Aquatic Biogeography in a Changing World

Aquatic Biogeography in a Changing World

Editor: Jonas Bailey

Callisto Reference

www.callistoreference.com

Callisto Reference,
118-35 Queens Blvd., Suite 400,
Forest Hills, NY 11375, USA

Visit us on the World Wide Web at:
www.callistoreference.com

ISBN: 978-1-63239-822-2 (Hardback)

Printed in China.

Cataloging-in-publication Data

Aquatic biogeography in a changing world / edited by Jonas Bailey.
p. cm.
Includes bibliographical references and index.
ISBN 978-1-63239-822-2
1. Aquatic animals--Geographical distribution. 2. Biogeography. 3. Aquatic biology. 4. Aquatic ecology.
5. Aquatic biodiversity. 6. Aquatic biodiversity conservation. I. Bailey, Jonas.
QL120 .A93 2017
577.6--dc23

Table of Contents

Preface

The study of distribution of plants and animals is called biogeography which is a branch of biology. Aquatic biogeography is defined as the study of aquatic ecosystem spatially distributed across globe. This book covers in detail some existent theories and innovative concepts revolving around aquatic biogeography. It aims to provide to its readers a wide variety of topics dealing with aquatic biogeography and its various aspects which are innovative and important. There has been rapid progress in this field and its applications are finding their way across multiple industries. The book is appropriate for students seeking detailed information in this area as well as for experts. It attempts to assist those with a goal of delving into the field of aquatic biogeography.

Significant researches are present in this book. Intensive efforts have been employed by authors to make this book an outstanding discourse. This book contains the enlightening chapters which have been written on the basis of significant researches done by the experts.

Finally, I would also like to thank all the members involved in this book for being a team and meeting all the deadlines for the submission of their respective works. I would also like to thank my friends and family for being supportive in my efforts.

Editor

1

Synergies for Improving Oil Palm Production and Forest Conservation in Floodplain Landscapes

Nicola K. Abram[1,2,3,4*], Panteleimon Xofis[5], Joseph Tzanopoulos[1], Douglas C. MacMillan[1], Marc Ancrenaz[2,4,6], Robin Chung[7], Lucy Peter[8], Robert Ong[9], Isabelle Lackman[2], Benoit Goossens[8,10,11], Laurentius Ambu[11], Andrew T. Knight[12,13]

1 Durrell Institute for Conservation and Ecology, School of Anthropology and Conservation, University of Kent, Canterbury, Kent, United Kingdom, **2** HUTAN/Kinabatangan Orang-utan Conservation Programme, Kota Kinabalu, Sabah, Malaysia, **3** Living Landscape Alliance, Reading, Berkshire, United Kingdom, **4** Borneo Futures initiative, People and Nature Consulting International, Jakarta, Indonesia, **5** Technological Education Institute of Kavala, Department of Forestry and Management of Natural Environment, Drama, Greece, **6** North England Zoological Society, Chester, Cheshire, United Kingdom, **7** C H Williams Talhar and Wong (Sabah) Sdn Bhd, Sandakan, Sabah, Malaysia, **8** Danau Girang Field Centre, c/o Sabah Wildlife Department, Kota Kinabalu, Sabah, Malaysia, **9** Forest Research Centre, Sabah Forestry Department, Sandakan, Sabah, Malaysia, **10** Organisms and Environment Division, School of Biosciences, Cardiff University, Cardiff, Wales, United Kingdom, **11** Sabah Wildlife Department, Kota Kinabalu, Sabah, Malaysia, **12** Department of Life Sciences, Imperial College London, Ascot, Berkshire, United Kingdom, **13** Department of Botany, Nelson Mandela Metrropolitan University, Port Elizabeth, Eastern Cape, South Africa

Abstract

Lowland tropical forests are increasingly threatened with conversion to oil palm as global demand and high profit drives crop expansion throughout the world's tropical regions. Yet, landscapes are not homogeneous and regional constraints dictate land suitability for this crop. We conducted a regional study to investigate spatial and economic components of forest conversion to oil palm within a tropical floodplain in the Lower Kinabatangan, Sabah, Malaysian Borneo. The Kinabatangan ecosystem harbours significant biodiversity with globally threatened species but has suffered forest loss and fragmentation. We mapped the oil palm and forested landscapes (using object-based-image analysis, classification and regression tree analysis and on-screen digitising of high-resolution imagery) and undertook economic modelling. Within the study region (520,269 ha), 250,617 ha is cultivated with oil palm with 77% having high Net-Present-Value (NPV) estimates (\$413/ha^{-yr}–\$637/ha^{-yr}); but 20.5% is under-producing. In fact 6.3% (15,810 ha) of oil palm is commercially redundant (with negative NPV of \$-299/ha^{-yr}-\$-65/ha^{-yr}) due to palm mortality from flood inundation. These areas would have been important riparian or flooded forest types. Moreover, 30,173 ha of unprotected forest remain and despite its value for connectivity and biodiversity 64% is allocated for future oil palm. However, we estimate that at minimum 54% of these forests are unsuitable for this crop due to inundation events. If conversion to oil palm occurs, we predict a further 16,207 ha will become commercially redundant. This means that over 32,000 ha of forest within the floodplain would have been converted for little or no financial gain yet with significant cost to the ecosystem. Our findings have globally relevant implications for similar floodplain landscapes undergoing forest transformation to agriculture such as oil palm. Understanding landscape level constraints to this crop, and transferring these into policy and practice, may provide conservation and economic opportunities within these seemingly high opportunity cost landscapes.

Editor: Francisco Moreira, Institute of Agronomy, University of Lisbon, Portugal

Funding: NKA thanks RICS Research Trust, U.S. Fish and Wildlife Service's Great Ape Fund and the Arcus Foundation for financial support; Planet Action and Trimble for providing satellite imagery and the Definiens eCognition License. ATK acknowledges support from the Department of Life Sciences at Imperial College London, UK, and ARC Centre of Excellence in Environmental Decisions, University of Queensland, Australia. The funders had no role in study design, data collection and analysis, decision to publish, or preparation of the manuscript.

Competing Interests: The authors have declared that no competing interests exist.

* E-mail: nicola_abram@hotmail.com

Introduction

Lowland tropical forest ecosystems contain some of the highest levels of species endemism and biological diversity globally [1,2]. However, tropical lowland forests are under increasing risk of conversion to oil palm (*Elaeis guineensis*) which drives biodiversity loss [2] and compromises forest functioning and maintenance of vital ecosystem services [3,4,5]. In 2012 over 17.1 million hectares of permanent cultivated cropland worldwide was comprised of oil palm agriculture compared with 9.97 million hectares in the year 2000 [6]. Oil palm establishment continues to increase within lowland areas of Malaysia and Indonesia, the two dominant producing countries, but also elsewhere in tropical Asia, Africa and central and South America [6,7]. Brazil alone has 2 million hectares of lowland forest identified as suitable for oil palm, and similarly the Democratic Republic of Congo, Indonesia, Peru and Columbia have an estimated 0.78 million, 0.61 million, 0.46 million and 0.42 million hectares respectively, of forest deemed as suitable for oil palm [8]. Expansion of this crop is largely due to its high opportunity costs driven by accelerating global demand for this versatile oil coupled with high palm production (i.e. yield) capacity [9].

Oil palm plantations are limited to low elevation areas and are in direct conflict with tropical lowland forests, including those found within riverine floodplains. Riverine floodplains are characterised by high levels of biodiversity and productivity [10]. These regions are low lying and subject to periodic inundation by associated rivers or streams [11]. Consequently, these landscapes present heterogenous environmental characteristics that dictate variable land suitability for oil palm cultivation, with unsuitable areas principally linked to seasonal and/or tidal inundation events. The palm *Elaeis guineensis* is flood intolerant with mortality occurring from root rot within two weeks of ground saturation and/or low oxygen levels or from saline water [12]. Palm mortalities reduce plantation yields and in-turn financial returns [13]. Nevertheless, simplistic biophysical criteria are often used by governments and agencies for agricultural zoning for oil palm that includes slope (<25°), elevation (<500 m) and soil types within suitable climatic zones [14]. These criteria may fail to capture regionalised constraints for this crop.

We investigated forest and oil palm dynamics within a floodplain system in the Malaysian context. In Malaysia it is estimated that between 1990–2005, 55–59% of Malaysia's oil palm extent replaced old-growth and secondary forests [15]. In Malaysia, Sabah (Borneo) is the largest oil palm planted state contributing 28.6% of the country's total oil palm extent [16]. In 2011, 1.43 million hectares (19.3%) of Sabah's terrestrial extent was under oil palm which could increase up to 2.1 million hectares by 2025, depending on land suitability [16,17]. It is likely that oil palm expansion will continue to target the eastern State floodplains areas that have very high yield potential. We employed the Lower Kinabatangan landscape, the largest floodplain in Sabah, as a 'model system' to understand spatial and economic components of forest conversion to oil palm in a riverine floodplain. We highlight the need for improving biophysical criteria used by governments in agricultural zoning for oil palm and emphasise the need for regional land-use planning of floodplain landscapes to promote synergies between agricultural development and biodiversity and ecosystem conservation goals. Understanding landscape level constraints to this crop, and transferring these into policy and practise, may provide conservation opportunities within these seemingly high opportunity cost landscapes.

Materials and Methods

Study Area

The study area comprised 520,269 ha of the Lower Kinabatangan floodplain region, in eastern Sabah (Figure 1(A)). The region experiences mean monthly temperatures of 21–34°C and average annual rainfall of 3,000 mm [18]. Forest types encompass those associated with mangrove, flooded forest and dry (humid) forest systems, which are threatened forest types [19]. Remaining forests have been heavily impacted from past commercial timber exploitation and extensive forest conversion resulting in significant forest loss, severe fragmentation and degradation within the forest ecosystem [20,21]. Yet, these forests remain important habitat for biodiversity harbouring 129 species of mammal, 314 species of birds, 101 species of reptiles, and 33 species of amphibians [22]. Many of these species are threatened IUCN (International Union for Conservation of Nature) Red List species such as the Bornean elephant [23], Bornean orangutan [24] and proboscis monkey [25]. These forests provide species habitat, are instrumental in connecting (at least in part) the fragmented protected areas network; as well as crucial for facilitating multiple ecological processes needed for the functioning of this ecosystem. Attempts to safeguard remaining forests has led to the gazetting of the Lower Kinabatangan Wildlife Sanctuary (LKWS, at 27,960 ha) in 2005. Other protected areas include protection forest reserves and virgin jungle forest reserves. Use-forests include the mangrove forest reserves and commercial (i.e. production) forest reserves; all herein referred to as protected areas (PA's). However, significant areas of unprotected forest remain outside the protected areas or production forests and on State land or land 'alienated' land for agriculture. Alienated land is leased (long-term) State land, under title and granted for a specific purpose such as oil palm [26]. Despite over a decade of conservation initiatives to secure these unprotected forests for conservation purposes they remain threatened largely with conversion to oil palm.

Forest Extent and Mapping Forest Types

Forest vegetation types were identified using SPOT5 10 m satellite images captured on 25/11/2007 and two on 19/06/2009; as well as two Landsat TM5 30 m images captured on 27/07/2006 and 11/08/2009 (downloaded from www.usgs.gov). Images were orthorectified and registered and indices calculated for: Normalized Difference Vegetation Index (NDVI); Normalized Difference Moisture Index (NDMI); and Soil Adjusted Vegetation Index (SAVI). Spectral transformations were performed including reflectance for SPOT5 and Tasseled Cap transformation on the Landsat TM5 Images [27]. We used a step-wise Object-Based Image Analysis (OBIA in eCognition Developer 8.7) for the classification that creates meaningful 'objects' by segmenting images into groups of spectrally similar pixels and spatial characteristics [28]. The eastern and western Kinagantagan areas were classified separately because different image capture dates prohibited the development of a common classification rule set. For training the classifier and testing the result, 1,938 ground validation points were collected in-field using a stratified random sample. The stratification was conducted on dominant regional forest types idenftified by local experts (for details see Table 1). Within a 20 m buffer around each validation point, data were recorded on: habitat type; degradation level (low, medium, high); percentage canopy cover; and dominant tree species with diameter at breast height (DBH) class. The ground truthed dataset was halved, one portion for training the classifier and the other for testing the result.

For image segmentation and object generation within eCognition, SPOT5 10 m resolution data were employed for delimitation of 'objects' within the images. Following trialing of numerous scale parameters, a scale of 15 was adopted. Colour/Shape and smoothness/compactness were set (0.9/0.1 and 0.5/0.5 respectively) [29]. All spatial data were used for ruleset development and the identification of features with the highest discrimination ability. The Support Vector Machine algorithm (SVM) within eCognition was employed to run the classification because it performed better in reproducing the training set than other trialled algorithms (i.e. Byes or classification and regression tree) [30,31]. We used a generic forest extent auxiliary layer, digitised from 2.5 m SPOT5 satellite images captured 12/12/2010 and 01/06/2011, to restrict the classification to areas of known forest (these images were not used in the classification due to inappropriate bands). Despite the high accuracy of the SVM classifications, classification refinement was undertaken on selected known misclassified objects using ground knowledge or neighborhood based features.

The eastern and western classifications were exported as shapefiles and 'merged' to form one continuous layer. An accuracy assessment was performed using an error confusion matrix method using the second portion of the ground truth dataset, and overall classification accuracy and kappa statistics were calculated for both portions [29].

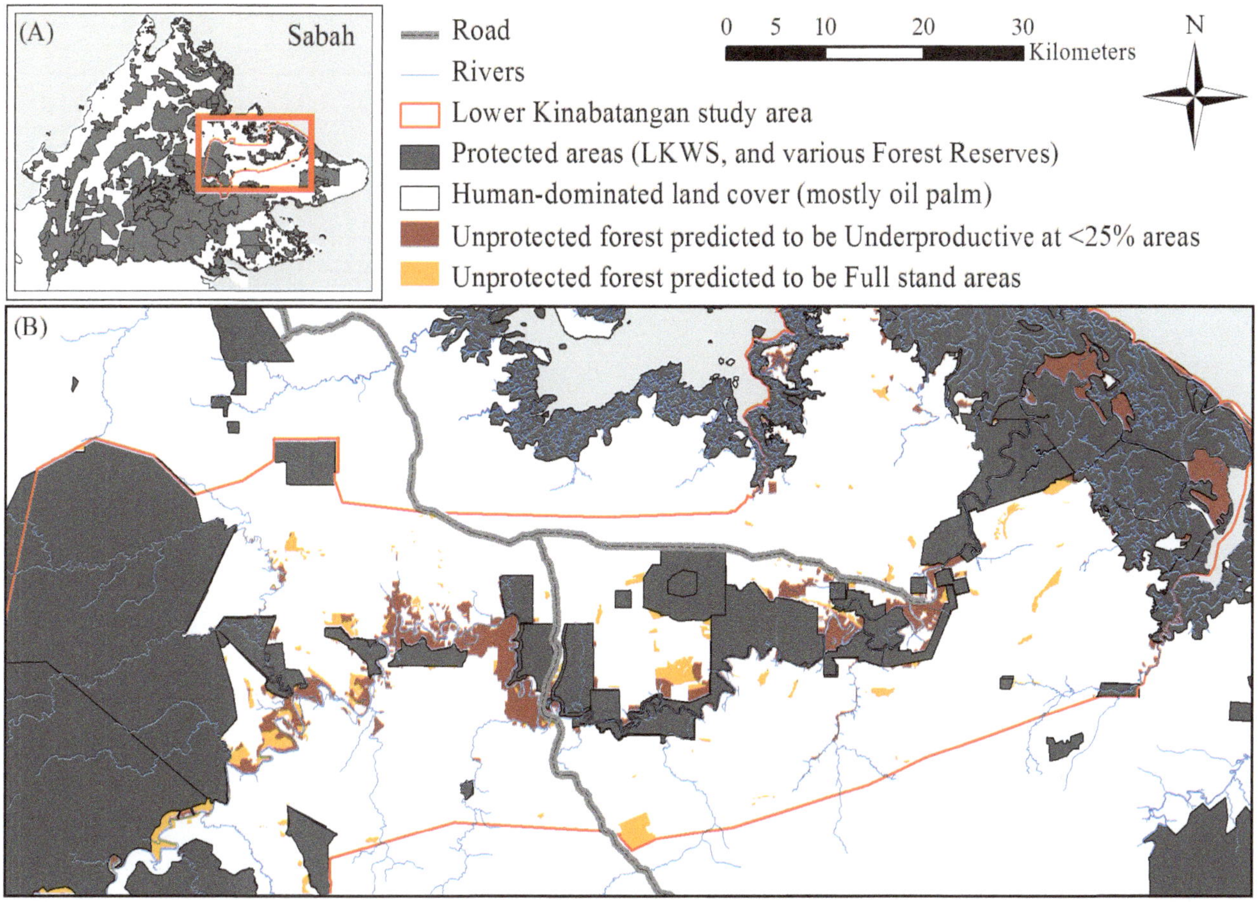

Figure 1. Study region and result from the CART analysis. (A) Map of the study region, the Lower Kinabatangan (red outline) within the context of Sabah, Malaysian Borneo. Protected areas (including all 'use' forests) are in grey. (B) Map of the Lower Kinabatangan study region with the spatial output from the Classification and Regression Tree analysis (CART) overlaid with the unprotected forest areas. Dark red demonstrates areas predicted as unsuitable for oil palm (i.e. areas that would be <25% palm capacity and commercially redundant); orange areas are predicted for full stands of oil palm (i.e. suitable areas).

Mapping Existing Oil Palm

Oil palm age and productivity classes were digitised using the 2.5 m SPOT5 images. Age classes were defined by characteristics viewable in the images that were associated with planting preparation and/or age associated features (see Table 2 for descriptions and Figure 2 for examples). To inform the visual characteristics, we used two geo-referenced maps (in ARcGIS 10) from large oil palm estates within the region (Sime Darby and Hilco Estates) that contained planting years for each oil palm 'block'. For productivity classes we based these on palm capacity. Optimal planting density for oil palm typically ranges from 128 to 148 palms per hectare depending on factors such as genotypes, environmental conditions, soil types and management practises [32]. In our study region 136 stands (i.e. palms) per hectare (SPH) is used, as recommended by the Malaysian Palm Oil Board (MPOB) and we used 136 SPH as our upper limit value for palm capacity (i.e., 136 SPH = 100% palm capacity). We classified the underproductive classes by defining the range of palm capacity within each class. Descriptions of each class can be seen in Table 2 and examples in Figure 2. In brief however, our oil palm categories were: (1) Cleared areas; (2) Planted out (0–2 years old); (3) Young mature (3–6 years, and ranging from 76–100% palm capacity); and (4) Prime mature at 'Full stand' (7–24 years, and ranging from 76–100% palm capacity); (5) Underproductive at 75% (areas that ranged from 51%–75% SPH capacity); (6) Underproductive at 50% (areas that had approximately 26–50% SPH capacity); and (7) Underproductive at ≤25% (areas that had approximately 25% SPH capacity or less).

Classification accuracy was determined through remotely obtained information as no ground-data was available. We used all recent (16/09/2009 and 24/08/2011) high-resolution tiles from Google Earth images (see http://earth.google.com) in our region to validate our oil palm age and productivity map. Although our satellite imagery was 2.5 m, which allowed individual palms to be seen, the higher resolution tiles in Google Earth enabled better interpretation ability and therefore permitted the validation of classes. We defined these tiles within a GIS (ArcGIS 10), generated 120 random points (see Figure S1 in File S1) with assigned unique identity numbers. We imported the validation points into Google Earth, assessed and allocated an oil palm class for each point. An error confusion matrix method was used to assess map accuracy and kappa statistic was calculated [29].

Economic Models for Oil Palm

We developed economic models to calculate the net present value (NPV) of oil palm for four oil palm class scenarios (see Tables S1 (a+b) to S4 (a+b) in File S2). These classes included; (1) 'Full

Table 1. Forest system and forest type classes found in the Kinabatangan with flooding periods and extents in hectares (ha) as calculated from the Object-Based Image Analysis classification.

Forest systems and forest type class	Annual flood-ing period	Total forest	Protected forest	Unprot-ected forest
		ha (%)	ha (%)	ha (%)
Mangrove forest:				
Beach forest: Occurs on sandy substrate along coastal areas. Dominant species includes *Casuarina equisetifolia.*	Tidal	5,327 (2)	4,672 (2)	655 (2)
Mangrove forest: Found in saline coastal sediments. Dominant species include *Rhizophora apiculata.*	Tidal	12,863 (5)	12,357 (6)	506 (2)
Nipah palm forest: Native type of palm (*Nypa fruticans*) found within the mangrove system either in mono-stands or mixed with *Rhizophora apiculata*	Tidal	26,618 (11)	25,399 (11)	1,219 (4)
Transitional forest: Occurs between mangrove and freshwater swamp forest. Brackish water. Dominant species of *Heritiera littoralis, Ilex cymosa, Excoecaria agallocha.*	Semi-tidal	13,849 (6)	10,567 (5)	3,282 (11)
Seasonally flooded forest:				
Freshwater swamp forest: Formed in backswamps and largely on poorly drained soil. Common species include *Dillenia excelsa, Croton oblongus, Mallotus muticus.*	>6 mths	22,284 (9)	16,721 (8)	5,563 (18)
Seasonal freshwater swamp forest: Heavy degradation thought to have occurred with many pioneer species. Common species include *Macaranga gigantea, Pterospermum elongatum, Cananga odarata.*	3–6 mths	12,501 (5)	8,253 (4)	4,248 (14)
Peat swamp forest: Oligotropic peat substrate, poorly drained forests exposed to flooding. Common treespecies include *Lophopetalum multinerviu, Baccaurea, Campnosperma coriaceum, Syzygium* and *Anisoptera costata.*	>6 mths	2,132 (1)	2,102 (1)	30 (<0)
Swamp: Open reed, swamp vegetation. Dominant tree species *Excoecaria indica*	>9 mths	2,750 (1)	2,048 (1)	702 (2)
Lowland dry forest:				
Lowland dry forest: Previous dipterocarp forest, secondary forest with species including, *Nauclea subdita, Neolamarckia cadamba, Glochidion rubrum*	<3 mths	39,008 (16)	30,693 (14)	8,315 (28)
Lowland dry dipterocarp forest: Logged lowland mixed dipterocarp forest, dominated with *Dipterocarp sp.*	Never/Rarely	101,878 (41)	100,866 (46)	1,012 (3)
Limestone forest: Gomantong substrate association of hill and ridge escarpments. Low human disturbance. Dominant species include *Dryobalanops lanceolata, Shorea pauciflora, Parashorea malaanonan* and *Dipterocarpus caudiferus.*	Never/Rarely	1,679 (1)	1,392 (1)	287 (1)
Mixed vegetation types:				
Severely degraded: areas of severe degradation with unknown previous forest types dominated by shrub/low lying vegetation.	Varied	10,511 (4)	6,159 (3)	4,352 (14)
Sum of hectares		251,400	436,295	30,173

stand' areas which incorporated Young mature and Prime mature oil palm classes that ranged from 76% (i.e., 103 SPH) to 100% (based on a maximum of 136 SPH) palm capacity (and combined enumerated 96.3% of this class extent). We also included Cleared areas and Planted out classes into 'full stand' under the assumption they will fall into this palm capacity range in the future or once planted (collectively these classes made up only 3.7% of the 'full stand' class extent); (2) 'Underproductive at 75%' which assumed 51% (69 SPH) to 75% (102 SPH) palm capacity; (3) 'Underproductive at 50%' at 26% (35 SPH) to 50% (68 SPH) palm capacity; and (4) 'Underproductive at ≤25%' which ranged from 0% (0 SPH) to 25% (34 SPH) palm capacity. We modelled the NPV of the lowest and highest palm capacity percentage range for the four classes.

Our economic models assumed a typical 25 year crop life as per industry standard [16]. For each model a yield curve was used

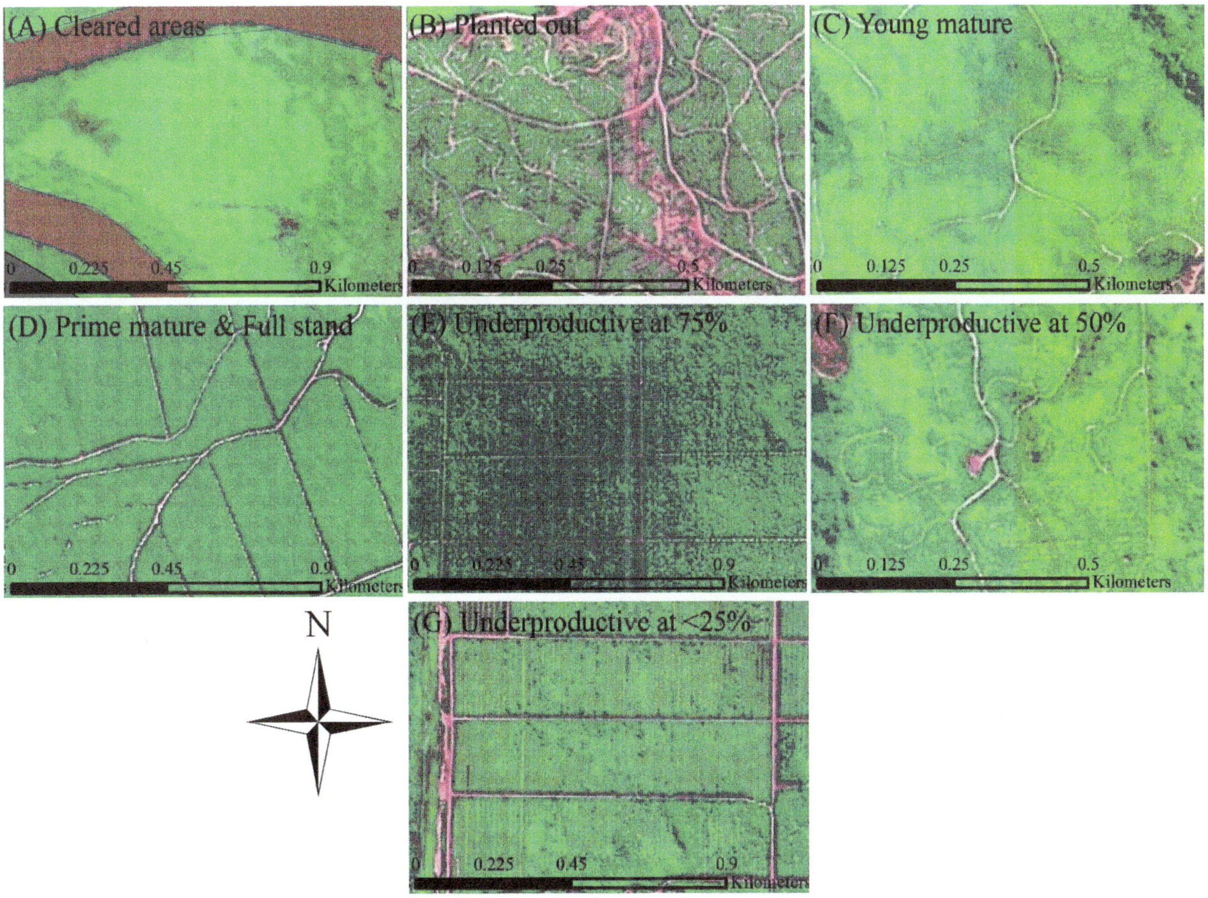

Figure 2. Examples of oil palm age and productivity classes. Images (A) to (D) show examples of the four age classes used for the oil palm mapping; and, images (E) to (G) show examples of underproductive classes, using the 2.5 m SPOT satellite images.

spanning the 25 year time frame (see oil palm yields in Figure S2 in File S2). Yields vary considerably over the lifespan of the crop with declines occurring after peak years due to factors such as palm mortality. To generate our yield estimates we used data based on actual yields from the region for full stand areas of 136 SPH (obtained from C H Williams). For the 'Full stand' at 136 SPH (100% palm capacity) NPV model, we used a maximum yield of 30 metric tons (t) of fresh fruit bunches (FFB) per hectare (ha) in peak years (years 8–11) declining to 17 t/FFB/ha in year 25, averaging at 21.92 t/FFB^{-25}. For all other models, we vary annual

Table 2. Oil palm age and productivity classes and descriptions, mapped using SPOT5 2.5 m satellite imagery.

Oil palm classes	Description
Cleared areas	Areas cleared of forest cover with bare earth or grass like vegetation cover. No roads laid nor ground prepared for oil palm planting (Figure 2A).
Planted out	Areas had roads cut and laid, ground prepared for planting, and in some areas palms planted out (palms would be approximately 2 years or younger) (Figure 2B).
Young mature	Young mature palms were visible but palm fronds did not overlap. Low-lying (leguminous) vegetation was visible. Palms in this category range from 3–6 years (Figure 2C).
Prime mature and Full stand	Palms in prime yield (7–24 years) and 'Full stand' (i.e., 76–100% palm capacity). Class had closed canopies i.e., overlapping fronds of neighbouring plants. Canopy closes at 7 years in areas with 136 palms per ha. Homogeneous texture (Figure 2D).
Underproductive at 75%	Class ranged from 51%–75% palm capacity. Class included: older palm (where mortality would naturally start to occur) or areas prone to water logging (dimpling effect can be seen in blocks) (Figure 2E).
Underproductive at 50%	Class ranged from 26–50% palms per ha capacity and were sometimes on slightly undulating areas or areas with water logging/ flooding issues (Figure 2F).
Underproductive at 25%	Class had ≤25% palms per ha and were largely associated with areas that experience annual flooding and/or daily inundation from tides (Figure 2G).

yields through the 25 year model timeframe, by calculating the proportion of yield (in t/FFB/ha) needed against our maximum 'Full stand' (100% at 136 SPH) yield values. In our models we assume a direct relationship between the number of SPH (or percentage thereof) and yield. We discuss the pitfalls and limitations of these assumptions in the discussion.

Within our models we retained FFB as our revenue unit rather than converting to crude palm oil (CPO) as for smallholders and commercial estates with no processing mills, revenue is derived from the sale of FFB. We based our model largely around 2011 data. As a result we used a constant FFB price of US$178 t/FFB, based on average 2011 values from the east coast of Sabah (calculated by C H Williams Talhar & Wong). We used a discount rate of 11% per annum in our main models, used by industry in 2011 in Sabah. However, to assess model robustness we undertook sensitivity analyses using discount rates of 5%, 8% and 14% per annum. Cost data were calculated from estates past actual costs, their budgeted future costs and typical industry average costs for 2011 (obtained from C H Talhar & Williams). These costs were cross referenced and supplemented with state wide data on costs from 2008 [33]. Costs were summarised in three categories: (1) General charges (or Joint Estate costs); (2) Field upkeep (weeding; manuring; pruning; pests and disease treatment; supplying; infrastructure etc); and (3) Harvesting and transport. Costs were constant in the four models except for 'supplying' which is the replanting of palms if palms die. We used a commercial estates approach (rather than smallholder approach) due to data availability and use in translating industry standards to build model assumptions. In regards to 'supplying' in commercial estates this is undertaken within the first two years of production to ensure homogeneous age blocks [33]. Values for supplying were adjusted for the three underproductive classes to reflect the proportion of palms needed for replanting within the initial two year period (these costs were relatively small, see Tables S1 (a+b) to S4 (a+b) in File S2). Costs excluded those associated with non-recurrent costs for establishing 'New Plantings' and those associated with costs of palm oil processing mills.

Modeling Oil Palm Suitability for Unprotected Forested Areas

To predict suitable and unsuitable areas for oil palm outside the protected areas, we used a Classification and Regression Tree (CART) analysis. The CART framework was developed by using data from existing oil palm blocks integrated with biophysical spatial information developed at 1 ha resolution. Predictor variables included: (1) Euclidian distance from main river (and main tributaries); (2) elevation; (3) aspect; (4) slope; and (5) a soil agriculture suitability layer using 4 suitability classes (unsuitable = 1; marginal = 2; suitable = 3; and very suitable = 4) [34]. Point training data were extracted randomly from known 'Full stand' and 'Underproductive at ≤25%' areas. 'Full stand' class comprised of 263 data points and 'Underproductive at ≤25%' had 211. We extracted raster values from the spatial predictors for each point using ArcGIS 10. Waikato Environment for Knowledge Analysis (WEKA) software was used to execute the CART analysis and a 10-fold cross-validation was used for tree pruning and the accuracy assessment. The software ArcPath was used to integrate the CART's classification decision tree with variables to produce a categorical map of oil palm suitability. This categorical output was imported into ArcGIS 10 and extracted for the unprotected forest extent.

Mapping Land Titles

Publicly available cadastral maps ($n = 14$) used by commercial land valuation agencies and industry were used to forecast future loss of unprotected forest by quantifying the extent of alienated land. Maps were undated but known to have been drafted in the 1990's by the Land and Survey Department. Although, these maps have been updated since the 1990's the extent of alienated land is likely underestimated. Maps were scanned at high resolution and geo-referenced to 2.5 m SPOT5 images. Land parcels were digitised, with title types and identity numbers recorded in a vector file.

Land title types included: (1) Native Title (NT), i.e., smallholdings alienated for agriculture (oil palm) for perpetuity and restricted to <40 ha in size [26]; (2) Country Land title (CL), i.e., 'alienated' State Land for commercial agriculture (oil palm) under a 99 year lease [26]; (3) Demarcated State land, with boundaries but no identity code (assumed to be under application but not alienated): and (iv) Undemarcated State land, assumed to have no title applications and not alienated (see Appendix S1 in File S1 for more details on land ordinance policies).

Results

Forest Extent and Forest Types

Overall classification accuracy was 72.2%, with a Kappa statistic of 0.65 using an error matrix method and 969 ground truth points (see Appendix S2 and Table S6 in File S1, for details), deemed as satisfactory e.g. kappa falls into 'fair' and 'good' according to differing scales outlined by Monserud [35] and Landis and Koch [36], respectively. Of the 520,269 ha study region, 48% (251,400 ha) is forested (inclusive of severely degraded forests) and comprised of 371 independent forest fragments (*mean* size = 679 ha; *sd* (standard deviation) = 5,231 ha). Of the forested extent, 12% (30,173 ha) occurs outside the protected areas (i.e. unprotected forest).

Within the unprotected forest, the Lowland dry forest system covered 32% (Table 1, Figure 3). The Seasonally flooded forest system was the largest (35%), harbouring: Freshwater swamp forest (18%); Seasonal freshwater swamp forest (14%); Peat swamp forest (<1%); and Swamp (2%) (Table 1, Figure 3). The Mangrove system comprised 19% of unprotected forest and contained: Beach forest (2%); Mangrove forest (2%); Nipah palm forest (4%); and Transitional forest (11%) (Table 1, Figure 3). These two forest systems enumerated 54% of the unprotected forest, are prone to inundation events and are likely unsuitable for oil palm. The severely degraded mixed vegetation type class totalled 14% and we predict that much of this area is prone to flooding and could also be unsuitable for oil palm.

Future Forest Loss

Of the unprotected forest, 64% had been alienated with 49% under Native or smallholder titles (i.e., 9,497 ha, either fully or partially under 1,175 titles); and 51% under Country Land or commercial titles (i.e., 9,732 ha spanning 163 titles); the remaining extent was State land (36%) (see Table 3). Of the forest systems, 75% of the Seasonally flooded forest system had been alienated for agriculture (of this NT = 46%: CL = 54%); 21% of the Mangrove forest system (of this NT = 87%: CL = 13%); 71% for the Lowland dry forest system (of this NT = 43%: CL = 57%); and 77% for the Mixed vegetation types (of this NT = 57%: CL = 43%) (see Table 4 for more details).

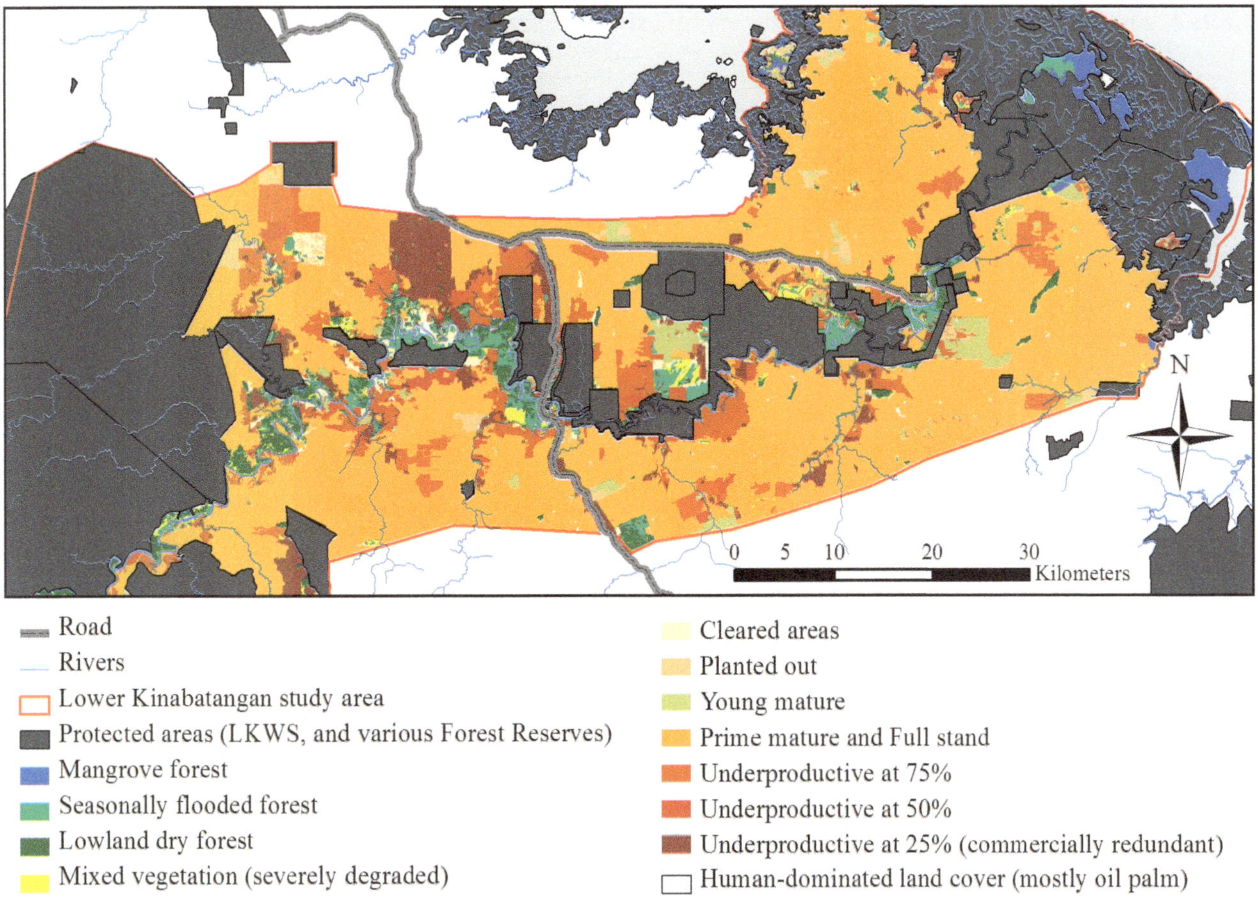

Figure 3. Map of forest systems and oil palm age and productivity classes. Map of the Lower Kinabatangan study area (red outline) showing the extent of the protected area network (grey) and the forest systems identified in the unprotected forest (as of 2010/2011 and generated through the Object-Based Image Analysis). Forest systems include: Mangrove forest (blue), Seasonal flooded forest (turquoise), Lowland dry forest (dark green) and Mixed vegetation that is severely degraded (yellow). Map also shows the oil palm age and productivity classes including: Cleared areas (cream), Planted out (beige), Young mature (olive), Prime mature and Full stand (orange), Underproductive at 75% (orange-red), Underproductive at 50% (red), and Underproductive at ≤25% (dark red), that are largely associated with areas proximal to the major river and its principal tributaries.

Oil Palm Landscape

Overall classification accuracy was 84% with a Kappa statistic of 0.816 (deemed 'very good' [35] to 'excellent' [36]) using an error matrix method with 116 test data, as four fell on cloud (see Table S7 in File S1, for details). Oil palm covered 250,617 ha (48%) of the study region (see Table 5 for ha level information and Figure 3). Seventy percent of oil palm is on commercial land titles (174,698 ha). Cleared areas covered 1.3% of the oil palm landscape; Planted out covered 1.6%; and Young mature and Prime mature and full stand collectively covered 77% of the oil palm extent. For the underproductive classes, Underproductive at 75% made up 11% and of this 39% were on NT and 59% on CL. This class largely included areas of old palm (>25 years) and/or areas of water logging both of which can increase palm mortality.

Table 3. Land title types, size in hectares (with percentages) and number of unique demarcated titles within the Kinabatangan, classified from cadastral maps for the unprotected forests.

	Unprotected forest	
Title type	**Total no. ha (%)**	**No. of unique titles**
Native Title	9,497 (31)	1,175
Country Land Title	9,732 (32)	163
State Land Demarcated	3,937 (13)	104
State Land Un-demarcated	7,009 (23)	-
TOTAL	30,173	1,338

Table 4. Land title type and extent (in ha and percentages) for: Forest systems; and, Modelled oil palm suitability (CART analysis) for suitable ('Full stand') and unsuitable ('Underproductive at ≤25%') areas.

	Land Title Type			
	Native title in ha (%)	Country Land title in ha (%)	State land (demarcated) in ha (%)	State land (un-demarcated) in ha (%)
Forest systems:				
Mangrove forest	1,016 (11)	147 (2)	321 (8)	4,179 (60)
Seasonally flooded forest	3,635 (38)	4,235 (44)	1,110 (28)	1,563 (22)
Lowland dry forest	2,934 (31)	3,917 (40)	1,747 (44)	1,017 (15)
Mixed vegetation types	1,912 (20)	1,433 (15)	758 (19)	250 (4)
Total number of ha	9,497	9,732	3,936	7,009
Oil palm suitability modelling:				
Full stand	3,665 (39)	3,952 (41)	1,723 (44)	1,003 (14)
Underproductive at ≤25%	5,832 (61)	5,780 (59)	2,213 (56)	6,006 (86)
Total number of ha	9,497	9,732	3,936	7,009

Underproductive at 50% totalled 3%, and spanned both NT and CL titles almost equally and were typical on slightly undulating areas or flood prone areas. Underproductive at ≤25% covered 6%, of which 28% was on NT and 66% on CL. These areas vary in age but knowledge of some areas e.g. Pontian Alljoy and Pendiros Pontian Estates, suggest conversion dates back 20 years and are largely associated with inundation.

Oil Palm Economic Models

The net present value of oil palm for each class ranged from: $413/ha^{-yr} to $637/ha^{-yr} for the 'Full stand' class; $179/ha^{-yr} to $403/ha^{-yr} for the Underproductive at 75% class; $-55/ha^{-yr} to $169/ha^{-yr} for the Underproductive at 50% class; and for those areas that were grossly underproductive at ≤25% had negative NPV of between $-299/ha^{-yr} to –$65/ha^{-yr} (see Table 6). We performed sensitivity analyses for varying discount rates for all NPV models (see Table 6; and Table S5 in File S2) and compared them to our 11% discounted base models. For Full stand areas, a 5% discount rate would increase NPV by approximately 100%, an 8% discount rate would seen a 40% approx increase in NPV; and around a 26% decrease in NPV under the 14% discount rate (Table 6). Similar percentage values were found for the Underproductive at 75% models (Table 6). For Underproductive at 50% areas the lower range models did not widely differ, but for the upper range models could see a 121% increase in NPV with a 5% discount rate, 48% NPV increase under an 8% discount rate; and decrease of about 33% for a discount rate of 14%. Models for the upper range of Underproductive at ≤25% were insensitive to changes in discount rates all having negative NPV (Table 6).

Oil Palm Suitability Modelling (CART)

The CART analysis identified distance to river, elevation and soil as strong predictors for suitable and unsuitable areas for oil palm. Aspect and slope variables were not significant. Overall classification accuracy was 75% and a Kappa statistic of 0.5, deemed as fair [35] to good agreement [36] according to differing scales. The CART decision tree identified three classification rules that dictate land feature thresholds for unsuitable areas for oil palm (Figure S3 in File S1): (1) areas ≤1,504 m from river, with elevation of ≤22.4 m above sea level (ASL); and (2) if over 1504 m from a river, ≤16 m ASL and marginal soil type; and (3) if over 1,947 m from a river, on very suitable soil and between 9.9 m–14 m ASL, with the latter two likely due to areas within lower lying depressions prone to periodic flooding in some years and/or water logged areas. Of the 30,173 ha of unprotected forest, the

Table 5. Land title type and extent (in ha and percentages) for existing oil palm age and productivity classes.

	Land Title Type			
Palm oil classes	Total in ha (%)	Native title in ha (%)	Country land title in ha (%)	Un-demarcated or State land in ha (%)
Cleared areas	3,290 (1)	1,729 (2)	1,218 (1)	343 (7)
Planted out	4,030 (2)	960 (1)	2,520 (1)	550 (11)
Young & Prime mature and full stand	191,832 (77)	49,328 (70)	140,157 (80)	2,347 (47)
Underproductive at 75%	28,081 (11)	10,959 (15)	16,563 (9)	560 (11)
Underproductive at 50%	7,575 (3)	3,489 (5)	3,754 (2)	332 (7)
Underproductive at ≤25%	15,810 (6)	4,481 (6)	10,486 (6)	843 (17)
Total area	250,617	70,945	174,698	4,974

Table 6. Range values for the main oil palm Net Present Value (NPV) (US $/ha/25 years) model discounted at 11% in the four suitability classes, as well as the outputs of the sensitivity analyses with variable discount rates of 5%, 8% and 14% (showing range NPV and percentage (%) difference from the main model at 11%).

	11% discount rate	5% discount rate	8% discount rate	14% discount rate
	lowest & highest NPV	lowest & highest NPV	lowest & highest NPV	lowest & highest NPV
Full stand	$413–$637	$813–$1,252 (101%–97%)	$579–$881 (40%–38%)	$300–$471 (−27%– −26%)
Underproductive at 75%	$179–$403	$392–$812 (119%–101%)	$263–$565 (47%–40%)	$122–$293 (−32%– −27%)
Underproductive at 50%	$-55–$169	$-48–$374 (−13%–121%)	$-52–$250 (−5%–48%)	$-57–$144 (4%– −33%)
Underproductive at ≤25%	$-299–$-65	$-505–$-66 (69%-2%)	$-380 $-65 (27%–0%)	$-242–$-64 (−19%– −2%)

CART model predicted 66% is unsuitable for oil palm (Figure 1(B)). Of these unsuitable lands, 59% has already been alienated for oil palm and 11% were probably under land applications, though may now be alienated (Table 4).

Discussion

It is likely that oil palm cultivations will continue to expand at large scales throughout the world [37], as it is estimated that global demand for palm oil will double by 2020 [38]. With such expansion careful land-use planning is needed so that multiple-benefits within landscapes can be attained and issues pertaining to biodiversity and ecosystem service loss can be mitigated [39]. We employed the Kinabatangan as a 'model system' to investigate spatial and economic patterns of land cover in a tropical forest system.

Heterogeneity of Oil Palm Landscapes

We considered the oil palm landscape to understand fine-scale variability of production within the wider floodplain extent to begin formulating ideas around conservation opportunity [40,41]. Oil palm production in many fertile floodplains may deliver high profits under optimal planting conditions [17]. Our economic models for full stand areas within the region (see Figure 3) estimated NPV from $\$413/ha^{-yr}$ to $\$637/ha^{-yr}$ using an 11% discount rate (potentially increasing by 100% with an 8% discount rate, see Table 6). However, floodplains display variable topography resulting in heterogeneous suitability for oil palm cultivation (Figure 3). Our study region is characterised by areas of inundation, particularly during the monsoon season (October to March). Our results indicated that 20.5% (51,466 ha) of the total 250,617 ha of oil palm plantations are under producing. At minimum 6.3% (15,810 ha) is likely to be commercially redundant with negative NPV estimates ranging from $\$\text{-}299/ha^{-yr}$ to $\$\text{-}65/ha^{-yr}$. These areas are largely associated with flooding in low-lying areas proximal to rivers, as predicted by our CART model (Figure 1). The economic models for the lower SPH range within the Underproductive at 50% class also demonstrated negative NPV suggesting that some of these areas could also be commercially redundant (Table 6).

Although flood mitigation measures can be implemented in flood prone areas, losses from inundation within our study region have proven to be largely ineffective and very costly [42] thereby posing a financial risk to industry. For example, in the year 2000 one company experienced palm mortalities in 5,000 hectares of immature palm, with estimated financial losses of US $3 million (equivalent to US $600/ha) due to high flood water (14 m ASL) [43]. Impacts of flood related financial losses is particularly pertinent for small scale farmers who often establish plantations using formal credit, borrow money through informal arrangements, or by investing a large proportion of their savings. Failed oil palm ventures therefore represent poor return on investment for small-scale producers [44]. Larger companies with processing mills are likely to have less associated financial risk in converting flood prone land as larger plantations may have a mosaic of land suitability thereby offsetting financial impacts. Moreover, major financial risks for some companies are generally associated with fixed costs of processing mills [45]. Nevertheless, large estates have social and environmental corporate responsibilities and conserving these forests may help companies achieve these requirements.

There were a number of assumptions in our economic models for oil palm. We assumed that the variability in yield is driven by the number of SPH. However, yield may vary due to a range of factors such as soil composition and other environmental factors [12], palm strains and management practises [32,46]; as well as pests and disease such as ganoderma [47]. In addition to factors that may affect actual yield, a number of other limitations are at play that may affect the harvesting of the fresh fruit bunches. Sabah is currently experiencing a foreign labour shortage notably of FFB harvesters, collectors and loose fruit collectors within the estates, resulting in substantial areas of oil palm remaining un-harvested [48]. Areas may also be un-harvestable if there is poor accessibility or potential health hazards such as areas with flooding. Such factors are not incorporated into our models due to a lack of data availability, as well as difficulties in up-scaling these factors to a landscape level. Nevertheless, our findings are based on sound, regional data using justifiable assumptions and provision needed values for understanding landscape economics for the region to start to formulating understanding of potential trade-offs.

Understanding Forest System Suitability for Oil Palm

Mapping natural forest systems at the agricultural frontier is fundamental to understanding opportunities for achieving biodiversity conservation goals [49]. This is particularly pertinent in forested areas zoned for oil palm that have low expected suitability. We estimated 54% (i.e., 16,207 ha of mangrove forest and flooded forests systems combined) to 68% (i.e., 20,555 ha with the inclusion of severely degraded forest) is unsuitable for oil palm production (Table 1, Figure 3), supported by our CART results. These areas experience significant inundation either daily (Mangrove system) or annually from Monsoon floods (Seasonally flooded forest system). If these forests are converted to oil palm, it is highly likely they will mirror the commercially redundant areas. Using our most conservative unsuitable forest extent estimate (16,209 ha) we estimate at minimum 56% was already alienated under smallholder (4,651 ha) and commercial (4,382 ha) titles. In parallel to existing commercially redundant areas, future investment of smallholders or companies may see no financial gain in these areas. If these areas are converted there could be over 32,000 ha of commercially redundant areas in the Kinabatangan region.

Further loss from the unprotected forests will be a significant blow for the Lower Kinabatangan ecosystem including for lowland forest types that are becoming increasingly threatened [50,51]. These forests provide a wide range of ecosystem function [52] and services such as above- and below- ground- carbon storage, essential in mitigating climate change [53], erosion control thereby mitigating severe runoff into the river system and flood attenuation, wildlife carrying capacity and species dispersal essential for the viability of globally threatened and endemic species found in the region [23,24,54,55]. Moreover the persistence of these forests and the populations of large mega-fauna species that reside in them are crucial to the flourishing international tourism industry in the region, essential to local livelihoods. Despite the value of these forests, significant forest extent is alienated and by law should be cultivated within several years of title acquisition [26]. The large extent of alienated forest suggests many landholders have not complied with these regulations. This could be due to a number of reasons including their understanding of these forests lack of suitability for oil palm. A window of opportunity for intervention schemes results from these unconverted forests. We discuss a number of plausible intervention schemes in the context of the State of Sabah.

State Policy/Management Interventions

For unprotected forests, the Sabah Government has commited to establishing corridors within the Kinabatangan to help promote the viability of State priority species e.g. the Bornean orangutan and Bornean elephant [56,57], two endangered species in the Kinabatangan. As an initial step, the Sabah Government should seek to excise State land to become incorporated into the Lower Kinabatangan Wildlife Sanctuary or as Class I Forest Reserves (i.e., protected forest status). For alienated forested land conversion may be imminent. A moratorium on conversion for these areas, regardless of forest type or level of degradation, would allow time for negotiations to be undertaken between government, the oil palm industry, and local people so that opportunities to secure these forests for conservation purposes can be sought. Solutions will need to consider land holders rights (local people and companies). Lands unsuitable for cultivation are likely to incur costs that potentially outweigh revenues and could be purchased for conservation by Government or by existing land purchase schemes by NGO's. Additionally, mechanisms such as reducing emissions from deforestation and forest degradation (REDD+) [58] may prove useful if State policies can be aligned with such mechanisms, i.e., to permit alienated land to retain standing forest (which is not permitted under current Land Ordinance Policy) [26].

For existing commercially redundant oil palm these areas could be excised to the protected area network and prioritised for restoration by either existing initiatives (e.g. HUTAN, Nestlé Project RiLeaf and MESCOT initiatives) or new reforestation projects (potentially under a carbon credit fund [59]). Companies could benefit through promoting corporate responsibility goals and securing benefits when applying for sustainable certification under the Roundtable on Sustainable Palm Oil (RSPO) [60]. Additionally, re-planting of old palm and improving yield of existing palms could help maximise returns from existing oil palm areas, offsetting need for further oil palm expansion.

Policy reform or creation of new policies could pave a more sustainable multi-use floodplain landscape. Firstly, biophysical criteria used by the State Government to demarcate areas for oil palm could be refined to account for more regional environmental constraints such as frequency and intensity of inundation. The Sabah Water Resources Enactment (1998) requires 'river reserves' to be retained in cultivated areas (up to 20 m from watercourse). This is rarely implemented by companies or enforced by authorities. River reserve widths are based on minimum needs to mitigate soil and bank erosion [61]. Revision of this policy to incorporate wildlife corridor needs or development of new 'wildlife corridor' policies should be considered. Although we strongly recommend fine-scale planning to be undertaken to promote best use of landscapes, we do suggest minimal corridor widths of 1 km for major floodplain rivers (such as the Kinabatangan which is 1.5 km wide at river mouth and 100 m wide at the most interior point of study region). This is because 'corridors' should be (multi) functional landscape features [62], resistant to issues such as edge effects that can compromise their long-term persistence [63]. Additionally, our findings suggests commercially redundant areas of oil palm within floodplains are largely within 1.5 km from (major) river banks, meaning that such corridors may not significantly impact industry profits.

In this study we highlighted prevalent issues pertaining to forest conversion to oil palm in a floodplain system. Our study provides an exemplary case study of how poor planning can result in unfavourable land cover whilst little serving either profitability for landholders or conservation agendas. Regional studies such as this one should be conducted for other floodplains undergoing land cover change. Greater understanding of opportunities and constraints in these landscapes is needed to promote informed trade-off decision making at multi-scale levels. As global palm oil demand increases, ensuring responsible utility of tropical landscapes is vital in synergising a balancing between agricultural and development needs with long-term biodiversity conservation and ecosystem functioning. This study has global significance beyond our study region as we hypothesis that biophysical criteria used by many governments and international agencies for targeting areas for oil palm cultivation are likely similar to those used in Sabah [64]. Revising biophysical criteria for agricultural zoning may better promote land use. Finally the long-term viability of the Kinabatangan lie in the hands of political will and landholders decisions. We hope that this study will provision information for better decision making within the context of the Kinabatangan landscape to ensure wise steps are made to secure this ecosystem long-term.

Supporting Information

File S1 Oil palm age and productivity map validation areas (Figure S1); Land ordinance policy under land titles (Appendix S1); Operational habitat map confusion matrix (Appendix S2); Confusion matrix table of the OBIA (Table S6); Confusion matrix table of the oil palm age and productivity (Table S7); Decision Tree from the CART analysis for oil palm suitability (Figure S3).

File S2 Excel economic models for four oil palm classes including: Full stand 100% palm capacity (136 SPH) (Table S1(a)); Full stand at 76% palm capacity (103 SPH) (Table S1(b)); Underproductive at 75% at 75% capacity (102 SPH) (Table S2(a)); Underproductive at 75% at 51% capacity (69 SPH) (Table S2(b)); Underproductive at 50% at 50% capacity (68 SPH) (Table S3(a)); Underproductive at 50% at 26% capacity (35 SPH) (Table S3(b)); Underproductive at 25% at 25% capacity (34 SPH) (Table S4(a));

Underproductive at 25% at 0% capacity (0 SPH) (Table S4(b)); Discounted summaries (Table S5); Yield curve (Figure S2). The authors are solely responsible for the content and functionality of these materials. Queries (other than absence of the material) should be directed to the corresponding author.

Acknowledgments

Thank you to D.Webber, RSPO Secretary General for comments and the staff of HUTAN/KOCP, especially A.R.Bin Saharon for his expertise, W. Bin Maharonl and H.Kler. The authors thank the two anonymous peer reviewers for their insightful comments which helped strengthen the manuscript.

Author Contributions

Conceived and designed the experiments: NKA PX JT DM MA RC IL BG LA ATK. Performed the experiments: NKA PX RC LP. Analyzed the data: NKA PX JT RC LP. Contributed reagents/materials/analysis tools: NKA PX JT MA RC LP RO IL BG. Wrote the paper: NKA PX JT DM MA RC RO IL BG ATK.

References

1. Sodhi NS, Koh LP, Brook BW, Ng PKL (2004) Southeast Asian biodiversity: an impending disaster. Trends in Ecology & Evolution 19: 654–660.
2. Fitzherbert EB, Struebig MJ, Morel A, Danielsen F, Brühl CA, et al. (2008) How will oil palm expansion affect biodiversity? Trends in Ecology & Evolution 23: 538–545.
3. Brown KA, Flynn DFB, Abram NK, Ingram JC, Johnson SE, et al. (2011) Assessing Natural Resource Use by Forest-Reliant Communities in Madagascar Using Functional Diversity and Functional Redundance Metrics. PLoS ONE 6: e24107.
4. Abram NK, Meijaard E, Ancrenaz M, Runting RK, Wells JA, et al. (2013) Spatially explicit perceptions of ecosystem services and land cover change in forested regions of Borneo. Ecosystem Services. Available: http://dx.doi.org/10.1016/j.ecoser.2013.11.004.
5. van der Werf GR, Morton DC, DeFries RS, Olivier JGJ, Kasibhatla PS, et al. (2009) CO_2 emmissions from forest loss. Nature Geoscience 2: 737–738.
6. FAO (2012) FAOSTAT online statistical service. Food and Agriculture Organization of the United Nations (FAO) Rome, Italy. Available: http://faostat.fao.org. Accessed 08-02-2014.
7. Gilbert N (2012) Palm-oil boom raises conservation concerns. Industry urged towards sustainable farming practices as rising demand drives deforestation. Nature News 12 July 2012.
8. Butler RA, Laurance WF (2009) Is oil palm the next emerging threat to the Amazon? Tropical Conservation Science 2: 1–10.
9. Feintrenie L, Chong WK, Levang P (2010) Why do Farmers Prefer Oil Palm? Lessons Learnt from Bungo District, Indonesia. Small-scale Forestry 9: 379–396.
10. Tockner K, Stanford JA (2002) Riverine flood plains: present state and future trends. Environmental Conservation 29: 308–330.
11. Junk WJ, Welcomme RL (1990) Floodplains. In: Wetlands and Shallow Continental Water Bodies, ed. B.C. Patten et al., 491–524. The Hague, The Netherlands: SPB Academic Publishers.
12. Mantel S, Wosten H, Verhagen J (2007) Biophysical Land Suitability for Oil Palm in Kalimantan, Indonesia. Wageningen, UR. Wageningen: ISRIC-World Soil Information, Alterra, Plant Research International.
13. Teoh CH (2000) Land Use and the Oil Palm Industry in Malaysia: Report of WWF Malaysia.
14. Carlson KM, Curran LM, Ratnasari D, Pittman AM, Soares-Filho BS, et al. (2012) Committed carbon emissions, deforestation, and community land conversion from oil palm plantation expansion in West Kalimantan, Indonesia. Proceedings of the National Academy of Sciences 109: 7559–7564.
15. Koh LP, Wilcove DS (2009) Oil palm: disinformation enables deforestation. Trends in Ecology & Evolution 24: 67–68.
16. MPOB (2012) Malaysian Oil Palm Statistics 2011, 31st Edition. Malaysian Palm Oil Board, Ministry of Plantation Industries and Commodities, Malaysia.
17. IDS (2007) Sabah Development Corridor, Socio-Economic Blueprint 2008–2025, Harnessing Unity in Diversity for Wealth Creation and Social Wellbeing. Institute for Development Studies, Sabah Kota Kinabalu, Sabah.
18. Ancrenaz M, Goossens B, Gimenez O, Sawang A, Lackman-Ancrenaz I (2004) Determination of ape distribution and population size using ground and aerial surveys: a case study with orang-utans in lower Kinabatangan, Sabah, Malaysia. Animal Conservation 7: 375–385.
19. Abram NK, Kandasamy Y, Sheppard S (2013) Spatial Planning for Conservation and Sustainable Development 'SPaCe': Report for Sbah Structure Plan 2033. Kota Kinabalu: Living Landscape Alliance, HUTAN, WWF-Malaysia.
20. McMorrow J, Talip MA (2001) Decline of forest area in Sabah, Malaysia: Relationship to state policies, land code and land capability. Global Environmental Change 11: 217–230.
21. Sabah Forestry Department (2007) Vegetation Assessment and Classification for Lower Kinabatangan. Kota Kinabalu: Sabah Forestry Department & WWF.
22. Lackman-Ancrenaz I, Manokaran K (2008) Kinabatangan Biosphere Reserve Nomination Dossier. HUTAN.
23. Estes J, Othman NG, Ismail S, Ancrenaz M, Goossens B, et al. (2012) Quantity and Configuration of Available Elephant Habitat and Related Conservation Concerns in the Lower Kinabatangan Floodplain of Sabah, Malaysia. PLoS ONE 7: e44601.
24. Goossens B, Chikhi L, Jalil MF, Ancrenaz M, Lackman-Ancrenaz I, et al. (2005) Patterns of genetic diversity and migration in increasingly fragmented and declining orang-utan (*Pongo pygmaeus*) populations from Sabah, Malaysia. Molecular Ecology 14: 441–456.
25. Sha JCM, Henry B, Nathan S (2008) Status and Conservation of Proboscis Monkeys (*Nasalis larvatus*) in Sabah, East Malaysia. Primate Conservation 23: 107–120.
26. Sabah Land Ordinance (ver2010) State of Sabah Land Ordinance (Sabah Cap.68). Land Ordinance.
27. Crist EP, Cicone RC (1984) Application of the Tasseled Cap Concept to simulated thematic mapper data. Photogrammetric Engineering and Remote Sensing 50: 343–352.
28. Benz UC, Hofmann P, Willhauck G, Lingenfelder I, Heynen M (2004) Multi-resolution, object-oriented fuzzy analysis of remote sensing data for GIS-ready information. ISPRS Journal of Photogrammetry and Remote Sensing 58: 239–258.
29. Heumann BW (2011) An Object-Based Classification of Mangroves Using a Hybrid Decision Tree-Support Vector Machine Approach. Remote Sensing 3: 2440–2460.
30. Foody GM, Mathur A (2006) The use of small training sets containing mixed pixels for accurate hard image classification: training on mixed spectral responses for classification by a SVM. Remote Sensing of Environment 103: 179–189.
31. Li HT, Gu HY, Han YS, Yang JH (2010) Object-orientated classification of high-resolution remote sensing imagery based on an improved colour structure and a support vector machine. International Journal of Remote Sensing 31: 1453–1470.
32. Mohamed Nazeeb AT, Tang MK, Loong SG, Syed Shahar SAB (2008) Variable density plantings for oil palms (*Elaeis guineensis*) in Peninsular Malaysia. Journal of Oil Palm Research: 61–90.
33. MPOB (2010) Palm Oil Cost of Production Malaysia 2008: A Report of the MPOB Palm Oil Cost of Production Survey 2009: MPOB.
34. Town and Regional Planning Department (unknown) Soil Suitability Map for Agriculture. Town and Regional Planning Department, Sabah.
35. Monserud RA (1990) Method for comparing global vegetation map. ASA, Laxenbury. 90–40.
36. Landis JR, Koch GC (1987) The measurement of observer agreement for catagorical data. Biometric 33: 159–179.
37. Wilcove DS, Koh LP (2010) Addressing the threats to biodiversity from oil-palm agriculture. Biodiversity Conservation 19: 999–1007.
38. UNEP (2011) Oil palm plantations: threats and oppertunities for tropical ecosystems. UNEP Global Environmental Alert Service (GEAS): Taking the pulse of the planet: connecting science with policy December 2011. www.unep.org/geas.
39. Garcia-Ulloa J, Sloan S, Pacheco P, Ghazoul J, Koh LP (2012) Lowering environmental costs of oil-palm expansion in Colombia. Conservation Letters 5: 366–375.
40. Knight AT, Cowling RM (2007) Embracing Opportunism in the Selection of Priority Conservation Areas. Conservation Biology 21: 1124–1126.
41. Knight AT, Cowling RM, Difford M, Campbell BM (2010) Mapping human and social dimensions of conservation opportunity for the scheduling of conservation action on private land. Conservation Biology 24: 1348–1358.
42. Hoh C, Ishak-Amin J (2001) Oil Palm Cultivation in the Lower Kinabatangan Floodplain: A Case Study. Forum on Making Land Use Sustainable in the Lower Kinabatangan Floodplain. Kota Kinabalu.
43. Hai TC, Ng A, Preduente C, Pang C, Yee JTC (2001) Balancing the Need for Sustainable Oil Palm Development and Conservation: The Lower ISP National Seminal 2001: Strategic Directions for the Sustainability of the Oil Palm Industry. Kota Kinabalu, Sabah, Malaysia.
44. Vermeulen S, Goad N (2006) Towards better practice in smallholder palm oil production. International Institute for Environment and Development. London, UK.
45. Levin J, Ng G, Fortes D, Garcia S, Lacey S, et al. (2012) Profitability and Sustainability in Palm Oil Production. Analysis of Incremental Financial Costs and Benefits of RSPO Compliance. WWF-US, FMO, CDC.
46. Ng PHC, Goh KJ, Mahamooth TN, Petronella GAT (2013) Challenges in fertiliser and cover crop management. Planter 89: 409–431.

47. Paterson RRM, Moen S, Lima N (2009) The Feasibility of Producing Oil Palm with Altered Lignin Content to Control Ganoderma Disease. Journal of Phytopathology 157: 649–656.
48. MPOB (2010) Oil Palm Industry Labour: Issues, Performance & Sustainability (PLIPS).
49. Cowling RM, Pressey RL (2003) Introduction to systematic conservation planning in the Cape Floristic Region. Biological Conservation 112: 1–13.
50. Butchart SHM, Scharlemann JPW, Evans MI, Quader S, Aricò S, et al. (2012) Protecting Important Sites for Biodiversity Contributes to Meeting Global Conservation Targets. PLoS ONE 7: e32529.
51. Lindermayer DB, Fischer J (2006) Habitat fragmentation and landscape change: an ecological and conservation synthesis. USA: Island Press.
52. Zavaleta ES, Pasari JR, Hulvey KB, Tilman GD (2010) Sustaining multiple ecosystem functions in grassland communities requires higher biodiversity. Proceedings of the National Academy of Sciences 107: 1443–1446.
53. Gillies CS, St Clair CC (2008) Riparian corridors enhance movement of a forest specialist bird in fragmented tropical forest. Proceedings of the National Academy of Sciences 105: 19774–19779.
54. Stark DJ (2009) Proboscis monkey (*Nasalis larvatus*) Population Viability Analysis: Reassessment and management for wild populations in Borneo.: Oxford Brookes University.
55. Ancrenaz M, Meijaard E, Ambu LN, Lackman I, Ahmad E, et al. (in press) Of pongo, palms, and perceptions – A multidisciplinary assessment of Bornean orangutans in an oil palm context. Oryx In press.
56. Sabah Wildlife Department (2012) Orangutan Action Plan 2012–2016. In: Sabah Wildlife Department MoT, Culture and Environment, editor. Kota Kinabalu.
57. Sabah Wildlife Department (2012) Elephant Action Plan 2012–2016. In: Sabah Wildlife Department MoT, Culture and Environment, editor. Kota Kinabalu.
58. den Besten JW, Arts B, Verkooijen P (2013) The evolution of REDD+: An analysis of discursive-institutional dynamics. Environmental Science & Policy.
59. Alexander S, Nelson CR, Aronson J, Lamb D, Cliquet A, et al. (2011) Opportunities and Challenges for Ecological Restoration within REDD+. Restoration Ecology 19: 683–689.
60. RSPO (2013) Roundtable For Sustainable Palm Oil: An International Multi Stakeholder Initiative Transforming Markets to Make Sustainable Palm Oil The Norm.
61. Sabah State Government (1998) Sabah Water Resources Enactment. Kota Kinabalu.
62. Roever CL, van Aarde RJ, Leggett K (2013) Functional connectivity within conservation networks: Delineating corridors for African elephants. Biological Conservation 157: 128–135.
63. Broadbent EN, Asner GP, Keller M, Knapp DE, Oliveira PJC, et al. (2008) Forest fragmentation and edge effects from deforestation and selective logging in the Brazilian Amazon. Biological Conservation 141: 1745–1757.
64. FAO (2009) FAOSTAT online statistical service. Food and Agriculture Organization of the United Nations (FAO) Rome, Italy.

Aboveground Allometric Models for Freeze-Affected Black Mangroves (*Avicennia germinans*): Equations for a Climate Sensitive Mangrove-Marsh Ecotone

Michael J. Osland[1]*, Richard H. Day[1], Jack C. Larriviere[2], Andrew S. From[2]

1 U.S. Geological Survey, National Wetlands Research Center, Lafayette, Louisiana, United States of America, **2** Five Rivers Services, LLC, U.S. Geological Survey, National Wetlands Research Center, Lafayette, Louisiana, United States of America

Abstract

Across the globe, species distributions are changing in response to climate change and land use change. In parts of the southeastern United States, climate change is expected to result in the poleward range expansion of black mangroves (*Avicennia germinans*) at the expense of some salt marsh vegetation. The morphology of *A. germinans* at its northern range limit is more shrub-like than in tropical climes in part due to the aboveground structural damage and vigorous multi-stem regrowth triggered by extreme winter temperatures. In this study, we developed aboveground allometric equations for freeze-affected black mangroves which can be used to quantify: (1) total aboveground biomass; (2) leaf biomass; (3) stem plus branch biomass; and (4) leaf area. Plant volume (i.e., a combination of crown area and plant height) was selected as the optimal predictor of the four response variables. We expect that our simple measurements and equations can be adapted for use in other mangrove ecosystems located in abiotic settings that result in mangrove individuals with dwarf or shrub-like morphologies including oligotrophic and arid environments. Many important ecological functions and services are affected by changes in coastal wetland plant community structure and productivity including carbon storage, nutrient cycling, coastal protection, recreation, fish and avian habitat, and ecosystem response to sea level rise and extreme climatic events. Coastal scientists in the southeastern United States can use the identified allometric equations, in combination with easily obtained and non-destructive plant volume measurements, to better quantify and monitor ecological change within the dynamic, climate sensitive, and highly-productive mangrove-marsh ecotone.

Editor: Ben Bond-Lamberty, DOE Pacific Northwest National Laboratory, United States of America

Funding: This project was supported by the Department of Interior Southeast Climate Science Center (doi.gov/csc/southeast/index.cfm). The funders had no role in study design, data collection and analysis, decision to publish, or preparation of the manuscript.

Competing Interests: JCL & ASF are employees of Fiver Rivers Services, LLC, who provide technical support via a contract with the USGS. Via this contract with the USGS, Five Rivers Services personnel "provide technical support in the following areas: biological, ecological, and hydrologic research; environmental chemistry; geography; river studies; geographic information systems and spatial analysis; information technology, and administrative services." More information about Five Rivers Services and this contract with USGS can be found at this link: http://www.fiveriversservices.com/2011/04/26/five-rivers-services-awarded-us-geological-survey-contract/. There are no patents, products in development or marketed products to declare. This does not alter the authors' adherence to all the PLOS ONE policies on sharing data and materials.

* Email: mosland@usgs.gov

Introduction

In response to climate change and land use change, species distributions are changing across the globe [1–3]. Some of these distributional changes are resulting in comparatively large ecological transformations, especially where traditional ecosystem boundaries (i.e., ecotones) are migrating [4–6]. In areas with shifting ecotones, ecologists and environmental managers seek information that will enable them to better understand the ecological implications of ecosystem transformations.

In coastal wetland ecosystems, abrupt ecotones are common and occur across multiple abiotic gradients (e.g., inundation, salinity, and macroclimatic gradients) [7–10]. Since tidal wetland ecotones involve foundation plant species, sensu [11–14], with divergent growth forms and functionality (e.g., graminoid plants, microbial mats, succulent plants, and woody mangrove plants), the ecological implications of shifting ecotones in coastal wetlands can be substantial.

In the southeastern United States and other tropical-to-temperate climatic transition zones, the mangrove-salt marsh ecotone is dynamic, visually striking, and highly productive [15,16]. Mangroves are woody foundation plant species that are sensitive to extreme winter temperatures and dominant in warmer climates [17,18]. In contrast, salt marsh graminoids and salt marsh succulent plant foundation species are dominant in cooler coastal reaches [8,19] where extreme winter temperature events lead to mangrove mortality and/or limit mangrove forest development, reproduction, and dispersal [10,20,21]. At the poleward mangrove-marsh ecotone, mangrove abundance and coverage is winter temperature-sensitive in that it increases and decreases in response to the absence or presence of extreme winter events, respectively [22–24]. In response to changing climatic conditions, a decrease in the frequency, duration, and/or intensity of extreme winter temperatures is expected which would facilitate poleward mangrove range expansion at the expense of salt marsh vegetation [10]. In the southeastern U.S., the northward expansion of mangroves is expected to occur in Texas, Louisiana, and parts of Florida.

The ecological implications of mangrove expansion are poorly understood. Mangroves and salt marsh ecosystems provide many important ecosystem goods and services including fish and wildlife habitat, carbon storage, nutrient and sediment retention, coastal protection, and maintenance of coastal food webs and fisheries [25]. Although salt marsh and mangrove forest ecosystems are both highly valuable and have many ecosystem goods and services in common, there are structural and functional differences between these two ecosystems as well as differences in the provision of some ecosystem goods and services. Like many other ecotones, structural and functional complexity is higher at the mangrove-marsh ecotone than in adjacent graminoid or mangrove-dominated ecosystems. As a result, it is likely that ecological resilience and the supply of some ecosystem goods and services within the mangrove-marsh ecotone is equivalent to or even greater than in adjacent marsh or mangrove-dominated areas. However, our understanding of the ecosystem goods and services that would be affected by mangrove expansion and replacement of salt marsh is limited, and coastal wetland ecologists seek information and tools that will enable them to better quantify ecological changes associated with mangrove expansion at the expense of salt marshes [26–31]. Here, we provide equations that can be used to quantify changes in important aspects of aboveground mangrove structure near and within the dynamic mangrove-marsh ecotone.

Aboveground biomass is an important and fundamental metric for quantifying change in coastal wetland ecosystems. Many important coastal wetland ecological functions and services are affected by changes in aboveground biomass, structure, and productivity, including resilience to sea level rise [32–34], carbon storage [18,35–37], coastal protection [18,38,39], recreation, and fish and avian habitat [40,41]. Allometric equations provide an approach for estimating plant biomass and structural attributes via easily-obtained and non-destructive measurements. Most allometric models for determining mangrove aboveground biomass have been developed in tropical climates in the absence of extreme freeze events [42]. Within the freeze-dependent mangrove-marsh ecotone, traditional metrics for quantifying mangrove forest aboveground biomass are not appropriate because the morphology and architecture of freeze-affected mangrove trees is different than tropical mangrove trees. In the absence of freeze events and where freshwater and nutrient resources are abundant, tropical mangrove plants can develop into comparatively tall and straight trees. In contrast, at the poleward range limit of mangroves, severe freeze events can either lead to total mangrove mortality or cause aboveground structural damage that is followed by vigorous basal resprouting (Figure 1). As a result, mangroves in freeze-prone areas are often shorter, wider, multi-stemmed, and more shrub-like relative to their tropical counterparts growing in resource-rich environments, but see [43]. Tree-focused allometric models for quantifying aboveground biomass (i.e., dbh and/or height-based models) [42,44–49] are not suitable for freeze-affected and shrubby mangrove individuals.

Our objective was to develop aboveground allometric models for freeze-affected black mangroves (*Avicennia germinans*). In the southeastern United States, *A. germinans* is the most cold-tolerant mangrove species and the most abundant mangrove species at the northern range limit of mangrove forests. Based upon studies from other shrub-dominated ecosystems [50–53] and models for stunted mangroves in New Zealand [54], southern Florida (USA) [55,56], and Iran [57], we expected that crown area and height measurements could be used to develop allometric models for quantifying total aboveground biomass, leaf area, and leaf biomass for freeze-stunted *A. germinans* individuals.

Figure 1. Photos of freeze-affected black mangroves (*Avicennia germinans*) in Louisiana (USA) near their northern range limit. The upper two photos show the shrub-like morphology. The middle left photo shows the size of the smallest individuals included in the analyses. The middle right photo shows leaf damage from an extreme winter temperature event in January, 2014. The lower two photos show the high stem density of freeze-affected individuals.

Materials and Methods

Study Area

Allometric equations were developed using mangroves from two tidal wetland sites near Port Fourchon, Louisiana (USA; 29°8′52″N, −90°14′38″W and 29°6′58″N, −90°11′26″W; Figure 2), which is within the Mississippi River Deltaic Plain. Tidal wetlands in this area are abundant and dominated primarily by *Spartina alterniflora* (smooth cordgrass) and/or *A. germinans*. For more information about the study area, see [26,29,58–62]. Port Fourchon is near the northern limit of *A. germinans* along the Gulf of Mexico coast and is periodically exposed to extreme winter events that can damage or kill *A. germinans* (see photos in Figure 1), reducing areal coverage and limiting structural development. Historical accounts of *A. germinans* in the region document freeze-induced mortality during the 1960's [63] and 1980's [24]. The aerial extent and abundance of *A. germinans* has expanded and contracted in response to extreme winter events [22,24,63,64].

Although most of the *A. germinans* stands in the Port Fourchon vicinity were less than 3 m in height at the time of this study, some individuals were between 4.5 and 5 m in height and some stands were in the 3–4 m height range (RHD and MJO, personal observation). This study was conducted, with permission, on private lands owned by The Louisiana Land and Exploration Company LLC (LL&E) and managed by the Conoco Phillips Company/LL&E office in Houma, Louisiana.

Allometric Measurements

In February and May 2013, we selected a total of 56 *A. germinans* individuals for allometric equation development. We selected 40 individuals in February and 16 individuals in May. For each individual, we measured the height and the basal diameter at 30 cm above the soil surface. We also measured crown diameter in two perpendicular directions, across the widest crown section (CD_1) and at the widest section perpendicular to the first section (CD_2). The two crown diameter measurements were used to calculate crown area via the equation for an ellipse: crown area = $[(CD_1)/2)*(CD_2)/2)]*\pi$. When an individual had multiple stems, height was measured only on the tallest stem and basal diameter was measured only on the thickest stem. Whereas height was measured to the base of the highest leaf, crown diameter measurements were made to the edge of the most horizontally-distant leaf tips. Crown area and height were integrated into a single estimate of volume via the following equation: volume = crown area*height. Once *in situ* measurements were recorded, we cut each individual at the soil surface and transported the samples back to the laboratories of the U.S. Geological Survey's National Wetlands Research Center in Lafayette, Louisiana.

In the laboratory, the February and May samples were processed in a slightly different fashion since the May samples were collected explicitly for the development of leaf area and leaf biomass allometric equations. Individuals collected in February were dried to a constant mass at 60°C to determine total aboveground biomass (i.e., leaves were not separated from stems and branches prior to drying and weighing). In contrast, the May samples were separated into leaves and non-leaves (i.e., stem+ branch) prior to drying, and the area of fresh leaves was determined using a Li-3000C (Licor, Inc., Lincoln, NE, USA). The separated leaf and non-leaf samples were then dried to a constant mass at 60°C to determine leaf biomass, stem+branch biomass, and total aboveground biomass. The data for this study can be found in Dataset S1.

Data Analyses

Linear regression models were developed for the following response variables: (1) total aboveground biomass; (2) leaf biomass; (3) stem plus branch biomass; and (4) leaf area. The independent variables (i.e., crown area, height, volume, and basal diameter) were evaluated alone and in combination via simple and multiple regression analyses. Response and independent variables were natural log transformed prior to analysis. For each equation, we calculated a correction factor (CF) sensu Sprugel [65], where $CF = \exp(SEE^2/2)$ and where SEE is the Standard Error of the Estimate. These correction factors are to be applied to the back-transformed estimates of the response variables in order to correct for a small underestimation associated with using log transformed data in the regression analyses [50,66]. For example, for an allometric equation of the form $\ln(y) = a+b * \ln(x)$, y should be calculated as follows: $y = CF * (\exp[a+b * \ln(x)])$. Data analyses were conducted in SAS Version 9.3 (SAS Institute, Cary, NC, USA) and Sigma Plot Version 12.5 (Systat Software, Inc., San Jose, CA, USA).

Results

We measured *A. germinans* individuals within a height range of 31–157 cm, a crown diameter range of 4–180 cm, and an aboveground biomass range of 1–2,452 g (Table 1). For the smallest plants, the measurements of crown diameter were essentially the lengths of two pairs of leaves on a single un-

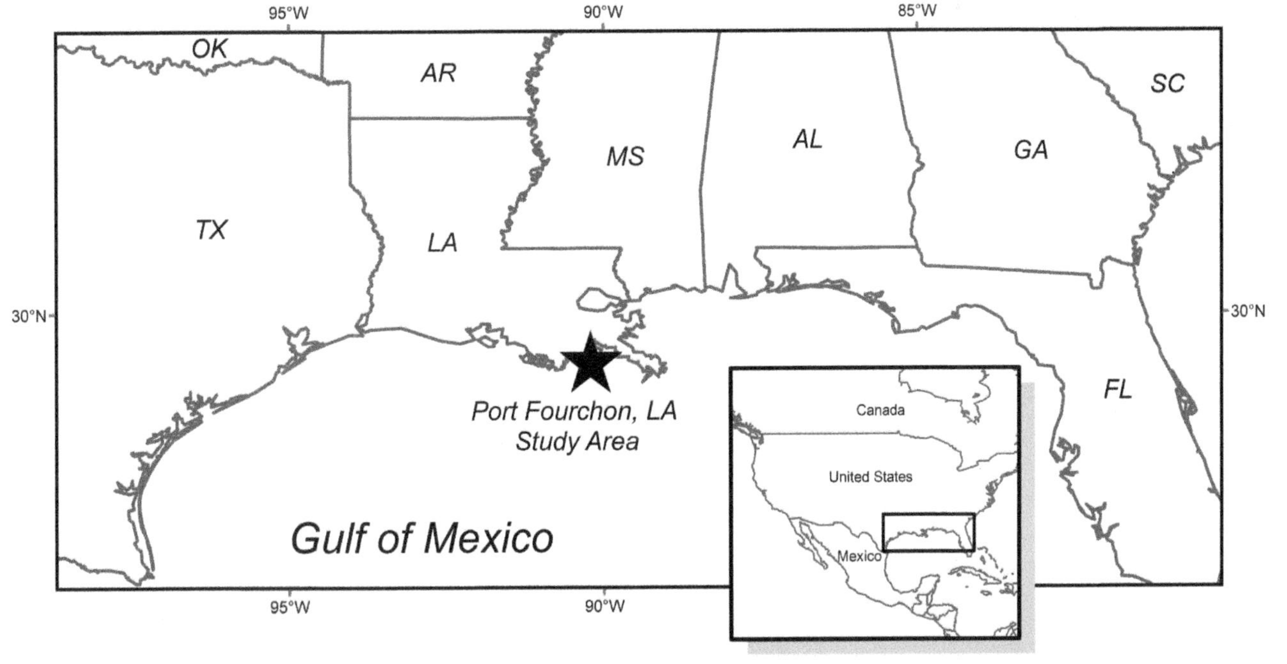

Figure 2. Map highlighting the mangrove-marsh ecotone where this study was conducted (Port Fourchon, Louisiana [USA]).

branched stem. In contrast, the largest plants contained multiple stems and branches extending to almost 1 m from a central base (Figure 1). We developed and compared multiple allometric equations using various combinations of predictor and response variables resulting in the selection of those shown in Table 2 and Table S1. The range limits presented in Table 1 identify boundaries for the application of the developed equations and should be used to avoid unwarranted extrapolation. Plant volume was selected as the best predictor of all of the response variables (i.e., leaf area, leaf biomass, stem plus branch biomass, and total above ground biomass) (Table 2, Figure 3, Table S1). Additional models based upon other combinations of measurements were also significant (Table S1). These additional models may be useful for researchers that want to determine biomass using alternative combinations of variables.

Discussion

In this study, our primary objective was to develop aboveground allometric equations for freeze-affected *A. germinans* individuals near and within the poleward mangrove-marsh ecotone in the southeastern United States. Allometric equations enable scientists to estimate plant community structural attributes via easily-obtained and non-destructive measurements. Although above-ground allometric models have been developed for mangrove species in different abiotic settings [42], most of these models are for trees growing in tropical wet climates. At the poleward range limit of mangrove forests, extreme winter temperature events affect the morphology of mangrove plants and result in plants that sometimes resemble shrubs (i.e., short, wide, and multi-stemmed individuals) rather than trees. To our knowledge, our study is the first to develop allometric models for freeze-affected and shrub-like *A. germinans* individuals at their northern range limit.

Mangrove shrub or dwarf morphologies can result from various abiotic conditions, including drought stress, freeze stress, salinity stress, nutrient limitation, and hydrologic isolation [44,67–71]. Shrub-based aboveground allometric models have been developed for some mangroves in some of these abiotic conditions. In New Zealand, near the southern latitudinal limit of mangroves and in an area with shrubby mangroves less than 1 m tall, Woodroffe [54] developed allometric equations that predict *A. marina* biomass using mean canopy width measurements. In an arid high-latitude mangrove forest in Iran, Parvaresh et al. [57] developed allometric equations that predict *A. marina* biomass from measurements of crown diameter. In a south Florida (USA) dwarf mangrove forest located in an irregularly-inundated basin setting, Ross et al. [55] developed allometric equations that predict *A. germinans*, *Rhizophora mangle*, and *Laguncularia racemosa* biomass from measurements of stem basal diameter and crown volume. In another south Florida dwarf mangrove forest, Coronado-Molina et al. [56] developed allometric equations that predict *Rhizophora mangle* biomass from measurements of height, crown diameter, and the number of prop roots. Collectively, these four studies along with our results show that crown area-focused measurements, sometimes in combination with other measurements (e.g., plant height, number of prop roots), can be used to develop allometric equations for estimating aboveground biomass of mangrove individuals with dwarf or shrub morphologies.

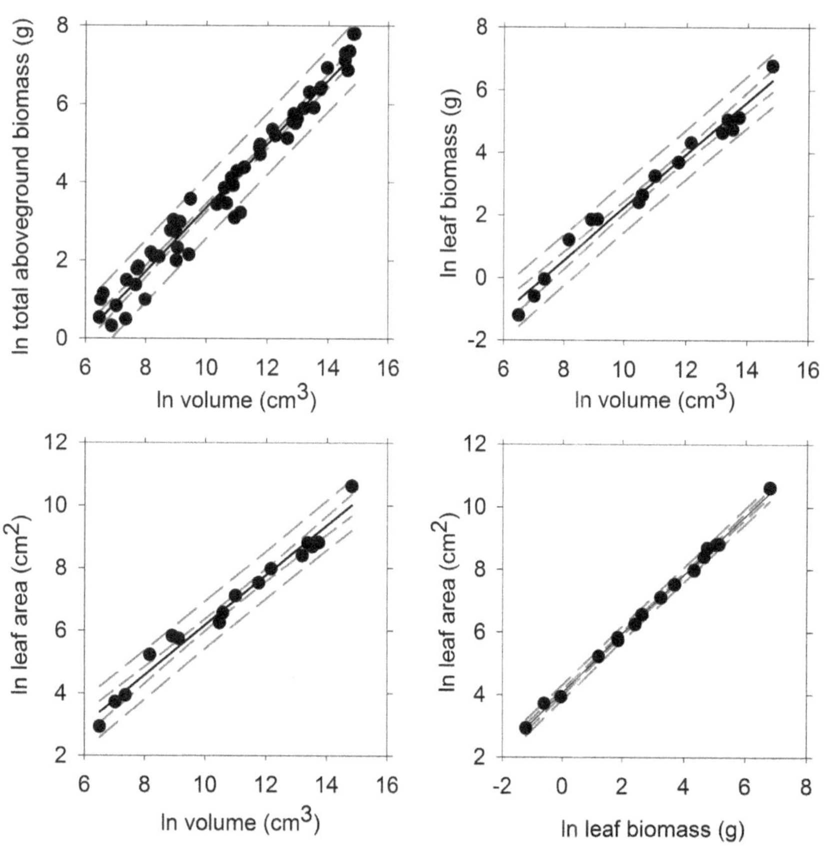

Figure 3. Aboveground allometric relationships for freeze-affected black mangrove (*Avicennia germinans*) individuals. The short and long dashed lines show the 95% confidence and prediction bands, respectively. Note the natural log scale on both axes.

Table 1. Sample size and measurement range of the variables used to develop allometric models for freeze-affected black mangroves (*Avicennia germinans*).

Variable	Sample Size	Minimum	Maximum
Height (cm)	56	31	157
Basal Diameter at 30 cm (mm)	56	1	31
Crown Diameter (cm)	56	4	180
Crown Area (cm^2)	56	19	23,629
Volume (cm^3)	56	641	2,811,792
Total Aboveground Biomass (g)	56	1	2,452
Leaf Biomass (g)	16	0.3	881
Stem and Branch Biomass (g)	16	2	1,556
Leaf Area (cm^2)	16	19	40,941

How do we envision that these equations will be used? In response to future changing winter temperature regimes and accelerated sea level rise, black mangroves are expected to migrate both poleward [10] and landward [72] along the northern Gulf of Mexico coast, in many cases at the expense of salt marshes, tidal freshwater systems, and/or upland forests. At many of these ecotones, ecologists will be challenged to characterize and monitor the structural changes that are already occurring or are expected to occur in the future. To the north and south of these ecotones (i.e., in salt marshes and large mangrove forests, respectively), methods for characterizing aboveground biomass and vegetation structure are well-established. Vegetation methods used in salt marshes are quite different than those used in mangrove forests. For example, whereas ground-layer measurements and vegetation samples (e.g., biomass clip plots and stem measurements from plants less than 2 m in height) are typically taken in small plots ($\sim$1-m^2) within marshes, tree-focused measurements (e.g., dbh and/or tree heights) are typically taken in large plots ($\sim$100-m^2) within tropical mangrove forests. Vegetation measurements in the freeze-affected and structurally-complex mangrove-marsh ecotone require a hybrid sampling approach that includes measurements tailored for the stunted morphology of freeze-affected mangroves (e.g., crown area and plant height measurements) as well as the multiple other vegetation strata that could be present ranging from graminoid and succulent marsh plants to large (i.e., 2–10 m tall) mangrove forest trees. See Osland et al. [36] for an example of the multiple plot sizes and stratum-specific measurements required to sample marshes and mangroves within a single study design. Sampling designs within mangrove-marsh ecotones often need to be customized to fit the range of plant species and morphologies present. Where conditions allow multi-stemmed shrubby mangroves to grow into large multi-stemmed trees, traditional tree-focused metrics can be applied to each individual stem contained within a tree [43].

Our findings and allometric equations provide a foundation for characterizing aboveground structural changes of freeze-affected black mangroves within mangrove-marsh ecotones. In combination with a robust sampling design, these equations can be used to quantify the temporal and spatial changes in plant community structure in response to the presence or lack of extreme climatic events including freeze events, drought, and hurricanes, as well as slower abiotic changes associated with sea level rise. Plant community structural information can be used to investigate the effects of ecosystem transformations and ecotone migration upon important ecosystem functions and services (e.g., sediment trapping, soil elevation change, nutrient cycling, fish and wildlife abundances, wave attenuation, carbon storage). When paired with carbon concentration data and information regarding below-ground processes and soil carbon accumulation, our equations can provide a foundation for quantifying the aboveground carbon present in freeze-affected black mangrove forests which is information that can contribute to estimates of carbon sequestration and inform best practices for carbon management and climate change mitigation [35,37]. The biomass equations can also be used to evaluate plant community development and functional equivalency after wetland restoration and/or management. The leaf area equations may be incorporated into ecophysiological and/or biogeochemical models, for example [60,73].

In summary, we developed aboveground allometric equations for freeze-affected black mangroves (*A. germinans*) which can be used to quantify the following: (1) total aboveground biomass; (2) leaf biomass; (3) stem plus branch biomass; and (4) leaf area. Plant

Table 2. Selected allometric equations for freeze-affected black mangrove (*Avicennia germinans*) individuals.

Response (y)	Predictor (x)	*a* (SE)	*b* (SE)	Adj-R^2	RMSE	CF	d.f.
Total Aboveground Biomass	Volume	−4.8045 (0.2250)	0.8157 (0.0204)	0.97	0.39	1.0781	56
Leaf Biomass	Volume	−6.2219 (0.3834)	0.8468 (0.0348)	0.98	0.36	1.0654	16
Stem plus Branch Biomass	Volume	−4.5075 (0.2971)	0.7682 (0.0269)	0.98	0.28	1.0388	16
Leaf Area	Volume	−1.8036 (0.3681)	0.7981 (0.0334)	0.97	0.34	1.0602	16
Leaf Area	Leaf Biomass	4.0626 (0.0437)	0.9418 (0.0121)	0.99	0.11	1.0057	16

These are all equations of the following form: $\ln(y) = a + b*\ln(x)$. CF is the correction factor sensu Sprugel [65]. Additional equations can be found in Table S1.

volume (i.e., a combination of crown area and plant height), was selected as the optimal predictor of these response variables. Coastal wetland scientists in the southeastern United States can use these equations to better quantify and monitor ecological changes within the dynamic and climate sensitive mangrove-marsh ecotone.

Supporting Information

Table S1 Additional allometric regression equations for freeze-affected black mangrove (*Avicennia germinans*) individuals.

Dataset S1 The data used to develop allometric equations for freeze-affected black mangrove (*Avicennia germinans*) individuals. An excel file that has two spreadsheets: (1) a data spreadsheet (spreadsheet entitled "data"); and (2) a spreadsheet that includes a description of column names and units contained within the data spreadsheet (spreadsheet entitled "legend").

Acknowledgments

We thank Erik Yando for assistance in the field, Ken Krauss for leaf area measurement recommendations, and the ConocoPhillips Company/ Louisiana Land and Exploration Company LLC for permission to conduct research on their land. The contents of the manuscript are solely the responsibility of the authors and do not necessarily represent the views of the Southeast Climate Science Center or the USGS. Any use of trade, product, or firm names is for descriptive purposes only and does not imply endorsement by the U.S. Government.

Author Contributions

Conceived and designed the experiments: MJO RHD JCL ASF. Performed the experiments: MJO RHD JCL ASF. Analyzed the data: MJO RHD. Wrote the paper: MJO RHD JCL ASF.

References

1. Parmesan C, Yohe G (2003) A globally coherent fingerprint of climate change impacts across natural systems. Nature 421: 37–42.
2. Millenium Ecosystem Assessment (2005) Ecosystems and human well-being: Synthesis. Washington, D.C.: Island Press. 137 p.
3. Chen IC, Hill JK, Ohlemüller R, Roy DB, Thomas CD (2011) Rapid range shifts of species associated with high levels of climate warming. Science 333: 1024–1026.
4. Allen CD, Breshears DD (1998) Drought-induced shift of a forest–woodland ecotone: rapid landscape response to climate variation. Proceedings of the National Academy of Sciences 95: 14839–14842.
5. Neilson RP (1993) Transient ecotone response to climatic change: some conceptual and modelling approaches. Ecological Applications: 385–395.
6. Risser PG (1995) The status of the science examining ecotones: A dynamic aspect of landscape is the area of steep gradients between more homogeneous vegetation associations. BioScience 45: 318–325.
7. Jiang J, Gao D, DeAngelis DL (2012) Towards a theory of ecotone resilience: Coastal vegetation on a salinity gradient. Theoretical Population Biology 82: 29–37.
8. Pennings SC, Bertness MD (2001) Salt marsh communities. In: Bertness MD, Gaines SD, Hay M, editors. Marine Community Ecology. Sunderland: Sinauer Associates. pp. 289–316.
9. Marani M, D'Alpaos A, Lanzoni S, Carniello L, Rinaldo A (2010) The importance of being coupled: Stable states and catastrophic shifts in tidal biomorphodynamics. Journal of Geophysical Research: Earth Surface 115: F04004.
10. Osland MJ, Enwright N, Day RH, Doyle TW (2013) Winter climate change and coastal wetland foundation species: salt marshes versus mangrove forests in the southeastern U.S. Global Change Biology 19: 1482–1494.
11. Ellison AM, Bank MS, Clinton BD, Colburn EA, Elliott K, et al. (2005) Loss of foundation species: consequences for the structure and dynamics of forested ecosystems. Frontiers in Ecology and the Environment 3: 479–486.
12. Angelini C, Altieri AH, Silliman BR, Bertness MD (2011) Interactions among foundation species and their consequences for community organization, biodiversity, and conservation. BioScience 61: 782–789.
13. Bruno JF, Bertness MD (2001) Habitat modification and facilitation in benthic marine communities. In: Bertness MD, Gaines SD, Hay J, editors. Marine Community Ecology. Sunderland: Sinauer Associates. pp. 201–218.
14. Dayton PK (1972) Toward an understanding of community resilience and the potential effects of enrichments to the benthos at McMurdo Sound, Antartica. In: Parker BC, editor. Proceedings of the Colloquium on Conservation Problems in Antartica. Lawrence: Allen Press. pp. 81–96.
15. McKee K, Rogers K, Saintilan N (2012) Response of salt marsh and mangrove wetlands to changes in atmospheric CO_2, climate, and sea level. In: Middleton BA, editor. Global Change and the Function and Distribution of Wetlands: Global Change Ecology and Wetlands. Dordrecht: Springer. pp. 63–96.
16. Saintilan N, Wilson NC, Rogers K, Rajkaran A, Krauss KW (2014) Mangrove expansion and salt marsh decline at mangrove poleward limits. Global Change Biology 20: 147–157.
17. Tomlinson PB (1986) The Botany of Mangroves. New York: Cambridge University Press. 413 p.
18. Alongi DM (2009) The Energetics of Mangrove Forests. New York: Springer. 216 p.
19. Adam P (1990) Saltmarsh Ecology. Cambridge: Cambridge University Press.
20. Cavanaugh KC, Kellner JR, Forde AJ, Gruner DS, Parker JD, et al. (2014) Poleward expansion of mangroves is a threshold response to decreased frequency of extreme cold events. Proceedings of the National Academy of Sciences 111: 723–727.
21. Pickens CN, Hester MW (2011) Temperature tolerance of early life history stages of black mangrove *Avicennia germinans*: implications for range expansion. Estuaries and Coasts 34: 824–830.
22. Giri C, Long J, Tieszen L (2011) Mapping and monitoring Louisiana's mangroves in the aftermath of the 2010 Gulf of Mexico Oil Spill. Journal of Coastal Research 27: 1059–1064.
23. Ross MS, Ruiz PL, Sah JP, Hanan EJ (2009) Chilling damage in a changing climate in coastal landscapes of the subtropical zone: a case study from south Florida. Global Change Biology 15: 1817–1832.
24. Woodroffe CD, Grindrod J (1991) Mangrove biogeography: the role of Quaternary environmental and sea-level change. Journal of Biogeography 18: 479–492.
25. Barbier EB, Hacker SD, Kennedy C, Koch EW, Stier AC, et al. (2011) The value of estuarine and coastal ecosystem services. Ecological Monographs 81: 169–193.
26. Perry CL, Mendelssohn IA (2009) Ecosystem effects of expanding populations of *Avicennia germinans* in a Louisiana salt marsh. Wetlands 29: 396–406.
27. Comeaux RS, Allison MA, Bianchi TS (2012) Mangrove expansion in the Gulf of Mexico with climate change: Implications for wetland health and resistance to rising sea levels. Estuarine, Coastal and Shelf Science 96: 81–95.
28. Chavez-Ramirez F, Wehtje W (2012) Potential impact of climate change scenarios on whooping crane life history. Wetlands 32: 11–20.
29. Henry KM, Twilley RR (2013) Soil development in a coastal Louisiana wetland during a climate-induced vegetation shift from salt marsh to mangrove. Journal of Coastal Research 29: 1273–1283.
30. Lewis DB, Brown JA, Jimenez KL (2014) Effects of flooding and warming on soil organic matter mineralization in *Avicennia germinans* mangrove forests and *Juncus roemerianus* salt marshes. Estuarine, Coastal and Shelf Science 139: 11–19.
31. Bianchi TS, Allison MA, Zhao J, Li X, Comeaux RS, et al. (2013) Historical reconstruction of mangrove expansion in the Gulf of Mexico: linking climate change with carbon sequestration in coastal wetlands. Estuarine, Coastal and Shelf Science 119: 7–16.
32. Krauss KW, McKee KL, Lovelock CE, Cahoon DR, Saintilan N, et al. (2014) How mangrove forests adjust to rising sea level. New Phytologist 202: 19–34.
33. Kirwan ML, Megonigal JP (2013) Tidal wetland stability in the face of human impacts and sea-level rise. Nature 504: 53–60.
34. Morris JT, Sundareshwar PV, Nietch CT, Kjerfve B, Cahoon D (2002) Responses of coastal wetlands to rising sea level. Ecology 83: 2869–2877.
35. Duarte CM, Losada IJ, Hendriks IE, Mazarrasa I, Marbà N (2013) The role of coastal plant communities for climate change mitigation and adaptation. Nature Climate Change 3: 961–968.
36. Osland MJ, Spivak AC, Nestlerode JA, Lessmann JM, Almario AE, et al. (2012) Ecosystem development after mangrove wetland creation: plant-soil change across a 20-year chronosequence. Ecosystems 15: 848–866.
37. Mcleod E, Chmura GL, Bouillon S, Salm R, Björk M, et al. (2011) A blueprint for blue carbon: toward an improved understanding of the role of vegetated coastal habitats in sequestering CO_2. Frontiers in Ecology and the Environment 9: 552–560.
38. Gedan KB, Kirwan ML, Wolanski E, Barbier EB, Silliman BR (2011) The present and future role of coastal wetland vegetation in protecting shorelines: answering recent challenges to the paradigm. Climatic Change 106: 7–29.
39. Feagin RA, Lozada-Bernard SM, Ravens TM, Möller I, Yeager KM, et al. (2009) Does vegetation prevent wave erosion of salt marsh edges? Proceedings of the National Academy of Sciences 106: 10109–10113.
40. Odum WE, McIvor CC, Smith TJ III (1982) The ecology of mangroves of south Florida: a community profile: U.S. Fish and Wildlife Service, Office of Biological Services, Washington, D.C. FWS/OBS-81/24. 144 p.

41. Caudill MC (2005) Nekton utilization of black mangrove (*Avicennia germinans*) and smooth cordgrass (*Spartina alterniflora*) sites in southwestern Caminada Bay, Louisiana. M.S. Thesis. Louisiana State University, Department of Oceanography and Coastal Sciences, Baton Rouge, LA.
42. Komiyama A, Ong JE, Poungparn S (2008) Allometry, biomass, and productivity of mangrove forests: A review. Aquatic Botany 89: 128–137.
43. Clough BF, Dixon P, Dalhaus O (1997) Allometric relationships for estimating biomass in multi-stemmed mangrove trees. Australian Journal of Botany 45: 1023–1031.
44. Cintrón G, Schaeffer Novelli Y (1984) Methods for studying mangrove structure. In: Snedacker SC, Snedaker JG, editors. The mangrove ecosystem: research methods. Paris: UNESCO. pp. 91–113.
45. Smith TJ III, Whelan KR (2006) Development of allometric relations for three mangrove species in South Florida for use in the Greater Everglades Ecosystem restoration. Wetlands Ecology and Management 14: 409–419.
46. Day JW, Conner WH, Ley-Lou F, Day RH, Navarro AM (1987) The productivity and composition of mangrove forests, Laguna de Términos, Mexico. Aquatic Botany 27: 267–284.
47. Fromard F, Puig H, Mougin E, Marty G, Betoulle JL, et al. (1998) Structure, above-ground biomass and dynamics of mangrove ecosystems: new data from French Guiana. Oecologia 115: 39–53.
48. Clough BF, Scott K (1989) Allometric relationships for estimating above-ground biomass in six mangrove species. Forest Ecology and Management 27: 117–127.
49. Clough B, Dixon P, Dalhaus O (1997) Allometric relationships for estimating biomass in multi-stemmed mangrove trees. Australian Journal of Botany 45: 1023–1031.
50. Cleary MB, Pendall E, Ewers BE (2008) Testing sagebrush allometric relationships across three fire chronosequences in Wyoming, USA. Journal of Arid Environments 72: 285–301.
51. Uresk DW, Gilbert RO, Rickard WH (1977) Sampling big sagebrush for phytomass. Journal of Range Management 30: 311–314.
52. Northup BK, Zitzer SF, Archer S, McMurtry CR, Boutton TW (2005) Above-ground biomass and carbon and nitrogen content of woody species in a subtropical thornscrub parkland. Journal of Arid Environments 62: 23–43.
53. McClaran MP, McMurtry CR, Archer SR (2013) A tool for estimating impacts of woody encroachment in arid grasslands: allometric equations for biomass, carbon and nitrogen content in *Prosopis velutina*. Journal of Arid Environments 88: 39–42.
54. Woodroffe CD (1985) Studies of a mangrove basin, Tuff Crater, New Zealand: I. Mangrove biomass and production of detritus. Estuarine, Coastal and Shelf Science 20: 265–280.
55. Ross MS, Ruiz PL, Telesnicki GJ, Meeder JF (2001) Estimating above-ground biomass and production in mangrove communities of Biscayne National Park, Florida (USA). Wetlands Ecology and Management 9: 27–37.
56. Coronado-Molina C, Day JW, Reyes E, Perez BC (2004) Standing crop and aboveground biomass partitioning of a dwarf mangrove forest in Taylor River Slough, Florida. Wetlands Ecology and Management 12: 157–164.
57. Parvaresh H, Parvaresh E, Zahedi G (2012) Establishing allometric relationship using crown diameter for the estimation of above-ground biomass of grey mangrove, *Avicennia Marina* (Forsk) Vierh in mangrove forests of Sirik, Iran. Journal of Basic and Applied Scientific Research 2: 1763–1769.
58. Patterson S, McKee KL, Mendelssohn IA (1997) Effects of tidal inundation and predation on *Avicennia germinans* seedling establishment and survival in a subtropical mangal/salt marsh community. Mangroves and Salt Marshes 1: 103–111.
59. Stagg CL, Mendelssohn IA (2011) Controls on resilience and stability in a sediment-subsidized salt marsh. Ecological Applications 21: 1731–1744.
60. Krauss KW, McKee KL, Hester MW (2014) Water use characteristics of black mangrove (*Avicennia germinans*) communities along an ecotone with marsh at a northern geographical limit. Ecohydrology 7: 354–365.
61. Alleman LK, Hester MW (2011) Refinement of the fundamental niche of black mangrove (*Avicennia germinans*) seedlings in Louisiana: applications for restoration. Wetlands Ecology and Management 19: 47–60.
62. McKee KL, Rooth JE (2008) Where temperate meets tropical: multi-factorial effects of elevated CO_2, nitrogen enrichment, and competition on a mangrove-salt marsh community. Global Change Biology 14: 971–984.
63. West RC (1977) Tidal salt-marsh and mangal formations of Middle and South America. In: Chapman VJ, editor. Wet coastal ecosystems. Amsterdam: Elsevier. pp. 193–213.
64. Michot TC, Day RH, Wells CJ (2010) Increase in black mangrove abundance in coastal Louisiana. Lousiana Natural Resource News. Newsletter of the Lousiana Association of Professional Biologists, January 2010. 4–5.
65. Sprugel D (1983) Correcting for bias in log-transformed allometric equations. Ecology 64: 209–210.
66. Beauchamp JJ, Olson JS (1973) Corrections for bias in regression estimates after logarithmic transformation. Ecology 54: 1403–1407.
67. Lugo AE, Snedaker SC (1974) The ecology of mangroves. Annual Review of Ecology and Systematics 5: 39–64.
68. Pool DJ, Snedaker SC, Lugo AE (1977) Structure of mangrove forests in Florida, Puerto Rico, Mexico, and Costa Rica. Biotropica 9: 195–212.
69. Feller IC, McKee KL, Whigham DF, O'Neill JP (2003) Nitrogen vs. phosphorus limitation across an ecotonal gradient in a mangrove forest. Biogeochemistry 62: 145–175.
70. Castañeda-Moya E, Twilley RR, Rivera-Monroy VH (2013) Allocation of biomass and net primary productivity of mangrove forests along environmental gradients in the Florida Coastal Everglades, USA. Forest Ecology and Management 307: 226–241.
71. Cintrón G, Lugo AE, Pool DJ, Morris G (1978) Mangroves of arid environments in Puerto Rico and adjacent islands. Biotropica 10: 110–121.
72. Doyle TW, Krauss KW, Conner WH, From AS (2010) Predicting the retreat and migration of tidal forests along the northern Gulf of Mexico under sea-level rise. Forest Ecology and Management 259: 770–777.
73. Lovelock CE, Ball MC, Choat B, Engelbrecht BM, Holbrook NM, et al. (2006) Linking physiological processes with mangrove forest structure: phosphorus deficiency limits canopy development, hydraulic conductivity and photosynthetic carbon gain in dwarf Rhizophora mangle. Plant, cell & environment 29: 793–802.

3

Are Regional Habitat Models Useful at a Local-Scale? A Case Study of Threatened and Common Insectivorous Bats in South-Eastern Australia

Anna McConville[1]*, Bradley S. Law[2], Michael J. Mahony[1]

1 School of Environmental and Life Sciences, University of Newcastle, Callaghan, New South Wales, Australia, 2 Forest Science Centre, Department of Primary Industries, Beecroft, New South Wales, Australia

Abstract

Habitat modelling and predictive mapping are important tools for conservation planning, particularly for lesser known species such as many insectivorous bats. However, the scale at which modelling is undertaken can affect the predictive accuracy and restrict the use of the model at different scales. We assessed the validity of existing regional-scale habitat models at a local-scale and contrasted the habitat use of two morphologically similar species with differing conservation status (*Mormopterus norfolkensis* and *Mormopterus* species 2). We used negative binomial generalised linear models created from indices of activity and environmental variables collected from systematic acoustic surveys. We found that habitat type (based on vegetation community) best explained activity of both species, which were more active in floodplain areas, with most foraging activity recorded in the freshwater wetland habitat type. The threatened *M. norfolkensis* avoided urban areas, which contrasts with *M.* species 2 which occurred frequently in urban bushland. We found that the broad habitat types predicted from local-scale models were generally consistent with those from regional-scale models. However, threshold-dependent accuracy measures indicated a poor fit and we advise caution be applied when using the regional models at a fine scale, particularly when the consequences of false negatives or positives are severe. Additionally, our study illustrates that habitat type classifications can be important predictors and we suggest they are more practical for conservation than complex combinations of raw variables, as they are easily communicated to land managers.

Editor: Danilo Russo, Università degli Studi di Napoli Federico II, Italy

Funding: Funding for this PhD project (A. McConville) was sourced from the following: The Tom Farrell Institute for the Environment; Donaldson Conservation Trust; Hunter-Central Rivers Catchment Management Authority; Royal Zoological Society of NSW (Ethel Mary Read Award); Wambo Coal Pty Ltd.; and Australian Geographic Society. B. Law is employed by Department Primary Industries (a government department). Whilst these organisations have generously provided funding for this research as part of an overall PhD, they have had no input into research design, direction, methodology or writing of this manuscript.

Competing Interests: Anna McConville received a small amount of funding ($2,000 AUD) for her PhD project from a commercial source (Wambo Coal Pty Ltd). Additionally, Anna McConville is employed as an ecological consultant and has worked for a company which was contracted to work for Wambo Coal. Anna McConville also had a personal relationship (friend) with the environmental officer at Wambo Coal when the funding was issued. The funding was used to assist with field expenses. There were no conditions attached to the funding, the company did not influence the design or direction of research and no comments were provided on written material. The project does not directly relate to the operations of Wambo Coal Pty Ltd and the authors do not stand to gain (future commissions or funding) from the publication of this manuscript. Additionally, this does not alter the authors' adherence to all the PLOS ONE policies on sharing data and materials.

* E-mail: anna.mcconville@uon.edu.au

Introduction

Habitat models [1,2] and predictive mapping [3,4] have been successful at classifying habitat of wide-ranging or lesser known species. Providing such models are validated [5], predictive maps can be used to direct conservation effort, identify areas for survey and research, and investigate disturbance effects [6–10]. Habitat selection operates at many spatial scales [11] and habitat modelling studies should select a scale that is appropriate to both the species and the model's intended use [12]. Habitat models prepared across the entire distribution of a species are likely to describe only broad patterns, whilst models based on a single population or family group are likely to contain location-specific variables that are unable to be extrapolated to broader scales [13]. Both broad- and fine-scale habitat models have value. For example, landscape-scale studies are useful to identify trends in habitat use by functional groups, such as the use of urban areas by insectivorous bats [14–16]. Additionally, studies that are undertaken at a fine-scale may identify factors that are important for conservation management that were unable to be identified from broader studies [17,18]. For land managers that may use habitat model predictions for land-use planning, it is essential to know whether models may be accurately applied to other spatial scales.

The factors that influence habitat use by a species are poorly understood for many species of conservation concern. Yet this information is fundamental to understanding threatening processes and to the development of adequate conservation strategies. *Mormopterus norfolkensis* Gray, 1839 is a hollow-roosting insectivorous bat species of which little is known. It occurs on the east coast of Australia and is listed as vulnerable under the New South Wales (NSW) *Threatened Species Conservation Act 1995* and as vulnerable C1 under the International Union for Conservation of Nature red list [19]. A morphologically similar species, *Mormopterus* species 2 [20], occurs in sympatry with *M. norfolkensis* in some parts of its range. Both species have morphological characteristics (high aspect ratio and high wing loading) and echolocation call designs (low frequency) indicative of species' that are adapted to foraging in open habitats [21–24]. Regional-scale habitat models based on systematically collected presence/absence data have been previously developed for both species (hereafter referred to as the 'regional-scale models') and indicate differences between the species in relation to the use of urban and floodplain habitats [25]. However, in a similar manner to the effect of false negatives on habitat models [26], it is possible that models created using presence/absence data inflate the importance of (and possibly over-predict) habitat for highly mobile species by assigning the same value to a site which had continuous foraging activity as a site where only a single animal passed by (a false positive).

The aim of this study was to determine if local-scale habitat associations, based on an index of habitat use (activity levels), are consistent with the predictions of regional-scale habitat models that were developed using presence/absence data. Additionally, we aimed to explore whether morphologically similar and sympatric species use different habitats by contrasting habitat models of *M. norfolkensis* with those of *M.* species 2. This was achieved with systematic local acoustic surveys in a variety of habitats, centred on a large roosting population of *M. norfolkensis* and nested within the extent of the previous regional-scale habitat modelling. We calculated different accuracy measures to explore model fit and predicted that if habitat use at a local-scale is consistent with regional-scale models [25], both species would be more likely to use floodplain habitats and *M. norfolkensis* would be less likely to occur in urban areas.

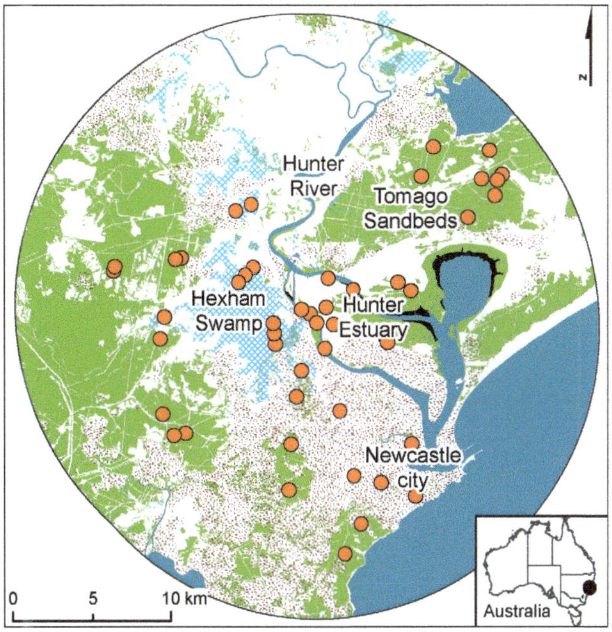

Figure 1. Study area location. Sample sites (orange circle), woody vegetation cover (green shading), mature mangrove forest (black shading), freshwater wetland (blue cross hatch), urban land-use (brown stipple) and major water bodies (blue shading) within 15 km.

Methods

Study area

The Hunter Estuary enters the Pacific Ocean at the Port of Newcastle (32°55'36"S 151°46'44"E) on the east coast of Australia (Figure 1). The area experiences a warm temperate climate (average monthly temperatures 8.4 – 25.6 °C) and average annual rainfall of 1134 mm [27]. The Hunter Estuary contains high quality estuarine vegetation communities, with the second largest area of mangroves (1600 ha) and the third largest area of saltmarsh (600 ha) in NSW [28]. Areas surrounding the Hunter Estuary have been highly modified by a long history of agriculture, coal mining and port-associated industries. The fertile floodplain of the Hunter River has been extensively cleared for agriculture with only small patches of native vegetation remaining and the construction of flood levies has altered the pattern of river flooding. Hexham Swamp, a large 2500 ha wetland dominated by freshwater vegetation (since the study, floodgates have been opened to re-establish tidal flushing and more estuarine conditions) and generally lacking trees, occurs in the west of the study area. Further west of Hexham Swamp the topography comprises low undulating hills with moderately nutrient-poor soils [29] that support remnant dry sclerophyll forest vegetation (Figure 1), which is typically young with few hollow-bearing trees (A. McConville pers. obs.). The most intact native vegetation (low open forest, woodland and heath) occurs to the north of the estuary on the flat, sandy and nutrient-poor soils of the Tomago sandbeds, which overlay a groundwater resource that is used locally. This area has been subject to disturbances such as sand mining and construction and the operation of military facilities and other infrastructure.

Study design and site selection

We used a 15 km buffer around two patches of mangroves in the Hunter Estuary that contained multiple maternity roosts and where over 700 *M. norfolkensis* have been trapped [30], to delineate the study area (approximately 76,500 ha; Figure 1), as the maximum recorded movement of *M. norfolkensis* over one night is 10 km (A. McConville, unpublished data 2012). We centred the study area on a large roosting population of *M. norfolkensis* to focus the study at a local-scale where we knew the species was active. This also assisted with obtaining sufficiently high activity levels to use as a response variable in statistical analyses. We classified the study area into 'habitat type' categories based on vegetation communities and human land-use. Habitat type categories were: freshwater wetland, dry sclerophyll forest, dry sclerophyll forest on sand; urban bushland; urban matrix; mature mangrove forest; and swamp oak forest. Small, distinct habitat types (e.g. coastal heath) and those that merged with other communities (e.g. saltmarsh) were not sampled. The freshwater wetland category included reed and rush vegetation that lacked tree cover and were dominated by *Typha bonariensis* or *Phragmites australis* with some open water. The dry sclerophyll forest habitat type consisted of open forest and woodland vegetation that was usually over 20 m in height, occurred on undulating slopes and was dominated by *Corymbia maculata* (spotted gum) and *Eucalyptus fibrosa* (broad-leaved ironbark) with *Eucalyptus punctata* (grey gum) occurring occasionally. The dry sclerophyll forest on sand habitat type was located on mostly flat, sandy plains and consisted of low (10 – 20 m height) open forest and woodland vegetation communities dominated by on flat terrain by *Eucalyptus haemastoma* (scribbly gum) and *Angphophora costata* (smooth-barked apple) or *Eucalyptus parramattensis* subsp. *decadens* (Earp's gum) and by *Eucalyptus pilularis* (blackbutt) on sandy rises. Urban bushland consisted of remnants (8 – 80 ha) of dry sclerophyll forest that were completely surrounded by the matrix of suburban and industrial development. Urban matrix habitat was located in backyards in well-established suburbs of Newcastle and contained scattered ornamental trees and shrubs. Mangrove forest was mostly a monoculture of *Avicennia marina* subsp. *australasica* (grey mangrove) and the swamp oak forest habitat type consisted of small remnant patches (0.4 – 5 ha) dominated by *Casuarina glauca* (swamp oak). A total of 47 sites were sampled, consisting of nine freshwater wetland, eight dry sclerophyll forest, eight dry sclerophyll forest on sand; five swamp oak forest, five urban bushland, five urban matrix, and seven mangrove forest habitat type sites. Sites were not selected randomly, but rather to achieve a spread of sites within the available habitat type and across the study area. Sites were separated by at least 250 m and were located on both private and public land.

Bat call sampling

Sampling was undertaken between25 January and 25 February 2011, following the maternity season (after weaning, when young are flying). One bat detector (Anabat II with CFZCAIM or Anabat SD1, Titley Electronics, Balina, Australia) was placed at each sample site to passively record for two entire nights. Up to seven sites were sampled on any one night and 10 different detectors were used during the study. Microphones were housed in a 1 m tall plastic pipe to provide some weather protection and were aimed up at a 45 ° angle. Detectors were set in the open, or aimed along flyways and away from vegetation in forested areas to minimise sound attenuation. Surveys were not conducted on windy or rainy nights or on a full moon to avoid any possible effect of these factors on bat activity [31]. Bat detectors were calibrated prior to sampling to ensure that detector sensitivity was similar across all detectors. The average minimum nightly temperature (sourced from the nearest weather station) was calculated for each site and included as candidate variables in analyses. Rainfall and wind speed data were also collected from nearby weather stations and were used to verify that substantial rainfall or high winds were not experienced during sampling.

We developed a filter in AnalookW [32] to reduce the number of call sequences for analysis by excluding most non-*Mormopterus* calls. The following filter parameters were used: smoothness = 30; body over 1000 ms; $F_c = 22–39$ kHz; $S_c = -15–20$ OPS; and $D_c = 3.5 – 20$ ms. We tested the performance of the filter on a reference library of *M. norfolkensis*, *M.* species 2 and *M.* species 4 calls (n = 104) and found that 95.4% were correctly identified. The reference calls that were excluded using the filter were reviewed and assessed to be poor quality recordings, mostly from old tape units. To test the filter performance on remotely recorded data, we then ran the filter through manually identified calls from one of the survey sites. In this test, the filter correctly identified 99.4% of calls (n = 639) as *Mormopterus* spp. and we considered this to be an acceptably low level of misclassification.

Bat calls that passed the *Mormopterus* filter were then manually identified using AnalookW to view calls and extract parameters. A key to the identification of *Mormopterus* spp. in the study area was developed based on a broader NSW guide [33] to make the process less subjective. Each pass (call sequence) was assigned to one of three categories - definite, probable or unidentified, according to the confidence with which an identification could be made [34]. During manual identification, feeding buzzes were identified by their characteristic shape [33]. Passes that did not contain any search phase pulses or contained fewer than three pulses were not considered. Definite and probable identifications were included in the analyses. We used total activity (number of passes) recorded for each species, at each site (n = 2 nights), to create habitat models. Additionally, we calculated average nightly bat activity (number of passes per night) to allow comparison with other studies and calculated foraging activity (percentage of passes that contained a feeding buzz) for each *Mormopterus* spp. at each site.

Landscape variables

We used ArcGIS (version 9.3, ESRI, Redlands, CA, USA) to derive a number of environmental variables describing the sample sites and surrounding area. The slope and elevation layers were derived from a 25 m digital elevation model (DEM), which was interpolated (drainage enforced) using 10 m contours and drainage lines and regions from 1:25,000 topographic maps. We created a vegetation coverage layer from regional vegetation mapping [35] with non-woody vegetation types removed (ie grassland and reeds). We also created a freshwater wetland layer consisting of communities dominated by reeds (both in a natural state and used for cattle grazing) as previous predictive mapping indicated that freshwater wetlands were suitable (typically > 0.5 probability of occurrence [25]). We excluded forested wetland types as they occupied only a relatively small portion of the study area. The urban land-use layer was based on land-use mapping (NSW Landuse) obtained from the NSW Office of Environment and Heritage and the soil type was based broadly on soil landscape mapping [29]. The line-work of all layers was revised following a review of recent aerial photography (ESRI, Redlands, CA, USA, 2001) and to ensure that all layers were consistent. We used ArcGIS with the Patch Analyst extension (version 4.0 [36]) to create a series of patch size and configuration variables from the vegetation dataset (Table S1) broadly following McGarigal and McComb [37]. A patch was considered to be > 10 trees together with a minimum area of 500 m^2. Core areas were considered to be > 100 m from the vegetation edge.

To determine the spatial scale at which habitat use was occurring, we created a series of concentric buffers at 250 m, 500 m, 1 km, 2.5 km and 5 km around each sample site. We then quantified the amount and configuration of environmental variables (Table S1) within each buffer for each site. Many of the environmental variables calculated from buffers had highly skewed distributions that could not be transformed to a normal distribution, so we converted these to ecologically relevant categories for use in the analyses (Table S1).

Statistical analyses

We used R [38] to prepare models and undertake other statistical analyses using the MASS [39], PresenceAbsence [40], verification [41] and visreg [42] packages. We use generalised linear models (GLMs) with a negative binomial distribution (glm.nb from MASS package) to model the activity of *M. norfolkensis* and *M.* species 2 against environmental variables (Table S1). We excluded the mangrove forest sites from statistical analyses as the very high activity levels were overly influential, most likely due to the use of mangroves as roosts [30]. We first constructed univariate models for each explanatory variable and species. We then ranked the univariate models by Akaike Information Criterion value corrected for small sample size (AICc) [43], with the highest ranking model having the lowest AICc. To reduce the number of variables offered to final candidate models, the variables within the four highest ranking univariate models were then selected for each species for consideration in the final candidate models. We avoided the use of correlated variables (Pearson's $r > 0.6$ or $r < -0.6$) in final candidate models by removing the variable which was the most difficult to interpret. Where landscape buffer variables of the same category were high ranking, only the highest-ranking buffer value was selected for consideration in the final models [44].The final model set contained all combinations of the four highest-ranking and non-correlated variables for each species. The habitat type variable was not combined with other variables to minimise the potential for model overparamaterisation.

We then calculated AICc and model weight (which is interpreted as the likelihood that the model is the best among the 95% confidence model set) [44] for each of the final candidate models. The best-fitting model for each species was considered to have the lowest AICc ranking and any model within two AICc points of the top model was considered to have strong support [44]. The variation explained by each model was assessed by calculating the Nagelkerke's R^2 value [45]. Additionally, to assess the predictive capacity of the models we calculated the area under the curve (AUC) for the receiver operating characteristic (ROC) [46,47] from the presence/absence records of each species. An AUC value of 0.5 suggests a completely random model and 1 indicates perfect discrimination. The traditional academic point system [48] was used as a rough guide for classifying the fit of each model with AUC values [49] where AUC values under 0.7 are considered poor, values of 0.7 to 0.8 are rated as fair, 0.8 to 0.9 as good and those over 0.9 as excellent. Model residuals were checked for spatial autocorrelation using Moran's I [50]. We plotted partial response curves for the best-fitting models using the conditional response option in the visreg package [42]. We used a chi-square goodness of fit test to compare percentage foraging activity between habitat types and we excluded habitat types with < 2 feeding buzzes from this analysis (urban matrix and urban bushland for *M. norfolkensis* and urban matrix, urban bushland and swamp oak forest for *M.* species 2). All results are reported as mean ± standard error.

Validation of regional-scale habitat models

We also used data collected during this study to assess the predictive ability of previous regional-scale habitat mapping undertaken for *M. norfolkensis* and *M.* species 2 within the study area [25]. The regional-scale habitat models were prepared from data collected using bat detectors set in a similar manner as this study (mostly for two nights and aimed along flyways and tracks), within an overlapping study area of 770,000 ha [25]. The regional predictive maps were created from the best-fitting generalised linear mixed models of presence/absence data [25]. See [25] for full methodology used to create regional habitat models.

We used the AUC for the ROC of regional-scale predictions against the local-scale dataset as a threshold-independent measure of predictive accuracy. To do this, we extracted the probability of occurrence value from the regional-scale predictive mapping [25] for each of our sample sites in ArcGIS and converted the activity levels to presence/absence records. Additionally, we used threshold-dependent confusion matrices to calculate the following accuracy measures to describe model predictive ability: sensitivity, specificity, prevalence, correct classification rate (CCR), kappa and true skill statistic [46,51]. We calculated the occurrence threshold for the confusion matrices as the probability of occurrence value from the regional-scale mapping where sensitivity was equal to specificity for the local-scale data. The kappa and TSS accuracy measures range from -1 to $+1$, where a value of 1 indicates perfect agreement and values of zero or less indicate a performance no better than random [51,52]. Kappa values of 0 – 0.4 are considered to indicate slight to fair model performance, values of 0.4 – 0.6 moderate, 0.6 – 0.8 substantial and 0.8 – 1.0 almost perfect [53] and this ranking was also applied to TSS values.

Ethics statement

This project was conducted under scientific licence (S132C under National Parks Act 1974; S12460) and as we used only indirect survey methods (ultrasonic detectors to record bat calls) we did not require animal ethics approval.

Results

In all, 5981 *M. norfolkensis* and 854 *M.* species 2 passes were recorded, including 402 *M. norfolkensis* (6.7% of total calls) and 58 *M.* species 2 feeding buzzes (6.8% of total calls) at 47 sites during the study. Site activity levels (63.6±31.7 average nightly passes for *M. norfolkensis* and 9.1±3.3 average nightly passes for *M.* species 2) ranged from 0 – 1272.5 average nightly passes for *M. norfolkensis* and 0 – 144.5 average nightly passes for *M.* species 2. Lower foraging activity was recorded for *M.* species 2 (0.31±0.11, range 0 – 4 feeding buzzes) than *M. norfolkensis* (2.14±1.31, range 0 – 60 feeding buzzes). *Mormopterus* species 2 was more widespread, occurring at 72.3% of sites compared to *M. norfolkensis*, which occurred at 57.4% of sites. *Mormopterus* species 2 was recorded at each of the different habitat types sampled and *M. norfolkensis* was recorded at all habitat types except for the urban matrix (Figure 2). We recorded very high levels of activity of *M. norfolkensis* at mangrove forest sites (403.4±170.7 average nightly passes) being over 50 times greater on average than other habitat types (freshwater wetland had the second highest activity with 7.9±2.1 average nightly passes; Figure 2).

Percentage foraging activity differed significantly between freshwater wetland, mangrove forest, swamp oak forest and dry sclerophyll forest on sand habitat types for *M. norfolkensis* (df = 3, $\chi 2 = 25.653$, $P < 0.001$) and also between freshwater wetland, mangrove forest, and dry sclerophyll forest on sand habitats for *M.* species 2 (df = 2, $\chi 2 = 33.868$, $P < 0.001$). The freshwater wetland habitat type had the greatest percentage foraging activity for both

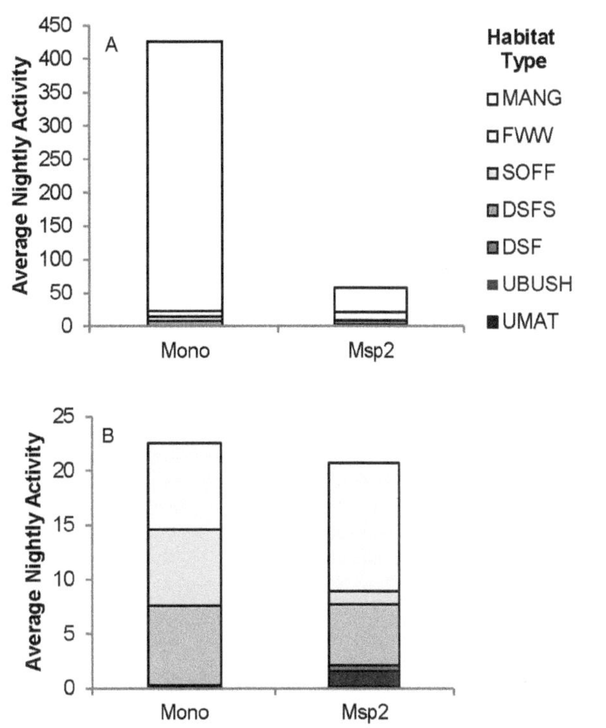

Figure 2. Average nightly bat activity. Average nightly bat activity (number of passes) recorded for each species in the study area across a) all seven landscape categories and b) landscape categories excluding mangrove forest. Abbreviations are: *Mormopterus norfolkensis* (Mono); *M.* species 2 (Msp2); FWW - freshwater wetland; DSF - dry sclerophyll forest and woodland; DSFS - dry sclerophyll forest and woodland on sand; UBUSH - urban bushland remnant; UMAT - urban matrix; MAN - mangrove forest; and SOFF - swamp oak forest.

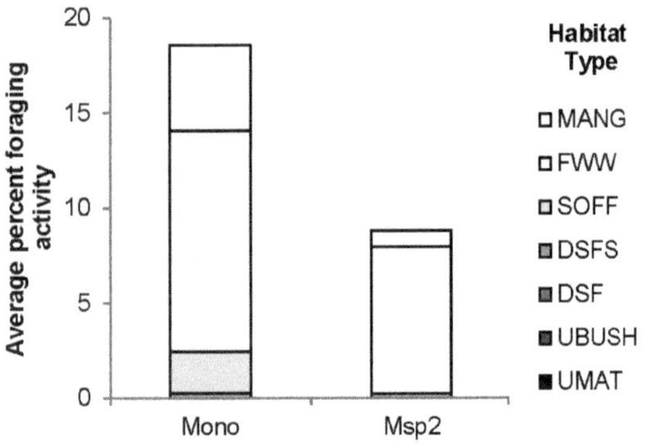

Figure 3. Average percentage foraging activity. Average percentage foraging activity recorded for *Mormopterus norfolkensis* (Mono) and *M.* species 2 (Msp2) across seven different habitat types. FWW - freshwater wetland; DSF - dry sclerophyll forest and woodland; DSFS - dry sclerophyll forest and woodland on sand; UBUSH - urban bushland remnant; UMAT - urban matrix; MAN - mangrove forest; and SOFF - swamp oak forest.

species (Figure 3). No foraging activity by either species was recorded in the urban habitat types and no *M.* species 2 foraging activity was recorded in swamp oak forest.

Local-scale habitat models

Habitat type best predicted the activity levels of both *M. norfolkensis* and *M.* species 2 in the best-fitting GLMs compared to other environmental variables sampled. *Mormopterus norfolkensis* was more likely to be active in freshwater wetland, swamp oak forest and dry sclerophyll forest on sand compared to dry sclerophyll forest ($P < 0.001$), in the best fitting model (Table 1; Figure 4). Urban habitats were less likely to be used by *M. norfolkensis* than dry sclerophyll forest in the best-fitting model, but this was not significantly different (Table 1; Figure 4). There were two other models with strong support (within 2 AICc points of the best-fitting model) and four models in the 95% confidence set for *M. norfolkensis*. In the first supported model, *M. norfolkensis* was likely to be more active at low elevations and in areas without urban land-use within 500 m (Table 1). In the second supported model, *M. norfolkensis* was likely to be more active in areas with no urban land-use within 500 m and that had sandplain and floodplain soils compared to soils on undulating slopes (Table 1). ROC plots indicated that the best-fitting model was an excellent fit to the data (AUC = 0.92; Table 1) and explained a considerable portion of the variation in the data (Nagelkerke's $R^2 = 0.78$). Other models were poorer fitting (AUC < 0.85) and explained less variation in the data (Nagelkerke's $R^2 < 0.70$; Table 1).

Mormopterus species 2 was more likely to occur in freshwater wetland and dry sclerophyll forest on sand compared to dry sclerophyll forest ($P < 0.001$; Table 1; Figure 4), in the best fitting model. There was no significant difference in the activity of *M.* species 2 in swamp oak forest, urban bushland and urban matrix compared to dry sclerophyll forest habitat types (Table 1; Figure 4). However, there was a trend for *M.* species 2 to be more active in swamp oak forest and urban bushland and less active in the urban matrix than the dry sclerophyll forest (Table 1; Figure 4). There was one other model in the 95% confidence set for *M.* species 2, but it did not have strong support. ROC plots indicated that the best-fitting model was a fair fit to the data (AUC = 0.76) and it explained considerable variation in the data (Nagelkerke's $R^2 = 0.77$; Table 1). There was no evidence of spatial autocorrelation in the residuals for models in the 95% confidence set for either of the study species ($P > 0.564$).

Validation of regional-scale habitat models

Regional-scale model accuracy measures were calculated using a probability of occurrence threshold (above which we classify a presence and below which we classify an absence) of 0.34 for both species, as determined from the value of the intersection of sensitivity and specificity measures (see Figure S1). The regional-scale models had a good fit to the local data for both species as measured by AUC (> 0.72; Table 2), while kappa and TSS accuracy measures were slight to fair (0.255 – 0.317; Table 2). Presence records occurred generally proportional to predictions, indicating a good level of model calibration for *M. norfolkensis* (Figure 5). *Mormopterus norfolkensis* records did tend to occur more often than predicted in areas mapped as having a low probability of occurrence (0 – 0.4 probability of occurrence; Figure 5). However, prevalence was not greatly under-estimated for *M. norfolkensis* overall (Figure 5). Figure 6 illustrates the predictive accuracy of the regional-scale mapping for *M. norfolkensis*, displaying false negatives in the north of the study area (which coincides with the dry sclerophyll forest on sand habitat type) and false positives in the low-lying parts of Newcastle city. *Mormopterus* species 2 was more prevalent than predicted by regional-scale models (Figure 5) and false negatives occurred in western parts of the study area, coinciding with dry sclerophyll forest habitat types (Figure 6).

Discussion

Few studies test the performance of habitat models against systematically collected and independent datasets [54–56]. Our findings suggest that whilst the regional-scale habitat models are useful to describe habitat for these poorly understood species, they should be used cautiously at a local-scale, particularly when the consequences of errors are severe. Habitat type categories best predicted the activity of both species, compared to other environmental variables sampled, confirming previous findings that both species are more active in floodplains and that the threatened *M. norfolkensis* avoids urban areas. Additionally, these habitat type categories are easily communicated to land managers, illustrating that they are not only good summaries of environmental variation at a local-scale, but also practical tools for conservation.

Local-scale habitat models

Both species were recorded more frequently and site activity levels were higher than recorded during previous studies in both urban and forested areas [14,16,57,58]. Activity levels were also considerably higher than those recorded for the regional-scale study, where mangroves were not sampled and freshwater wetlands were under-sampled [25]. This allowed habitat use to be modelled using activity levels, which are preferable to presence/absence data that may inflate the importance of some habitats for mobile insectivorous bats due to false positives. The activity of both *M. norfolkensis* and *M.* species 2 was best explained by habitat type categories, which were based on vegetation communities that reflect environmental conditions such as soil type, climate, geology and topography. Habitat types may also incorporate other aspects such as vegetation structure, condition, disturbance history and the presence of important components such as hollow-bearing trees or rock outcrops, depending on the target species and the project aims. In this way, habitat types are likely to better represent the environmental variation in a

Table 1. Summary from habitat analysis of a) *Mormopterus norfolkensis* and b) *M.* species 2 surrounding a large *M. norfolkensis* maternity roost.

Variable (s)	Estimate	Standard Error	Z	p	Log(L)	AICc	Wi	△AICc	ROC	R^2
a) Mormopterus norfolkensis										
Model 1 - Best-fitting model					-89.17	192.89	0.38	0	0.92	0.76
Intercept	-0.98	0.76	-1.29	0.20						
DSF versus FWW	3.75	0.90	4.18	<0.001*						
DSF versus SOFF	3.62	0.99	3.65	<0.001*						
DSF versus DSFS	3.66	0.91	4.01	<0.001*						
DSF versus URB	-1.32	1.33	-0.99	0.322						
Model 2					-92.19	193.53	0.19	0.64	0.83	0.66
Intercept	-0.65	0.70	-0.93	0.353						
Low elevation	3.41	0.74	4.58	<0.001*						
Urban land-use presence within 500 m	-1.75	0.65	-2.69	0.007*						
Model 3					-91.26	194.29	0.18	1.41	0.85	0.70
Intercept	-0.70	0.69	-1.02	0.307						
Urban land-use presence within 500 m	-1.69	0.67	-2.53	0.011*						
Undulating Slope versus Sandplain	3.29	0.82	4.00	<0.001*						
Undulating Slope versus Floodplain	3.60	0.77	4.66	<0.001*						
b) Mormopterus species 2					-101.76	221.01	0.93	0	0.76	0.77
Intercept	0.00	0.52	0	1.000						
DSF versus FWW	3.16	0.63	4.99	<0.001*						
DSF versus SOFF	0.88	0.76	1.15	0.251						
DSF versus DSFS	2.42	0.65	3.72	<0.001*						
DSF versus UBUSH	1.03	0.76	1.36	0.173						
DSF versus UMAT	-0.92	1.00	-0.92	0.359						

*denotes significant coefficients at $\alpha = 0.05$.
FWW - freshwater wetland; DSF - dry sclerophyll forest and woodland; DSFS - dry sclerophyll forest and woodland on sand; UBUSH - urban bushland remnant; UMAT - urban matrix; MAN - mangrove forest; and SOFF - swamp oak forest.
Total activity was modelled using generalised linear models with a negative binomial distribution. Data presented includes the coefficient estimates, Nagelkerke's R^2 and ROC AUC values for the best-fitting and supported (within 2 AICc) models.

particular area, compared to the combination of a small number of raw variables (such as elevation, aspect, etc). The classification of landscapes into different habitat types by experts is difficult at large spatial scales and has been previously found to offer little benefit over raw environmental variables [59]. However, we found that the creation of habitat type spatial layers can be efficiently compiled based on vegetation mapping combined with local expert knowledge for small study areas. Additionally, broad habitat type categories are easily communicated to conservation managers and authorities, which is beneficial compared to interpreting complex relationships with multiple raw variables.

Mormopterus norfolkensis and *M.* species 2 were more active in mangrove forest than any other habitat type sampled. This is most likely due to the abundant hollow resources in the Hunter Estuary mangrove forests, which have been found to be important roosting habitat for *M. norfolkensis* maternity colonies [30]. The percentage foraging activity in the mangrove forest was relatively low for *M. norfolkensis* and radio-tracking of bats indicated that they quickly move out of mangrove forests upon exiting roosts [30]. Additionally, very little *M. norfolkensis* activity has been recorded near hollow-depauperate mangrove forests elsewhere in NSW [60,61]. As such, we suggest that mangrove forests within the study area represent roosting habitat for *M. norfolkensis* and that the foraging activity recorded during this study was opportunistic as bats exited and returned to roosts. Mangrove forests may not be as important for *M.* species 2 with less activity recorded and almost 45 times fewer *M.* species 2 individuals were captured during trapping within the Hunter Estuary mangrove forests [30].

When the mangrove forest sites were excluded from analyses, we found that both *M. norfolkensis* and *M.* species 2 were more active in freshwater wetland habitats and percentage foraging activity was also greatest in freshwater wetland for both species. Riparian habitats often support high levels of bat activity, which is likely to be due to high insect prey abundance [62–66]. Additionally, more productive soils (as are often associated with floodplains) have also been found to be related to high bat activity levels [14,67]. These productive low-lying floodplain areas were previously predicted to be high habitat quality by regional habitat

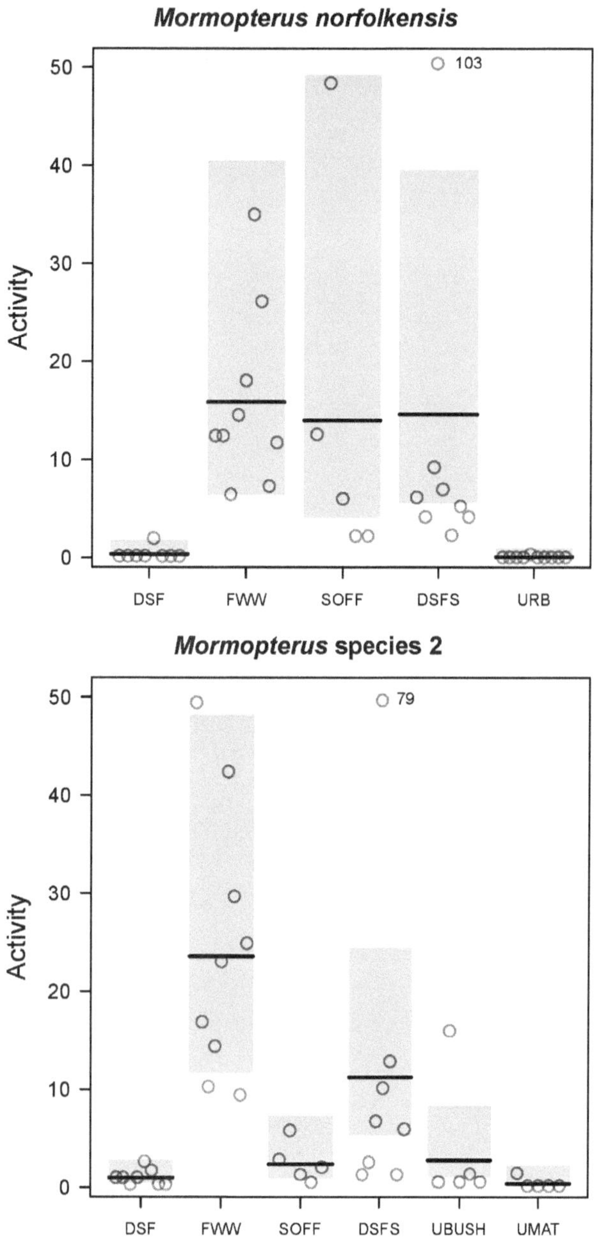

Figure 4. Habitat model response graphs. Partial conditional response graph of the relationship between the relative activity (number of passes) and habitat types included in the best-fitting negative binomial generalised linear models for *Mormopterus norfolkensis* and *Mormopterus* species 2. The grey shading represents 95% confidence intervals. Outliers are indicated by a number next to the point which specifies the value. FWW - freshwater wetland; DSF - dry sclerophyll forest and woodland; DSFS - dry sclerophyll forest and woodland on sand; UBUSH - urban bushland remnant; UMAT - urban matrix; URB - urban sites; MAN - mangrove forest; and SOFF - swamp oak forest.

mapping for *M. norfolkensis* and *M.* species 2 [25,68] and our study confirms the importance of these habitats.

Both *M. norfolkensis* and *M.* species 2 were more active in dry sclerophyll forest on sand compared to the dry sclerophyll forest on low undulating hills. The Tomago sandbeds (where this habitat type occurs) have characteristics (sandy soil and low vegetation height) that are typically considered to be associated with nutrient-

Table 2. Predictive accuracy of regional-scale models [25] to local-scale data collected during this study.

Accuracy measure	*Mormopterus norfolkensis*	*Mormopterus* species 2
AUC	0.757	0.719
Kappa	0.313	0.255
TSS	0.317	0.292
CCR	0.660	0.660
Sensitivity	0.667	0.676
Specificity	0.65	0.615
Prevalence	0.574	0.723

AUC – area under the curve calculated from the receiver operating characteristic; TSS – true skill statistic; CCR – correct classification rate. The confusion matrix was constructed using a probability threshold of 0.34 for both species as determined by the intersection of sensitivity and specificity values.

poor environments. This contrasts with the high activity of both *M. norfolkensis* and *M.* species 2 recorded in productive floodplain areas of the study area. It is possible that the high activity is attributed to the presence of hollow-bearing trees that provide roosting habitat for these species. However, koalas (*Phascolarctos cinereus*) which have been linked to high fertility and clay soils in Victoria and Queensland [69,70], also prefer these sandy soils within the study area [70,71]. This occurrence of preferred koala habitat on the sandbeds has been attributed to the historical clearing of more suitable floodplain habitats [70,72]. However, it is possible that the Tomago sandbeds are actually more productive than the sandy soil suggests, perhaps due to the high water table. Further investigation into the factors that drive productivity, particularly in relation to ground water would be valuable.

Differential habitat use

Whilst floodplain habitats were used by both morphologically similar species, there were also differences in habitat use between the two species. Most notably, *M. norfolkensis* activity was negatively associated with urban areas, whereas urban bushland was positively associated with *M.* species 2 activity levels. *Mormopterus norfolkensis* was never recorded in the urban matrix and rarely in urban bushland remnants and this confirms previous assertions that *M. norfolkensis* avoids urban habitats [14,25]. Habitat mapping from regional-scale models indicated that urban habitats were only marginal habitat for *M.* species 2 [25]. This is also consistent with our local study where *M.* species 2 was recorded at only one urban matrix site and whilst it was recorded relatively frequently in urban bushland, activity was low and foraging was rarely recorded. *Mormopterus norfolkensis* may be particularly sensitive to disturbance associated with urbanisation such as noise and artificial lighting (in contrast with *M.* species 2 which has been found to be positively associated with urban lighting [73]), but this requires explicit testing. Alternatively, these subtle differences in habitat use by *M. norfolkensis* and *M.* species 2 may represent spatial segregation to either prevent competitive interactions, or in response to them [74,75]. Interspecific interactions are poorly understood for bat communities and further research into competition and niche partitioning is required.

Validation of regional-scale habitat models

The descriptions of habitat use from regional-scale models were consistent with those from local-scale models. Additionally, the

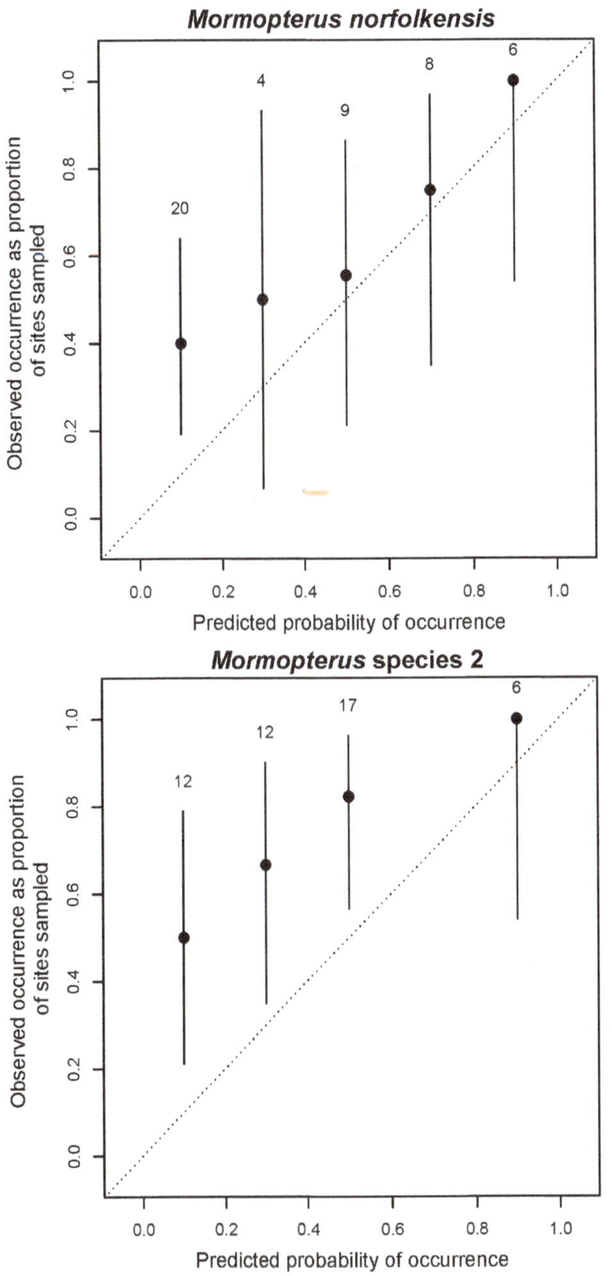

Figure 5. Regional model calibration plots. Calibration plots of the proportion of observed occurrence (collected during this study), against the predicted values of regional-scale probability of occurrence mapping [25] with 95% confidence intervals, for *Mormopterus norfolkensis* and *Mormopterus* species 2. Sample size in each bin is given by the numbers above. A close fit to the diagonal indicates good model calibration.

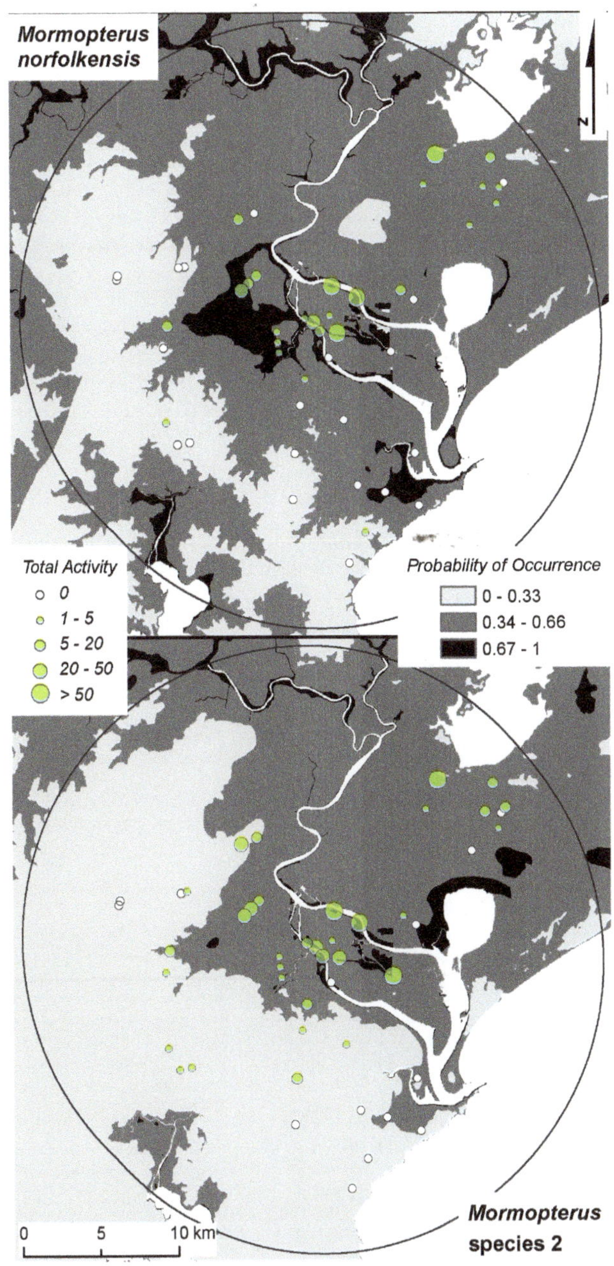

Figure 6. Spatial visualisation of prediction errors. Regional-scale probability of occurrence maps [25] presented at the local-scale of the study for *Mormopterus norfolkensis* and *Mormopterus* species 2 with activity levels from the local study overlaid. The probability of occurrence of each species from the regional-scale models was classified into three categories to aid visualisation of prediction errors. The sensitivity = specificity occurrence threshold (0.34) obtained from the local-scale study was used to define the lowest category.

discrimination ability of regional-scale mapping, as assessed by AUC was fair to good. However, threshold-based accuracy measures (kappa and TSS) indicated that the regional models performed poorly with both false positive and false negative occurrences for *M. norfolkensis* and *M.* species 2 at a local-scale. Additionally, model calibration for *M.* species 2 suggested a systematic bias in which the regional-scale model underestimates species prevalence [12]. The broad level of agreement between these models suggests that they are useful to describe general habitat for these poorly understood species. However, our findings also illustrate that regional-scale models should not be used without testing for fine-scale prediction of habitat use where high levels of certainty are required. This is particularly the case when the consequences of errors are severe, such as using regional-scale models to allocate land for development or conservation.

Management implications

The high level of use of floodplain areas and the avoidance of urban areas by *M. norfolkensis* suggest that its threatened conservation status is warranted. Floodplains are subject to high anthropogenic pressures from agriculture and urban development, with less than 30% of floodplain vegetation communities remaining in NSW [76]. Whilst wetland restoration has been found to have a positive effect on local bat communities elsewhere [77], it is important to monitor how particular management actions such as draining, insect spraying and installation of tidal floodgates affect insect and bat communities and these factors would benefit from further investigation. The findings of both regional and local-scale habitat models suggest that conservation efforts should focus on conserving and restoring floodplains, the sensitive management of riparian areas on private land and limiting urbanisation in these areas.

Supporting Information

Figure S1 Sensitivity and probability of occurrence threshold. Effect of the probability of occurrence threshold on the sensitivity (dotted line), specificity (dashed line) and kappa (black line) accuracy measures for regional-scale models [25] tested against independent local-scale data collected for *Mormopterus norfolkensis* and *Mormopterus* species 2.

Table S1 Description of candidate variables used in the analyses. Values are mean ± se (range) or frequency of ordinal values. A patch was defined as > 10 trees occurring together and patch sizes > 500 m^2 were mapped.

Acknowledgments

We would like to thank the land managers that provided property access for this project. Trent Penman assisted with general statistical discussions, Leroy Gonsalves, Evan Pickett and two referees provided helpful comments on this manuscript.

Author Contributions

Conceived and designed the experiments: AM BL MM. Performed the experiments: AM. Analyzed the data: AM. Contributed reagents/materials/analysis tools: AM BL MM. Wrote the paper: AM BL MM.

References

1. Guisan A, Zimmermann NE (2000) Predictive habitat distribution models in ecology. Ecological Modelling 135: 147–186.
2. Elith J, Graham CH, Anderson RP, Dudik M, Ferrier S, et al. (2006) Novel methods improve prediction of species' distributions from occurrence data. Ecography 29: 129–151.
3. Jaberg C, Guisan A (2001) Modelling the distribution of bats in relation to landscape structure in a temperate mountain environment. Journal of Applied Ecology 38: 1169–1181.
4. Wintle B, Elith J, Potts J (2005) Fauna habitat modelling and mapping: A review and case study in the Lower Hunter Central Coast region of NSW. Austral Ecology 30: 719–738.
5. Elith J, Burgman MA, Regan HM (2002) Mapping epistemic uncertainties and vague concepts in predictions of species distribution. Ecological Modelling 1575: 313–329.
6. Greaves GJ, Mathieu R, Seddon PJ (2006) Predictive modelling and ground validation of the spatial distribution of the New Zealand long-tailed bat (*Chalinolobus tuberculatus*). Biological Conservation 132: 211–221.
7. Ferrier S, Watson G, Pearce J, Drielsma M (2002) Extended statistical approaches to modelling spatial pattern in biodiversity in northeast New SouthWales. I. Species-level modelling. Biodiversity and Conservation 11: 2275–2307.
8. Rhodes JR, Wiegand T, McAlpine CA, Callaghan J, Lunney D, et al. (2006) Modeling species' distributions to improve conservation in semiurban landscapes: Koala case study. Conservation Biology 20: 449–459.
9. Gibson LA, Wilson BA, Cahill DM, Hill J (2004) Spatial prediction of rufous bristlebird habitat in a coastal heathland: a GIS-based approach. Journal of Applied Ecology 41: 213–223.
10. Manning AD, Lindenmayer DB, Barry SC, Nix HA (2007) Large-scale spatial and temporal dynamics of the vulnerable and highly mobile superb parrot. Journal of Biogeography 34: 289–304.
11. Krebs CJ (2001) Ecology. San Francisco: Benjamin Cummings.
12. Franklin J (2009) Mapping species distributions: spatial inference and prediction. New York: Cambridge University Press.
13. Razgour O, Hanmer J, Jones G (2011) Using multi-scale modelling to predict habitat suitability for species of conservation concern: The grey long-eared bat as a case study. Biological Conservation 144: 2922–2930.
14. Threlfall C, Law B, Penman TD, Banks PB (2011) Ecological processes in urban landscapes: mechanisms influencing the distribution and activity of insectivorous bats. Ecography 34: 814–826.
15. Avila-Flores R, Fenton M (2005) Use of spatial features by foraging insectivorous bats in a large urban landscape. Journal of Mammalogy 86: 1193–1204.
16. Hourigan C, Johnson C, Robson S (2006) The structure of a micro-bat community in relation to gradients of environmental variation in a tropical urban area. Urban Ecosystems 9: 67–82.
17. Brambilla M, Casale F, Bergero V, Crovetto GM, Falco R, et al. (2009) GIS-models work well, but are not enough: Habitat preferences of *Lanius collurio* at multiple levels and conservation implications. Biological Conservation 142: 2033–2042.
18. Watrous KS, Donovan TM, Mickey RM, Darling SR, Hicks AC, et al. (2006) Predicting minimium habitat characteristics for the Indiana Bat in the Champlain Valley. Journal of Wildlife Management 70: 1228–1237.
19. IUCN (2009) IUCN red list of threatened species. Vienna: International Union for Conservation of Nature and Natural Resources (IUCN).
20. Adams M, Reardon T, Baverstock P, Watts C (1988) Electrophoretic resolution of species boundaries in Australian Microchiroptera. IV. The Molossidae (Chiroptera:Vespertilionidae). Australian Journal of Biological Science 41: 315–326.
21. Adams MD, Law BS, French KO (2009) Vegetation structure influences the vertical stratification of open- and edge-space aerial-foraging bats in harvested forests. Forest Ecology and Management 258: 2090–2100.
22. Rhodes M (2002) Assessment of sources of variance and patterns of overlap in microchiropteran wing morphology in southeast Queensland, Australia. Canadian Journal of Zoology 80: 450–460.
23. Norberg U, Rayner J (1987) Ecological morphology and flight in bats (Mammalia; Chiroptera): Wing adaptations, flight performance foraging strategy and echolocation Philosophical Transactions of the Royal Society of London Series B, Biological Sciences 316: 335–427.
24. McKenzie N, Rolfe J (1986) Structure of bat guilds in the Kimberley mangroves, Australia. Journal of Animal Ecology 55: 401–420.
25. McConville A, Law B, Penman TD, Mahony M (2013) Contrasting habitat use of morphologically similar bat species with differing conservation status in south-eastern Australia. Austral Ecology Early View: DOI: 10.1111/aec.12047.
26. Tyre AJ, Tenhumberg B, Field SA, Niejalke D, Parris K, et al. (2003) Improving precision and reducing bias in biological surveys: estimating false-negative error rates. Ecological Applications 13: 1790–1801.
27. Bureau of Meteorology (2012) Climate data online. Available: http://www.bom.gov.au/climate/data/index.shtml. Accessed 2012 Jun 30.
28. Geoscience Australia (2012) OzCoasts: Australian Online Coastal Information. Available: http://www.ozcoasts.org.au/search_data/estuary_search.jsp. Accessed 2012 Jun 30.
29. Mattei LE (1995) Soil landscapes of the Newcastle 1:100 000 sheet. Sydney: Depatment of Land and Water Conservation.
30. McConville A, Law B, Mahony M (2013) Mangroves as maternity roosts for a colony of the rare east-coast free-tailed bat (*Mormopterus norfolkensis*) in south-eastern Australia. Wildlife Research 40: 318–327.
31. de Oliveira M (1998) Anabat System Practical Guide. Brisbane: Department of Natural Resources.
32. Corben C (2008) AnalookW. 3.5m Available: http://www.hoarybat.com. Accessed 2009 Jan 15.
33. Pennay M, Law B, Reinhold L (2004) Bat calls of NSW, region based guide to the echolocation calls of Microchiropteran bats. Hurstville: NSW Department of Environment and Conservation.
34. Mills D, Norton T, Parnaby H, Cunningham R, Nix H (1996) Designing surveys for microchiropteran bats in complex forest landscapes- a pilot study from south-east Australia. Forest Ecology and Management 85: 149–161.
35. House S (2003) Lower Hunter and Central Coast Regional Biodiversity Conservation Strategy, Updated Extant Vegetation Community Map. Callaghan, NSW: Lower Hunter and Central Coast Regional Environmental Management Strategy.
36. Rempel RS, Kaukinen D, Carr AP (2011) Patch Analyst and Patch Grid extension for ArcMap. 4.0 ed. Thunder Bay, Ontario: Ontario Ministry of Natural Resources Centre for Northern Forest Ecosystem Research.

37. McGarigal K, McComb WC (1995) Relationships between landscape structure and breeding birds in the Oregon Coast Range. Ecological Monographs 65: 235–260.
38. R Development Core Team (2012) R: a language and environment for statistical computing. 2.14.2 ed: The R Foundation for Statistical Computing.
39. Venables WN, Ripley BD (2002) Modern Applied Statistics with S. New York: Springer.
40. Freeman EA, Moisen GG (2008) PresenceAbsence: an R package for presence-absence model analysis. Journal of Statistical Software 23: 1–31.
41. NCAR - Research Application Program (2012) verification: forecast verification utilities. 1.32. Available: http://CRAN.R-project.org/package=verification. Accessed 2012 Jan 10.
42. Breheny P, Burchett W (2012) Visualizing regression models using visreg. Available: http://cran.r-project.org/web/packages/visreg/visreg.pdf. Accessed 2012 Feb 10.
43. Akaike H (1973) Information theory and an extension of the maximum likelihood. In: Petrov BN, Cs'aki F, editors. Akademia Kaido, Budapest. pp. 267–281.
44. Burnham KP, Anderson DR (2002) Model selection and multimodel inference: a practical information-theoretic approach. New York: Springer-Verlag.
45. Nagelkerke NJD (1991) A note on a general definition of the coefficient of determination. Biometrika 78: 691–692.
46. Fielding AH, Bell JF (1997) A review of methods for the assessment of prediction errors in conservation presence/absence models. Environmental Conservation 24: 38–49.
47. Pearce J, Ferrier S (2000) Evaluating the predictive performance of habitat models developed using logistic regression. Ecological Modelling 133: 225–245.
48. Swets KA (1988) Measuring the accuracy of diagnostic systems. Science 240: 1285–1293.
49. Thuiller W, Araujo MB, Lavorel S (2003) Generalized models vs. classification tree analysis: predicting spatial distributions of plant species at different scales. Journal of Vegetation Science 14: 669–680.
50. Moran PAP (1950) Notes on continuous stochastic phenomena. Biometrika 37: 17–23.
51. Allouche O, Tsoar A, Kadmon R (2006) Assessing the accuracy of species distribution models: prevalence, kappa and the true skill statistic (TSS). Journal of Applied Ecology 43: 1223–1232.
52. Cohen J (1960) A coefficient of agreement of nominal scales. Educational and Psychological Measurement 20: 37–46.
53. Landis JR, Koch GG (1977) The measurements of observer agreement for categorical data. Biometrics 33: 159–174.
54. Mouton AM, De Baets B, Goethals PLM (2010) Ecological relevance of performance criteria for species distribution models. Ecological Modelling 221: 1995–2002.
55. Manel S, Williams HC, Ormerod SJ (2001) Evaluating presence-absence models in ecology: the need to account for prevalence. Journal of Applied Ecology 38: 921–931.
56. Araujo MB, Guisan A (2006) Five (or so) challenges for species distribution modelling. Journal of Biogeography 33: 1677–1688.
57. Basham R, Law B, Banks P (2010) Microbats in a 'leafy' urban landscape: are they persisting, and what factors influence their presence? Austral Ecology 36: 663–678.
58. Lloyd A, Law B, Goldingay R (2006) Bat activity on riparian zones and upper slopes in Australian timber production forests and the effectiveness of riparian buffers. Biological Conservation 129: 207–220.
59. Pearce JL, Cherry K, Drielsma M, Ferrier S, Whish G (2001) Incorporating expert opinion and fine-scale vegetation mapping into statistical models of faunal distribution. Journal of Applied Ecology 38: 412–424.
60. Gonsalves L, Law B, Webb C, Monamy V (2012) Are vegetation interfaces important to foraging insectivorous bats in endangered coastal saltmarsh on the Central Coast of New South Wales. Pacific Conservation Biology 18: 282–292.
61. Laegdsgaard P, Monamy V, Saintilan N (2004) Investigating the presence of threatened insectivorous bats on coastal NSW saltmarsh habitat. Wetlands (Australia) 22: 29–41.
62. Grindal SD, Morissette JL, Brigham RM (1999) Concentration of bat activity in riparian habitats over an elevational gradient. Canadian Journal of Zoology 77: 972–977.
63. Hayes JP (1997) Temporal variation in activity of bats and the design of echolocation monitoring studies. Journal of Mammalogy 78: 541–524.
64. Racey PA, Swift SM (1985) Feeding ecology of *Pipistrellus pipistrellus* (Chiroptera: Vespertilionidae) during pregnancy and lactation. I. foraging behaviour. Journal of Animal Ecology 54: 205–215.
65. Walsh AL, Harris S (1996) Foraging habitat preferences of Vespertilionid bats in Britain. Journal of Applied Ecology 33: 508–518.
66. Flaquer C, Puig-Montserrat X, Goiti U, Vidal F, Curco A, et al. (2009) Habitat selection in Nathusius' pipistrelle (*Pipistrelle nathusii*): the importance of wetlands. Acta Chiropterologica 11: 149–155.
67. Law B, Chidel M, Penman TD (2011) Do young eucalypt plantations benefit bats in an intensive agricultural landscape? Wildlife Research 38: 173–187.
68. Reside AE, Lumsden LF (2011) Resource partitioning by two closely-related sympatric freetail bats, *Mormopterus* spp. In: Law B, Eby P, Lunney D, Lumsden L, editors. The Biology and Conservation of Australian Bats. Mosman, NSW: Royal Zoological Society of NSW. pp. 155–166.
69. Januchowski SR, McAlpine CA, Callaghan JG, Griffin CB, Bowen M, et al. (2008) Identifying multiscale habitat factors influencing koala (*Phascolarctos cinereus*) occurrence and management in Ballarat, Victoria, Australia. Ecological Management & Restoration 9: 134–142.
70. McAlpine CA, Rhodes JR, Bowen ME, Lunney D, Callaghan JG, et al. (2007) Can multiscale models of species' distribution be generalized from region to region? A case study of the koala. Journal of Applied Ecology 45: 558–567.
71. Lunney D, Phillips SJ, Callaghan J, Coburn D (1998) Determining the distribution of Koala habitat across a shire as a basis for conservation: a case study from Port Stephens, New South Wales. Pacific Conservation Biology 4: 186–196.
72. Knott T, Lunney D, Coburn D, Callaghan J (1998) An Ecological History of Koala Habitat in Port Stephens Shire and the Lower Hunter on the Central Coast of New South Wales, 1801–1998. Pacific Conservation Biology 4: 354–368.
73. Threlfall C (2011) Conserving biodiversity in urban landscapes: mechanisms influencing the distribution, community assembly and resource use of insectivorous bats in Sydney, Australia. Sydney: University of New South Wales.
74. Arlettaz R (1999) Habitat selection as a major resource partitioning mechanism between the two sympatric sibling bat species *Myotis myotis* and *Myotis blythii*. Journal of Animal Ecology 68: 460–471.
75. Nicholls B, Racey PA (2006) Habitat selection as a mechanism of resource partitioning in two cryptic bat species *Pipistrellus pipistrellus* and *Pipistrellus pygmaeus*. Ecography 29: 697–708.
76. Keith DA (2004) Ocean shores to desert dunes: the native vegetation of New South Wales. Sydney: NSW Department of Environment and Conservation.
77. Menzel JM, Menzel MA, Kilgo JC, Ford WM, Edwards JW (2005) Bat response to Carolina Bays and wetland restoration in the southeastern U.S. coastal plain. WETLANDS 25: 542–550.

4

Changes in Fish Assemblages following the Establishment of a Network of No-Take Marine Reserves and Partially-Protected Areas

Brendan P. Kelaher[1,2]*, Melinda A. Coleman[1], Allison Broad[3], Matthew J. Rees[3], Alan Jordan[1], Andrew R. Davis[3]

1 New South Wales Fisheries, Department of Primary Industries, PO Box 4321, Coffs Harbour 2450, New South Wales, Australia, **2** National Marine Science Centre & Centre for Coastal Biogeochemistry Research, School of Environment, Science and Engineering, Southern Cross University, Coffs Harbour, New South Wales, Australia, **3** Institute for Conservation Biology & Environmental Management, School of Biological Sciences, University of Wollongong, New South Wales, Australia

Abstract

Networks of no-take marine reserves and partially-protected areas (with limited fishing) are being increasingly promoted as a means of conserving biodiversity. We examined changes in fish assemblages across a network of marine reserves and two different types of partially-protected areas within a marine park over the first 5 years of its establishment. We used Baited Remote Underwater Video (BRUV) to quantify fish communities on rocky reefs at 20–40 m depth between 2008–2011. Each year, we sampled 12 sites in 6 no-take marine reserves and 12 sites in two types of partially-protected areas with contrasting levels of protection (n = 4 BRUV stations per site). Fish abundances were 38% greater across the network of marine reserves compared to the partially-protected areas, although not all individual reserves performed equally. Compliance actions were positively associated with marine reserve responses, while reserve size had no apparent relationship with reserve performance after 5 years. The richness and abundance of fishes did not consistently differ between the two types of partially-protected areas. There was, therefore, no evidence that the more regulated partially-protected areas had additional conservation benefits for reef fish assemblages. Overall, our results demonstrate conservation benefits to fish assemblages from a newly established network of temperate marine reserves. They also show that ecological monitoring can contribute to adaptive management of newly established marine reserve networks, but the extent of this contribution is limited by the rate of change in marine communities in response to protection.

Editor: Arga Chandrashekar Anil, CSIR- National institute of Oceanography, India

Funding: The study was funded by the NSW Government. Independent university-based scientists and NSW Fisheries scientists were responsible for the data collection and analysis, decision to publish, and preparation of the manuscript.

Competing Interests: The authors have declared that no competing interests exist.

* E-mail: brendan.kelaher@scu.edu.au

Introduction

Human activities, such as catchment development, overfishing, pollution and maritime industries, have degraded marine and estuarine environments [1,2]. Global concern for the health of marine systems has driven an unprecedented increase in marine protected area establishment over the last decade [3]. A small percentage of these marine protected areas are marine reserves where extraction of living marine resources is not permitted [4]. Many published studies have evaluated the responses of marine ecosystems to reserve establishment [5]. These include highlighting the types of species that do and do not benefit (e.g. [6,7]) cascading trophic responses (e.g. [8,9]), their influence on surrounding areas (e.g. [10,11]), their influence on invasive species (e.g. [12]) and the enforcement effort required for significant changes to occur [13,14].

While individual marine reserves provide conservation benefits, social and economic considerations often limit their size to a fraction of the bioregion whose biodiversity they are often designed to represent [4]. A limitation of most marine reserves is that they are not large enough to be completely self-sustaining because their size is less than the average dispersal distance of key species [15]. Although this issue can be resolved by establishing much larger marine reserves, socio-economic pressures are likely to prevent this, particularly on densely populated coasts. In an attempt to scale up the benefits of individual marine reserves to broader regions, networks of marine reserves are increasingly being established [4,16]. Effective networks of marine reserves require adequate connectivity, such that each reserve can contribute and receive sufficient adults and larvae from adjacent reserves [4,17]. Theoretical models suggest that a network of marine reserves may synergistically increase conservation benefits relative to the sum of the benefits from unconnected individual reserves [15,18–20]. However, published data on changes in marine communities across marine reserve networks is limited relative to research on individual marine reserves and rigorous empirical tests of theoretical models optimizing marine reserve network designs are still in their infancy [4,21].

Partially-protected areas are typically marine protected areas with less restrictive regulations than marine reserves [22,23]. Depending on local objectives, they usually involve restrictions on particular activities, gear types, user groups, target species, or extraction periods [23]. Partially-protected areas may also be used to limit foreshore developments that require marine infrastructure (e.g. marinas or discharge outlets) thereby further reducing environmental threats [24]. Relative to marine reserves, there is much less published information about ecological changes associated with the establishment of partially-protected areas [23]. A meta-analysis of 20 studies found that partially-protected areas maintain higher biomasses, density and richness of marine organisms than areas with less regulation, but do not provide the same level of ecological benefits as no-take marine reserves [23]. These conclusions are, however, limited by (i) major differences in fishing restrictions in partially-protected areas among the different studies and (ii) most comparisons within a region being based on a single marine reserve or partially-protected area (but see [25]).

The establishment of multiple-use marine parks with replicated, closely spaced marine reserves, partially-protected areas and open access areas provides the opportunity to test hypotheses about networks of marine reserves and make rigorous comparisons of change in marine communities associated with different levels of environmental protection [26]. Over the last decade, six such multi-zoned marine parks containing 115 individual marine reserves (i.e. no-take sanctuary zones) have been established in the state waters of New South Wales, Australia [27]. Built into the legislation administering these marine parks are statutory requirements to review and, if necessary, adaptively manage the zoning arrangements 5 years after establishment. Some species can display significant changes after only a few years of protection (e.g. <5 years [28]), while others may take decades [29,30]. Significant changes in marine community structure may take well in excess of 15 years [28,31]. It is uncertain, therefore, the extent to which marine environmental monitoring will contribute to evidence-based adaptive management of marine park zoning arrangements at a 5 year review.

To assess the recovery trajectory of a newly established marine reserve network, we tested the hypothesis that reef-associated fish assemblages in reserves will change significantly relative to fished areas within 5 years of establishment. Concurrently, we also tested the hypothesis that reef-associated fish assemblages vary with different levels of environmental protection by including partially-protected areas. We also evaluated the performance of individual reserves within the network and related this to reserve size and enforcement actions.

Materials and Methods

Study Area

This study was undertaken in the Batemans Marine Park, a ~85000 Ha multi-use marine park on the NSW South Coast, Australia (northern boundary = 35°31.086'S and southern boundary = 36°22.290'S) encompassing waters from the mean high tide mark to the limit of state waters (ca. 3 nm from land). The zoning plan for the marine park commenced in June 2007, after which activities (e.g. fishing, recreation, foreshore development, boating, pollution discharge, etc.) were regulated by the NSW Marine Parks Act (1998) and Regulations (1999, 2009), as well as a range of other legislation (e.g. Fisheries Management Act 1994, Coastal Protection Act 1979, Protection of Environmental Operations Act 1997, Threatened Species Conservation Act 1995, etc.). Marine park legislation specifically prohibits dredge and demersal trawling, mining and long-lining throughout the entire park.

As part of the objective to achieve conservation of biodiversity, the Marine Park was zoned into 4 types of areas: sanctuary zones, habitat protection zones, general use zones and special purpose zones, which represented 19.1%, 43.3%, 37.2% and 0.4% of the entire park, respectively. The different zone types are interspersed throughout the marine park creating a network of marine reserves and partially-protected areas. Special purpose zones were not included in the hypotheses tested because they only represented <0.5% of the marine park and were created for a range of specific management purposes (e.g. oyster farming, foreshore development and cultural resource use).

Sanctuary zones are strict no-take marine reserves that allow for non-extractive activities. Habitat protection zones are partially-protected areas where the species that can be harvested and the fishing methods that can be used are prescribed by legislation. For example, lawful recreational fishing methods are allowed in habitat protection zones with a few exceptions, but commercial purse seining, lift netting, mesh-netting, estuary prawn and haul netting are not permitted. With only the overall park-wide prohibitions enforced, general use zones are the least restrictive partially protected areas in NSW Marine Parks. Lawful commercial and recreational fishing methods other than trawling and long-lining are permitted in general use zones in the Batemans Marine Park. More specific details about prohibited activities can be found in Read and West [24].

Sampling Methodology

Baited Remote Underwater Video (BRUV) was used to test hypotheses about changes in fish assemblages across the network of marine reserves and partially-protected areas. In many situations, BRUV units are preferred over other sampling techniques because they can be deployed in environments unsuitable for conventional diver based assessments [32], they are able to detect diver-shy species [33], they provide usable estimates of the relative abundance of economically-important species [33] and they provide a permanent visual record of surveys [32]. BRUV was particularly suitable for our study because it is a non-destructive sampling technique appropriate for high conservation areas (e.g. no-take marine sanctuaries) and survey depths often exceeded 30 m. Like all fish survey methods, BRUV only samples a subset of the fish community with a tendency towards sampling more predatory species than other methods on shallow reefs (e.g. underwater visual census [34]). The observed fish assemblage with BRUV systems can also be influenced by the presence of large predatory species (e.g. sharks [35]). Importantly, these issues did not systematically vary among zone types and, as such, did not influence the hypotheses that were tested here.

BRUV units were deployed on rocky reef at 12 sites in sanctuary zones, 6 habitat protection and 6 general use zones (Fig. 1). This design allowed for planned balanced comparisons between no-take and fished areas (12 sites vs 12 sites) and between the two types of partially protected areas (6 sites vs 6 sites) (see design below). The sanctuary zones included were between 2 km to 14 km apart, which is likely within the range of either larval or adult movements for many common reef fish species (e.g. [36] and references within), especially considering the East Australia Current [37,38]. Sites were haphazardly interspersed throughout the Marine Park from Brush Island (35°31.086'S) to Potato Point (35°06.172'S) (Fig. 1). Each site was dominated by rocky reef and was sampled in 2008, 2009, 2010 and 2011. In 2008, sampling occurred from January to May and for the following years it occurred from June to August. This change in timing was related to the implementation of a state-wide monitoring program and was not a major consideration for interpretation of results because there is often no clear seasonal signal in demersal fish assemblages in this region [39]. This likely stems from substantial spatial-temporal variation and relatively mild winters. Moreover, the key

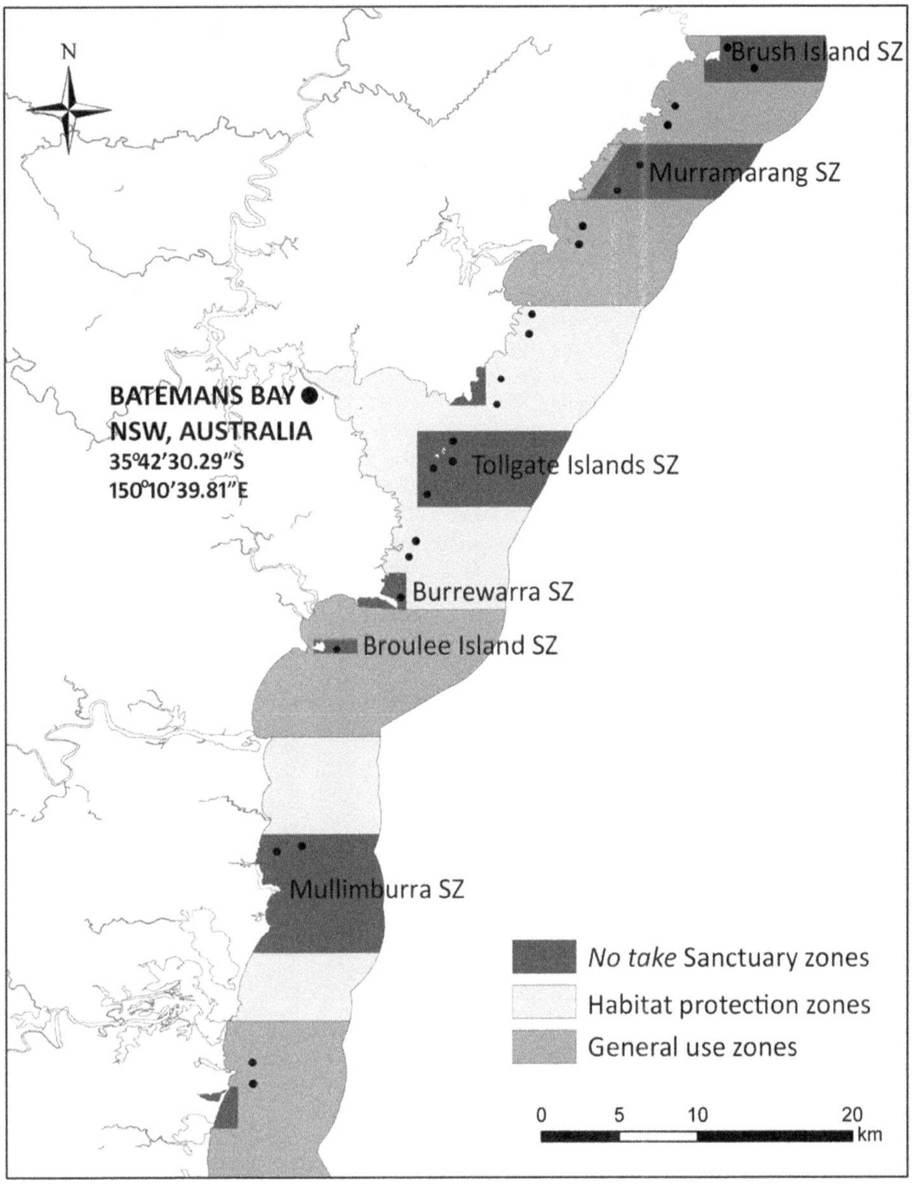

Figure 1. Map showing the configuration of zones in the part of the Batemans Marine Park (NSW, Australia) included in our study. The map highlights spatial arrangement of the network of no-take marine sanctuaries. ● indicates the location of each BRUV sites.

hypotheses of this study focused on differences between sanctuary zones and partially protected areas rather than temporal variation.

In each site, 4 BRUV units were deployed at approximately 200 m intervals onto reef habitat. The mean (±S.E.) depth of deployments was 26.0 (1.3) m, 26.3 (1.6) m and 26.6 (0.5) m for sanctuary, habitat protection and general use zones, respectively, and did not differ significantly among zones (PERMANOVA, $pseudo\text{-}F_{2,21} = 1.48$, $P = 0.13$). Each BRUV unit was constructed as per Malcolm et al. [40], which included a galvanized metal frame containing a video camera (mini DV SONY) pointed at a bait bag mounted horizontally at the end of a 1.5 m long bait arm. Cameras were housed within high-pressure polyvinyl chloride pipe with flat acrylic end-ports yielding a field of view of 110°. For each BRUV deployment, the bait bag was replenished with ~500 g of chopped pilchards (*Sardinops* spp.) and each BRUV unit was left on the bottom for 30 minutes. This bait type was determined to yield the most consistent outcomes compared to others previously tested (e.g. abalone viscera or crushed urchin [41]). This bottom time was considered appropriate for sampling reef fish between 20–40 m because there is no significant differences among fish assemblages and the *max* N of many common species when deployment times of 30, 60, and 90 minutes were compared (D. Harasti, unpublished data). Furthermore, the replication levels of sites and camera deployments within sites provide adequate power to reliably detect significant differences between fish species richness and total *max* N in sanctuary zones compared to fished areas with mean differences of 30% and 100%, respectively (B. Kelaher, unpublished data).

Videos were analyzed in the laboratory using a field of view 2 m behind the bait bag, which represented a standardized area of 9.4 m^3 [40]. For each replicate BRUV deployment, we determined species richness, total *max* N, and *max* N of each fish species. *Max* N for a species was the maximum number of individuals in any frame and total *max* N was the sum of *max* N's for each deployment [32]. When the abundances of families of fishes were analyzed, the *max* N value used for each replicate was the summed *max* N of each fish species in that family. Analyses were restricted to fin fish to avoid complications associated with extra protection of all but two species of elasmobranchs (i.e. *Mustelus antarcticus* and *Galeorhinus galeus*) in habitat protection zones.

Comparisons across a Network of Marine Reserves and Partially-protected Areas

Hypotheses about changes in fish assemblages across the network of no-take marine reserves and partially-protected areas were tested using 2 factor analyses with *zone type* (3 levels, orthogonal and fixed) and *years* since the commencement of the zoning plan (4 levels, orthogonal and fixed), with analyses based on site averages. To test for differences in fish assemblages between no-take sanctuary zones and fished areas a contrast was included to compare sanctuary zones against zones where fishing was allowed. To test for differences in fish assemblages between the two types of partially-protected zones a contrast was included comparing fish assemblages in habitat protection and general use zones.

Hypotheses were based on multivariate comparisons of fish assemblage structure and univariate comparisons of fish species richness and total *max* N. Hypotheses were also tested using the total *max* N of four numerically-dominant families, Carangidae, Kyphosidae, Labridae and Monacanthidae, which represented 17%, 29%, 14% and 7% of the overall total *max* N, respectively. Analyses were also conducted on fish species with a summed *max* N that totaled more than 300 individuals and are commonly-caught in NSW waters. Each species is currently assessed as either moderately fished, fully fished, growth overfished or overfished indicating that they are each under fishing pressure. These taxa were *Pagrus auratus* [snapper, growth overfished], *Pseudocaranx georgianus* [silver trevally, growth overfished], *Scorpis lineolata* [silver sweep, moderately to fully fished], *Ophthalmolepis lineolatus* [southern maori wrasse, moderately fished], *Trachurus novaezelandiae* [yellow tail scad, fully fished] and *Nemadactylus douglasii* [grey morwong, overfished] (see [42] for details). In Batemans Marine Park each of the above species is caught recreationally, as well as in the commercial ocean trap and line fishery. However, *T. novaezelandiae* is mostly caught in purse seine nets [42], which cannot be used in habitat protection zones.

Hypotheses about changes in fish assemblages and individual families and species were tested with non-parametric multivariate analysis of variance (PERMANOVA [43]). These non-parametric procedures are robust to variable ecological data commonly obtained from marine communities [44]. All univariate analyses were done using Euclidean distance to create similarity matrices. All multivariate analyses used the Bray-Curtis similarity coefficient [45]. Non-metric multidimensional scaling (nMDS) [46] was used to generate two-dimensional ordination plots which graphically illustrated multivariate patterns in fish assemblages.

Comparisons of the Performance of Individual Marine Reserves

The 12 sanctuary zone sites were located within six of the 10 offshore sanctuary zones in the Batemans Marine Park. From north to south, these were Brush Island, Murramarang, Tollgate Islands, Burrewarra Point, Broulee Island and Mullimburra (GPS co-ordinates of boundaries included in the NSW Marine Parks (Zoning Plan) Regulation 1999). These zones encompassed the smallest and largest offshore sanctuary zones in the marine park. As well as size, these sanctuary zones varied across a range of marine park planning criteria (see Table 1 for details). To compare the individual performance of these 6 sanctuary zones since the commencement of the zoning plan, a ratio was established with $(x_{SZ}+1)/(X_{FA}+1)$, where x_{SZ} was the response variable from each sanctuary zone BRUV deployment and X_{FA} was the average of the closest two sites in areas where fish could be legally caught. This sanctuary zone/fished area ratio (hereafter called SZ/FA ratio) provided an indication of relative changes in fish assemblages in no-take and fished zones in a local area around individual sanctuary zones rather than across the network of reserves and partially-protected areas.

To test whether the performance of individual sanctuary zones was variable, a two factor PERMANOVA analysis was carried out on overall fish species richness, total *max* N and the total *max* N of four numerically-dominant families: Carangidae, Kyphosidae, Labridae and Monacanthidae with the factors *sanctuary zones* (*SZ*, 6 levels orthogonal and random) and *years* since zoning plan commencement (4 levels, orthogonal and fixed). These univariate analyses used Euclidean distance to create similarity matrices and were based on individual BRUV deployments.

The average direction of change of the six key fish outlined above (*Pagrus auratus*, *Pseudocaranx georgianus*, *Scorpis lineolata*, *Ophthalmolepis lineolatus*, *Trachurus novaezelandiae* and *Nemadactylus douglasii*) in each sanctuary zone was determined by calculating Pearson's correlation coefficient for the average SZ/FA ratio vs years since commencement of the zoning plan. These correlation coefficients were then averaged to determine a generalized direction of change (r_{av}) for each individual sanctuary zone with r_{av} being a value between -1 and 1 with positive and negative values indicative of positive and negative associations between SZ/FA ratio and time since establishment, respectively. To evaluate potential explanations for variation in individual sanctuary zone performance, the r_{av} for individual sanctuary zones were correlated using Pearson's correlation coefficient with the number of enforcement actions by marine park staff from 1 July 2009 to 30 June 2011 and the size of the sanctuary zone. To control for Type 1 error, the significance level of these correlations was corrected with sequential Bonferroni's technique [47]. Qualitative comparisons were also made between individual sanctuary zone performance and other important aspects of individual sanctuary zones, including whether they (i) terminated at the 3 nm limit maximizing cross shelf diversity, (ii) were directly linked to no-take estuarine areas facilitating connectivity, (iii) were buffered by habitat protection zones limiting accidental damaging activities or (iv) were adjacent to terrestrial reserves reducing land-based impacts (e.g. urban run-off).

Ethics Statement

This study was conducted with the permission of the NSW Marine Parks Authority and the NSW Department of Primary Industries. BRUV work was done under the auspices of the University of Wollongong animal ethics committee (approval number AE12/07). The study complied with the current laws of Australia.

Results

Comparisons across a Network of Marine Reserves and Partially-protected Areas

In total, 17,681 individuals from 89 species of fin fish were identified from the 384 BRUV deployments from 2008–2011. The structure of fish assemblages in no-take marine reserves (i.e sanctuary zones) differed significantly from fished areas (Table 2,

Table 1. Average direction of association for the six abundant and commonly-caught fish species (see methods) since the commencement of the zoning plan and potential explanatory variables in six offshore sanctuary zones in the Bateman Marine Park.

	Brush Island	Murramarang	Tollgate Islands	Burrewarra	Broulee Island	Mullimburra	*r*	*P*
Direction of association since the establishment of the zoning plan								
Average Pearson's *r* for fish response variables	−0.07	−0.46	0.64	0.06	−0.19	−0.20		
Potential explanatory variables								
Enforcement actions	8	7	104	11	8	26	0.91	**<0.01**
Sanctuary zone area (ha)	1709	2449	3291	312	172	4542	0.12	0.82
Terminates at 3 nm boundary	Yes	Yes	Yes	No	No	Yes		
Directly Linked to estuarine sanctuary zones	No	No	No	No	No	Yes		
Buffered by Habitat Protection Zone	No	No	Yes	Part	No	Yes		
Adjacent to mainland National Park/Nature reserve	Yes	Yes	No	No	Yes	Yes		

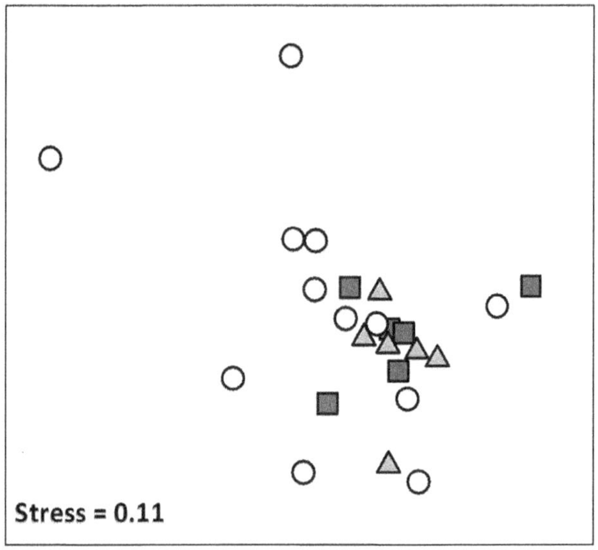

Figure 2. nMDS ordination of fish assemblages represented as centroids for each site within sanctuary (white circles), habitat protection (light grey triangles) and general use (dark grey squares) zones since the commencement of the zoning plan for the Batemans Marine Park. As there were no significant interactions between years and main effect contrasts (Table 1), points indicate site centroids averaged across years since establishment.

Fig. 2). In contrast, the structure of fish assemblages in habitat protection zones did not differ significantly from general use zones (Table 2, Fig. 2).

The richness of fish species was significantly greater in general use zones than in habitat protection zones, but did not differ significantly between sanctuary zones and fished areas (Table 2, Fig. 3). In contrast, the total *max* N of fishes was 37% greater in no-take marine reserves (i.e. sanctuary zones) compared to fished areas, which was significant (Table 2, Fig. 2). There was a trend towards more fish in general use zones compared to habitat protection zones ($P = 0.053$, Fig. 3). Of the numerically-dominant families examined, the mean *max N* of kyphosids was significantly higher in no-take sanctuary zones compared to fished areas (Table 3, Fig. 3). For monacanthids, however, the differences in mean *max* N were significant between fished zones (Table 3, Fig. 3). The total *max* N of carangids and labrids did not differ significantly among zone types (Table 3).

In general, the *max* N's of individual species were more variable than univariate community measures (i.e. species richness and total *max* N) or family groups, leading to fewer significant results. There was, however, a trend towards more *P. auratus* (snapper) in sanctuary zones than fished areas in 2008, 2010 and 2011 (Fig. 4). The *max* N of *P. georgianus* (silver trevally), *S. lineolata* (silver sweep), *O. lineolatus* (southern maori wrasse), *T. novaezelandiae* (yellow tail scad) and *N. douglasii* (grey morwong) did not vary significantly among zone types since the zoning plan's establishment (Table 4, Fig. 4). The average *max* N of *S. lineolata* increased in sanctuary zones with years since establishment.

Comparisons of the Performance of Individual Marine Reserves

There were significant differences in the SZ/FA ratio in the richness and total *max* N of fish assemblages among individual sanctuary zones (Table 5, Fig. 5). These ratios indicated a trend for

Table 2. PERMANOVA analyses comparing the structure of fish assemblages using multivariate data on the richness and abundance (total *max* N) of fishes among zone types and among years since the commencement of the zoning plan.

		(a) Fish assemblages			(b) Species richness			(c) Total *max* N		
	df	MS	*p-F*	*P*	MS	*p-F*	*P*	MS	*p-F*	*P*
Zone type	2	3926.20	2.16	**<0.01**	21.17	2.34	0.10	3210.50	5.61	**<0.01**
SZ vs FA	1	5352.00	2.96	**<0.01**	5.16	0.56	0.48	5188.40	9.22	**<0.01**
HPZ vs GUZ	1	2500.40	1.41	0.17	37.19	7.57	**<0.01**	1232.70	4.42	0.05
Years	3	4194.00	2.30	**<0.01**	47.52	5.26	**<0.01**	1978.90	3.46	**<0.05**
Zone type × Years	6	1058.90	0.58	0.99	3.86	0.43	0.86	228.44	0.40	0.88
Years × SZ vs FA	3	894.56	0.49	0.99	4.73	0.52	0.67	363.55	0.65	0.60
Years × HPZ vs GUZ	3	1223.20	0.69	0.92	2.98	0.61	0.60	93.32	0.33	0.79
Residual	84	1820.70			9.03			571.82		

Fish assemblages (a) used Bray-Curtis similarity measures following square root transformation while species richness (b) and total *max* N (c) used Euclidean distance to generate similarity matrices. Contrasts were included to compare sanctuary zones (SZ) with fished areas (FA) and habitat protection zones (HPZ) with general use zones (GUZ).

p-F = pseudo *F* ratio generated by PERMANOVA.

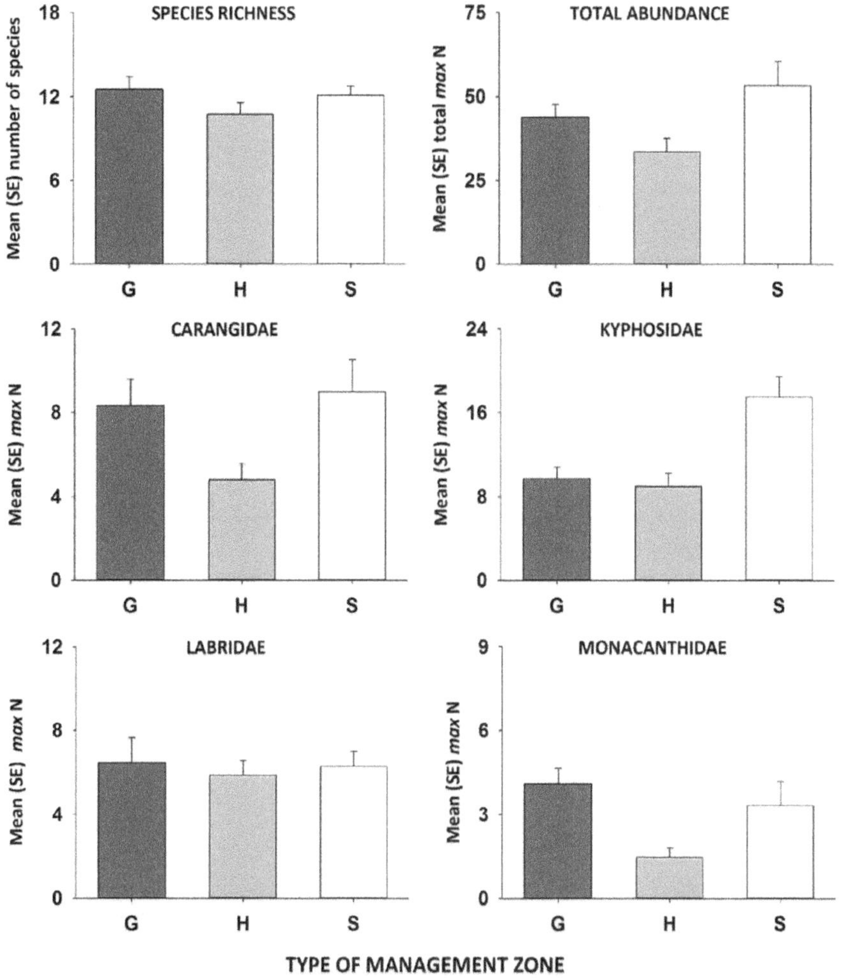

Figure 3. Mean (±1 SE) richness and total *max* N of fish assemblages and numerically-dominant family groups in general use (dark grey bars), habitat protection (light grey bars) and sanctuary (white bars) zones since the commencement of the zoning plan for the Batemans Marine Park.

Table 3. PERMANOVA analyses comparing the total *max* N of the numerically-dominant families among zone types and among years since the commencement of the zoning plan using Euclidean distance.

		(a) Carangidae			(b) Kyphosidae		
	df	MS	p-F	P	MS	p-F	P
Zone type	2	146.12	1.32	0.29	803.43	6.03	**<0.01**
SZ vs FA	1	142.59	1.32	0.26	1600.70	12.37	**<0.01**
HPZ vs GUZ	1	149.64	2.11	0.17	6.20	0.14	0.70
Years	3	92.76	0.84	0.50	48.76	0.37	0.77
Zone type × Years	6	19.05	0.17	0.98	72.47	0.54	0.78
Years × SZ vs FA	3	15.91	0.15	0.93	81.74	0.63	0.58
Years × HPZ vs GUZ	3	22.19	0.31	0.84	63.20	1.38	0.26
Residual	84	110.57			133.24		
		(c) Labridae			**(d) Monacanthidae**		
	df	MS	p-F	P	MS	p-F	P
Zone type	2	2.27	0.42	0.65	44.87	5.12	**<0.01**
SZ vs FA	1	0.32	0.06	0.82	7.04	0.74	0.40
HPZ vs GUZ	1	4.23	0.99	0.32	82.69	9.94	**<0.01**
Years	3	63.84	11.71	**<0.01**	15.89	1.82	0.15
Zone type × Years	6	2.18	0.40	0.89	9.64	1.10	0.37
Years × SZ vs FA	3	1.14	0.21	0.88	12.15	1.27	0.31
Years × HPZ vs GUZ	3	3.21	0.75	0.53	7.14	0.86	0.46
Residual	84	5.45			8.76		

Contrasts were included to compare sanctuary zones (SZ) with fished areas (FA) and habitat protection zones (HPZ) with general use zones (GUZ).
p-F = pseudo *F* ratio generated by PERMANOVA.

greater *max* N of fish in sanctuary zones (i.e. the probability (P) of 6 ratios greater than $1 = 0.059$) and substantially richer fish assemblages in two of the six sanctuary zones sampled (i.e. where the mean plus standard error bar is greater than 1 on Fig. 4). However, the average number of fish species in the Mullimburra sanctuary zone plus one standard error was less than 1, indicating fewer fish species in this sanctuary zone relative to the surrounding fished area (i.e. mean plus standard error are less than 1 on Fig. 5).

The SZ/FA ratio of total *max* N for carangids, kyphosids and labrids also varied significantly among individual sanctuary zones demonstrating variation in individual reserve performance (Table 5, Fig. 5). For four out of six sanctuary zones, the SZ/FA ratio was close to one. For labrids, however, the average total *max* N plus one standard error was less than 1 in the Murramarang reserve indicating fewer of these fishes in this sanctuary zone than in the surrounding fished areas (Fig. 4). The SZ/FA ratio of total *max* N for carangids and monacanthids interacted significantly among sanctuary zones and years since establishment (Table 5). Post hoc tests (PHT) revealed that the patterns of average SZ/FA ratio for these fish taxa varied significantly among sanctuary zones in some years but not others (PHT: $P<0.05$). For example, the SZ/FA ratio of carangids did not vary among individual sanctuary zones in 2008 and 2011, but was significantly greater at Mullimburra than other sanctuary zones in 2009 (PHT: $P<0.05$) and significantly smaller than other sanctuary zones in 2010 (PHT: $P<0.05$). The average SZ/FZ ratio for monacanthids and carangids indicated more of these fishes in 5 out of 6 and 3 out of 6 sanctuary zones than the surrounding fished areas, respectively (Fig. 4). The average SZ/FA ratio plus one standard error for monacanthids at the Mullimburra reserve was less than 1 for each year of sampling (2008 SZ/FA ratio [SE] = 0.41 [0.08], 2009 = 0.32 [0.04], 2010 = 0.21 [0.03], 2011 = 0.51 [0.085]). Similarly, the average SZ/FA ratio plus one standard error of carangids in the Murramarang reserve, was less than 1 for three of the four years of sampling (2008 SZ/FA ratio [SE] = 0.67 [0.10], 2009 = 0.28 [0.06], 2010 = 0.11 [0.02], 2011 = 1.17 [0.61]).

The r_{av} for the six abundant species considered to be important for commercial and recreational fishing in NSW waters varied substantially among individual sanctuary zones (Table 1). The value of 0.64 for the Tollgate Island sanctuary zone was strongly positive with each of the six species having a positive association between SZ/FA ratio and the years since the zoning plan's commencement (Table 1). In contrast, the average direction of change at Murramarang ($r_{av} = -0.46$) and Mullimburra ($r_{av} = -0.20$) suggested limited performance in these marine reserves for the species we considered. For the remaining three sanctuary zones, there was no strong average directional association between SZ/FA ratio and the years since the zoning plan's commencement ($0.20 > r_{av} < -0.20$).

After P-values were corrected using sequential Bonferroni's technique, there was a significant correlation between the r_{av} for the six key fish species and the number of enforcement actions undertaken (Table 1), indicating a positive association between individual reserve performance and compliance activity. In contrast, there were no significant correlations between average directional association of the SZ/FA ratio for the six key fish species since the park's establishment (r_{av}) and the size of sanctuary zones (Table 1). With respect to key reserve attributes (Table 1),

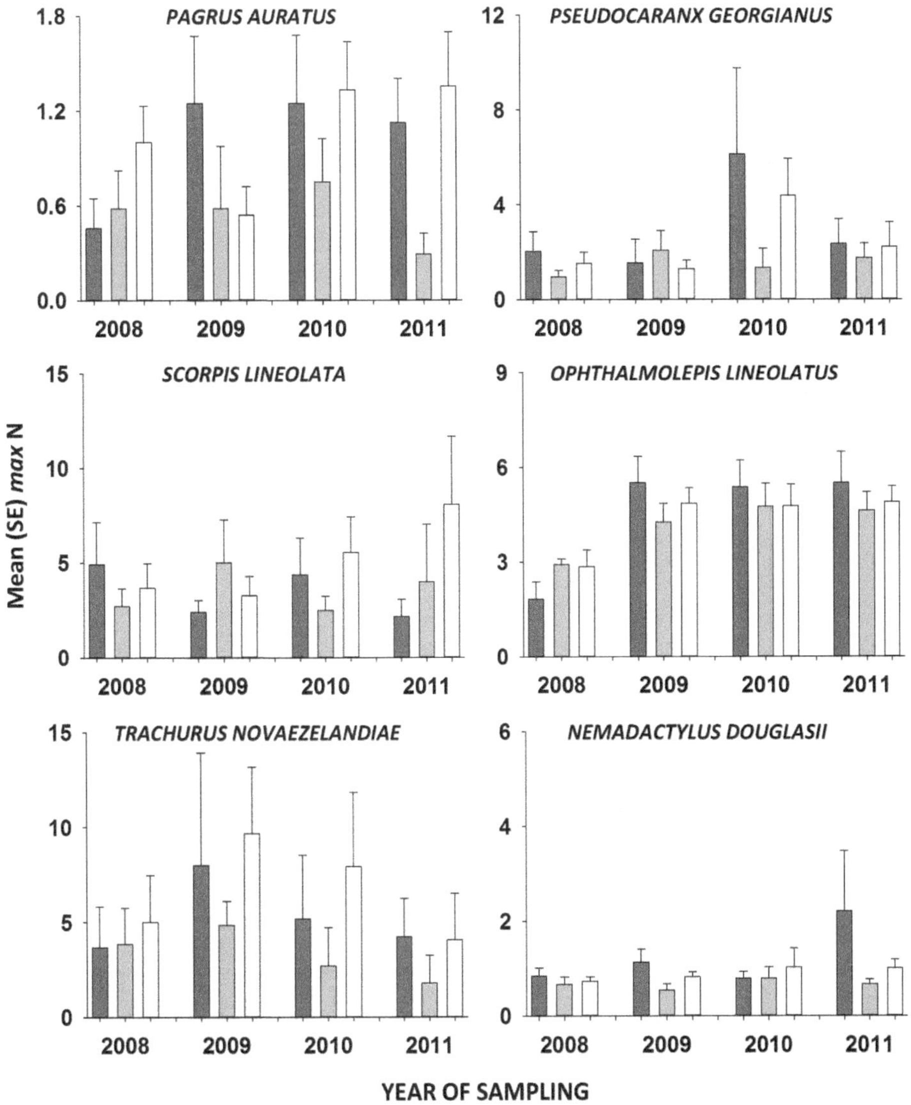

Figure 4. Mean (±1 SE) *max* N of fish species important to recreational and commercial fisheries in general use (dark grey bars), habitat protection (light grey bars) and sanctuary (white bars) zones in each year since the commencement of the zoning plan for the Batemans Marine Park.

the large reserve at Mullimburra was the only one to have full coverage from the shore to the 3 nm limit of state waters, to be directly linked to estuarine sanctuary zones, to be buffered by habitat protection zones and to be adjacent to mainland National Parks. In contrast, the only key attribute possessed by the relatively small Broulee Island Reserve was that it was adjacent to a Nature Reserve. Moreover, the small Burrewarra reserve only had partial buffering from habitat protection zones. The reserves at Brush Island and Murramarang each terminated at the 3 nm limit and were adjacent to National Parks. The Tollgate Island Reserve ran out to the 3 nm limit of state waters and was buffered by habitat protection zones. As it commenced approximately 1 km offshore (Fig. 1), the Tollgate Island reserve could not link directly to estuarine sanctuary zones. It was, however, directly adjacent to the Clyde River Estuary that included several substantial no-take estuarine sanctuaries. Although the Tollgate Islands Reserve was adjacent to urban development, its distance from shore (Fig. 1) provided a substantial buffer from land-based impacts. The Tollgate Islands themselves are Nature Reserves not accessible to the general public.

Discussion

On average there were 38% more fish in the network of marine reserves than in fished areas of the Batemans Marine Park. The largest contribution to this effect came from Kyphosids (drummers). Compared to global averages for individual reserves (e.g. 166%, n = 124 reserves [5]) the elevated fish abundances across the network of marine reserves was modest, but well within the spectrum of positive responses. This may, in part, be due to the marine park only being in place for 5 years (e.g. [30,48]) and previous fishing pressure being regulated by conventional fisheries management [42]. It may also stem from the fished areas being partially-protected such that even the most unprotected places in the marine park (general use zones) were free from potentially damaging activities such as demersal trawling and long-lining [24].

Table 4. PERMANOVA analyses comparing the *max* N of key fish species among zone types and among years since the commencement of the zoning plan.

		(a) *P. auratus*			(b) *P. georgianus*			(c) *S. lineolata*		
	df	MS	p-F	P	MS	p-F	P	MS	p-F	P
Zone type	2	2.20	2.95	0.07	13.20	1.06	0.37	31.73	0.81	0.45
SZ vs FA	1	1.76	2.32	0.15	0.15	0.01	0.92	63.38	1.67	0.22
HPZ vs GUZ	1	2.64	4.51	**<0.05**	26.26	1.96	0.18	0.08	0.01	0.95
Years	3	0.74	0.99	0.41	27.44	2.20	0.09	5.88	0.15	0.94
Zone type × Years	6	0.87	1.16	0.34	8.70	0.70	0.70	28.19	0.72	0.62
Years × SZ vs FA	3	1.21	1.59	0.19	1.38	0.11	0.95	37.98	1.00	0.41
Years × HPZ vs GUZ	3	0.53	0.90	0.44	16.02	1.19	0.31	18.40	0.96	0.44
Residual	84	0.75			12.48			39.20		
		(d) *O. lineolatus*			(e) *T. novaezelandiae*			(f) *N. douglasii*		
	df	MS	p-F	P	MS	p-F	P	MS	p-F	P
Zone type	2	1.04	0.31	0.74	92.33	1.07	0.36	1.86	1.87	0.13
SZ vs FA	1	<0.01	<0.01	1.00	137.16	1.65	0.19	0.07	0.06	0.85
HPZ vs GUZ	1	2.08	0.70	0.43	47.50	0.95	0.38	3.66	2.64	0.07
Years	3	31.49	9.37	**<0.01**	70.34	0.81	0.48	1.33	1.34	0.27
Zone type × Years	6	1.94	0.58	0.76	9.65	0.11	0.99	1.02	1.03	0.39
Years × SZ vs FA	3	0.69	0.21	0.90	12.79	0.15	0.93	0.51	0.49	0.77
Years × HPZ vs GUZ	3	3.20	1.08	0.40	6.52	0.13	0.95	1.53	1.11	0.33
Residual	84	3.36			86.55			0.99		

Contrasts were included to compare sanctuary zones (SZ) with fished areas (FA) and habitat protection zones (HPZ) with general use zones (GUZ).
p-F = pseudo *F* ratio generated by PERMANOVA.

Despite the total *max* N of fishes being significantly greater in marine reserves than in fished areas, there were no significant differences in the richness of reef fishes across the network of no-take reserves compared to fished areas. There was, however, greater richness of reef fish species in Brush Island and Burrewarra reserves compared to the adjacent fished areas. Similar to richness, there were also no significant differences in the abundances of some family groups and commonly-caught fish species among zone types. Large variation in the measurement of fish populations contributed to these results. For example, although there was 37% more Carangids in marine reserves than in fished areas, this comparison was not close to being significant due to substantial variation among zones and sites in fished areas. Nonetheless, it was to be expected that only the very large changes in fish assemblages would be detected because power analyses demonstrated that, for the levels of replication used, effects of 30% and 100% were required to reliably detect significant differences in the richness and max *N* between reserves and fished areas, respectively.

Another consideration for the non-significant results was the influence of time since reserve establishment. In comparisons of other temperate Australian marine reserves to fished areas from before to three years after establishment, Edgar et al [49] demonstrated few changes in the abundance of fish and invertebrates in the marine reserves compared to fished areas. They concluded that the three-year period studied after reserve commencement may have been insufficient to generate clear trends in fish population recoveries. The results from our study suggest that 5 years may also not be sufficient to detect change of some fish species whose abundances have been demonstrated to recover in much older marine reserves (e.g. *Pagrus auratus*, snapper [28]). Similar conclusions were reached about fish populations on shallow subtidal reefs sampled using underwater visual census over the first five years following the establishment of the Batemans Marine Park [50].

An important consideration for interpreting positive effects of marine reserves on fish abundances is whether marine reserves were deliberately placed in areas with more fish. For the Batemans Marine Park this was not the case because, although some data was collected prior to the parks' establishment [31,51], detailed regionally specific data on reef fish assemblages and reef extent and complexity were not available to marine park planners prior to the marine parks establishment. Furthermore, there are two lines of evidence to support positive reserve effects: (i) the abundances of some species increased in sanctuary zones over time (e.g. *S. lineolata*, silver sweep) and (ii) sampling conducted prior to the zoning plan's establishment indicated that there were similar if not fewer fish in marine reserves compared to fished areas [48].

Given that all levels in the factor *zone type* were replicated with multiple sites and BRUV deployments were haphazardly located on reefs of similar structure, our study was of the form of a standard ecological field experiment where the manipulation was the implementation and enforcement of marine park regulations. While this sampling was sufficient for testing the proposed hypotheses about reserve effects, similar to most published field experiments, the ability to attribute treatment effects to the manipulation (e.g. conservation measures in this case) rather than site selection could be improved by the incorporation of pre-establishment data into comparisons (e.g. BACI-type experimental

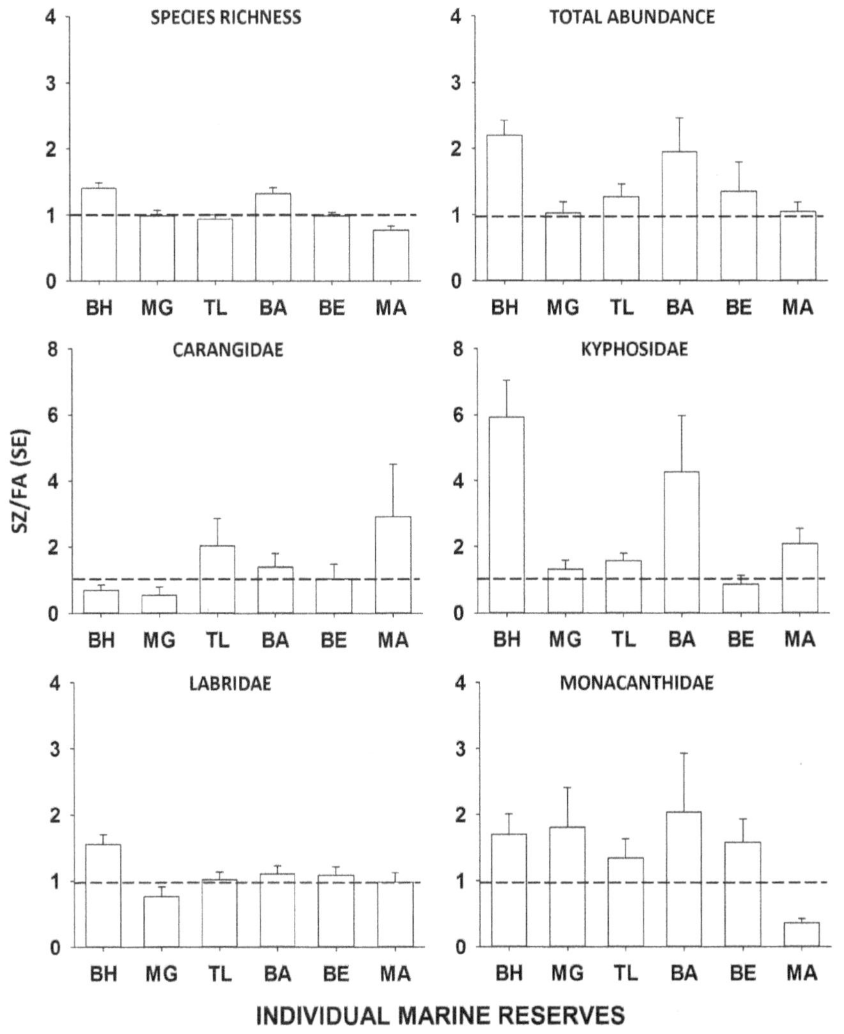

Figure 5. Mean (±1 SE) sanctuary zone to local fished areas ratio (SZ/FA) for richness and total *max* N of fish assemblages and numerically-dominant family groups in the Brush Island (BH), Murramarang (MG), Tollgate Islands (TL), Burrewarra (BA), Broulee Island (BE) and Mullimburra (MA) sanctuary zones. Bars represent the main effects of sanctuary zone averaged across years since establishment.

designs [52]). The marine park planning process is, however, not always conducive to implementation of robust BACI experimental designs. For the Batemans Marine Park, there was around 14 months between the declaration of the Park and the implementation of the zoning plan [27]. Most of this period was taken up with planning and public consultation, leaving only a few months between when the locations of the marine reserves were finalized and the zoning plan coming into effect. Consequently, there was insufficient time to collect the inter-annual pre-establishment data required for a temporally-replicated BACI-style experimental design.

In general, fish assemblages either did not differ between the partially-protected areas with different levels of protection or there were more species in general use zones than habitat protection zones. There was, therefore, no evidence that the additional restrictions associated with habitat protection zones, such as removal of commercial purse seining, lift netting and set lining, improved conservation outcomes for reef fish assemblages on offshore reefs after 5 years. Given that trawling and long-lining were removed from the entire Batemans Marine Park, the removal of other less damaging commercial fishing activities from habitat protection zones probably had limited additional influence on fish communities. Furthermore, the designation of habitat protection zones could have attracted some increased recreational fishing effort, therefore reducing differences between the two different zones. This is because habitat protection zones are often promoted as enhancing recreational fishing opportunities through reduced commercial fishing effort, with similar types of areas in NSW estuaries (e.g. Recreational Fishing Havens) being perceived by recreational anglers as improving catch rates [53]. Management strategies that result in shifting recreational fishing effort towards partially-protected areas may limit the conservation benefits of these areas.

Although there was a general increase in overall fish abundance in marine reserves across the network, there was significant variation among the performance of individual reserves. The six commonly caught fish species in marine reserves at the Tollgate Islands showed the strongest positive trend over the 5 years of reserve protection (Pearson's $r = 0.64$). Although quantitative data on fishing effort was not collected consistently across the Batemans

Table 5. PERMANOVA analyses comparing the sanctuary zone to local fished area ratio (SZ/FA) for univariate measures of fish assemblages, numerically-dominant families and key fish species among sanctuary zones (SZ) and among years since the commencement of the zoning plan for the Batemans Marine Park.

		(a) Species richness			(b) Total abundance		
	df	MS	*p-F*	*P*	MS	*p-F*	*P*
SZ	5	1.70	16.15	**<0.01**	6.97	4.57	**<0.01**
Years	3	0.18	1.24	0.33	4.39	2.25	0.13
SZ × Years	15	0.15	1.47	0.13	2.03	1.33	0.21
Residual	168	0.11			1.52		
		(c) Carangidae			**(d) Kyphosidae**		
	df	**MS**	***p-F***	***P***	**MS**	***p-F***	***P***
SZ	5	26.72	3.38	**<0.05**	114.14	8.60	**<0.01**
Years	3	2.64	0.12	0.96	30.17	2.04	0.14
SZ × Years	15	25.65	3.24	**<0.01**	15.05	1.13	0.35
Residual	168	7.91			13.27		
		(e) Labridae			**(f) Monacanthidae**		
	df	**MS**	***p-F***	***P***	**MS**	***p-F***	***P***
SZ	5	2.18	6.77	**<0.01**	9.98	4.28	**<0.01**
Years	3	1.10	2.30	0.12	15.18	3.29	**<0.05**
SZ × Years	15	0.51	1.57	0.10	5.04	2.16	**<0.05**
Residual	168	0.32			2.33		

p-F = pseudo *F* ratio generated by PERMANOVA.

Marine Park prior to its establishment, it was well known that the Tollgate Islands were heavily targeted by boat-based fishers prior to the enforcement of marine park regulations, as the islands are adjacent to the largest town and boating facilities in the region. This offshore reserve was also one of the most commonly patrolled because of its central location and proximity to a relatively safe ocean bar crossing. The substantial reduction in fishing effort combined with the greatest compliance effort would have contributed to the Tollgate Islands reserve showing the strongest positive responses to protection over the first 5 years.

Effective compliance cannot be underestimated in achieving positive marine conservation outcomes [13,14]. As expected, enforcement actions were positively associated with individual reserve performance in the Batemans Marine Park. As well as active enforcement, the Batemans Marine Park operational plan included priority actions aimed at increasing voluntary compliance. This included local education and awareness activities, programs to improve signage and zone markers as well as proactively restricting potentially harmful activities through permitting. In response to these strategies, we contend that public knowledge of the marine park zoning arrangements improved substantially since the parks establishment. For example, marine park awareness by tourists increased from 47% (n = 203) to 72% (n = 36) from 2008 to 2011 (Eurobodalla Shire Council and NSW Marine Park Authority, unpublished data). Public knowledge and support for marine reserves increases voluntary compliance, which can both improve the effectiveness of marine reserves and reduce the costs of enforcement [54,55]. Greater consideration of compliance planning during establishment and adaptive management of marine reserve networks can enhance voluntary compliance and improve conservation outcomes [24].

Marine reserve size is generally regarded as a fundamental principle in effective marine reserve design with larger reserves often having greater conservation benefits [56]. By this criterion, the largest marine reserve in our study, Mullimburra, did not perform as well as smaller reserves in the network. Consequently, factors other than reserve size must have been driving this result. Importantly, the Mullimburra reserve had many characteristics considered important for effective reserve design (Table 1). Mullimburra marine reserve was adjacent to the Eurobodalla National Park minimizing potential land-based threats to the marine ecosystem [57,58]. It was also directly linked to no-take estuarine reserves ensuring undisrupted connectivity between juvenile and adult habitats [59,60]. Mullimburra marine reserve had cross-shelf coverage from the shore to the 3 nm limit of NSW state waters, maximizing reef habitat representation [61], which is known to be extensive in inner- and mid-shelf waters in the region [51]. It was also surrounded by extensive partially-protected areas (i.e. habitat protection zones) buffering it from unintentional commercial fishing activities [62].

Given all these key reserve attributes, it is not clear why the large reserve at Mullimburra did not perform as well as some smaller reserves, although it should be noted that the BRUV sites in fished areas adjacent to this reserve were further away than they were for other reserves. A more likely explanation is the influence of compliance levels because the least effective reserves, Murramarang and Mullimburra, also had the lowest number of enforcement actions per unit area. A review of compliance related issues from the Great Barrier Reef Marine Park suggests that even a small amount of poaching can have major ecological consequences [26]. Although there are no data available to discriminate between compliance efficacy and the amount of illegal fishing activity in the Batemans Marine Park, the significant relationship between enforcement actions and reserve performance suggests that quantitative monitoring of compliance and illegal activities should be prioritized to facilitate adaptive management to maximize marine conservation outcomes.

It is not possible from our results to determine whether the performance of individual marine reserves within the first 5 years will be a useful predictor of long-term reserve performance. This raises important questions about how much park-specific ecological monitoring can contribute to evidence-based adaptive management of marine park zoning arrangements at a 5 year review, as is currently required in NSW. Certainly, clear advice can be given that the network of marine reserves in the Batemans Marine Park had a positive influence on the abundance of fishes, particularly kyphosids, despite differences in the performance of the individual marine reserves we examined. In contrast, there were no consistent effects to validate the efficacy of habitat protection zones. Ongoing enforcement will also be required to maintain reserve efficacy and extra compliance attention should be focused on the large marine reserves at Mullimburra and Murramarang, which appear to be underperforming given their attributes (see Table 1).

Apart from this general advice, 5 years of ecological monitoring was insufficient to provide scientific evidence that would justify changing the current network of marine reserves and partially-protected areas in the Batemans Marine Park to improve long-term conservation of biodiversity. Nonetheless, the broader scientific literature about marine reserves will still have an invaluable contribution to the review process, with rigorous assessments of reserve attributes (e.g. size, habitat linkages, buffering) from much older marine reserve networks being particularly informative. After the initial 5 year review, NSW marine parks are reviewed every 10 years. At the 15 year review, sufficient time should have passed for substantial changes in the structure of marine communities to have occurred [28,30]. At this point, the results from local ecological monitoring and other complimentary research will be in a much stronger position to drive evidence-based adaptive management to enhance long-term conservation objectives.

In conclusion, few studies have examined changes in fish assemblages across a network of marine reserves relative to fished areas with different levels of environmental protection. We show that after 5 years of protection, fish abundances were 37% greater across the network of marine reserves compared to partially-protected areas, although not all individual reserves performed equally and performance was temporally variable. These changes are relatively modest compared to some reserve networks (e.g. [63]), but still add to the growing weight of evidence that conservation outcomes from planned networks of marine reserves are greater than those from individual reserves [4,26]. Our results also provide insight into factors (e.g. past fishing effort and compliance) that promote early conservation benefits to fish in temperate marine reserves and thus should be carefully considered in marine reserve establishment and management. As coastal population growth and associated development increases stress on marine environments, it is critical that networks of marine reserves are designed and adaptively managed to maximise their conservation objectives. Although local environmental monitoring can contribute to adaptive management of newly established marine reserve networks, the extent of this contribution will be limited by the rate of change in marine communities in response to protection. The adaptive management processes of newly established marine reserve networks could, therefore, be enhanced by rigorous assessment of the efficacy of ecological attributes and planning principles from much older networks.

Acknowledgments

We thank Ian Osterloh, Jess Klages, James Wraith and Jessica Rugge for assistance with field work and sample processing. This is contribution no. 305 from the Ecology and Genetics Group at this University of Wollongong.

Author Contributions

Conceived and designed the experiments: BPK MC AB MR AJ ARD. Performed the experiments: BPK MC AB MR AJ ARD. Analyzed the data: BPK MC AB MR AJ ARD. Contributed reagents/materials/analysis tools: BPK MC AB MR AJ ARD. Wrote the paper: BPK MC AB MR AJ ARD.

References

1. Lotze HK, Lenihan HS, Bourque BJ, Bradbury RH, Cooke RG, et al. (2006) Depletion, degradation, and recovery potential of estuaries and coastal seas. Science 312: 1806–1809.
2. Crain CM, Halpern BS, Beck MW, Kappel CV (2009) Understanding and managing human threats to the coastal marine environment. In: Ostfeld RS, Schlesinger WH, editors. Year in Ecology and Conservation Biology 2009. 39–62.
3. Halpern BS, Lester SE, McLeod KL (2010) Placing marine protected areas onto the ecosystem-based management seascape. Proceedings of the National Academy of Sciences of the United States of America 107: 18312–18317.
4. Gaines SD, White C, Carr MH, Palumbi SR (2010) Designing marine reserve networks for both conservation and fisheries management. Proceedings of the National Academy of Sciences of the United States of America 107: 18286–18293.
5. Lester SE, Halpern BS, Grorud-Colvert K, Lubchenco J, Ruttenberg BI, et al. (2009) Biological effects within no-take marine reserves: a global synthesis. Marine Ecology Progress Series 384: 33–46.
6. Mosqueira I, Cote IM, Jennings S, Reynolds JD (2000) Conservation benefits of marine reserves for fish populations. Animal Conservation 3: 321–332.
7. Barrett NS, Edgar GJ, Buxton CD, Haddon M (2007) Changes in fish assemblages following 10 years of protection in Tasmanian marine protected areas. Journal of Experimental Marine Biology and Ecology 345: 141–157.
8. Shears NT, Babcock RC (2003) Continuing trophic cascade effects after 25 years of no-take marine reserve protection. Marine Ecology Progress Series 246: 1–16.
9. Mumby PJ, Harborne AR, Williams J, Kappel CV, Brumbaugh DR, et al. (2007) Trophic cascade facilitates coral recruitment in a marine reserve. Proceedings of the National Academy of Sciences of the United States of America 104: 8362–8367.
10. Harrison HB, Williamson DH, Evans RD, Almany GR, Thorrold SR, et al. (2012) Larval Export from Marine Reserves and the Recruitment Benefit for Fish and Fisheries. Current Biology 22: 1023–1028.
11. Russ GR, Alcala AC (2011) Enhanced biodiversity beyond marine reserve boundaries: The cup spillith over. Ecological Applications 21: 241–250.
12. Burfeind DD, Pitt KA, Connolly RM, Byers JE (2013) Performance of non-native species within marine reserves. Biological Invasions 15: 17–28.
13. Byers JE, Noonburg EG (2007) Poaching, enforcement, and the efficacy of marine reserves. Ecological Applications 17: 1851–1856.
14. Guidetti P, Milazzo M, Bussotti S, Molinari A, Murenu M, et al. (2008) Italian marine reserve effectiveness: does enforcement matter? Biological Conservation 141: 699–709.
15. Hastings A, Botsford LW (2003) Comparing designs of marine reserves for fisheries and for biodiversity. Ecological Applications 13: 65–70.
16. Roberts CM, Andelman S, Branch G, Bustamante RH, Castilla JC, et al. (2003) Ecological criteria for evaluating candidate sites for marine reserves. Ecological Applications 13: S199–S214.
17. Coleman MA, Chambers J, Knott NA, Malcolm HA, Harasti D, et al. (2011) Connectivity within and among a Network of Temperate Marine Reserves. Plos One 6: 10.1371/journal.pone.0020168.
18. Botsford LW (2005) Potential contributions of marine reserves to sustainable fisheries: recent modeling results. Bulletin of Marine Science 76: 245–259.
19. Hastings A, Botsford LW (2006) Persistence of spatial populations depends on returning home. Proceedings of the National Academy of Sciences of the United States of America 103: 6067–6072.
20. Moffitt EA, Botsford LW, Kaplan DM, O'Farrell MR (2009) Marine reserve networks for species that move within a home range. Ecological Applications 19: 1835–1847.
21. Planes S, Jones GP, Thorrold SR (2009) Larval dispersal connects fish populations in a network of marine protected areas. Proceedings of the National Academy of Sciences of the United States of America 106: 5693–5697.
22. Denny CM, Babcock RC (2004) Do partial marine reserves protect reef fish assemblages? Biological Conservation 116: 119–129.
23. Lester SE, Halpern BS (2008) Biological responses in marine no-take reserves versus partially protected areas. Marine Ecology Progress Series 367: 49–56.
24. Read AD, West RJ (2010) Qualitative risk assessment of multiple-use marine park effectiveness - a case study from NSW, Australia. Ocean & Coastal Management 53: 636–644.
25. Tuya FC, Soboil ML, Kido J (2000) An assessment of the effectiveness of Marine Protected Areas in the San Juan Islands, Washington, USA. Ices Journal of Marine Science 57: 1218–1226.
26. McCook LJ, Ayling T, Cappo M, Choat JH, Evans RD, et al. (2010) Adaptive management of the Great Barrier Reef: a globally significant demonstration of the benefits of networks of marine reserves. Proceedings of the National Academy of Sciences of the United States of America 107: 18278–18285.
27. Banks SA, Skilleter GA (2010) Implementing marine reserve networks: a comparison of approaches in New South Wales (Australia) and New Zealand. Marine Policy 34: 197–207.
28. Babcock RC, Shears NT, Alcala AC, Barrett NS, Edgar GJ, et al. (2010) Decadal trends in marine reserves reveal differential rates of change in direct and indirect effects. Proceedings of the National Academy of Sciences of the United States of America 107: 18256–18261.
29. Roberts CM (2005) Marine protected areas and biodiversity conservation. In: Norse E, Crowder LB, editors. Marine conservation biology: the science of maintaining the sea's biodiversity. Washington DC: Island Press. 265–279.
30. Molloy PP, McLean IB, Cote IM (2009) Effects of marine reserve age on fish populations: a global meta-analysis. Journal of Applied Ecology 46: 743–751.
31. Edgar GJ, Stuart-Smith RD (2009) Ecological effects of marine protected areas on rocky reef communities-a continental-scale analysis. Marine Ecology Progress Series 388: 51–62.
32. Gladstone W, Lindfield S, Coleman M, Kelaher B (2012) Optimisation of baited remote underwater video sampling designs for estuarine fish assemblages. Journal of Experimental Marine Biology and Ecology 429: 28–35.
33. Lowry M, Folpp H, Gregson M, Suthers I (2012) Comparison of baited remote underwater video (BRUV) and underwater visual census (UVC) for assessment of artificial reefs in estuaries. Journal of Experimental Marine Biology and Ecology 416: 243–253.

34. Colton MA, Swearer SE (2010) A comparison of two survey methods: differences between underwater visual census and baited remote underwater video. Marine Ecology Progress Series 400: 19–36.
35. Klages J, Broad A, Kelaher BP, Davis AR (2013) The influence of gummy sharks, Mustelus antarcticus, on observed fish assemblage structure. Environmental Biology of Fishes. doi: 101007/s10641-013-0138-2.
36. Curley BG, Jordan AR, Figueira WF, Valenzuela VC (2013) A review of the biology and ecology of key fishes targeted by coastal fisheries in south-east Australia: identifying critical knowledge gaps required to improve spatial management. Reviews in Fish Biology and Fisheries.
37. Coleman MA, Kelaher BP (2009) Connectivity among fragmented populations of a habitat-forming alga, Phyllospora comosa (Phaeophyceae, Fucales) on an urbanised coast. Marine Ecology Progress Series 381: 63–70.
38. Coleman MA, Roughan M, Macdonald HS, Connell SD, Gillanders BM, et al. (2011) Variation in the strength of continental boundary currents determines continent-wide connectivity in kelp. Journal of Ecology 99: 1026–1032.
39. Gray CA, Otway NM (1994) Spatial and temporal differences in assemblages of demersal fishes on the innner continental-shelf off Sydney, South-Eastern Australia. Australian Journal of Marine and Freshwater Research 45: 665–676.
40. Malcolm HA, Gladstone W, Lindfield S, Wraith J, Lynch TP (2007) Spatial and temporal variation in reef fish assemblages of marine parks in New South Wales, Australia - baited video observations. Marine Ecology Progress Series 350: 277–290.
41. Wraith J, Lynch T, Minchinton TE, Broad A, Davis AR (2013) Bait type affects fish assemblages and feeding guilds observed at baited remote underwater video stations. Marine Ecology Progress Series 477: 189–199.
42. Rowling K, Hegarty A, Ives M (2010) Status of fisheries resources in NSW 2008/09. SydneyAustralia: NSW Industry & Investment. 392 p.
43. Anderson MJ (2001) A new method for non-parametric multivariate analysis of variance. Austral Ecology 26: 32–46.
44. Kelaher BP (2003) Changes in habitat complexity negatively affect diverse gastropod assemblages in coralline algal turf. Oecologia 135: 431–441.
45. Bray JR, Curtis JT (1957) An ordination of the upland forest communities of southern Wisconsin. Ecological Monographs 27: 326–349.
46. Field JG, Clarke KR, Warwick RM (1982) A practical strategy for analyzing multispecies distribution patterns. Marine Ecology Progress Series 8: 37–52.
47. Rice WR (1989) Analysing tables of statistical tests. Evolution 43: 223–225.
48. Edgar GJ, Barrett NS, Stuart-Smith RD (2009) Exploited reefs protected from fishing transform over decades into conservation features otherwise absent from seascapes. Ecological Applications 19: 1967–1974.
49. Edgar GJ, Barrett NS (2012) An assessment of population responses of common inshore fishes and invertebrates following declaration of five Australian marine protected areas. Environmental Conservation 39: 271–281.
50. Coleman MA, Palmer-Brodie A, Kelaher BP (2013) Conservation benefits of a network of marine reserves and partially protected areas. Biological Conservation 167: 257–264.
51. Jordan A, Davies P, Ingleton T, Mesley E, Neilson J, et al. (2010) Seabed habitat mapping of continental shelf waters of NSW. Sydney: NSW Department of Environment, Climate Change and Water Occasional Paper Series.
52. Underwood AJ (1993) The mechanics of spatially replicated sampling programs to detect environmental impacts in a variable world. Australian Journal of Ecology 18: 99–116.
53. Steffe AS, Murphy JJ, Chapman DJ, Barrett GP, Gray CA (2005) An assessment of changes in the daytime, boat-based, recreational fishery of the Tuross Lake estuary following the establishment of a 'Recreational Fishing Haven' Sydney, Australia: NSW Department of Primary Industries. 70 p.
54. Davis BC, Morett GS (2005) Enforcing U.S. marine protected areas: synthesis report. Silver SpringMaryland, USA: National Marine Protected Areas Center in Cooperation with the National Oceanic and Atmospheric Administration Coastal Services Center. 69 p.
55. Read AD, West RJ, Haste M, Jordan A (2011) Optimizing voluntary compliance in marine protected areas: A comparison of recreational fisher and enforcement officer perspectives using multi-criteria analysis. Journal of Environmental Management 92: 2558–2567.
56. Claudet J, Osenberg CW, Benedetti-Cecchi L, Domenici P, Garcia-Charton JA, et al. (2008) Marine reserves: size and age do matter. Ecology Letters 11: 481–489.
57. Stoms DM, Davis FW, Andelman SJ, Carr MH, Gaines SD, et al. (2005) Integrated coastal reserve planning: making the land-sea connection. Frontiers in Ecology and the Environment 3: 429–436.
58. Alvarez-Romero JG, Pressey RL, Ban NC, Vance-Borland K, Willer C, et al. (2011) Integrated Land-Sea Conservation Planning: The Missing Links. In: Futuyma DJ, Shaffer HB, Simberloff D, editors. Annual Review of Ecology, Evolution, and Systematics, Vol 42. 381–409.
59. Olds AD, Connolly RM, Pitt KA, Maxwell PS (2012) Habitat connectivity improves reserve performance. Conservation Letters 5: 56–63.
60. Mumby PJ, Edwards AJ, Arias-Gonzalez JE, Lindeman KC, Blackwell PG, et al. (2004) Mangroves enhance the biomass of coral reef fish communities in the Caribbean. Nature 427: 533–536.
61. Malcolm HA, Smith SDA, Jordan A (2010) Using patterns of reef fish assemblages to refine a Habitat Classification System for marine parks in NSW, Australia. Aquatic Conservation-Marine and Freshwater Ecosystems 20: 83–92.
62. NSW Marine Parks Authority (2009) Solitary Islands Marine Park: zoning plan review report. SydneyAustralia: NSW Marine Parks Authority. 117 p.
63. Russ GR, Cheal AJ, Dolman AM, Emslie MJ, Evans RD, et al. (2008) Rapid increase in fish numbers follows creation of world's largest marine reserve network. Current Biology 18: 514–515.

5

Salmon Migration Patterns Revealed the Temporal and Spatial Fluctuations of the Radiocesium Levels in Terrestrial and Ocean Environments

Takaomi Arai*

Institute of Oceanography and Environment, Universiti Malaysia Terengganu, Kuala Terengganu, Terengganu, Malaysia

Abstract

The disabling of the Fukushima Daiichi Nuclear Power Plant (F1NPP) resulted in the release of radionuclides, including ^{134}Cs and ^{137}Cs, into the air and the ocean. The unpredicted nuclear accident is of global concern for human health and the ecosystem. Although investigations of radionuclides in environments were performed shortly after the accident started, the temporal and spatial impacts and fluctuations on the releasing radionuclides to natural environment remain unclear. I focused on salmon, which migrate from inland to the open ocean globally, to reveal the three-year (May 2011 to February 2014) fluctuations and accumulations of ^{134}Cs and ^{137}Cs from terrestrial to open ocean environments after the F1NPP accident. The ^{134}Cs and ^{137}Cs concentrations in six salmonids exhibited lower temporal variations for three years after the F1NPP accident, suggesting that these radionuclides are widely distributed and these radionuclides remain in the natural environment globally with less convergence. The accumulation patterns were significantly different among the different salmon species. Fluvial (freshwater residence) type salmons exhibited significantly higher accumulation in ^{134}Cs (25.3–40.2 Bq kg^{-1} in mean) and ^{137}Cs (41.4–51.7 Bq kg^{-1} in mean) than did the anadromous (sea-run) type salmons (0.64–8.03 Bq kg^{-1} in mean ^{134}Cs and 0.42–10.2 Bq kg^{-1} in mean ^{137}Cs) suggesting widespread contamination in terrestrial environments versus the coastal and open ocean environments. Salmonids are the most highly migratory animals and are characterised by their strong tendency to return home to their natal site for reproduction. Salmonids have a potential to be a good indicator as an effective monitoring animal.

Editor: Georges Chapouthier, Université Pierre et Marie Curie, France

Funding: This author has no support or funding to report.

Competing Interests: The author has declared that no competing interests exist.

* Email: arai@umt.edu.my

Introduction

The Great East Japan Earthquake of magnitude 9.0 on Friday 11 March 2011 caused considerable damage in the region, and the large tsunami it created caused much more. Following a major earthquake, a 15-m tsunami disabled the power supply and the cooling of three Fukushima Daiichi reactors, causing a nuclear accident at that date. Radioactive contamination following the Fukushima Daiichi Nuclear Power Plant (F1NPP) accident was the most significant artificial radioactive liquid release into the environment ever known, on a short time and space scale basis [1–5]. The amount of radionuclides discharged from the F1NPP immediately after the accident was estimated to be 5–10 PBq to the atmosphere and 3–6 PBq to the sea, the latter being caused by direct leakage of the contaminated cooling water [5–8]. Afterwards, trace quantities of radioactive particles from the incident have since been detected around the world, including ^{131}iodine and ^{134}Cs and ^{137}Cs [1,3]. Fukushima-derived ^{134}Cs and ^{137}Cs were found throughout a 150,000 km^2 area of the Pacific Ocean of Japan [4]. In Europe, the first signs of the release were detected seven days later, while the first peak of the activity level was observed between 28 March and 30 March 2011 [3].

Among the radionuclides discharged by the accident, the effects of ^{134}Cs and ^{137}Cs are the most serious because of their high levels and long decay periods. These radionuclides are considered to most substantially contribute to the contamination of various organisms due to the contamination of their environments. Prior to the accident, the concentrations of ^{137}Cs in the surface water of the Pacific Ocean were in the range of 1–4 Bq m^{-3} [9–11], which mainly resulted from global fallout due to atmospheric nuclear weapon tests, reprocessing plants, fallout from the accident of the Chernobyl nuclear power plant, and solid waste dumping at sea [12]. Recently, Bailly du Bois et al. [12] proposed that 12–41 PBq of ^{137}Cs was discharged to the Pacific Ocean between March 25 and July 28, 2011 and predicted an increase of approximately 6 Bq m^{-3} after dilution in the surface ocean. In addition to this direct release into the ocean, the accident also dispersed approximately 15 PBq of ^{137}Cs into the atmosphere, dominated by a leak from 12 March to 6 April 2011 [2]. These nuclides, once discharged into the atmosphere, can have a global effect through fast atmospheric circulation. These phenomena seriously suggest that the contamination by the F1NPP is a global issue and has spread widely throughout the water, soil, air and biota.

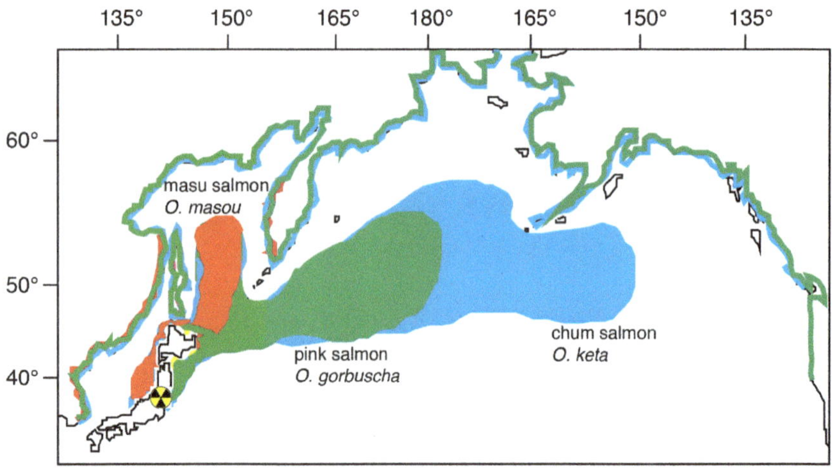

Figure 1. Migratory route and scale from Japanese habitats and the distribution of anadromous salmonids in the Pacific region used in the present study. Chum salmon (blue) have the largest natural range of any Pacific salmon, and undergo the longest migrations within the genus *Oncorhynchus*. Chum are found around the north Pacific, in the waters of Korea, Japan, and the Okhotsk and Bering seas, British Columbia in Canada, and from Alaska to California in the United States. The native range of pink salmon (green) is from the Arctic and Pacific drainages from Mackenzie River delta, the Northwest Territories, to the Sacramento River drainage, California (occasionally as far south as La Jolla, southern California) and in the west from the Lena River in Siberia to Korea. Populations in Asia occur as far south as Honshu in Japan. Masu (cherry) salmon (red) found in the Western Pacific Ocean along East Asia, range from the Kamchatka, Kuril Islands, Sakhalin and Primorsky Krai south through Korea, Taiwan and Japan.

Several agencies have conducted investigations to monitor the radioactive quantities of water, soil, air and biota after the F1NPP accident, and considerable data have been reported to understand the present status of radionuclide pollution levels [1–4]. Although it has been three years since the disaster by the F1NPP accident, the radionuclide pollution issues have not converged yet, and there is little information about radionuclide behaviour in the natural environment and ecosystem globally. A recent study found that the Pacific blue tuna transported Fukushima-derived radionuclides from Japan to California, and such highly migratory animals transport radionuclides globally as a vector [13], leading to the cause of the distribution of the radionuclides globally. In such a situation, systematic and comprehensive investigations for radionuclides in natural environments are required. However, a number of animals are living in a certain restricted area, e.g., aquatic animals generally can survive in either freshwater or seawater throughout their lives, and thus, it was quite difficult to understand the radionuclide behaviour systematically and comprehensively in various environments from inland water to open ocean environments.

Salmonids are widely distributed in cool and cold waters of all northern continents occurring from the Arctic drainages of Europe, Asia and North America, south as far as the Mediterranean and northern Africa [14]. All of salmonids spawn in

Table 1. Information of life history in the salmonids examined between May 2011 and February 2014 in the present study.

Species	Migration scale (migration type)	Freshwater life period before descending to sea	Marine life period before spawning
S. leucomaenis	freshwater	whole life	n/a
(whitespotted charr)	(fluvial)		
O. masou	freshwater	whole life	n/a
(masu salmon)	(fluvial)		
O. nerka	freshwater	whole life	n/a
(sockeye salmon)	(fluvial)		
O. masou	freshwater-coastal	1–2 year(s)	1–2 year(s)
(masu salmon)	(anadromous)		
O. gorbuscha	freshwater-coastal-open ocean	immediately after hatching	1–2 year(s)
(pink salmon)	(anadromous)		
O. keta	freshwater-coastal-open ocean	immediately after hatching	2–8 years
(chum salmon)	(anadromous)		

n/a: not applicable.

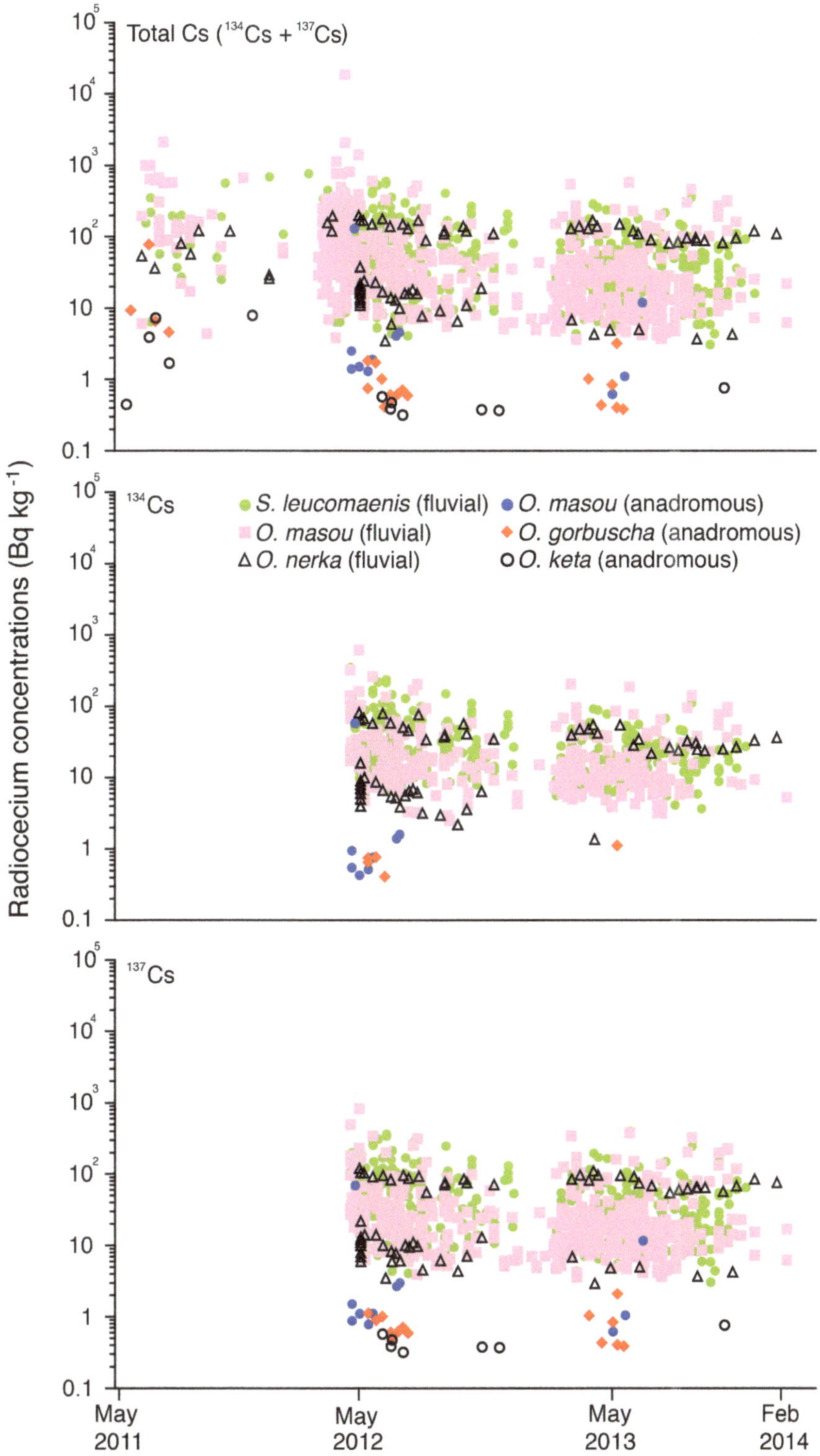

Figure 2. Fluctuations of the radiocesium levels in the total Cs (top) between May 2011 and February 2014, ^{134}Cs (middle) and ^{137}Cs (bottom) between April 2012 and February 2014 after the F1NPP accident. No significant temporal trends were found for all types of salmon suggesting the continued widespread distribution of these radionuclides with less convergence and indicating that these radionuclides remain in the natural environment for both terrestrial and marine ecosystems. All radiocesium data in the salmonids was from the information published by the Fisheries Agency of the Japanese Government [16] between May 2011 and February 2014. nd: not detectable due to levels below the detection limits.

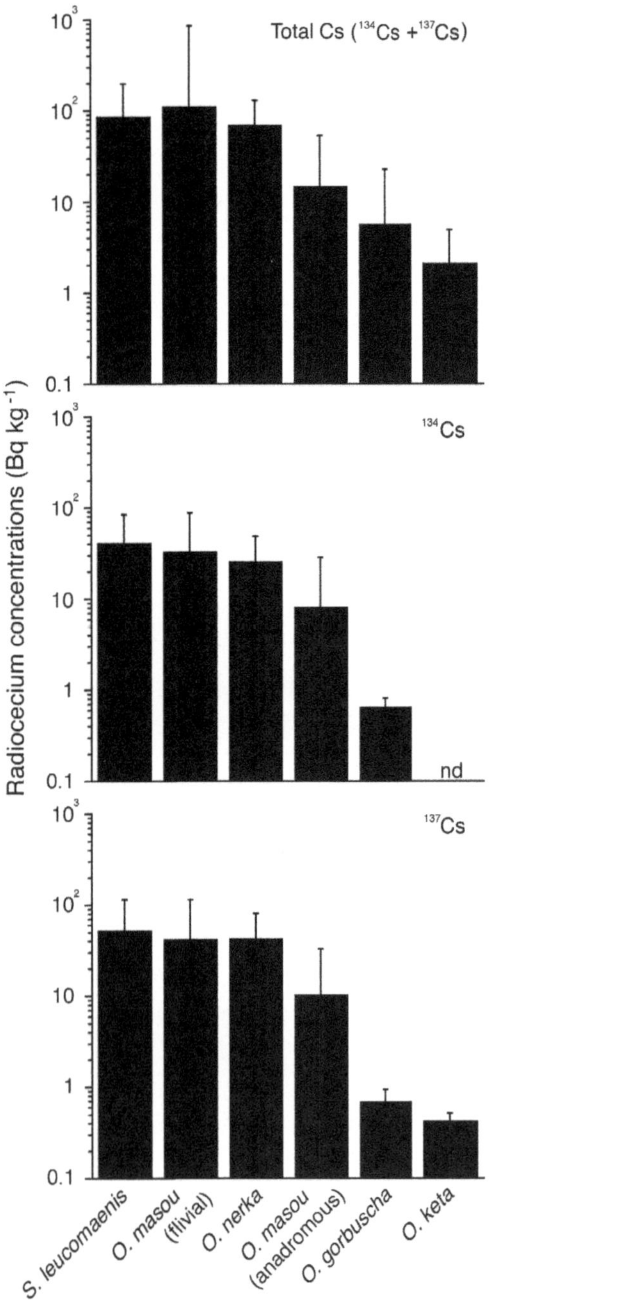

Figure 3. Differences in the radiocesium accumulations among the salmon species between May 2011 and February 2014 after the F1NPP accident. Fluvial salmons exhibited significantly higher concentrations in each radiocesium than those of the anadromous salmon. The detectable rates were also higher in fluvial salmon than in anadromous salmon as presented in Table 2.

freshwater [15] but anadromy (sea-run) is strongly and widely represented. The animal is one of the most studied due to their great importance for food and angling. Thus, the biology and ecology are well understood. Some salmonids highly migrate to far offshore locations, such as the Pacific and Atlantic oceans (Fig. 1); however, all salmons come back to freshwater environments for spawning. Thus, this animal can be useful indicator to monitor for the spread of radionuclides from inland water to open ocean environments without allocation of a significant catching effort and time because a large number of salmons return to their natal rivers.

A number of salmonid species exist in Japan, and the Fisheries Agency of the Japanese Government has examined radionuclides in the animals (Table 1) [16]. In this study, I compared radiocesium (^{134}Cs, ^{137}Cs and these total) levels among salmonid species using a total of 3,024 samples between May 2011 and February 2014. The levels of accumulation were significantly different in accordance with their life patterns; fluvial (freshwater residence) salmons exhibited the highest level, influenced by the higher radionuclide levels in terrestrial environments, a coastal migrant exhibited a moderate level because the salmon migrate between freshwater and coastal water, and the long-distance migrants to the open ocean exhibited the lowest level, being less impacted by the radiocesium because they live in the open ocean for almost all of their lives after hatching. The present study indicates that migration patterns of the different types of salmon can reveal temporal and spatial fluctuations of radiocesium globally, ranging from terrestrial to open ocean environments.

Materials and Methods

All radiocesium data in the salmonids was from the information published by the Fisheries Agency of the Japanese Government [16] between May 2011 and February 2014 in the eastern and northern parts of Japan from 3 to 15 prefectures. The radiocesium concentrations were determined as a total between May 2011 and March 2012, while thereafter each ^{134}Cs and ^{137}Cs concentration was determined in each sample (Fig. 2). All data were shown in the wet weight basis. Each radiocesium concentration among the salmon species was compared using the Kruskal-Wallis test. Significance of the correlation coefficient and regression slope were tested by Fisher's Z-transformation.

Results

The radiocesium concentrations in six salmonids (five different species and a different migratory type in a same species) (Table 1) exhibited less temporal variations for three (total Cs) or 2 (^{134}Cs and ^{137}Cs) years after the F1NPP accident (Fig. 2). No significant temporal trends were found for all of the types of salmon (Fisher's Z-transformation, $p>0.05$), suggesting a continued widespread distribution of these radionuclides with less convergence and that these radionuclides remain in natural environment for both terrestrial and marine ecosystems. Fluvial salmons, such as *Salvelinus leucomaenis*, *Oncorhynchus masou* and *O. nerka*, only reside in a freshwater environment, and thus they accumulate these radionuclides from the terrestrial environment. In contrast, anadromous (sea-run) salmons *O. masou*, *O. gorbuscha* and *O. keta* accumulate radionuclides from terrestrial, coastal and open ocean environments depending on their migration scale (Table 1). The highest concentration of 18,700 Bq/kg in total cesium was found in fluvial *O. masou* in March 2012, one year after the F1NPP accident, and there were still detectable amounts in salmon of over 100 Bq kg^{-1} (Japanese safety limit) of ^{134}Cs and ^{137}Cs, even three years after the F1NPP incident (Fig. 2).

However, the accumulation patterns were significantly different among the salmon species. In general, fluvial-type salmon, such as *S. leucomaenis*, *O. masou* and *O. nerka*, exhibited significantly higher accumulation of ^{134}Cs (25.3–40.2 Bq kg^{-1} in mean) and ^{137}Cs (41.4–51.7 in mean) than the anadromous-type salmon of *O. masou*, *O. gorbuscha* and *O. nerka* (0.64–8.03 Bq kg^{-1} in mean ^{134}Cs and 0.42–10.2 Bq kg^{-1} in mean ^{137}Cs) (Table 2, Fig. 3) ($p<0.005$–0.0001). Among the anadromous salmon, the highest concentration was found in *O. masou*, the second highest was in *O. gorbuscha*

Table 2. Radiocesium data information and radiocesium concentrations (above the detection limit) in the salmonids examined between May 2011 and February 2014 in the present study.

Species	Number of the data above detection limit (Ratio of detectable (%))			Mean ± SD (Bq/kg) (minimum-maximum (Bq/kg))		
	Total Cs*	^{134}Cs	^{137}Cs	Total Cs*	^{134}Cs	^{137}Cs
S. leucomaenis	524	317	429	84.7±112	40.2±43.5	51.7±62.8
(whitespotted charr)	(70.6)	(45.6)	(61.3)	(3.10–840)	(5.63–350)	(3.07–490)
O. masou	670	327	493	109±740	32.4±54.3	41.4±72.2
(masu salmon)	(58.8)	(31.6)	(48.2)	(3.60–18700)	(2.50–610)	(3.57–820)
O. nerka	85	68	74	68.4±60.1	25.3±22.7	42.2±38.4
(sockeye salmon)	(88.0)	(73.0)	(82.0)	(3.50–200)	(1.37–82)	(2.94–120)
O. masou	11	8	11	14.6±38.4	8.03±20.2	10.2±22.3
(masu salmon)	(11.7)	(10.5)	(18.0)	(0.62–130)	(0.43–58)	(0.62–69)
O. gorbuscha	20	5	14	5.60±16.9	0.64±0.16	0.69±0.25
(pink salmon)	(71.6)	(20.2)	(48.4)	(0.38–76.7)	(0.41–1.11)	(0.38–2.07)
O. keta	12	0	7	2.06±2.82	nd	0.42±0.09
(chum salmon)	(1.6)	(0)	(1.3)	(0.32–8.00)		(0.32–0.77)

*Total Cs = ^{134}Cs+^{137}Cs.
Total Cs was available between May 2011 and February 2014.
^{134}Cs+^{137}Cs were available between April 2012 and February 2014.
nd: all samples were below the detection limit.

and the lowest was in *O. keta* (Table 2, Fig. 3). Furthermore, the detectable numbers were also different among either fluvial- or anadromous- type salmon. There were 529–772 samples of *O. keta* investigated for three years; however, all of the samples were below the detection limit in ^{134}Cs, and only seven samples were found to be above the limit. In contrast, radionuclide levels in fluvial salmon could be determined, being generally over 50% (Table 2).

Discussion

The present study clearly demonstrated that accumulations of radiocesium were different depending on the migratory patterns and migration scale of the different types of salmon. The highest accumulation was found in the fluvial species (type); their concentrations were similar to those of the other freshwater fish, such as Ayu (*Plecoglossus altivelis*) [17], used to investigate the radiocesium contamination levels in the freshwater ecosystem after the F1NPP accident. Thus, these salmons can also be used to document the ^{134}Cs and ^{137}Cs contamination levels in the terrestrial environments in East Japan. In contrast to the fluvial salmon, ^{134}Cs and ^{137}Cs concentrations were quite low in the anadromous salmon, such as *O. masou*, *O. gorbuscha* and *O. nerka*. The detectable ratios of the cesium isotopes were the lowest in *O. keta*, which is the species that has the longest distance of migration, spending almost all of their lives in the North Pacific Ocean (Fig. 1). Among the anadromous salmon, *O. masou* accumulated higher ^{134}Cs and ^{137}Cs concentrations than those of the other types of salmon. The species remain in freshwater for one to two year(s) after hatching and thereafter go downstream to the sea to growth. In addition, the salmon tends to migrate to a coastal area along the northern Japan for growth (Fig. 1). Thus, anadromous *O. masou* probably have a higher risk for accumulation of radionuclides compared to *O. gorbuscha* and *O. keta*. *O. gorbuscha* and *O. keta* migrate to sea immediately after hatching, and further migrate along the coastal to the open ocean for growth (Fig. 1); as a result, these salmon probably exhibit a lower risk to accumulate these radionuclides in their lives.

Although the higher levels of radiocesium in fluvial salmons are caused by the serious contamination in the terrestrial environments of East Japan after the F1NPP accident, marine fish are usually approximately 100 times lower in ^{137}Cs than are freshwater fish because potassium, which is more abundant in seawater, blocks the uptake of cesium by marine organisms [18]. Thus, the differences in physiological mechanism, especially ionic uptake and elimination of the related osmoregulation, might also accelerate higher ^{137}Cs accumulation in the fluvial salmons. The F1NPP accident released large amounts of radioactive substances into the environment and contaminated the soil of Tohoku and the Kanto (East) districts in Japan, and the radiocesium levels in water, soil and biota in terrestrial environment are much more higher than those of the coastal and marine environments [4,19]. The natural habitats of fluvial salmon, such as *Salvelinus leucomaenis* and *O. masou* are widespread in East Asian countries, as they reside streams, ponds and lakes; as a result, these salmons can be useful indicators for radiocesium pollutions in terrestrial environments.

After the F1NPP accident, the ^{137}Cs concentrations in a 30-km perimeter around the plant exceeded 10 Bq L^{-1} [12]. Chum salmon *O. keta* used in the present study might have migrated in the North Pacific at the time of the F1NPP accident because their oceanic migration in North Pacific takes two to eight years. In fact, the *O. keta* ^{134}Cs and ^{137}Cs concentrations were below the detection limit and 0.42±0.09 Bq kg^{-1} (mean ± SD), respectively, for the 529 samples. These levels were much lower than the highly migratory marine animals of the Pacific blue fin tuna in the North Pacific Ocean (4.0±1.4 Bq kg^{-1} in ^{134}Cs and 6.3±1.5 Bq kg^{-1} in ^{137}Cs) [13]. The tissue concentrations of radiocesium in the tuna might vary depending on the time spent near the Japanese coast, foraging strategies and the timing of the migration [13]. During open-ocean migration by chum salmon for growth, the salmon has never returned to the Japanese coast and are migrating in the

North Pacific Ocean (Fig. 1); and thus the salmon might not accumulate radiocesium of F1NPP accident origin. Nevertheless, trace amounts of radiocesium might have a potential to accumulate during their open ocean migration in the North Pacific Ocean, it could metabolite and eliminate during spawning migration to Japanese coasts. These suggest that the present data of radiocesium levels in chum salmon might be used as a back ground level, and continuous investigations might reveal the fluctuation of radionuclide pollution levels in North Pacific Ocean. The current distribution of chum salmon extends from approximately 45–70°N to 140°E–125°W longitude (Fig. 1). The species has been recorded globally from Russia, United States, Canada, Japan and Korea. Chum salmon has higher homing rate to the natal river, and thus each country could use the salmon to monitor radionuclide levels during their open ocean migration.

The life history, ecology and evolution of salmonid fishes (salmon, trout and char) are dominated by their strong tendency to return home to their natal site for reproduction. Homing characterises the family but is best studied in the anadromous (i.e., 'sea run') forms of salmon. Typically, salmon spawn in streams or lakes, and after a variable period of freshwater residence (zero to three years), depending on the species and population, the offspring migrate to the ocean. Salmon remain in the ocean until they begin to mature, and then they return to their natal site to spawn. Thus, salmon can have a great potential to monitor the fluctuation of radionuclides in inland, coastal and open ocean environments as we do not need to catch salmons in coastal and off shore regions because anadromous salmons return to their natal river mouth in spite of their migration scale. Anadromous salmon as well as fluvial salmon are also useful to monitor the radionuclide levels considering the migration scale. In contrast to anadromous salmons, fluvial salmons are suitable indicators for investigating terrestrial environments to reveal the fluctuations of local pollution patterns from the F1NPP event. The present study indicates that the levels of ^{134}Cs and ^{137}Cs pollution after the F1NPP accident did not converge during the three subsequent years (Fig. 2). Furthermore, Pacific Bluefin tuna, which are highly migratory marine animals, transported Fukushima-derived radionuclides from Japan to California [13], suggesting the widespread distribution of radionuclide pollution globally. Thus, continuous systematic and comprehensive monitoring, taking advantage of the characteristics of salmon migration patterns, might be needed to reveal the temporal and spatial fluctuations of the radionuclides in various environments in the future.

Author Contributions

Conceived and designed the experiments: TA. Performed the experiments: TA. Analyzed the data: TA. Contributed reagents/materials/analysis tools: TA. Contributed to the writing of the manuscript: TA.

References

1. Buesseler K, Aoyama M, Fukasawa M (2011) Impacts of the Fukushima nuclear power plants on marine radioactivity. Environ Sci Technol 45: 9931–9935.
2. Chino M, Nakayama H, Nagai H, Terada H, Katata G (2011) Preliminary estimation of release amount of ^{131}I and ^{137}Cs accidentally discharged from the Fukushima Daiichi Nuclear Power Plant into the atmosphere. J Nucl Sci Technol. 48: 1129–1134.
3. Masson O, Baeza A, Bieringer J, Brudecki K, Bucci S, et al. (2011) Tracking of airborne radionuclides from the damaged Fukushima Dai-Ichi nuclear reactors by European networks. Environ Sci Technol 45: 7670–7677.
4. Buesseler KO, Jayne SR, Fisher NS, Rypina II, Baumann H, et al. (2012) Fukushima-derived radionuclides in the ocean and biota off Japan. Proc Natl Acad Sci USA 109: 5984–5988.
5. Estournel C, Bosc E, Bocquet M, Ulses C, Marsaleix P, et al. (2012) Assessment of the amount of Cesium-137 released into the Pacific Ocean after the Fukushima accident and analysis of its dispersion in Japanese coastal waters. J Geophys Res 117: C11. doi:10.1029/2012JC007933.
6. Kawamura H, Kobayashi T, Furuno A, In T, Ishikawa Y, et al. (2011) Preliminary numerical experiments on oceanic dispersion of ^{131}I and ^{137} Cs discharged into the ocean because of the Fukushima Daiichi nuclear power plant disaster. J Nucl Sci Technol 48: 1349–1356.
7. Miyazawa Y, Masumoto Y, Varlamov SM, Miyama T, Takigawa M, et al. (2012) Inverse estimation of source parameters of oceanic radioactivity dispersion models associated with the Fukushima accident. Biogeosci Discuss 9: 13783–13816.
8. Tsumune D, Tsubono T, Aoyama M, Hirose K (2012) Distribution of oceanic ^{137}Cs from the Fukushima Dai-ichi nuclear power plant simulated numerically by a regional ocean model. J Environ Radioact 111: 100–108.
9. Ikeuchi Y (2003) Temporal variations of ^{90}Sr and ^{137}Cs concentrations in Japanese coastal surface seawater and sediments from 1974 to 1998. Deep Sea Res 50: 2713–2726.
10. Nakanishi T, Aono T, Yamada M, Kusakabe M (2010) Temporal and spatial variations of ^{137}Cs in the waters off a nuclear fuel reprocessing facility in Rokkasho, Aomori, Japan. J Radioanal Nucl Chem 283: 831–838.
11. Povinec, Breier R, Coppola L, Groening M, Jeandel C, et al. (2004) Spatial distribution of ^{3}H, ^{90}Sr, ^{137}Cs and $^{239,\ 240}$Pu in surface waters of the Pacific and Indian Oceans–GLOMARD database. J Environ Radioact 76: 113–137.
12. Bailly du Bois P, Laguionie P, Boust D, Korsakissok I, Didier D, et al. (2012) Estimation of marine source-term following Fukushima Dai-ichi accident. J Environ Radioact 114: 2–9.
13. Madigan DJ, Baumann Z, Fisher NS (2012) Pacific bluefin tuna transport Fukushima-derived radionuclides from Japan to California. Proc Natl Acad Sci USA 109: 9483–9486.
14. McDowall RM (1988) Diadromy in fishes. Croom Helm. 308 p.
15. Nelson JS (1976) Fishes of the world. Wiley-Interscience. 416 p.
16. Japanese Ministry of Agriculture, Forestry, and Fisheries (2014) Results of the inspection on radioactivity materials in fisheries products. Available: http://www.jfa.maff.go.jp/e/inspection/index.html. Accessed 2014 Feb 25.
17. Mizuno T, Kubo H (2013) Overview of active cesium contamination of freshwater fish in Fukushima and Eastern Japan. Sci Rep 3: 1742 doi:10.1038/srep01742.
18. Harte J, Holdren C, Schneider R, Shirley C (1991) Toxics A to Z: A Guide to. Everyday Pollution Hazards. University of California Press. 576 p.
19. Yasunari TJ, Stohl A, Hayano RS, Burkhart JF, Eckhardt S (2011) Cesium-137 deposition and contamination of Japanese soils due to the Fukushima nuclear accident. Proc Natl Acad Sci USA 108: 19447–19448.

6

Extensive Behavioural Divergence following Colonisation of the Freshwater Environment in Threespine Sticklebacks

Carole Di-Poi[1], Jennyfer Lacasse[1], Sean M. Rogers[2], Nadia Aubin-Horth[1]*

1 Département de Biologie & Institut de Biologie Intégrative et des Systèmes (IBIS), Université Laval, Québec, Québec, Canada, 2 Department of Biological Sciences, University of Calgary, Calgary, Alberta, Canada

Abstract

Colonisation of novel environments means facing new ecological challenges often resulting in the evolution of striking divergence in phenotypes. However, little is known about behavioural divergence following colonisation, despite the predicted importance of the role of behavioural phenotype-environment associations in adaptive divergence. We studied the threespine stickleback (*Gasterosteus aculeatus*), a model system for postglacial colonisation of freshwater habitats largely differing in ecological conditions from the ones faced by the descendants of the marine ancestor. We found that common-environment reared freshwater juveniles were less social, more active and more aggressive than their marine counterparts. This behavioural divergence could represent the result of natural selection that acted on individuals following freshwater colonisation, with predation as a key selection agent. Alternatively, the behavioural profile of freshwater juveniles could represent the characteristics of individuals that preferentially invaded freshwater after the glacial retreat, drawn from the standing variation present in the marine population.

Editor: Walter Salzburger, University of Basel, Switzerland

Funding: Funding was provided to C. Di-Poi by the Government of Canada Post-Doctoral Research Fellowship, to N. Aubin-Horth and S. Rogers through Natural Sciences and Engineering Research Council of Canada Discovery grants and an Alberta Innovates Technology Futures New Faculty Award to S. Rogers. The funders had no role in study design, data collection and analysis, decision to publish, or preparation of the manuscript.

Competing Interests: The authors confirm that co-author Sean Rogers is a PLOS ONE Editorial Board member. This does not alter the authors' adherence to PLOS ONE Editorial policies and criteria.

* E-mail: Nadia.Aubin-Horth@bio.ulaval.ca

Introduction

The novel ecological challenges that individuals face when colonising new environments have been shown in many instances to result in adaptive trait divergence. Morphology has often been shown to diverge following colonisation as the result of natural selection [1]. Remarkable examples include the diversification of leg and tail lengths in anole lizards that use different microhabitats in trees and on land [2] and trophic morphology associated with different feeding ecology in African cichlid fishes [3]. Much less information is available on whether behaviours that could influence fitness in the new ecological context also evolve. This is true even for very well-known model systems of phenotypic divergence following colonisation of a novel environment. Yet, evolutionary changes in behaviour may be integral in initiating adaptive shifts [4]. It has been shown that behavioural divergence can be associated with morphological divergence in a new ecological context, whereby populations differ in the behavioural traits used to exploit different resources, such as in the association between beak shape and foraging behaviour in Darwin's finches [5]. The specific ecological constraints acting in the novel environment could directly affect the benefits associated with expressing a given behaviour. For example, when there are no substantial benefits associated with group life, such as in low-productivity freshwater habitats where predation pressure is weaker and intraspecific competition is the most relevant process affecting fitness, the advantageous strategy might be to adopt the solitary lifestyle and to increase foraging activity, boldness and aggressiveness towards conspecifics [6–8]. Boldness, exploratory behaviours and sociability (schooling) have been shown to diverge when individuals face new ecological conditions (poeciliid *Brachyraphis episcopi* [9]; *Poecilia reticulata* [10,11]). Such behavioural divergence has been shown to enhance the probability of ecological speciation [12]. However, quantitative measurements of behavioural divergence between the ancestral population and the population in a new environment, potentially as the result of selection, remain scarce compared to other traits. Understanding the ecological context of behavioural divergence will help draw a complete picture of integrated phenotypes evolving in response to adaptive peak shifts [13].

To address this problem, we used a well-studied model system of colonisation of new environments, the threespine stickleback (*Gasterosteus aculeatus*). In the last 20,000 years, marine threespine sticklebacks have successfully colonised many freshwater systems with widely divergent ecological conditions in the Northern hemisphere [14]. A distinctive feature of this system is that the ancestral marine population is still extent, although it has also potentially diverged from the marine ancestor during this time period, allowing comparison with populations that have colonised new freshwater environments. Comparative studies between marine and freshwater populations have shown that colonisation of freshwater has resulted in rapid evolutionary changes in numerous traits in morphology (diminution of lateral armour and spine development, lower number of iridophores and presence of barred flanks [15,16]), life history (younger and smaller at reproduction [17]), and physiology (lower levels of thyroid hormones, lower critical swimming speed [18,19]). Furthermore, freshwater juveniles show higher territoriality [20] and a lower schooling tendency [21] than juvenile marine fish (representing the ancestral population). However, whether other ecologically relevant behaviours diverged following colonisation of the freshwater habitat by threespine sticklebacks has not been thoroughly studied.

We hypothesised that behaviour would have diverged between populations of sticklebacks that inhabit contrasting marine and freshwater environments [13,22] since the costs and benefits of expressing a particular behaviour may vary according to ecological demands [21,23]. We focused on genetically based differences in behaviour, rather than on phenotypic plasticity by studying common-environment reared juveniles originating from two populations of threespine sticklebacks representing the marine ancestor and the result of freshwater colonisation. We quantified boldness, aggressiveness, sociability, activity and exploratory behaviours, as they are relevant to fitness and ultimately to the ecology and evolution of populations [24]. Previous studies comparing two freshwater populations of sticklebacks differing in predation pressure have shown divergence in behaviour in sociability [25], activity [26,27], aggressiveness [27,28], and boldness [26,28,29]. Therefore, the potential behavioural divergence observed between marine and freshwater sticklebacks may be the result of natural selection, with predation as a key selection agent. Thus we predicted that sticklebacks from a freshwater lake population that experienced a presumed lower predation pressure based on their lower number of lateral armour plates would exhibit a lower sociability tendency, higher activity and exploratory behaviours, as well as a higher propensity to be aggressive towards a conspecific and to be bolder when foraging in a risky situation, compared to the individuals originating from the marine ancestor population that face a high-predation level environment.

Methods

Ethics statement

Wild animals were sampled under the Department of Fisheries and Ocean Permit #XR 315 2011 and the BC Ministry of Environment NA-SU-PE-10-63485 permit. The research adheres to the ASAB/ABS Guidelines for the Use of Animals in Research and was approved by the Comité de Protection des Animaux de l'Université Laval (permit #2010012-1).

Sampling and rearing

Adult threespine sticklebacks from marine and freshwater populations were collected using minnow traps in British Columbia (Canada) in two breeding sites, which differ in fundamental ecological characteristics [13,22]. Marine adult fish were collected from a small and shallow inlet off Malaspina Strait (Oyster Lagoon, 49°36′48″N 124°01′47″W) where salinity ranges from 20 ppt in winter to 32 ppt in summer. The surface area of the lagoon is 1.8 ha and the maximum depth is 3 m. Marine sticklebacks in Oyster Lagoon face a wide variety of predators [22]. Birds, such as great blue herons, mergansers and kingfishers, are the primary piscivores in the lagoon. Mammals (river otters, racoons, mink), and reptiles (e.g. Pacific garter snakes) are also potential predators of sticklebacks. Sculpins also feed on adult and juvenile sticklebacks in the lagoon [22]. The marine fish from Oyster Lagoon have a high-plated phenotype [30], which has been associated with high predation by vertebrate predators [31]. Freshwater adult fish were collected from Hoggan Lake on Gabriola Island (49°09′08″N 123°49″W). The lake is not connected to the ocean, and this population has been isolated from marine ancestors for approximately 10,000 years. It covers a surface area of 19.7 ha and has a perimeter of 2.2 km. The lake does not contain sculpin, a known predator of sticklebacks, but fish stocking records indicate that it was stocked with rainbow trout (*Oncorhynchus mykiss*) and coastal cutthroat trout (*Oncorhynchus clarkii clarkii*) in 1927 [32] with both of these species being intermittently observed in this environment, indicating that either stocked fish established a self-sustaining population, or that the lake has always contained a natural population. There is no known benthic-limnetic phenotypic divergence in this population. The fish from this population have a low number of lateral armour plates [13], which has been associated with low predation regimes [31].

Adult threespine sticklebacks were crossed to produce pure lines of F1 individuals, resulting in 14 full-sub F1 families, seven from each population. Juveniles were kept in 110 L-tanks under a 12L:12D cycle, with temperature at 18°C in the Life and Science Animal Research Centre at the University of Calgary and fed frozen bloodworms to satiation twice daily. Salinity was maintained at 5 ppt to promote health and maintain proper development in both marine and freshwater stickleback (see [33] for a similar rearing protocol of marine and freshwater populations). Juveniles (8-month old) were air-shipped to the "Laboratoire de Recherche en Sciences Aquatiques" at Université Laval. Juveniles were held in separate 80 L-tanks (stocking density: 0.11 g/L) under similar rearing conditions with the exception that all fish were acclimated to 15°C with a 8L:16D photoperiod (winter conditions). Juveniles from all families originating from a given environment (marine or freshwater) were kept together during air transport and later rearing, such that the information about the family of origin of individuals is not available. All fish were in a non-reproductive state (body length: average of 46.4±0.6 mm for fish from Hoggan Lake; 42.5±0.5 mm for fish from Oyster Lagoon) and sex was determined using a genetic sex marker [34].

Behavioural series

Behavioural experiments began after at least two weeks of acclimatisation. Each individual fish ($n = 36$ fish/population) was exposed to five behavioural tests over a 5-day period (sociability, exploration, aggressiveness, activity and boldness, see Fig. 1A for a timeline). After the sociability assay, the focal individual was transferred to an individual 45 L-experimental tank with plants and shelter as enrichment. The focal individual stayed in this tank for the rest of the experiment. In total, there were 6 series of 5 days (6 fish per population tested per series; 12 fish per series). All tests were done at the same time of day for all fish. Repeatability of these behaviours has already been demonstrated in the threespine stickleback [35,36]. Each behavioural assay was filmed with a digital camera (JVC model GZ-MS120) mounted in front of the tank. The tanks were drained, cleaned and filled with water again between each series.

Sociability. We used an assay previously described in [21] that ensures repeatable stimulus conditions and the absence of confounding effects of the behaviour of the conspecifics in the school. A motorised mobile reproducing a school of five fish with the same colour as juvenile sticklebacks and of a smaller body length than the average size of the populations (36 mm) was constructed. The artificial school was immersed 10 cm from the water surface and from the tank bottom in an 80 L-tank. The focal fish was added in the tank in the presence of the motionless school for a 15-minute acclimatisation period. The rotating motor

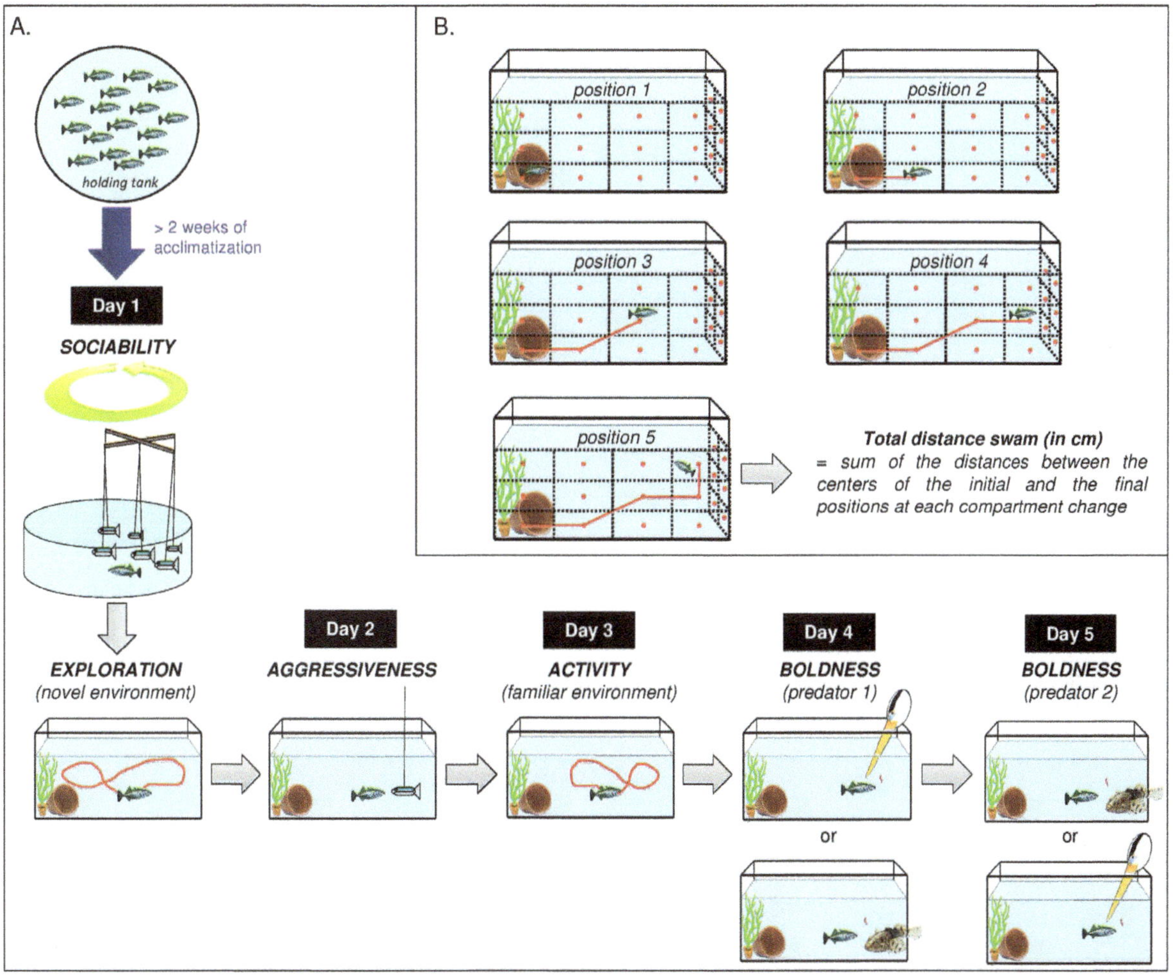

Figure 1. Illustrations of (A) the set-ups and the timeline of the experiment, and (B) the method to calculate the total distance swam of each fish in the exploration and the activity trials.

connected to the mobile then moved the fish school around the tank for 15 minutes. The focal fish's schooling behaviour was filmed from above. The time spent swimming within one body length of the group in the same direction and at the same speed (in seconds) was quantified as sociability.

Exploration. Immediately after the sociability test, the focal fish was moved into its novel individual experimental tank, where it remained throughout the rest of the behavioural experiment. The fish was placed in a 7 cm-diameter opaque cylinder with an open end, which was then introduced into the tank. The exploratory behaviour in this novel environment was quantified using three measures: (1) latency to exit the cylinder (in seconds) was measured (if the fish was still inside the cylinder after ten minutes following introduction, the cylinder was removed and it was given the maximum possible latency score, i.e. 600 sec), (2) after cylinder removal, time spent swimming (in seconds) in the new environment, and (3) total length swam (in cm) were measured for 5 minutes. The swimming activity of each fish was observed on the video recording and the total distance swam quantified manually using a 3-D grid divided into 4×2×3 compartments. Each time the fish swam from a compartment to another, the distance between the centers of the initial and the final compartments was measured (Fig. 1B). All distances were summed-up over the 5-min period.

Aggressiveness. On the second day, a model juvenile stickleback fish (body length: 36 mm, colour: olive green and grey) was placed as an intruder in the observation tank and the resulting interactions were quantified for a period of 15 minutes. The number of bites made by the focal fish against the intruder was used as a measure of aggressiveness [37]. The percentage of individuals exhibiting at least one biting event was also quantified.

Activity. Activity is a measure of the general level of swimming activity in its familiar environment, in contrast to exploration, which is measured in a novel environment. On the third day when the fish was familiarised with its experimental tank, the time spent moving (in seconds), as well as the total length swam (in cm) were measured, as previously explained for the exploration assay, for each subject during a 5-minute period, at the same time of day as in the exploration test.

Boldness. The response of an individual in a risky situation was used as a measure of boldness. On the fourth and fifth day, during the feeding session, a simulated predator attack was manually performed by the experimenter on the focal fish using a replica of the head of a great blue heron or a flexible lure in the shape of a sculpin fish, two natural predators of sticklebacks. The feeding session started with provision of a single bloodworm, followed by a 30-second wait and a second bloodworm. The predator attack was simulated at the third bloodworm, when the

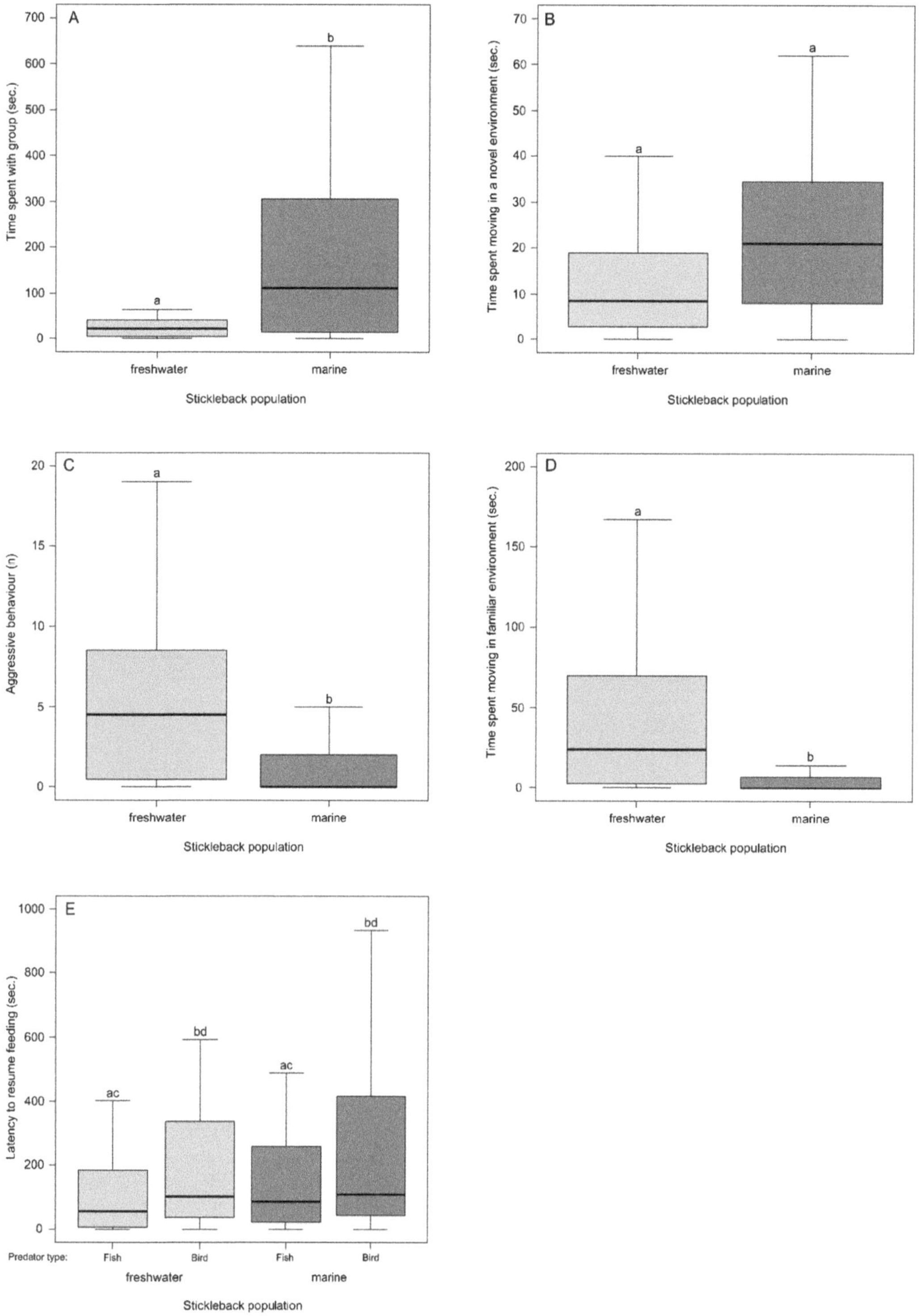

Figure 2. Behaviour (median ± quartiles) of juvenile threespine sticklebacks from a marine and a freshwater population. (A) Sociability, (B) exploratory behaviour in a novel environment, (C) aggressiveness, (D) activity in a familiar environment, and (E) boldness. Different letters indicate a significant difference between populations.

focal fish approached within at least one body length of the food. Striking the water with the beak directly over the area where food was distributed simulated the sudden overhead attack of a heron searching for prey. The beak was plunged about 5 cm below the water surface and moved in the water for 10 seconds. To simulate a large sculpin attack, the sculpin replica was suddenly placed outside the front of the aquarium and moved along the length of the tank for 10 seconds. The experimenter was hidden behind a black curtain, such that no human movement was perceptible to the fish. Latency to resume feeding after the attack (in seconds) was used as a measure of boldness. Each focal fish was tested for each predator on two days (one predator per day). The order of appearance of the two predators was randomised for each fish to minimise carry over effects.

Data availability

The entire dataset used is available as Table S1.

Data analysis

All statistical analyses were performed using R software 3.0.2 (http://cran.r-project.org/bin/windows/base/). Normal distribution of data and homogeneity of variances were assessed using Shapiro-Wilk and Bartlett tests respectively; however, the data did not meet the assumption of normality. Because the data was not normally distributed, we used a linear model for each behaviour, with p-values obtained by permutation tests (lmPerm package, Permutation tests for linear models in R). These models were used to test for behavioural differences between the two populations including population, sex and length of the fish as fixed effects. For boldness, the significance of interactions between the effects of population, predator type, day of tests, as well as fish identification number (ID) were tested in order to correct for pseudo-replication (as individuals were tested twice on the fourth and fifth days). Only population effects are reported in the text as no effect of sex, fish length, day and fish ID was found on any behaviour tested. Median ± quartiles are presented and $p<0.05$ was used as the statistical significance threshold (see Table S1 for the dataset).

For each population separately, Spearman's rank correlations were calculated between body length of fish, and each of the eight behavioural parameters measured. No correlation was found with body length and these terms were subsequently removed from the analysis. Correlations between behaviours were also quantified using Spearman's rank correlations. As multiple correlations were performed, sequential Benjamini and Hochberg corrections [38] were applied to reduce type I errors by controlling for the false discovery rate. Correlations were found between different behaviour components within a population. However, after correcting for multiple testing, the only correlations that were significant were found between different measures of the same behaviour (see Table S2).

Results

Average behavioural divergence between populations

Sociability. The two populations exhibited different schooling tendencies under identical social circumstances (Table 1 & Fig. 2A). The model school assay elicited schooling behaviour for 85% of individuals in the freshwater population and for 90% of the individuals in the marine fish. During the entire 15 min-session, freshwater fish spent an average of 4.4% of their time schooling (mean = 39.4 sec; SEM: ±9.9) while marine fish spent 21% of their time grouping with the school (mean = 188.9 sec; SEM: ±35.5; Fig. 2A). Accordingly, freshwater sticklebacks showed a significantly lower schooling behaviour than the marine fish (Linear model: population term, $p = 0.002$, $n = 35$ marine fish, $n = 35$ freshwater fish).

Exploration. Exploration in a novel environment was not significantly different between populations for the three measures observed (Linear model: population term, $p>0.05$ for latency to exit a refuge, $n = 32$ marine fish, $n = 34$ freshwater fish; total distance swam, $n = 33$ marine fish, $n = 31$ freshwater fish; and time spent moving, $n = 35$ marine fish, $n = 32$ freshwater fish; Table 1 & Fig. 2B).

Aggressiveness. About 75% and 39% of the fish exhibited aggressiveness in the freshwater and marine population respectively, whereas the rest did not show a reaction to the intruder (i.e. no approach within one body length). When comparing all individuals tested, the number of bites was four times higher in the freshwater than in the marine population (Linear model: population term, $p = 0.002$, $n = 26$ marine fish, $n = 32$ freshwater fish; Table 1 & Fig. 2C).

Activity. Activity in a familiar environment measured as time spent moving was five times higher for freshwater individuals than marine ones (Linear model: population term, $p = 0.0006$ for time spent moving, $n = 36$ marine fish, $n = 36$ freshwater fish; Table 1 & Fig. 2D) and a significant difference was also found for total distance swam ($p = 0.010$, $n = 36$ marine fish, $n = 36$ freshwater fish; Table 1).

Boldness. In the boldness under predation risk test, 94% of freshwater fish responded to the heron predator attacks and 79% to the sculpin by hiding in the shelter, while 97% of marine fish showed this reaction for both predators. The two populations did not differ in boldness; all fish exhibited the same average latency to resume feeding after a predator stimulus (Linear model: population term, $p = 0.961$, $n = 32$ marine fish, $n = 35$ freshwater fish for the fish predator, $n = 33$ marine fish, $n = 35$ freshwater fish for the bird predator; Table 1 & Fig. 2E). Fish responded differently to the type of predator in both populations (Linear model: predator type term, $p = 0.041$; no significant population X predatory type interaction). The bird-shaped predator induced the longest latency in both populations (Fig. 2E).

Discussion

We found that common-environment reared and predator-naïve juvenile threespine sticklebacks originating from a freshwater habitat were less social, more active in a familiar environment and more aggressive towards conspecifics than their marine conspecifics. Given that all adult fish were bred in the laboratory and that juveniles were raised in the same setting without parental care, our results suggest that a genetic difference may underlie the measured divergence in behaviour; although further investigations are needed to confirm this hypothesis. It has been shown that boldness [28,39], aggressiveness [27,28,39], exploration in an unfamiliar environment [28,39] and activity in a familiar environment [27,28,39] among others show genetic variation between freshwater populations in threespine stickleback, while schooling behaviour has been shown to differ between marine and freshwater populations of this species [21]. We thus uncover important behavioural differences between individuals belonging to the ancestral marine population and the freshwater one for two previously under- or unstudied ecologically-relevant behaviours, i.e. activity and aggressiveness, in addition to confirming divergence in sociability, in line with what has been found previously in another population pair [21].

Our results suggest that the behavioural divergence observed could represent the result of natural selection that acted on individuals after freshwater colonisation [40]. However, the hypothesis that this phenotypic divergence could result from other causes than selection cannot be excluded. Neutral processes acting during and after colonisation on genetic diversity of the population, e.g. genetic drift, could explain the divergence [41], although the findings that sociability diverges in the same direction as what was found previously for other stickleback populations [21] and other species [11] tends to support the hypothesis of the effects of natural selection. The evidence gathered demonstrating the parallel evolution of various traits (morphology, physiology)

Table 1. Summary of linear model statistical tests conducted to compare the two juvenile threespine stickleback populations for five behaviours.

Behaviours	Variable	Estimate	P-value	Population trends
Sociability	population	−70.782	0.011*	marine > freshwater
	sex	3.410	0.922	
	length	−1.912	0.784	
Exploration – latency to exit cylinder	population	13.159	0.902	marine = freshwater
	sex	10.506	0.638	
	length	6.331	0.302	
Exploration – time spent swimming	population	−3.626	0.745	marine = freshwater
	sex	3.285	0.510	
	length	−0.878	0.317	
Exploration – total length swam	population	0.332	0.961	marine = freshwater
	sex	−1.132	0.961	
	length	1.921	0.312	
Aggressiveness	population	2.297	0.002**	marine < freshwater
	sex	−0.970	0.270	
	length	−0.220	0.265	
Activity – time spent swimming	population	19.573	0.002**	marine < freshwater
	sex	−3.572	0.388	
	length	−0.522	0.638	
Activity – total length swam	population	40.187	0.013*	marine < freshwater
	sex	−18.725	0.171	
	length	−0.002	1.000	
Boldness	population	−46.004	0.961	marine = freshwater
	sex	−0.838	1.00	
	length	1.965	1.00	
	day	−31.193	0.863	
	fish_ID	−0.419	0.863	
	predator type	−53.608	0.041*	

Estimate and p-value probability obtained using the lmp function of the lmPerm package in R. Asterisk (*) denotes significant difference.

occurring in hundreds of geographically and genetically isolated freshwater populations of threespine stickleback after colonisation from the marine environment speaks in favour of natural selection [42]. Our study provides additional evidence supporting the hypothesis of repeated phenotypic divergence when facing similar ecological challenges, as we find the same significantly lower sociability tendency in freshwater fish compared to marine fish as previously found for entirely different marine (Pacific Ocean in Japan) and freshwater populations (benthic population from Paxton Lake in Canada) of this species [21]. This repeated strong behavioural difference associated with colonisation of the freshwater environment supports the hypothesis that this divergence in behaviour enhances fitness in the new ecological conditions. Of course, other stickleback marine-freshwater population pairs must be characterised for their behavioural phenotypes in both habitat types in order to confirm that behavioural differences measured in this study between a marine and freshwater population were due to divergent natural selection rather than being merely difference between any two populations, regardless of ecological pressures.

Predation pressure is one of the main ecological factors that act as agents of selection, which can cause phenotypic divergence among prey populations [1]. In threespine stickleback, contrasting predation pressures between marine and freshwater populations are correlated with divergence in lateral plate armour development [31]. The marine and freshwater habitats from which breeding adults were sampled to create the two juvenile laboratory lines studied differed in predation regime (based on armour morphology, see methods) thus suggesting that predation could be a selection agent acting differentially on behaviour in the marine and the freshwater population, similar to what is found for predator defence morphology [31]. Indeed, studies comparing freshwater populations showed an association between predatory regime differences and divergence in behaviour in sticklebacks in sociability [25], activity [26,27], aggressiveness [27,28], and boldness [26,28,29]. The strength of our results, although based on only two populations, is that they are in accordance with the predictions of behavioural divergence in sociability, aggressiveness, and activity between marine and freshwater fish we made *a priori* based on these studies of freshwater populations. For example, in a high-predation habitat such as the marine environment, schooling provides fish with several benefits such as a better foraging efficiency and a greater antipredator defence [43]. It has been hypothesised to lead to a reduction of aggression between individuals because high levels of fighting might offset the benefits of schooling [44]. However, we keep in mind that marine and freshwater habitats differ in several other ecological aspects, such as habitat structure and stability [26,45], food resource availability [46], intraspecific competition [47] and temperature [30], which could also explain the behavioural divergence observed. The fact that boldness did not differ between populations contrary to our prediction, and that both populations reacted more to the bird predator, suggests that other factors must also be considered. The behaviour of freshwater juveniles could also represent heritable behavioural characteristics of the marine individuals that preferentially invaded freshwater after the glacial retreat to form the new lake population. A greater swimming activity and a higher boldness could be associated with a higher propensity to use novel habitats [48,49]. The freshwater individuals were significantly more active than marine fish, although they did not differ in boldness. This is in line with the hypothesis that the freshwater behaviour stems from standing genetic variation in the marine population, as very bold individuals could also be found in the marine population, as observed in previous studies for morphological traits in threespine stickleback [50]. In this case, natural selection acting on these individuals once in the new freshwater environment is also possible.

Our study suggests that mean behaviours in a population are associated with their own costs and benefits modulated according to the habitat. Taken together, our results provide an integrated view of the divergence in mean behaviour of individuals colonising a novel environment, potentially resulting from the new ecological constraints encountered, in contrast with the behaviour of the ancestral population, information which are of crucial importance considering the role of behavioural phenotype-environment associations in adaptive divergence.

Supporting Information

Table S1 Dataset used for the study by Di Poi et al.

Table S2 Statistically significant correlations found between behaviours in each population. Correlation coefficients and p-values calculated before and after corrections for multiple testing (Benjamini-Horchberg) are presented.

Acknowledgments

The authors gratefully acknowledge the personnel of the Laboratoire Régional des Sciences Aquatiques (LARSA) at Université Laval for their help in fish rearing, Matthew Morris for help with fish and Nancy Diplo for chromosomal sex determination of sticklebacks.

Author Contributions

Conceived and designed the experiments: NAH CDP JL. Performed the experiments: CDP JL. Analyzed the data: CDP NAH. Contributed reagents/materials/analysis tools: SMR. Wrote the paper: CDP SMR NAH.

References

1. Schluter D (2000) The ecology of adaptive radiation. Oxford, UK: Oxford Series in Ecology and Evolution, 296.
2. Losos JB (2009) Lizards in an evolutionary tree: ecology and adaptive radiation of Anoles. Berkeley and Los Angeles, CA, USA: University of California Press, 528.
3. Cooper WJ, Parsons K, McIntyre A, Kern B, McGee-Moore A, et al. (2010) Bentho-pelagic divergence of cichlid feeding architecture was prodigious and consistent during multiple adaptive radiations within African rift-lakes. PloS ONE 5(3): e9551.
4. Rogers SM, Gagnon V, Bernatchez L (2002) Genetically based phenotype-environment association for swimming behavior in lake whitefish ecotypes (*Coregonus clupeaformis Mitchill*). Evolution 56: 2322–2329.
5. Grant PR, Grant BR (2011) How and why species multiply: the radiation of Darwin's finches. Princeton, NJ, USA: Princeton University Press, 218.
6. Budaev SV (1997) "Personality" in the guppy (*Poecilia reticulata*): a correlational study of exploratory behavior and social tendency. J Comp Psychol 111: 399–411.
7. Herczeg G, Välimäki K (2011) Intraspecific variation in behaviour: effects of evolutionary history, ontogenetic experience and sex. J Evol Biol 24: 2434–2444.
8. Herczeg G, Gonda A, Kuparinen A, Merilä J (2012) Contrasting growth strategies of pond versus marine populations of nine-spined stickleback (*Pungitius pungitius*): a combined effect of predation and competition? Evol Ecol 26: 109–122.
9. Brown C, Jones F, Braithwaite V (2005) *In situ* examination of boldness-shyness traits in the tropical poeciliid, *Brachyraphis episcopi*. Anim Behav 70: 1003–1009.
10. Magurran AE, Seghers BH, Carvalho GR, Shaw PW (1992) Behavioural consequences of an artificial introduction of guppies (*Poecilia reticulata*) in N. Trinidad: evidence for the evolution of antipredator behaviour in the wild. Proc R Soc B 248: 117–122.
11. Huizinga M, Ghalambor CK, Reznick DN (2009) The genetic and environmental basis of adaptive differences in shoaling behaviour among populations of Trinidadian guppies, *Poecilia reticulata*. J Evol Biol 22: 1860–1866.
12. Chamberlain NL, Hill RI, Kapan DD, Gilbert LE, Kronforst MR (2009) Polymorphic butterfly reveals the missing link in ecological speciation. Science 326: 847–850.
13. Rogers SM, Tamkee P, Summers B, Balabahadra S, Marks M, et al. (2012) Genetic signature of adaptive peak shift in threespine stickleback. Evolution 66: 2439–2450.
14. Bell MA, Foster SA (1994) The evolutionary biology of the threespine stickleback. Oxford, UK: Oxford University Press, 571.
15. Cresko WA, McGuigan KL, Phillips PC, Postlethwait JH (2007) Studies of threespine stickleback developmental evolution: progress and promise. Genetica 129: 105–126.
16. Greenwood AK, Jones FC, Chan YF, Brady SD, Absher DM, et al. (2011) The genetic basis of divergent pigment patterns in juvenile threespine sticklebacks. Heredity 107: 155–166.
17. Snyder RJ, Dingle H (1989) Adaptive, genetically based differences in life history between estuary and freshwater threespine sticklebacks (*Gasterosteus aculeatus* L.). Can J Zool 67: 2448–2454.
18. Kitano J, Lema SC, Luckenbach JA, Mori S, Kawagishi Y, et al. (2010) Adaptive divergence in the thyroid hormone signaling pathway in the stickleback radiation. Curr Biol 20: 2124–2130.
19. Dalziel AC, Vines TH, Schulte PM (2012) Reductions in prolonged swimming capacity following freshwater colonization in multiple threespine stickleback populations. Evolution 66: 1226–1239.
20. Bakker TCM, Feuth-de Bruijn E (1988) Juvenile territoriality in stickleback *Gasterosteus aculeatus* L. Anim Behav 36: 1556–1558.
21. Wark AR, Greenwood AK, Taylor EM, Yoshida K, Peichel CL (2011) Heritable differences in schooling behavior among threespine stickleback populations revealed by a novel assay. PloS ONE 6: e18316.
22. Saimoto RK (1993) Life history of marine threespine stickleback in Oyster Lagoon, British Columbia. M. Sc. thesis, University of British Columbia, Vancouver, 97.
23. Godin JGJ, Smith SA (1988) A fitness cost of foraging in the guppy. Nature 333: 69–71.
24. Réale D, Reader SM, Sol D, McDougall PT, Dingemanse NJ (2007) Integrating animal temperament within ecology and evolution. Biol Rev 82: 291–318.
25. Nomakuchi S, Park PJ, Bell MA (2009) Correlation between exploration activity and use of social information in three-spined sticklebacks. Behav Ecol 20: 340–345.
26. Brydges NM, Colegrave N, Robert JP, Braithwaite VA (2008) Habitat stability and predation pressure affect temperament behaviours in populations of three-spined sticklebacks. J Anim Ecol 77: 229–235.
27. Lacasse J, Aubin-Horth N (2012) A test of the coupling of predator defense morphology and behavior variation in two threespine stickleback populations. Curr Zool 58: 53–65.
28. Bell AM (2005) Behavioural differences between individuals and two populations of stickleback (*Gasterosteus aculeatus*). J Evol Biol 18: 464–473.
29. Dingemanse NJ, Wright J, Kazem AJN, Thomas DK, Hickling R, et al. (2007) Behavioural syndromes differ predictably between 12 populations of three-spined stickleback. J Anim Ecol 76: 1128–1138.
30. Barrett RDH, Paccard A, Healy TM, Bergek S, Schulte PM, et al. (2011) Rapid evolution of cold tolerance in stickleback. Proc R Soc B 278: 233–238.
31. Marchinko KB (2009) Predation's role in repeated phenotypic and genetic divergence of armor in threespine stickleback. Evolution 63: 127–138.
32. Ministry of Environment – Government of British Columbia website. Available: http://a100.gov.bc.ca/pub/fidq/fishDistributionsQuery.do. Accessed 2014 May 14.
33. Greenwood AK, Wark AR, Yoshida K, Peichel CL (2013) Genetic and neural modularity underlie the evolution of schooling behavior in threespine sticklebacks. Curr Biol 23: 1884–1888.

34. Peichel CL, Ross JA, Matson CK, Dickson M, Grimwood J, et al. (2004) The master sex-determination locus in threespine sticklebacks is on a nascent Y chromosome. Curr Biol 14: 1416–1424.
35. Dzieweczynski TL, Crovo JA (2011) Shyness and boldness differences across contexts in juvenile three-spined stickleback *Gasterosteus aculeatus* from an anadromous population. J Fish Biol 79: 776–788.
36. Aubin-Horth N, Deschênes M, Cloutier S (2012) Natural variation in the molecular stress network correlates with a behavioural syndrome. Horm Behav 61: 140–146.
37. Huntingford FA (1976) The relationship between anti-predator behaviour and aggression among conspecifics in the three-spined stickleback. Anim Behav 24: 245–260.
38. Benjamini Y, Hochberg Y (1995) Controlling the false discovery rate: a practical and powerful approach to multiple testing. J R Stat Soc B 57: 289–300.
39. Giles N, Huntingford FA (1984) Predation risk and inter-population variation in antipredator behaviour in the three-spined stickleback, *Gasterosteus aculeatus* L. Anim Behav 32: 264–275.
40. Endler JA (1986) Natural selection in the wild. Monographs in population biology. Princeton, New Jersey, USA: Princeton University Press, 354.
41. Knowles LL, Richards CL (2005) Importance of genetic drift during Pleistocene divergence as revealed by analyses of genomic variation. Mol Ecol 14: 4023–4032.
42. Schluter D, Marchinko KB, Barrett RDH, Rogers SM (2010) Natural selection and the genetics of adaptation in threespine stickleback. Phil Trans R Soc B 365: 2479–2486.
43. Pitcher TJ, Parrish JK (1993) Functions of shoaling behaviour in teleosts. In: Behaviour of Teleost Fishes (ed. Pitcher, T.J.). 2nd edn. Chapman & Hall, London, UK, 363–439.
44. Magurran AE, Seghers BH (1991) Variation in schooling and aggression amongst guppy (*Poecilia reticulata*) populations in Trinidad. Behaviour 118: 214–234.
45. Braithwaite VA, Girvan JR (2003) Use of water flow direction to provide spatial information in a small-scale orientation task. J Fish Biol 63: 74–83.
46. Laskowski KL, Bell AM (2013). Competition avoidance drives individual differences in response to a changing food resource in sticklebacks. Ecol Lett 16: 746–753.
47. Svanbäck R, Bolnick DI (2007) Intraspecific competition drives increased resource use diversity within a natural population. Proc R Soc B 274: 839–844.
48. Fraser DF, Gilliam JF, Daley MJ, Le AN, Skalski GT (2001) Explaining leptokurtic movement distributions: intrapopulation variation in boldness and exploration. Amer Nat 158: 124–135.
49. Barrett RDH, Vines TH, Bystriansky JS, Schulte PM (2009) Should I stay or should I go? The Ectodysplasin locus is associated with behavioural differences in threespine stickleback. Biol Lett 5: 788–791.
50. Colosimo PF, Peichel CL, Nereng K, Blackman BK, Shapiro MD, et al. (2004) The genetic architecture of parallel armor plate reduction in threespine sticklebacks. PLoS Biology 2: E109.

7

Millimeter-Sized Marine Plastics: A New Pelagic Habitat for Microorganisms and Invertebrates

Julia Reisser[1,2,3]*, Jeremy Shaw[4], Gustaaf Hallegraeff[5], Maira Proietti[6], David K. A. Barnes[7], Michele Thums[2,8], Chris Wilcox[3,9], Britta Denise Hardesty[3,9], Charitha Pattiaratchi[1,2]

1 School of Environmental Systems Engineering, University of Western Australia, Perth, Australia, **2** Oceans Institute, University of Western Australia, Perth, Australia, **3** Wealth from Oceans Flagship, Commonwealth Scientific and Industrial Research Organisation, Perth, Australia, **4** Centre for Microscopy, Characterisation and Analysis, University of Western Australia, Perth, Australia, **5** Institute for Marine and Antarctic Studies, University of Tasmania, Hobart, Australia, **6** Instituto de Oceanografia, Universidade Federal do Rio Grande, Rio Grande, Brazil, **7** British Antarctic Survey, Natural Environment Research Council, Cambridge, United Kingdom, **8** Australian Institute of Marine Science, The UWA Oceans Institute, Perth, Australia, **9** Marine and Atmospheric Research, Commonwealth Scientific and Industrial Research Organisation, Hobart, Australia

Abstract

Millimeter-sized plastics are abundant in most marine surface waters, and known to carry fouling organisms that potentially play key roles in the fate and ecological impacts of plastic pollution. In this study we used scanning electron microscopy to characterize biodiversity of organisms on the surface of 68 small floating plastics (length range = 1.7–24.3 mm, median = 3.2 mm) from Australia-wide coastal and oceanic, tropical to temperate sample collections. Diatoms were the most diverse group of plastic colonizers, represented by 14 genera. We also recorded 'epiplastic' coccolithophores (7 genera), bryozoans, barnacles (*Lepas* spp.), a dinoflagellate (*Ceratium*), an isopod (Asellota), a marine worm, marine insect eggs (*Halobates* sp.), as well as rounded, elongated, and spiral cells putatively identified as bacteria, cyanobacteria, and fungi. Furthermore, we observed a variety of plastic surface microtextures, including pits and grooves conforming to the shape of microorganisms, suggesting that biota may play an important role in plastic degradation. This study highlights how anthropogenic millimeter-sized polymers have created a new pelagic habitat for microorganisms and invertebrates. The ecological ramifications of this phenomenon for marine organism dispersal, ocean productivity, and biotransfer of plastic-associated pollutants, remains to be elucidated.

Editor: Adrianna Ianora, Stazione Zoologica Anton Dohrn, Naples, Italy

Funding: This project was funded by University of Western Australia (http://www.uwa.edu.au) and Commonwealth Scientific and Industrial Research (http://www.csiro.au). It has also been supported by Australia's Marine National Facility, Austral Fisheries, Australian Institute of Marine Science, CSIRO's Flagship postgraduate scholarship (JR), and the Shell social investment program (BDH and CW). The funders had no role in study design, data collection and analysis, decision to publish, or preparation of the manuscript.

Competing Interests: The authors have the following interests. This study was partly funded by Austral Fisheries and the Shell social investment program. There are no patents, products in development or marketed products to declare. This does not alter their adherence to all the PLOS ONE policies on sharing data and materials, as detailed online in the guide for authors.

* E-mail: jureisser@gmail.com

Introduction

Millimeter-sized plastics resulting from the disintegration of synthetic products (known as 'microplastics' if smaller than 5 mm) are abundant and widespread at the sea surface [1–7]. These small marine plastics are a toxic hazard to food webs since they can contain harmful compounds from the manufacturing process (e.g. Bisphenol A), as well as contaminants adsorbed from the surrounding water (e.g. polychlorinated biphenyls) [8–11]. These substances can be carried across marine regions and transferred from plastics to a wide range of organisms, from zooplankton and small fish to whales [8,12–19]. Furthermore, they can physically damage suspension- and deposit-feeding fauna (e.g. internal abrasions and blockages after ingestion) [20], and alter pelagic and sediment-dwelling biota by modifying physical properties of their habitats [21]. Finally, these small marine plastics can transport rafting species [22–27], potentially changing their natural ranges to become non-native species and even invasive pests.

Apart from providing long-lasting buoyant substrata that allow many organisms to widely disperse [28–38], marine plastics may also supply energy for microbiota capable of biodegrading polymers and/or associated compounds [27,39–43], and perhaps for invertebrates capable of grazing upon plastic inhabitants. The hydrophobic nature of plastic surfaces stimulates rapid formation of biofilm, which drives succession of other micro- and macro-organisms. This 'epiplastic' community appears to influence the fate of marine plastic pollution by affecting the degradation rate [27,44], buoyancy [3,45,46], and toxicity level [43] of plastics. Moreover, epiplastic microbiota could have impacts on the microflora of its consumers, and infectious organisms may reach their hosts through plastic ingestion [27,43,47].

Although epiplastic organisms may play an important role in determining the fate and ecological impacts of plastic pollution, little research has been directed to such study, particularly on the inhabitants of the widely dispersed and abundant millimeter-sized marine plastics [43]. In 1972, two papers first reported the occurrence of organisms (diatoms, hydroids, and bacteria) on small

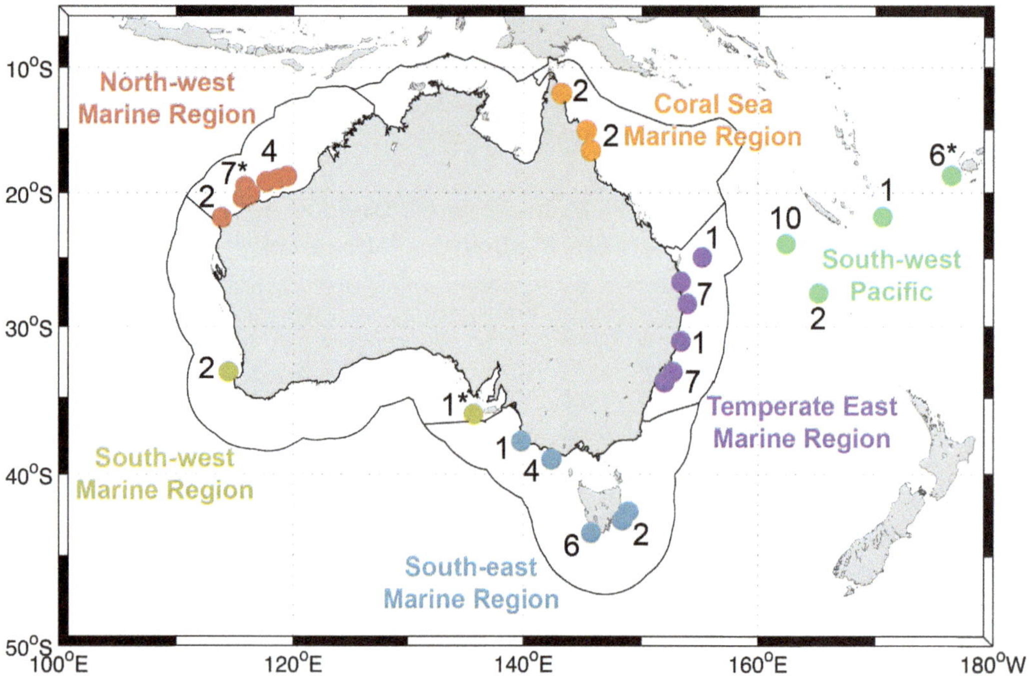

Figure 1. Sampling locations of the 68 plastics analyzed in this study. Black lines delimit marine regions of Australia (environment.gov.au/topics/marine/marine-bioregional-plans); dots indicate areas where the analyzed plastics were collected; numbers represent how many plastics were taken for scanning electron microscopy analyses at these locations. Samples collected were fragments of hard plastic (N = 65), except at locations marked with an asterisk: one piece of Styrofoam cup in Fijian waters, one pellet in South Australia, and one piece of soft plastic in the Australia's North-west marine region.

plastics (0.1–5 mm long) collected by plankton nets [22,23]. Further at-sea studies focusing on microplastic fouling biota only emerged in the 2000's [21,27,48]. Zettler et al. (2013) conducted the first comprehensive characterisation of epiplastic microbial communities, which they coined the "Plastisphere" [27]. These authors used scanning electron microscopy (SEM) and next-generation sequencing to analyze three polyethylene and three polypropylene plastic pieces (approx. 2–20 mm long) from offshore waters of the North Atlantic. This pioneer study revealed a unique, diverse, and complex microbial community that included diatoms, ciliates, and bacteria.

Here, we used SEM to examine types of organisms inhabiting the surface of 68 small marine plastics (length range = 1.7–24.3 mm, median = 3.2 mm) from inshore and offshore waters from around the Australian continent (Figure 1). We contributed many new records of taxa associated with millimeter-sized marine plastics and imaged a variety of marine plastic shapes and surface textures resulting from the interaction of polymers with environments and organisms.

Materials and Methods

Ethics Statement: Permits to conduct field research within the Great Barrier Reef area were obtained from the Great Barrier Reef Marine Park Authority (GBRMPA: permit G11/34378.1). No other special permit was required since sampling was limited to marine debris.

Buoyant plastics were collected using surface net tows in waters around Australia (see details in [4,49]) and preserved in 2.5% glutaraldehyde buffered in filtered seawater. Prior to analysis with a scanning electron microscope, plastics were dehydrated through a series of increasing ethanol concentrations (up to 100%), critical-point dried using CO_2, mounted on aluminum stubs with carbon tape, and sputter coated with a 20–30 nm layer of gold. We used a Zeiss 1555 VP-FESEM scanning electron microscope operated at 10–20 kV, 11–39 mm working distance, and 10–30 μm aperture. We randomly selected 65 hard plastics among those small enough to fit onto SEM stubs (<10 mm) and large enough to be easily handled (>1 mm). For comparison, a piece of (1) soft plastic, (2) industrial plastic pellet, and (3) expanded polystyrene (Styrofoam) were also examined, totaling 68 plastic pieces examined with SEM. These plastics were collected from offshore waters of the South-west Pacific (N = 19) and from different Australian marine regions (environment.gov.au/topics/marine/marine-bioregional-plans): North-west (N = 13), South-west (N = 3), South-east (N = 13), Temperate East (N = 16), and Coral Sea (N = 4; Figure 1).

The different types of organisms detected on each plastic piece were imaged, measured using ImageJ (length and width, http://rsb.info.nih.gov/ij/), classified into taxonomic/morphological groups, and the frequency of occurrence (FO) for each type was calculated. We used online resources (e.g. marinespecies.org, westerndiatoms.colorado.edu), primary taxonomic literature (e.g [50–54]), and expert consultation (see acknowledgments section) to identify the organisms at the lowest possible taxonomic level. Long filaments were very common but were excluded from the analysis due to difficulty in determining if they were organisms or mucilage.

For each plastic piece observed, an image of the entire piece was taken at 50×magnification. These images were uploaded to ImageJ to measure plastic particles' size parameters (length, area, perimeter, aspect ratio) and shape parameters (circularity and solidity indexes [55,56]). Surface fractures, pits and grooves [57,58] were also observed, recorded, and imaged while examining the entire surface of the plastics at magnifications of 100–

Figure 2. Overall appearance of marine plastics, as shown by scanning electron micrographs. Dot color indicates the marine region where the piece was sampled (see legend and Figure 1). Pieces are hard plastic fragments, with the exception of the soft plastic fragment (red dot), pellet (yellow dot), and Styrofoam fragment (green dot) shown at the bottom of the diagram and marked with a white asterisk. All images are at the same magnification (see scale bar at lower right).

500×. Other peculiar microtextures observed at higher magnifications, such as those suggesting interactions with biota, were also recorded and imaged. After SEM analyses, plastics were washed with distilled water and submitted to Fourier Transform Infrared spectrometry (FT-IR) for polymer identification. Two plastic pieces were destroyed while being cleaned for FT-IR; as such, we identified the polymer of 66 out of the 68 plastics examined using SEM.

Results

We examined 65 hard plastic fragments with lengths ranging from 1.7 to 8.9 mm (median = 3.2 mm), one 4 mm-wide plastic pellet, one 8.7 mm portion of a 15 mm long soft plastic fragment, and one 7 mm piece of a 24.3 mm Styrofoam cup fragment. Apart from the Styrofoam cup fragment (expanded polystyrene), plastics were made of polyethylene (N = 54) and polypropylene (N = 11).

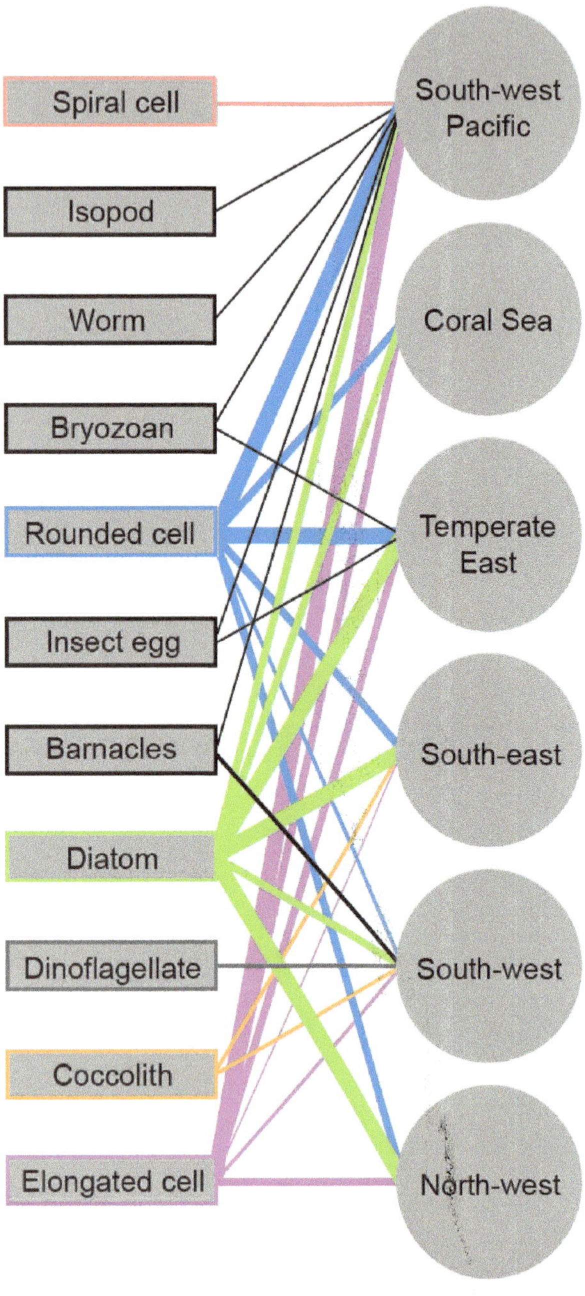

Figure 3. Types of epiplastic organisms detected at each of the marine regions sampled in this study (see Figure 1). Lines connect types of organisms (squares) to the marine regions (circles) where they were observed. Line color indicates type of organism, with black lines representing invertebrates. Line thickness is proportional to the organism's frequency of occurrence (FO = <25%, 25–50%, 50–75%, >75%).

Hard plastics had a diverse range of shapes (solidity index = 0.87–0.98, circularity index = 0.28–0.83; Figure 2) and types of surface microtextures, including linear fractures, pits, and scraping marks (Figure S1). Diatoms and bacteria (rounded, and elongated cells) were by far the most frequently observed organisms, being detected in all sampled marine regions (Figure 3). Plastics' FT-IR spectra, 1143 SEM micrographs, and a matrix containing information from collection sites, plastics characteristics, and organism/microtexture presence-absence data are available in [59].

Diatoms were the most abundant, widespread, and diverse group of plastic colonizers (Figures 3 and 4). These organisms were frequently observed (FO = 78%, N = 68 plastics) and included symmetrical biraphids/naviculoids (*Navicula* subgroup lineatae, *Mastogloia* sp., *Haslea* sp.; Figure 4a–c), Nitzschioids (*Nitzschia* spp., *Nitzschia longissima*; Figure 4d–f), monoraphids (*Cocconeis* spp., *Achnanthes* sp.; Figure 4g–i), centrics (*Minidiscus trioculatus*, *Thalassiosira* sp.; Figure 4j), araphids (*Thalassionema nitzschioides* var. *parva*, *Microtabella* spp., *Licmophora* spp., *Grammatophora* sp.; Figure 4k,l,o), and asymmetrical biraphids (*Amphora* spp., *Cymbella* sp.; Figure 4m,n). Most diatoms were growing flat on the surface (adnate and motile diatoms), but some were erect, attached to plastics by mucous pads or stalks/peduncles. The genus *Nitzschia* was the most frequent diatom (FO = 42.6%), followed by *Amphora* (13.2%), *Licmophora* (11.8%), *Navicula* (8.8%), *Microtabella* (5.9%), *Cocconeis* (4.4%), *Thalassionema* (2.9%), and *Minidiscus* (2.9%). The other six genera were only detected on a single plastic piece (FO = 1.5%). These frequencies of occurrence are likely to be underestimated, as many diatoms could not be identified from girdle-view positions (FO unidentified diatoms = 45.6%).

Calcareous coccolithophores were observed only on plastics from southern Australia (South-east and South-west marine regions; FO = 37.5%, N = 16 plastics; Figure 3, Figure 5a–h). The species identified included *Calcidiscus leptoporus* (Figure 5a), *Emiliania huxleyi* (Figure 5b,c), *Gephyrocapsa oceanica* (Figure 5d), *Umbellosphaera tenuis* (Figure 5e), *Umbilicosphaera hulburtiana* (Figure 5f), *Coccolithus pelagicus* (Figure 5g), and *Calciosolenia* sp. (Figure 5h). Many of these observations related to detached coccolith scales held in place by mucilage and chitin filaments resembling those produced by diatoms (e.g. *Thalassiosira*; Figure 5b,f). However, intact coccospheres were also present (Figure 5c,d,f). Additionally, one specimen of the dinoflagellate *Ceratium* cf. *macroceros* was present on a 8.2 mm plastic from South-west Australia (Figure 3, Figure 5i).

We found several unidentified organisms of various morphotypes and sizes, mostly resembling bacterial, cyanobacterial, and fungal cells (Figure 6). After diatoms, rounded/oval cells (length-width ratio <1.5; Figure 6a–c,i–m) were the most frequently observed morphotype (FO = 72%, N = 68 plastics; Figure 3). Rounded/oval cells with widths <1 μm and ≥1 μm had an overall FO of 38.2% and 54.4%, respectively. Elongated cells (length-width ratio ≥1.5; Figure 6e–h) were also frequently observed, being detected on 59% of the plastics examined (Figure 3). Those with widths <1 μm and ≥1 μm had an overall FO of 51.5% and 11.7%, respectively. Spiral cells (Figure 6d) had similar appearances (resembling spirochaete bacteria) and sizes (0.2–0.3 μm width), and were only observed in the South-west Pacific region (FO = 31.6%, N = 19; Figure 3). Several plastic pits and grooves contained bacteria-like cells closely resembling their shape (Figure 6i–m). They were particularly common on plastics covered by large rounded cells (Figure 6k).

A few invertebrates were observed on the millimeter-sized plastics (FO = 16.2%, N = 68 plastics; Figures 3 and 7). Colonies of encrusting bryozoans were the most common epiplastic animal (FO = 8.8%; Figure 7a–d). They occurred on two fragments from the Temperate East marine region and on four fragments from oceanic waters of the South-west Pacific (plastic length = 3.2–5.4 mm). Four of these bryozoan colonies were hosting abundant diatom assemblages dominated by *Licmophora* sp., *Nitzschia longissima* (Figure 7a), *Amphora* sp. (Figure 7c), and *Nitzschia* sp. (Figure 7d). Additionally, lepadomorph barnacles (*Lepas* spp.; Figure 7e,f) were attached to the 24.3 mm Styrofoam cup

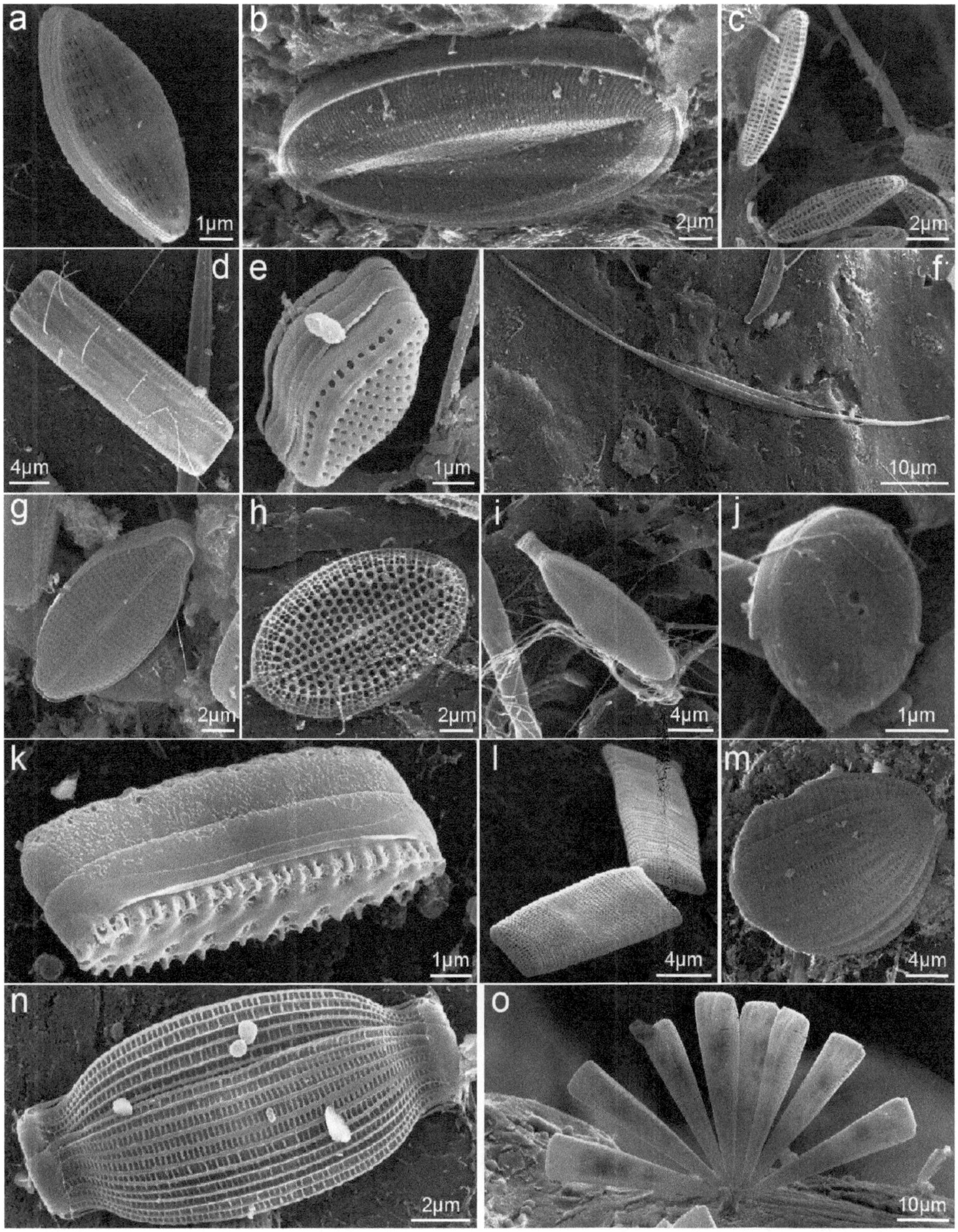

Figure 4. Examples of epiplastic diatoms. a: *Navicula* sp.; b: *Mastogloia* sp.; c: small naviculoids; d: *Nitzschia* sp.; e: *Nitzschia* sp.; f: *Nitzschia longissima*.; g: *Cocconeis* sp.; h: *Cocconeis* sp.; i: *Achnanthes* sp.; j: *Thalassiosira* sp.; k: *Thalassionema nitzschioides*; l: *Microtabella* sp.; m: *Amphora* sp.; n: *Amphora* sp.; o: *Licmophora* sp.

fragment and to a 8.2 mm-long hard plastic; an Asellote isopod (Figure 7g) was found on the Styrofoam cup fragment; eggs of the marine insect *Halobates* sp. (Figure 7h) were observed on two plastics (4.6 and 5.5 mm long); and a unidentified marine worm (Figure 7i,j) was found on a 6 mm hard plastic fragment.

Figure 5. Examples of epiplastic coccoliths and dinoflagellate. a: Calcidiscus leptoporus; b, c: Emiliania huxleyi; d: Gephyrocapsa oceanica; e: Umbellosphaera tenuis; f: Umbilicosphaera hulburtiana; g: Coccolithus pelagicus; h: Calciosolenia sp.; i: Ceratium cf. macroceros dinoflagellate.

Discussion

There now exists a large body of evidence that millimeter-sized plastics are abundant and widespread in marine environments [1–6,22,23] and our study significantly adds to this by conclusively demonstrating that they are colonized by a wide range of biota, particularly diatom and bacteria species (Table 1, [3,22,24–27]). We more than doubled the number of known diatom genera inhabiting millimeter-sized marine plastics and provide the first identifications of coccolithophore genera attached to these floating plastic particles. We also recorded a few invertebrate species living on these small plastics. As such, our findings provide further evidence that not only large debris [28–38] serve as vehicles for organism dispersal. Abundant 'microplastics' are equally providing a new pelagic habitat to many microorganism and a few invertebrate taxa.

We observed fouling diatoms to be diverse and widespread on marine plastics. These diatoms seemed to firmly attach to the plastic, resisting water turbulence and wave action. All the identified diatom genera are well known to form biofilms on estuarine and marine sediments and rocks (epilithic), vegetation (epiphytic), and animals (epizoic) [60–65]; marine plastics thus create a novel, long-lasting and abundant floating habitat for 'benthic' diatoms, in a light and nutrient-filled environment that is stable and beneficial to these organisms. Future epiplastic diatom research should focus on the quantitative contribution of these organisms to enhancing primary and secondary productivity of different marine regions, such as within subtropical gyres where productivity tends to be low but plastic pollution level high [1–3,66]. Because of their rapid growth and production of extracellular substances [67], epiplastic diatoms may provide an important food source for invertebrate grazers. As plastic debris can contain harmful substances [8,10–12,19], it remains unclear if such grazer-plastic relationships would have a positive or negative impact on the populations involved in this new type of food web.

A significant number of coccolithophore species were present on millimeter-sized marine plastics. These planktonic organisms are not commonly recognized as fouling or rafting organisms [36], although their occasional occurrence on marine plastics was briefly mentioned in recent studies [26,27]. Some of our observations were of clusters of mixed coccolith species, resembling zooplankton fecal pellets, and of solitary coccoliths, likely detached from living coccospheres and stuck to clingy parts of the plastic biofilm. However, entire coccolithophores were also seen attached to plastics, suggesting that these organisms could be using ocean plastics as 'floating devices'. We only observed coccoliths on plastics from southern Australia; as such, additional studies in these temperate waters may help better understand this potential coccolith-plastic relationship. Another atypical organism detected was the planktonic dinoflagellate *Ceratium* cf. *macroceros*. Recent studies have found plastics heavily fouled by dinoflagellates, including individuals and cysts of the potentially harmful species *Ostreopsis* sp., *Coolia* sp., and *Alexandrium* spp. [27,31], but here we only detected a single specimen of this group.

Several unidentified organisms (rounded, oval, elongated, and spiral) resembling bacterial cells were flourishing on millimeter-sized marine plastics. This supports previous studies that describe well established bacterial populations growing on plastic fragments [26,27]. Many of these unidentified cells were apparently interacting with the plastic surface by forming pits and grooves. Within this group of "pit-formers", colonies of rounded cells (around 5 micron in diameter) covered large areas of the plastic surface. They were similar to some previously unidentified epiplastic organisms from the North Atlantic [27]. These SEM

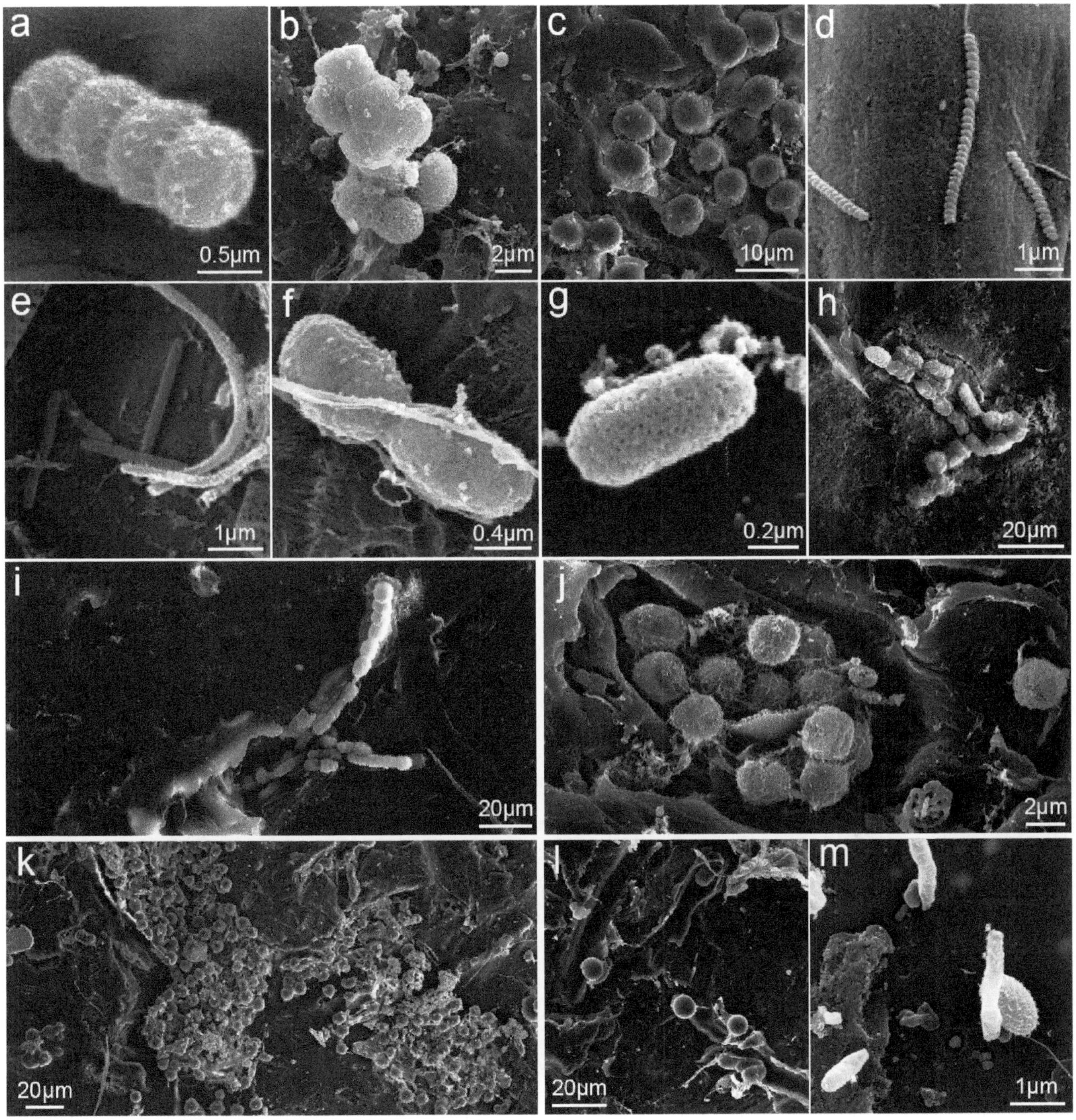

Figure 6. Examples of epiplastic rounded, elongated and spiral cells. a, b, c: rounded cells; d: spiral "spirochaete" cell; e, f, g, h: elongated cells.; i, j, k, l, m: pits and grooves on plastics with rounded cells.

observations, along with detections of hydrocarbon-degrading bacteria genes on marine plastics [27] and experiments demonstrating that marine bacteria can biodegrade polymers [27,39–43], strongly suggest that plastic biodegradation is occurring at the sea surface. Such process could partially explain why quantities of millimeter-sized marine plastics are not increasing as much as expected [2,7]. Studies of the "Plastisphere" from different marine regions worldwide will prove invaluable for extending our knowledge on epiplastic marine microbial communities, and may support the development of biotechnological solutions for better plastic waste disposal practices [68–70].

A number of invertebrates inhabited the small plastics examined here: bryozoans, barnacles *Lepas* spp., an Asellota isopod, a marine worm, and eggs of the marine insect *Halobates* sp. Even though microplastic-associated animals are rare and less diverse when compared to those associated with macroplastics [28–38], ecological implications of this phenomenon may be significant (e.g. [48]), given the large quantities and wide distribution ranges of millimeter-sized plastics in the marine environment [1–6,22,23]. Among the effects plastic associates may have is to shape 'epiplastic' microbiota by hosting unique epizoic assemblages on their bodies. For instance, the bryozoan colonies examined here covered a large proportion of their plastic-host, with some of them

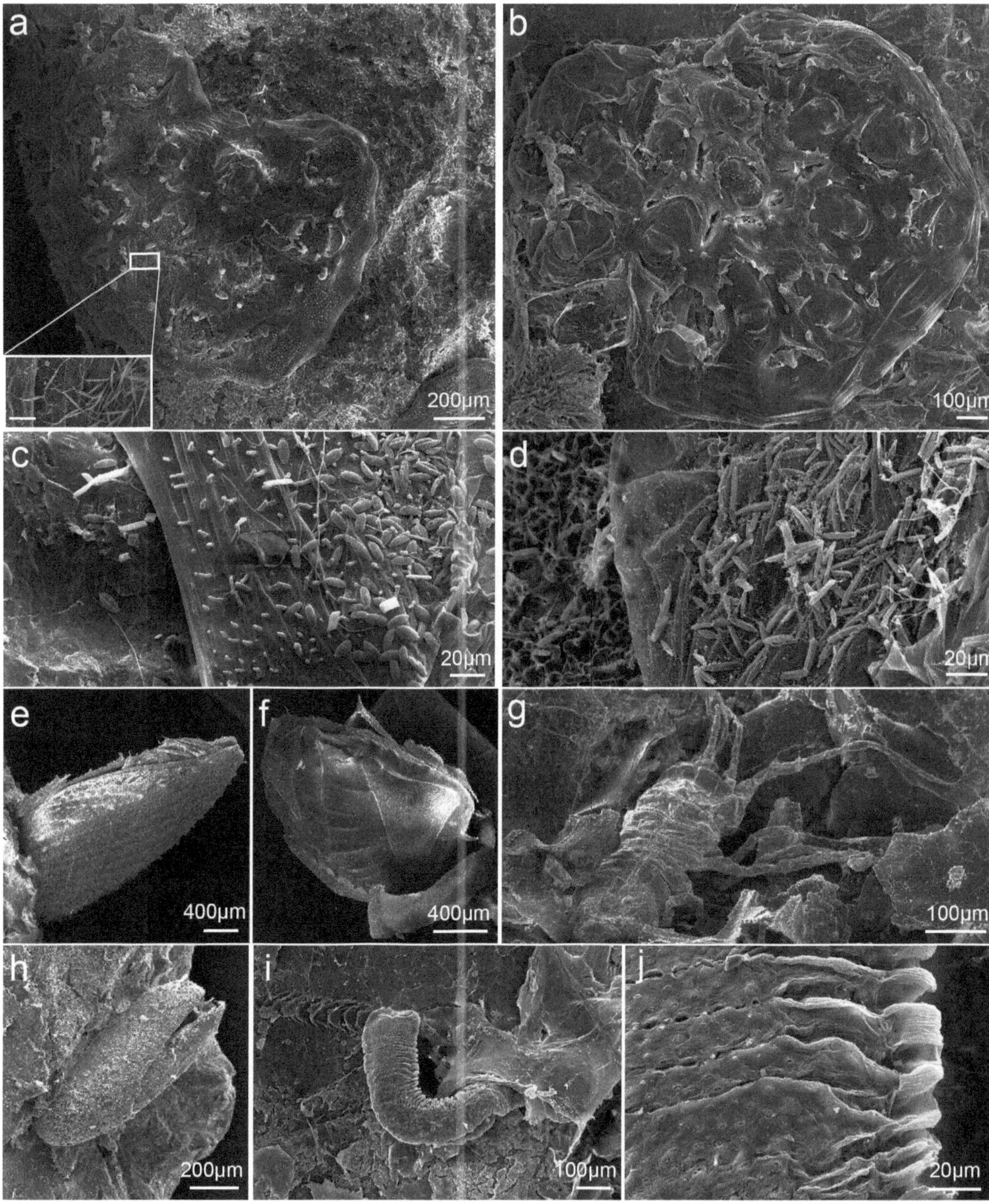

Figure 7. Examples of epiplastic invertebrates. a: Bryozoan colony harboring an abundant assemblage of *Nitzschia longissima* (zoomed image shows part of this assemblage, scale bar = 20 µm); b: bryozoan colony relatively free of fouling; c: bryozoan-plastic interface displaying an abundant epizoic assemblage of *Amphora* sp.; d: bryozoan-plastic interface displaying an abundant epizoic assemblage of *Nitzschia* sp.; e, f: barnacles (*Lepas* spp.); g: Asellota isopod; h: egg of the marine insect *Halobates* sp.; i: marine worm; j: zoom on the surface of the unidentified marine worm.

harboring unique diatom-dominated assemblages. Previous studies have shown that bryozoans do not represent neutral surfaces for microbial colonizers [71,72], with some species offering a favorable habitat for diatoms when compared to the surrounding

Table 1. List of known genera occurring on millimeter-sized pelagic plastics.

Group	Abundance/FO	Genera
Bacteria[a,b,c,d,j]	[d]1833 per mm^{-2}	*Acinetobacter*[b], *Albidovulum*[b], *Alteromonas*[b], *Amoebophilus*[b], *Bacteriovorax*[b], *Bdellovibrio*[b], *Blastopirellula*[b], *Devosia*[b], *Erythrobacter*[b], *Filomicrobium*[b], *Fulvivirga*[b], *Haliscomenobacter*[b], *Hellea*[b], *Henriciella*[b], *Hyphomonas*[b], *Idiomarina*[b], *Labrenzia*[b], *Lewinella*[b], *Marinoscillum*[b], *Microscilla*[b], *Muricauda*[b], *Nitrotireductor*[b], *Oceaniserpentilla*[b], *Parvularcula*[b], *Pelagibacter*[b], *Phycisphaera*[b], *Phormidium*[b], *Pleurocapsa*[b], *Prochlorococcus*[b], *Pseudoalteromonas*[b], *Pseudomonas*[b], *Psychrobacter*[b], *Rhodovulum*[b], *Rivularia*[b], *Roseovarius*[b], *Rubrimonas*[b], *Sediminibacterium*[b], *Synechococcus*[b], *Thalassobius*[b], *Thiobios*[b], *Tenacibaculum*[b], *Thalassobius*[b], *Vibrio*[b]
Diatoms[a,b,c,d,f]	[a]77.9%[d] 1188 per mm^{-2}	***Amphora***[a], ***Achananthes***[a], *Chaetoceros*[b], ***Cocconeis***[a], *Cyclotella*[c], ***Cymbella***[a], ***Grammatophora***[a], ***Haslea***[a], ***Licmophora***[a], *Mastogloia*[a,c], ***Microtabella***[a], ***Minidiscus***[a], *Navicula*[b], *Nitzschia*[a,b], *Pleurosigma*[c], *Sellaphora*[b], *Stauroneis*[b], ***Thalassionema***[a], ***Thalassiosira***[a]
Coccoliths[a,d,b]	[a]8.8%	***Calcidiscus***[a], ***Emiliania***[a], ***Gephyrocapsa***[a], ***Umbellosphaera***[a], ***Umbilicosphaera***[a], ***Coccolithus***[a], ***Calciosolenia***[a]
Bryozoa[a,e,f]	[a]8.8%	*Membranipora*[f], *Jellyella*[e], *Bowerbankia*[e], *Filicrisia*[e]
Hydroids[c,e]	–	*Clytia*[c], *Gonothyraea*[c], *Obelia*[e]
Polychaete[g]	–	*Spirorbis*[g], *Hydroides*[g]
Dinoflagellates[a,b,d]	[a]1.5%	*Alexandrium*[b], *Ceratium*[a]
Insect eggs[a,h,i]	[a]2.9%	*Halobates*[a,h,i]
Barnacles[a]	[a]2.9%	***Lepas***[a]
Rhodophyta[b,g]	–	*Fosliella*[g]
Foraminifera[g]	–	*Discorbis*[g]
Radiolaria[b,d]	–	*Circorrhegma*[d]
Ciliate[b]	–	*Ephelota*[b]

Organism groups (first column), their abundance and/or frequency of occurrence (when available; second column), and genera (third column). References are indicated by superscript letters and given at the bottom of the table, along with approximate length range of plastics examined. Genera in bold indicate those first detected in this study.
[a]This study (1.7–24.3 mm),
[b]Zettler et al. 2013 (2–20 mm) [27],
[c]Carpenter and Smith 1972 (2.5–5 mm) [22],
[d]Carson et al. 2013 (1–10 mm) [26],
[e]Goldstein et al. 2014 (4–5 mm) [38],
[f]Gregory 1978 (2–5 mm) [24],
[g]Gregory 1983 (1–5 mm) [25],
[h]Majer et al. 2012 (2–5 mm) [74],
[i]Goldstein et al. 2012 (1.2–6.5 mm) [48],
[j]Carpenter et al. 1972 (0.1–2 mm) [23].

substratum (e.g. by protecting against predators and supplying nutrients through flow generated by polypids [73]). Further studies focusing on both epiplastic microorganisms and invertebrates have the potential to further elucidate symbiotic and/or competitive relationships between inhabitants of this new type of pelagic habitat.

In summary, this study showed that millimeter-sized marine plastics are providing a new niche for several types of microorganisms and some invertebrates. This phenomenon has considerable ecological ramifications and deserves further research. As discussed here, additional observational and experimental studies on the inhabitants of these small plastic fragments may better elucidate several key plastic pollution processes that remain poorly assessed, such as at-sea polymer degradation and mineralisation, impacts of epiplastic communities on their consumers, and changes in the distributional range of species by plastic rafting.

Supporting Information

Figure S1 Examples of marine plastics' surface textures. a, d: polypropylene plastics with linear fractures and pits; b, c: higher magnification of the plastic surface shown in 'a' (note very similar pits – one empty and one with a cell conforming its shape); e: higher magnification of the plastic surface shown in 'd' (note three equally spaced deep pits); f: polyethylene soft plastic with linear fractures, producing squared microplastics; g: higher magnification of the plastic surface shown in 'f' (note shallow pits likely formed by *Cocconeis* sp.); h: rounded scrape mark similar to the ones found close to the worm-like animal (see Figure 6i); i,k: sub-parallel scrape marks; j: large plastic pit likely formed by an egg of *Halobates* sp.

Acknowledgments

We thank CSIRO Marine National Facility, Australian Institute of Marine Science (AIMS), and Austral Fisheries for proving us with sea time onboard their vessels, as well as the staff and crew of RV Southern Surveyor, RV Solander, and Comac Enterprise for logistic support during the voyages. The authors also acknowledge the UWA School of Chemistry and Biochemistry, and the Centre for Microscopy, Characterisation and Analysis for their facilities, scientific and technical assistance. A special thank you to Steve Rogers, Susana Agusti, Luana Lins, Piotr Kuklinski, Paco Cardenas, Pat Hutchings, Anja Schulze, Christopher Boyko, George (Buz) Wilson, Marilyn Schotte, John Hooper, Christine Schoenberg, Jean

Vacelet, Andrzej Pisera, Alexander Muir, and John Taylor for help with identification of organisms. We also acknowledge Martin Thiel for valuable suggestions on the manuscript.

Author Contributions

Conceived and designed the experiments: JR JS GH MP DKAB MT CW BDH CP. Performed the experiments: JR JS GH MP DKAB CW. Analyzed the data: JR JS GH MP DKAB. Contributed reagents/materials/analysis tools: JR JS MP MT CW BDH CP. Wrote the paper: JR JS GH MP DKAB.

References

1. Eriksen M, Maximenko N, Thiel M, Cummins A, Lattin G, et al. (2013) Plastic pollution in the South Pacific subtropical gyre. Marine Pollution Bulletin 68: 71–76.
2. Law KL, Morét-Ferguson S, Maximenko NA, Proskurowski G, Peacock EE, et al. (2010) Plastic accumulation in the North Atlantic subtropical gyre. Science 329: 1185–1188.
3. Moore CJ, Moore SL, Leecaster MK, Weisberg SB (2001) A comparison of plastic and plankton in the North Pacific central gyre. Marine Pollution Bulletin 42: 1297–1300.
4. Reisser J, Shaw J, Wilcox C, Hardesty BD, Proietti M, et al. (2013) Marine plastic pollution in waters around Australia: characteristics, concentrations, and pathways. PLOS ONE 8(11): e80466.
5. Hidalgo-Ruz V, Gutow L, Thompson RC, Thiel M (2012) Microplastics in the marine environment: a review of the methods used for identification and quantification. Environmental Science & Technology 46: 3060–3075.
6. Morét-Ferguson S, Law KL, Proskurowski G, Murphy EK, Peacock EE, et al. (2010) The size, mass, and composition of plastic debris in the western North Atlantic Ocean. Marine Pollution Bulletin 60: 1873–1878.
7. Law KL, Moret-Ferguson S, Goodwin DS, Zettler ER, DeForce E, et al. (2014) Distribution of surface plastic debris in the eastern Pacific Ocean from an 11-year dataset. Environmental Science & Technology 48: 4732–4738.
8. Teuten EL, Saquing JM, Knappe DR, Barlaz MA, Jonsson S, et al. (2009) Transport and release of chemicals from plastics to the environment and to wildlife. Philosophical Transactions of the Royal Society B: Biological Sciences 364: 2027–2045.
9. Rochman CM, Browne MA, Halpern BS, Hentschel BT, Hoh E, et al. (2013) Policy: Classify plastic waste as hazardous. Nature 494: 169–171.
10. Mato Y, Isobe T, Takada H, Kanehiro H, Ohtake C, et al. (2001) Plastic resin pellets as a transport medium for toxic chemicals in the marine environment. Environmental Science & Technology 35: 318–324.
11. Rios LM, Moore C, Jones PR (2007) Persistent organic pollutants carried by synthetic polymers in the ocean environment. Marine Pollution Bulletin 54: 1230–1237.
12. Gassel M, Harwani S, Park J-S, Jahn A (2013) Detection of nonylphenol and persistent organic pollutants in fish from the North Pacific Central Gyre. Marine Pollution Bulletin 64: 2374–2379.
13. Basheer C, Lee HK, Tan KS (2004) Endocrine disrupting alkylphenols and bisphenol-A in coastal waters and supermarket seafood from Singapore. Marine Pollution Bulletin 48: 1161–1167.
14. Boerger CM, Lattin GL, Moore SL, Moore CJ (2010) Plastic ingestion by planktivorous fishes in the North Pacific Central Gyre. Marine Pollution Bulletin 60: 2275–2278.
15. Choy CA, Drazen JC (2013) Plastic for dinner? Observations of frequent debris ingestion by pelagic predatory fishes from the central North Pacific. Marine Ecology Progress Series 485: 155–163.
16. Cole M, Lindeque P, Fileman E, Halsband C, Goodhead RM, et al. (2013) Microplastic ingestion by zooplankton. Environmental Science & Technology 47: 6646–6655.
17. Fossi MC, Panti C, Guerranti C, Coppola D, Giannetti M, et al. (2012) Are baleen whales exposed to the threat of microplastics? A case study of the Mediterranean fin whale (Balaenoptera physalus). Marine Pollution Bulletin 64: 2374–2379.
18. von Moos N, Burkhardt-Holm P, Köhler A (2012) Uptake and effects of microplastics on cells and tissue of the blue mussel *Mytilus edulis* L. after an experimental exposure. Environmental Science & Technology 46: 11327–11335.
19. Rochman CM, Hoh E, Kurobe T, Teh S (2013) Ingested plastic transfers hazardous chemicals to fish and induces hepatic stress. Nature 3: 3263.
20. Wright SL, Thompson RC, Galloway TS (2013) The physical impacts of microplastics on marine organisms: A review. Environmental Pollution 178: 483–492.
21. Carson HS, Colbert SL, Kaylor MJ, McDermid KJ (2011) Small plastic debris changes water movement and heat transfer through beach sediments. Marine Pollution Bulletin 62: 1708–1713.
22. Carpenter EJ, Smith K (1972) Plastics on the Sargasso Sea Surface. Science 175: 1240–1241.
23. Carpenter EJ, Anderson SJ, Harvey GR, Miklas HP, Peck BB (1972) Polystyrene spherules in coastal waters. Science 178: 749–750.
24. Gregory MR (1978) Accumulation and distribution of virgin plastic granules on New Zealand beaches. New Zealand Journal of Marine and Freshwater Research 12: 399–414.
25. Gregory MR (1983) Virgin plastic granules on some beaches of eastern Canada and Bermuda. Marine Environmental Research 10: 73–92.
26. Carson HS, Nerheim MS, Carroll KA, Eriksen M (2013) The plastic-associated microorganisms of the North Pacific Gyre. Marine Pollution Bulletin 75: 126–132.
27. Zettler ER, Mincer TJ, Amaral-Zettler LA (2013) Life in the 'Plastisphere': Microbial communities on plastic marine debris. Environmental Science & Technology 47: 7137–7146.
28. Winston JE (1982) Drift plastic–An expanding niche for a marine invertebrate? Marine Pollution Bulletin 13: 348–351.
29. Jokiel PL (1990) Long-distance dispersal by rafting: reemergence of an old hypothesis. Endeavour 14: 66–73.
30. Barnes DK (2002) Biodiversity: invasions by marine life on plastic debris. Nature 416: 808–809.
31. Masó M, Garcés E, Pagès F, Camp J (2003) Drifting plastic debris as a potential vector for dispersing Harmful Algal Bloom (HAB) species. Scientia Marina 67: 107–111.
32. Barnes DK, Fraser KP (2003) Rafting by five phyla on man-made flotsam in the Southern Ocean. Marine Ecology Progress Series 262: 289–291.
33. Barnes D (2004) Natural and plastic flotsam stranding in the Indian Ocean. The Effects of Human Transport on Ecosystems: Cars and Planes, Boats and Trains Davenport, D & Davenport, J(Eds) Royal Irish Academy, Dublin: 193–205.
34. Gregory MR (2009) Environmental implications of plastic debris in marine settings–entanglement, ingestion, smothering, hangers-on, hitch-hiking and alien invasions. Philosophical Transactions of the Royal Society B: Biological Sciences 364: 2013–2025.
35. Fortuño J, Masó M, Sáez R, De Juan S, Demestre M (2010) SEM microphotographs of biofouling organisms on floating and benthic plastic debris. Rapp Comm int Mer Médit 39: 358.
36. Thiel M, Gutow L (2005) The ecology of rafting in the marine environment. II. The rafting organisms and community. Oceanography and Marine Biology: An Annual Review 43: 279–418.
37. Aliani S, Molcard A (2003) Hitch-hiking on floating marine debris: macrobenthic species in the Western Mediterranean Sea. Migrations and Dispersal of Marine Organisms: Springer. 59–67.
38. Goldstein MC, Carson HS, Eriksen M (2014) Relationship of diversity and habitat area in North Pacific plastic-associated rafting communities. Marine Biology 161: 1441–1453.
39. Balasubramanian V, Natarajan K, Hemambika B, Ramesh N, Sumathi C, et al. (2010) High-density polyethylene (HDPE)-degrading potential bacteria from marine ecosystem of Gulf of Mannar, India. Letters in Applied Microbiology 51: 205–211.
40. Artham T, Doble M (2009) Fouling and degradation of polycarbonate in seawater: field and lab studies. Journal of Polymers and the Environment 17: 170–180.
41. Sudhakar M, Trishul A, Doble M, Suresh Kumar K, Syed Jahan S, et al. (2007) Biofouling and biodegradation of polyolefins in ocean waters. Polymer Degradation and Stability 92: 1743–1752.
42. Harshvardhan K, Jha B (2013) Biodegradation of low-density polyethylene by marine bacteria from pelagic waters, Arabian Sea, India. Marine Pollution Bulletin 77: 100–106.
43. Harrison JP, Sapp M, Schratzberger M, Osborn AM (2011) Interactions between microorganisms and marine microplastics: a call for research. Marine Technology Society Journal 45: 12–20.
44. Andrady AL (2011) Microplastics in the marine environment. Marine Pollution Bulletin 62: 1596–1605.
45. Lobelle D, Cunliffe M (2011) Early microbial biofilm formation on marine plastic debris. Marine Pollution Bulletin 62: 197–200.
46. Ye S, Andrady AL (1991) Fouling of floating plastic debris under Biscayne Bay exposure conditions. Marine Pollution Bulletin 22: 608–613.
47. Pham P, Jung J, Lumsden J, Dixon B, Bols N (2012) The potential of waste items in aquatic environments to act as fomites for viral haemorrhagic septicaemia virus. Journal of Fish Diseases 35: 73–77.
48. Goldstein MC, Rosenberg M, Cheng L (2012) Increased oceanic microplastic debris enhances oviposition in an endemic pelagic insect. Biology Letters 8: 817–820.
49. Reisser J, Shaw J, Wilcox C, Hardesty BD, Proietti M, et al. (2013) Data from: Marine Plastic Pollution in waters around Australia: characteristics, concentrations, and pathways. Figshare.
50. Cheng L (1976) Marine Insects. University of California: 581p.
51. Hallegraeff GM, Bolch C, Hill D, Jameson I, LeRoi J, et al. (2010) Algae of Australia: phytoplankton of temperate coastal waters. Australia. 421 p.
52. Kensley B (1994) Redescription of *Iais elongata* Sivertsen & Holthuis, 1980 from the South Atlantic Ocean (Crustacea: Isopoda: Asellota). Proceedings of the Biological Society of Washington 107: 274–282.

53. Magalhães W, Bailey-Brock J (2012) Capitellidae Grube, 1862 (Annelida: Polychaeta) from the Hawaiian Islands with description of two new species. Zootaxa 3581: 1–52.
54. Weise W, Rheinheimer G (1977) Scanning electron microscopy and epifluorescence investigation of bacterial colonization of marine sand sediments. Microbial Ecology 4: 175–188.
55. Ferreira T, Rasband W (2012) ImageJ User Guide. 198.
56. Paulrud S, Mattsson JE, Nilsson C (2002) Particle and handling characteristics of wood fuel powder: effects of different mills. Fuel Processing Technology 76: 23–39.
57. Cooper DA, Corcoran PL (2010) Effects of mechanical and chemical processes on the degradation of plastic beach debris on the island of Kauai, Hawaii. Marine Pollution Bulletin 60: 650–654.
58. Corcoran PL, Biesinger MC, Grifi M (2009) Plastics and beaches: A degrading relationship. Marine Pollution Bulletin 58: 80–84.
59. Reisser J, Shaw J, Hallegraeff G, Proietti M, Barnes D, et al. (2014) Data from: Millimeter-sized Marine Plastics: A New Pelagic Habitat for Microorganisms and Invertebrates. Figshare.
60. Congestri R, Albertano P (2011) Benthic Diatoms in Biofilm Culture. The Diatom World: Springer. 227–243.
61. Tiffany MA (2011) Epizoic and Epiphytic Diatoms. The Diatom World: Springer. 195–209.
62. Carpenter EJ (1970) Diatoms attached to floating *Sargassum* in the western Sargasso Sea 1. Phycologia 9: 269–274.
63. Reisser J, Proietti M, Sazima I (2010) First record of the silver porgy (*Diplodus argenteus*) cleaning green turtles (*Chelonia mydas*) in the south-west Atlantic. Marine Biodiversity Records 3: e75.
64. Romagnoli T, Totti C, Accoroni S, De Stefano M, Pennesi C (2014) SEM analysis of the epibenthic diatoms on *Eudendrium racemosum* (Hydrozoa) from the Mediterranean Sea. Turkish Journal of Botany 38.
65. Totti C, Poulin M, Romagnoli T, Perrone C, Pennesi C, et al. (2009) Epiphytic diatom communities on intertidal seaweeds from Iceland. Polar Biology 32: 1681–1691.
66. Polovina JJ, Howell EA, Abecassis M (2008) Ocean's least productive waters are expanding. Geophysical Research Letters 35.
67. Kawamura T, Saido T, Takami H, Yamashita Y (1995) Dietary value of benthic diatoms for the growth of post-larval abalone *Haliotis discus hannai*. Journal of Experimental Marine Biology and Ecology 194: 189–199.
68. Sangale M, Shahnawaz M, Ade A (2012) A review on biodegradation of polythene: the microbial approach. J Bioremed Biodeg 3.
69. Sivan A (2011) New perspectives in plastic biodegradation. Current Opinion in Biotechnology 22: 422–426.
70. Webb HK, Arnott J, Crawford RJ, Ivanova EP (2012) Plastic degradation and its environmental implications with special reference to poly (ethylene terephthalate). Polymers 5: 1–18.
71. Scholz J, Hillmer G (1995) Reef-bryozoans and bryozoan-microreefs: control factor evidence from the Philippines and other regions. Facies 32: 109–143.
72. Kittelmann S, Harder T (2005) Species-and site-specific bacterial communities associated with four encrusting bryozoans from the North Sea, Germany. Journal of Experimental Marine Biology and Ecology 327: 201–209.
73. Wuchter C, Marquardt J, Krumbein WE (2003) The epizoic diatom community on four bryozoan species from Helgoland (German Bight, North Sea). Helgoland Marine Research 57: 13–19.
74. Majer A, Vedolin M, Turra A (2012) Plastic pellets as oviposition site and means of dispersal for the ocean-skater insect *Halobates*. Marine Pollution Bulletin 64: 1143–1147.

8

Tiny Is Mighty: Seagrass Beds Have a Large Role in the Export of Organic Material in the Tropical Coastal Zone

Lucy G. Gillis[1]*, Alan D. Ziegler[2], Dick van Oevelen[3], Cecile Cathalot[4], Peter M. J. Herman[1], Jan W. Wolters[5], Tjeerd J. Bouma[1]

1 Spatial Ecology Department, Royal Netherlands Institute for Sea Research (NIOZ), Yerseke, Zealand, The Netherlands, **2** Geography Department, National University of Singapore (NUS), Singapore, Singapore, **3** Ecosystems Studies Department, Royal Netherlands Institute for Sea Research (NIOZ), Yerseke, Zealand, The Netherlands, **4** Laboratoire Environnement Profond (LEP), French Research Institute for Exploitation of the Sea, Polouzane, Brittany, France, **5** Department of Biology, University of Antwerp, Antwerp, Flanders, Belgium

Abstract

Ecosystems in the tropical coastal zone exchange particulate organic matter (POM) with adjacent systems, but differences in this function among ecosystems remain poorly quantified. Seagrass beds are often a relatively small section of this coastal zone, but have a potentially much larger ecological influence than suggested by their surface area. Using stable isotopes as tracers of oceanic, terrestrial, mangrove and seagrass sources, we investigated the origin of particulate organic matter in nine mangrove bays around the island of Phuket (Thailand). We used a linear mixing model based on bulk organic carbon, total nitrogen and $\delta^{13}C$ and $\delta^{15}N$ and found that oceanic sources dominated suspended particulate organic matter samples along the mangrove-seagrass-ocean gradient. Sediment trap samples showed contributions from four sources oceanic, mangrove forest/terrestrial and seagrass beds where oceanic had the strongest contribution and seagrass beds the smallest. Based on ecosystem area, however, the contribution of suspended particulate organic matter derived from seagrass beds was disproportionally high, relative to the entire area occupied by mangrove forests, the catchment area (terrestrial) and seagrass beds. The contribution from mangrove forests was approximately equal to their surface area, whereas terrestrial contributions to suspended organic matter under contributed compared to their relative catchment area. Interestingly, mangrove forest contribution at 0 m on the transects showed a positive relationship with the exposed frontal width of the mangrove, indicating that mangrove forest exposure to hydrodynamic energy may be a controlling factor in mangrove outwelling. However we found no relationship between seagrass bed contribution and any physical factors, which we measured. Our results indicate that although seagrass beds occupy a relatively small area of the coastal zone, their role in the export of organic matter is disproportional and should be considered in coastal management especially with respect to their importance as a nutrient source for other ecosystems and organisms.

Editor: David William Pond, Scottish Association for Marine Science, United Kingdom

Funding: This research was funded by Ecoshape, Building with Nature (SDWI/Eco-Shape/Building with Nature grant R-303-001-020-414). The funders had no role in study design, data collection and analysis, decision to publish, or preparation of the manuscript.

Competing Interests: The authors have declared that no competing interests exist.

* Email: lucygwen.gillis@nioz.nl

Introduction

The tropical coastal zone is composed of ecosystems emerging from the open ocean, extending over coral reefs and seagrass beds towards the tidal zone, which may include mangrove forests. This zone forms a well-structured and gradual interface between the land and the sea that contains some of the most productive and biogeochemically active ecosystems in the world [1]. Part of this productivity is maintained by particulate and dissolved organic matter (POM and DOM) inputs from both terrestrial and oceanic sources [2].

The importance of external inputs to the coastal zone and the exchanges between coastal ecosystems depends on various factors. For example, the quantity of terrestrial organic matter (OM) input to the coastal zone has been shown to depend on the size of the catchment area and the land-use within [3]. Tidally-dominated estuaries have large exchanges of water with the ocean, and therefore, receive ocean-derived DOM and POM [4–6]. The recognized mangrove outwelling hypothesis [7–8] postulates that export of mangrove-derive OM supports adjacent ecosystems and food webs.

Another potentially important OM source and sink within the tropical coastal zone are seagrass beds. Seagrass plants trap suspended POM originating from both external sources and from leaf shedding inside the seagrass bed [9–10]. Trapping results directly from the physical structure of the bed and from settling induced by changes in the hydrodynamic conditions - both are related in part to plant density and leaf characteristics [11–12]. Once POM is deposited within seagrass beds, the matter is 'protected' from re-suspension, and consequently, most POM degrades within the bed, releasing dissolved nutrients that may in turn be taken up by the seagrass [13–15] or released to the water column [9,16]. Seagrass beds however also export particulate nutrients via leaf shedding enhanced during strong hydrodynamic events and via marine herbivore consumption [17].

Due to the different ecosystems in the seascape, the high production of ecosystems and the potential for import and export from a variety of sources, it is challenging to constrain the contribution of the different resources to the exchange of or OM in the tropical coastal zone. One classical approach is to use a chemical tracer mixing model based on a combination of carbon to nitrogen ratios (C:N) with carbon and nitrogen stable isotope ratios (δ^{13}C and δ^{15} N, respectively) [18–19]. Each of these tracers has a specific signature for each source (oceanic, seagrass, mangrove and terrestrial), which can be used to identify the contribution of the different sources in a mixture [18–19].

Investigations of source contributions in Southern Thailand and found that seagrasses contributed between 36–42% of OM to sediment trap samples [4]. Other studies in Gazi Bay, Kenya discovered that mangrove and seagrass matter dominated suspended POM in the water column [17,20]. These studies indicate that contributions from individual sources likely depend on the local conditions. In this paper, we build upon the foundations of previous work to investigate the origin of POM in trap and suspended sediment samples in the coastal zone in Phang-nga Bay in southern Thailand. We studied nine sites with contrasting physical attributes of bays (width and length), of ecosystems (aerial extent and width) and of catchment areas (aerial extent and land use). This allowed us to gain a greater understanding of export of POM from different ecosystems at the tropical seascape scale. Specifically, we compared the amount of OM originating from seagrass beds versus mangrove, oceanic, and terrestrial sources across a number of different sub-habitats in the region, comparing the contributions to their respective surface areas.

Methods

Ethics statement

The work was conducted in collaboration with Rajabhat University (Phuket) on public lands. No permit was required for sampling. No live samples were collected.

Study Area and Physical attributes

Within the Andaman Sea, along the southwest coast of Thailand is the province of Phuket, which borders Phang-nga bay (Figure 1). Our sampling sites are located in Phang-nga Bay, which is 68 km long (head to mouth) and has a surface area of 3000 km^2. Mean tidal range is 1.8 m (Figure 1). The southern part of the bay is open to the Andaman Sea and the northwest area is open to the sea via the Pak Pra Inlet, which separates Phuket Island from the mainland (Figure 1). Estuarine salinity conditions dominate in the north, whilst marine conditions dominate in the south [21]. Circulation in the bay changes depending on the dry (May to October) or wet season (November to April), which alters the wind direction from northeast in the dry season to southwest in the wet season. Mean annual rainfall is about 2300 mm and mean temperature is 28°C.

Land-use around Phang-nga Bay changed from natural forest to initially tin mining (1600–1800) [21]. Other land-uses have gained greater importance such as rubber and palm oil plantations as well as shrimp (located on land) and fish farms (situated in coastal zones) [21]. In recent decades, rapid urbanisation related to tourism has occurred in Phuket.

Samples were collected from all the fully accessible mangrove sites located in the west of Phang-nga Bay, on the islands of Phuket, Yao Yai and Yao Noi. A total of 9 sites were sampled (Figure 1) during March 2011. All sites were outside the reach of direct river discharge and experienced a strong tidal exchange (tidal range 0.5–2.5 m) with the ocean. Therefore, all sites are classified as tidal mangroves [22].

Sample collection and data processing

At each site we established transects beginning at the edge of the mangrove, extending towards the ocean through an adjacent intertidal seagrass patch. Samples were taken, when logistically possible, at distances of 0 (seaward edge of mangrove forest), 50, 100, 200, 300, 400, 500, 1000, 1500 and 2000 m along the transect, perpendicular to the seaward edge of the mangrove forest towards the open sea. Two types of samples were collected: trap samples of suspended matter above the bed (0.05 m) and suspended particulate matter in the water column. At least half of the sampling points (for trap and water samples) were located in the seagrass beds; the other half were taken at points landward or seaward of the bed, depending on the physical constraints of the different sites. Sediment traps at each point were secured to steel rods and anchored to the substrate at a height of 0.05 m above the sediment. They were emptied once after approximately 24 hours. On the day of trap installation, 2-liter surface water samples were taken at each point above the traps at the water surface during high tide.

Four POM sources were considered in this study: terrestrial vegetation, mangrove leaves, seagrass leaves and oceanic plankton. For the terrestrial vegetation, three replicates of 3–4 leaves from rice paddies, rubber trees (*Hevea sp.*) and native vegetation (*Delonix sp.* and *Terminalia sp.*) were collected on the island of Koh Yai. These were considered to be representative for terrestrial vegetation of the province. This matter represents the source value for the terrestrial OM used in the mixing model. Mangrove and seagrass leaves were collected during the trap deployment. Plankton and suspended particulate matter (SPM) samples were collected at a point believed to have the majority of oceanic influence (7°52.573′N and 98°35.635′E) and taken as proxy for oceanic matter. Plankton samples were collected with weighted nets of mesh sizes 400 µm and SPM with a five 1-litre water sample. The mean of the C and N content and δ^{13}C and δ^{15} N value of each POM source was taken as the respective end-member to determine the contribution of each source to the OM along the transect.

All solid samples (sediment and leaves) were placed in separate sample bags, stored in a cooling box and transported immediately to the field laboratory where they were dried at 60°C for 48 hours. Water samples for SPM and plankton were filtered through pre-combusted glass fiber filters (GF/F, 0.7 µm pore diameter). The GF/F filters were dried at 60°C for 48 hours. Both solid material (sediment and leaves) and GF/F filters were packed in airtight containers and transported to the laboratory of Royal Netherlands Institute for Sea Research (NIOZ) for elemental and isotope analysis.

Elemental and isotope analysis

Sediment and leaf material samples were ground for homogenisation. The trap, plankton and SPM samples were acidified to remove carbonates [23]; the leaf samples were not acidified. All samples were analysed for total organic carbon (TOC), total nitrogen (TN), and the isotopes δ^{13}C and δ^{15}N by means of elemental analysis isotope ratio mass spectrometry (EA-IRMS) using a Thermo Finnigan Flash 1112.

Stable isotope ratios are expressed as δ values (‰) relative to conventional standards (VPDB limestone for C and atmospheric N_2 for N):

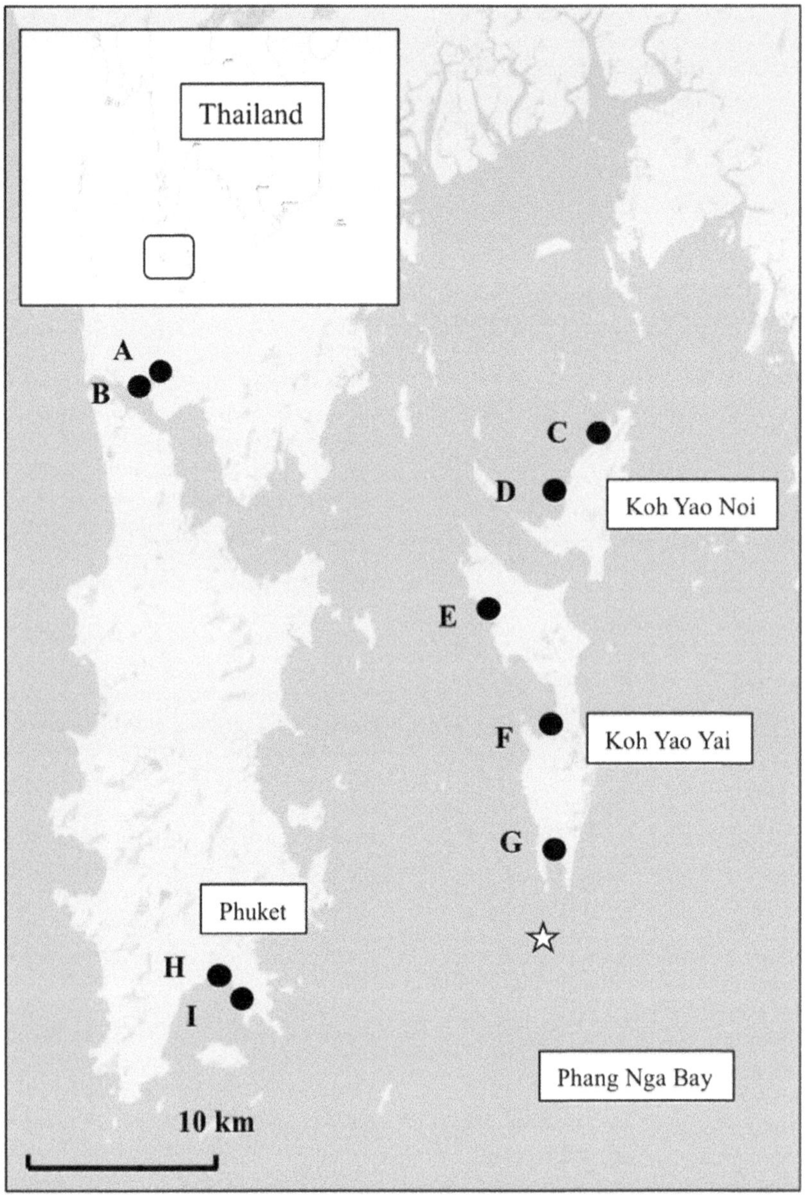

Figure 1. Map showing location of the nine sampling sites (black circles) in Phang Nga Bay, southern Thailand. The white star represents the point oceanic samples where collected.

$$\delta X = ((X_{sample} - X_{standard})/X_{standard}) * 1000(‰) \quad (1)$$

where δX is either $\delta^{13}C$ or $\delta^{15}N$; X_{sample} is the $^{13}C/^{12}C$ or $^{15}N/^{14}N$ ratio of the sample; $X_{standard}$ is the $^{13}C/^{12}C$ or $^{15}N/^{14}N$ ratio of the standards (0.01118 for C and 0.00368 for N).

Mixing model

We considered the following potential sources (Table 1): terrestrial vegetation (*Delonix sp.*, *Terminalia sp.* and *Hevea sp.*), mangrove leaves (*Rhizophora sp.*), seagrass leaves (*Enhalus sp.*, *Halodule sp.* and *Halophile sp.*) and oceanic production (plankton and SPM). Mean values were taken as end member for each source (Table 1). Only SPM and trap samples were analysed with the mixing model (as discussed below).

A linear mixing model was used to determine the contributions of the different sources to the SPM and trap samples [24–25]. The mixing model uses the means of $\delta^{13}C$, $\delta^{15}N$ and the C:N ratios (corrected for differences in %C and %N between sources) of individual trap or SPM samples to determine the contribution (f) of each of the four sources (Table 1) to the mixture. The variation between replicates is small justifying the use of the means. The mixing model has the following four equations that are solved simultaneously to recover a unique solution of the four source contributions (f_{oce}, f_{ter}, f_{man}, f_{sea}):

$$f_{oce} + f_{ter} + f_{man} + f_{sea} = 1 \quad (2)$$

Table 1. Mean and standard error values of isotopes carbon isotope ($\delta^{13}C$), nitrogen isotope ($\delta^{15}N$), bulk organic carbon (C) and total nitogen (N) for terrestrial trees, mangroves trees, seagrass plants and oceanic sources.

Sources	Species	$\delta^{13}C$		$\delta^{15}N$		C		N		n
Group		Mean (°/₀₀)	SE	Mean (°/₀₀)	SE	Mean	SE	Mean	SE	
Terrestrial	*Havea sp.*	−31.8	0.5	0.4	0.001	46.8	0.2	4.2	0.04	3
	Delomix sp.	−30.1	0.5	0.6	0.001	46.9	0.3	2.6	0.2	3
	Terminalia sp.	−31.2	0.5	1.9	0.001	44.8	0.2	3.6	0.006	3
Overall terrestrial	-	**−31.0**	**0.4**	**1.0**	**0.0004**	**46.1**	**0.2**	**4.01**	**0.08**	**3**
Overall Mangrove trees	***Rhizophora sp.***	**−26.9**	**0.6**	**3.6**	**0.2**	**37.5**	**0.6**	**1.2**	**0.07**	**45**
Seagrass plants	*Halodule sp.*	−11.4	0.7	2.9	0.15	21.4	2.6	1.4	0.2	7
	Halophile sp.	−11.6	0.5	2.8	0.9	21.4	2.1	1.3	0.2	19
	Enhalus sp.	−10.1	0.3	5.2	0.3	25.6	1.6	1.9	0.1	17
Overall seagrass plants	-	**−11.0**	**0.6**	**3.6**	**0.2**	**21.7**	**1.4**	**1.5**	**0.1**	**3**
Oceanic	SPM	−24.1	0.13	18.3	3	1.2	0.07	0.2	0.01	3
	Plankton	−24	0.04	7.9	0.03	4.6	0.5	0.8	0.1	3
Overall oceanic	-	**−24.1**	**0.06**	**13.1**	**3.5**	**2.9**	**0.8**	**0.5**	**0.1**	**6**

The values for each source (highlighted) are the means of the different contributions (species for terresrtial, mangrove and seagrasses & SPM/plankton for oceanic). These values are defined as the end members (organic matter sources), which are used in the mixing model to determine the different mixture of fractions for organic matter. The organic matter was collected in the core, sediment trap and suspended sediment matter samples. *n*, is the number of samples and SE is standard error.

$$\delta^{13}C_{mix} = \delta^{13}C_{oce} \cdot f_{oce} + \delta^{13}C_{ter} \cdot f_{ter} + \delta^{13}C_{man} \cdot f_{man} + \delta^{13}C_{sea} \cdot f_{sea} \quad (3)$$

$$\delta^{15}N_{mix} = \delta^{15}N_{oce} \cdot f_{oce} + \delta^{15}N_{ter} \cdot f_{ter} + \delta^{15}N_{man} \cdot f_{man} + \delta^{15}N_{sea} \cdot f_{sea} \quad (4)$$

$$0 = (\%C_{oce} - CN_{mix} \cdot \%N_{oca}) \cdot f_{oce} + (\%C_{ter} - CN_{mix} \cdot \%N_{ter}) \cdot f_{ter} + (\%C_{man} - CN_{mix} \cdot \%N_{man}) \cdot f_{man} + (\%C_{sea} - CN_{mix} \cdot \%N_{sea}) \cdot f_{sea} \quad (5)$$

Where oce = oceanic, ter = terrestrial, man = mangrove trees and sea = seagrass plants.The model was implemented and solved in R [26], using the *lsei* function available in the LIM package [27]. Using the contribution for each source as returned by the mixing model, we calculated if each source was under-contributing or over-contributing compared to its relative surface:

$$F_x / A_{fx} \quad (6)$$

This index was not calculated for POM derived from oceanic sources, because no representative surface area for the 'ocean' could be defined. The contribution (F_x; %) per source in each sample was then divided by the percent of total surface area (A_{fx}) occupied by the particular system (catchment area-terrestrial, mangrove forest, seagrass bed). This gives a dimensionless number, for which 1 implies that the ecosystem contributes proportionally to its relative surface area; values below or above 1 indicates that the proportional contribution is lower or higher, respectively.

Statistical analysis

From the results of equation 6 we used step-wise regression analysis to determine if the contribution of each resource was related to relevant physical aspects of the nine sites. For this analysis we only used the mixing model results at 0 and 100 m along the transect, as data for these distances were available across all nine sites. Contributions from terrestrial, mangrove and seagrass to SPM and trap samples were correlated with physical attributes of each site. The mangrove contribution was tested against area of mangrove forest (m^2), area of bay (m^2), urbanisation in the catchment area (%) and width of mangrove forest (m) (Table 2). Terrestrial contribution was analysed against catchment area (m^2), area of bay (m^2) and urbanisation in the catchment area (%) (Table 2). Finally, seagrass was tested with area of the seagrass bed (m^2), area of bay (m^2) and urbanisation in the catchment area (%) (Table 2).

The physical characteristics (bay area, bay width/length), surface areas of the mangrove forests and seagrass beds were determined with ground truthing during the sampling campaign. Land use in the catchment area, width of the bay, and length of the bay were determined by analysing Quickbird, WorldView-1 and WorldView-2 satellite imagery. All statistical analysis was completed in R [26], where probabilities (p) <0.05 were considered significant.

Results

All samples (both trap and SPM) along the gradient showed different contributions of the different sources of OM (Figure 2). At sites A and B, the signatures for the majority of samples were between those of terrestrial, mangrove and oceanic sources (Figure 2). The trap sample signatures from sites C and I were spread from seagrass to the oceanic end members (Figure 2). At sites D and F, the trap and SPM samples indicated a pattern from terrestrial and mangrove sources to the mid-point between all the sources (Figure 2). The majority of site E trap samples similarly

Table 2. Physical description of the sites and the physical attributes of the marine ecosystems for each of the nine sites (A to I) and their associated bay.

Physical attributes	Site								
	A	B	C	D	E	F	G	H	I
Urbanized area (km^2)	1.4	1.4	0.9	0.4	0.1	0.1	0.6	0.7	5.8
Mangrove forest area (km^2)	3.3	3.3	10	0.6	7.5	3.2	2	9	4.7
Catchment area (km^2)	29.6	29.6	3	0.2	0.6	0.7	1.5	2.1	0.4
Seagrass bed area (km^2)	1	0.5	0.1	0.3	0.8	0.04	0.8	0.9	0.8
Width of bay (km)	0.97	0.97	NA	1.77	1.44	3.76	1.22	2	2
Length of bay (km)	1.47	1.47	NA	1.11	0.72	3.31	2.94	1.06	1.06
Width of mangrove forest (km)	3.4	3.4	0.06	0.71	1.18	0.89	1.84	0.96	0.31

NA means that these sites are situated in an open coast, all other sites are bays.

plotted in the centre of the sources; and sites C, E and I SPM samples followed this pattern too (Figure 2). Terrestrial sources dominated trap samples from sites G and H, but in the case of SPM samples they where spaced between terrestrial and mangrove end numbers and the oceanic end number (Figure 2).

The mixing model determined the fraction of each source end-member (mangrove and seagrass plants, terrestrial and oceanic) in the mean of 2–3 individual sediment traps and SPM samples for the three sites (Table 3). The majority of SPM samples (for sites B, C, D, F, G, H and I) showed a spatial pattern for oceanic sources, where the oceanic source contribution increased from 0 m to the end of the transect (Table 3). This pattern (for oceanic sources) was similarly repeated in six of the trap samples (sites A, B, C, D, F and I) (Table 3). The bulk of mangrove plant contributions for SPM samples showed a spatial pattern where the source influence decreased with distance from 0 m to the end of the transect (B, C, D, E, F, G, I). This same pattern was seen in the trap samples (A, B, C, D, E, H and I) (Table 3). Only approximately half of terrestrial plant contributions for SPM samples showed a gradient through the transect (C, D, H and I) (Table 3). However, over half of trap samples (A, B, C, D, G and I) did show a spatial pattern, where terrestrial plant contribution was higher nearer the landward side of the transect (Table 3). The seagrass contribution was highest in the seagrass beds for SPM samples at sites A, B, F and for trap samples at sites E and F (Figures 3, 4 and Table 3). The remainder of the trap and SPM samples showed no spatial pattern in the seagrass plant contribution (Figures 3, 4 and Table 3). For SPM samples, variation of contributions across the transect was very small (0.8–6%); and this was repeated for trap samples (0.5–12%) (Table 3).

Step-wise regression indicated only two significant correlations. A negative correlation between terrestrial trap contributions (y) and catchment area (x) at 100 m ($y = -0.78x+27.2$, $R^2 = 0.5$, $p = 0.04$) and a positive correlation were shown for mangrove contribution to trap samples at 0 m contributions with width of mangrove forest ($y = 12.8x+2.8$, $R^2 = 0.6$, $p = 0.03$).

Discussion

This study was designed to clarify the origin and exchange of different OM sources within tropical coastal waters. Despite the high local primary production in mangrove and seagrass stands, we found that oceanic sources dominated trap and SPM samples along the entire transect. The mixing model results however showed that mangrove forests and terrestrial sources did contribute, especially to the trap samples. Interestingly we found a correlation between the width of the mangrove forest and their contribution, which provides new insights into physical processes (such as exposure to hydrodynamics) associated with mangrove outwelling.

However when taking into account the relative size of the ecosystem (mangrove forest, catchment area and seagrass beds) we found that seagrass beds, although occupying the smallest area of all the potential sources, contributed significantly to the OM in stations along land- to seaward transects. This provides further evidence of their importance as a nutrient (dissolved and particulate) source in the tropical coastal seascape. Before moving to the implications of these findings, we first discuss the sampling and modeling approach that we undertook.

Source contributions to organic matter

Mixing models are often used to estimate contributions to a mixture [24–25]. Some samples could not be solved with the model, which indicates that they violate some of the implicit assumptions of the linear mixing model formulation, such as imprecise measurement of end-members values, the existence of unidentified sources, or degradation processes that have altered the isotope or C:N ratio of the OM. For our study site, we are confident that we included the dominant POM sources in our design and the end-member values were consistent and estimated with a low uncertainty (Table 1). We therefore believe that the few samples that fell outside the mixing polygons (Figure 2) had already undergone substantial biogeochemical modification that lead to a decrease of $\delta^{15}N$ values and/or N content through mineralization and denitrification [4,28]. We think that the trap samples that fell out of the end-member polygon contained predominantly re-suspended bottom sediment (not unlikely given the close proximity to the sediment). Re-suspension of bottom sediment is a potential source of matter caught in the trap that had been subjected to biogeochemical modification.

There was a strong oceanic signal in the trap samples for the nine sites, suggesting that oceanic-derived OM settled out (Figure 2). The SPM and trap samples indicated a substantial oceanic input, which is in agreement with previously studies on coastal wetlands [29–30]. The OM of the trap samples at sites A, B, G and H showed a strong contribution from mangrove sources. These sites have large channels coming from the inner mangrove area, which may explain the increased contribution of mangrove OM. Sites G, H and I trap samples within the seagrass bed show an strong inclination towards the seagrass plant end-member and

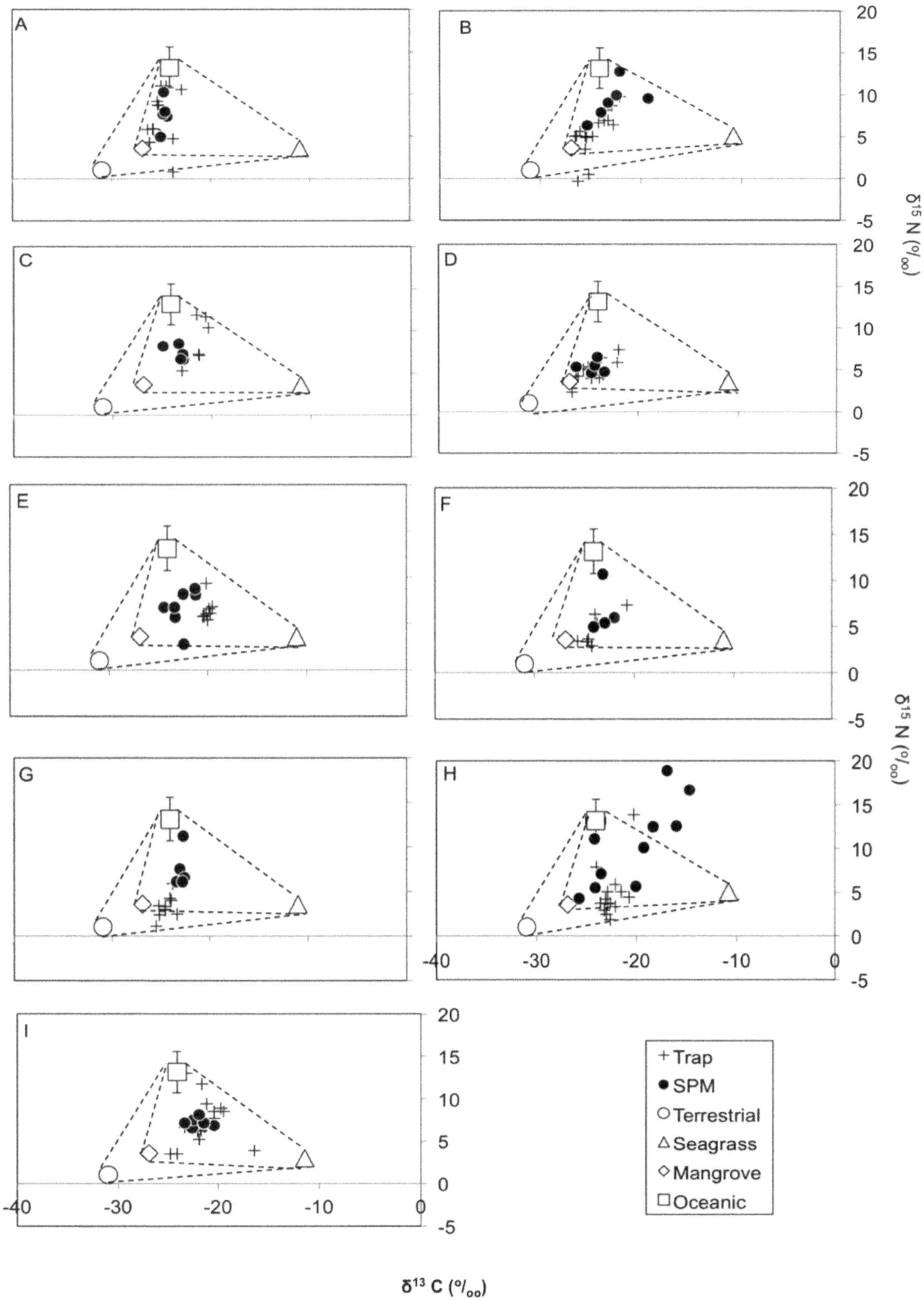

Figure 2. Carbon ($\delta^{13}C$ °/oo) and nitrogen ($\delta^{15}N$ °/oo) isotopic composition for organic matter from trap and suspended particulate matter (SPM) samples. The end members shown are: white circles represent the terrestrial value; white squares represent oceanic sources; white triangles represent seagrass plants; and white diamonds are mangrove trees. The trap and SPM samples are represented by crosses and black circles, respectively. Values for the end members are means (+/− SE). For the end members of the oceanic, terrestrial, seagrass plants and mangrove plants sources, n = 6, 9, 44 and 45, respectively. Note that this is a 2-dimensional representation of isotope values only. Table 3 shows results using all three components, i.e. carbon, nitrogen isotopes and C:N ratios.

Table 3. End member contributions of each source (terrestrial, mangrove, seagrass and oceanic) to each sediment sample (SPM & trap), as determined from the mixing model.

Site	Sample	Distance (m)	0	50	100	200	300	400	500	1000	1500
		Source	**Contribution (%)**								
A	SPM	Mangrove	17	3	1		1^	1			
		Oceanic	66	86	90		83^	85			
		Terrestrial	12	7	5		12^	4			
		Seagrass	5	4	4		4^	10			
	Trap	Mangrove	33	6	1		0.9^	2.8^			
		Oceanic	60	90	97		98^	95^			
		Terrestrial	5	3	1		0.5^	0.8^			
		Seagrass	2	1	1		0.6^	0.9^			
B	SPM	Mangrove	2	2	8	18	11^	5^	ns	ns	ns
		Oceanic	93	94	86	73	84^	88^	ns	ns	ns
		Terrestrial	1	1	3	4	3^	4^	ns	ns	ns
		Seagrass	3	3	3	5	2^	3^	ns	ns	ns
	Trap	Mangrove	65	ns	21	6	6^	1^	14	ns	ns
		Oceanic	22	ns	65	86	91^	95^	80	ns	ns
		Terrestrial	9	ns	9	4	1^	2^	2	ns	ns
		Seagrass	4	ns	5	4	2^	1^	4	ns	ns
C	SPM	Mangrove	13	10	3	11	12^				
		Oceanic	65	78	77	67	78^				
		Terrestrial	10	5	13	11	6^				
		Seagrass	12	7	7	11	4^				
	Trap	Mangrove	5	0.01	3	0.01	1^				
		Oceanic	64	78	60	98	80^				
		Terrestrial	27	20	33	1	17^				
		Seagrass	4	2	4	0.9	2^				
D	SPM	Mangrove	1	0.3^	0.2^	0.1^	0.6^				
		Oceanic	75	85^	79^	75^	83^				
		Terrestrial	23	14^	20^	24^	16^				
		Seagrass	1	0.7^	0.8^	0.9^	0.4^				
	Trap	Mangrove	11	6^	5^	2^	0.2^				
		Oceanic	52	57^	59^	67^	80^				
		Terrestrial	36	35^	33^	29^	19^				
		Seagrass	1	2^	3^	2^	0.8^				
E	SPM	Mangrove	7	1	1	0.2^	ns	ns			
		Oceanic	79	69	34	76^	ns	ns			
		Terrestrial	11	28	63	23^	ns	ns			
		Seagrass	3	1	2	0.8^	ns	ns			
	Trap	Mangrove	3	3	1	1^	2^	3^			
		Oceanic	77	65	68	72^	82^	70^			
		Terrestrial	17	29	29	25^	13^	25^			
		Seagrass	3	3	2	2^	3^	2^			
F	SPM	Mangrove	1	0.5	1	0.3^	0.8^				
		Oceanic	74	72	81	75^	72^				
		Terrestrial	24	27	18	24^	26^				
		Seagrass	1	1	1	1^	2^				
	Trap	Mangrove	7	3	0	7^					
		Oceanic	43	65	80	30^					
		Terrestrial	47	30	18	59^					
		Seagrass	3	2	2	4^					

Table 3. Cont.

Site	Sample	Distance (m)	0	50	100	200	300	400	500	1000	1500
		Source	**Contribution (%)**								
G	SPM	Mangrove	1.0	ns	1.6	ns	ns	1.3^			1^
		Oceanic	74	ns	80	ns	ns	94^			71^
		Terrestrial	17	ns	12	ns	ns	2^			19^
		Seagrass	8	ns	6	ns	ns	3^			9^
	Trap	Mangrove	26	30	24	32	38^				
		Oceanic	62	34	50	41	40^				
		Terrestrial	5	17	10	9	6^				
		Seagrass	7	19	16	18	16^				
H	SPM	Mangrove	1	6	9	11	7	15	22^	15^	
		Oceanic	68	58	54	56	75	71	67^	74^	
		Terrestrial	23	25	24	20	7	4	3^	1^	
		Seagrass	8	11	13	13	11	10	8^	10^	
	Trap	Mangrove	20	ns	20	ns	ns	ns	16^	8^	7
		Oceanic	48	ns	23	ns	ns	ns	31^	58^	40
		Terrestrial	17	ns	34	ns	ns	ns	29^	19^	34
		Seagrass	15	ns	23	ns	ns	ns	24^	15^	19
I	SPM	Mangrove	6	4	5	1	6	ns	4^	1^	
		Oceanic	69	77	71	72	72	ns	76^	82^	
		Terrestrial	14	11	11	14	10	ns	13^	10^	
		Seagrass	11	8	13	13	13	ns	7^	7^	
	Trap	Mangrove	22	ns	9	10	3	4^	2^	3^	
		Oceanic	59	ns	68	62	71	71^	84^	77^	
		Terrestrial	6	ns	10	12	15	8^	5^	8^	
		Seagrass	13	ns	13	16	11	17^	9^	12^	

Distance (m) represents the distance along the transect from land (0 m) to the ocean for each site. SPM and trap are sample type, SPM is suspended particulate matter taken from the water column. (ns) means that the model did not find a solution. (^) represents sediment trap and SPM samples located in seagrass beds. Blank spaces indicate that no samples were collected.

this, indicates that the seagrass canopy traps seagrass OM [10,31–32].

Landscape patterns of organic matter fluxes

Terrestrial-derived organic material showed an under contribution relative to the size of the catchment area. Different land uses may donate particulate and dissolved nutrients to downstream mangrove forests, but it is plausible that mangrove forests trap much of the nutrients that are flowing into them [33–34]. The negative correlation between catchment area and terrestrial OM is a data artefact. This has occurred due to sites A and B having catchment areas that are an order of magnitude bigger than the others, conversely sites A and B display a minor terrestrial contribution thus these two sites skew the correlation.

Mangrove derived organic matter contributed in proportion to their surface area. The latter finding may be surprising because mangrove forests are known to have a high productivity [7–8]. However, mangrove forests are also known to retain much of their POM and associated nutrients within the forests due to retention mechanisms as supported by other studies [33–35]. Our results indicate that the magnitude of outwelling of OM at the mangrove boundary (0 m) is correlated with the sea-facing width of the mangrove forest and therefore the extent that the forest is exposed to waves. Waves have been shown to decrease the trapping capacity of mangrove roots and therefore increase the transport of POM through mangrove roots [34]. The shape of the mangrove forest may have a strong influence on retention of POM. This suggests that mangrove forests that are distributed thinly along a coastline or large delta would export more POM because of their extensive frontal width, which is exposed to greater levels of hydrodynamic energy. In contrast, mangrove forests that extend inland or within protected bays may retain much of their POM and other sources of nutrients (dissolved and particulate) flowing into them. Our study sites are a mixture of these two categories category (Figure 1).

If one compares the potential amount of influx of OM sources in relation to their areal extent, seagrass is "relatively" a much more important source than mangrove forests or terrestrial sources (Figures 3 and 4). If seagrass beds are an important source of OM they may donate this material to other ecosystems and organisms. For example, stony corals have been found to assimilate particulate seagrass as a food source [36] and seagrass leaves are known to be an important food source for herbivore organisms [40]. No relationships were found between seagrass plant contribution and physical factors: hence there may be another controlling factor that has not been measured in the present study. Local hydrodynamics are a potential one. Indeed, as turbulence and wave energy can be a structuring factor in seagrass habitats. For instance, waves with an overlaying current have been found to

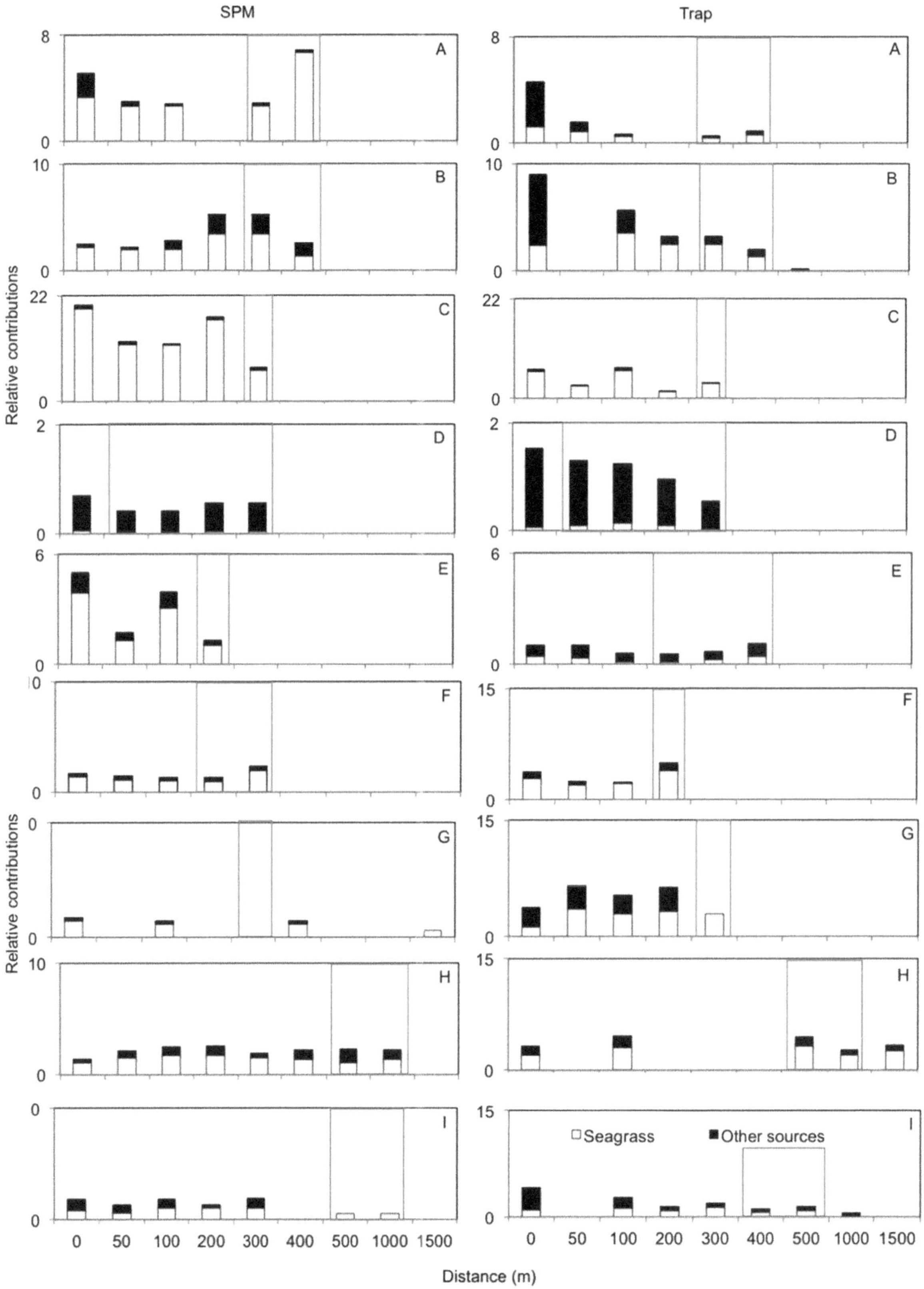

Figure 3. A dimensionless index indicating the relative contribution of seagrass plants and other sources (mangrove trees + terrestrial) for suspended particulate matter (SPM) and sediment trap samples (trap). The values were determined by dividing the mixing model contribution result (Table3) with the per cent of total surface area occupied by a particular ecosystem at each site. Distance (m) represents the distance along the length of the transect from the land (0 m) to the ocean for each site. SPM and trap are sample type, SPM is suspended particulate matter taken from the water column. Whilst trap represents sediment samples captured on the sediment floor over a 48-hour period. Each chart represents a site location; see letter at the top left right side of chart (Figure1). White areas represent seagrass beds, black areas symbolize other sources (mangrove trees + terrestrial). Green boxes around columns indicate when samples were taken in a seagrass bed. This index was not calculated for POM derived from oceanic sources because a representative surface area for the 'ocean' could not be defined.

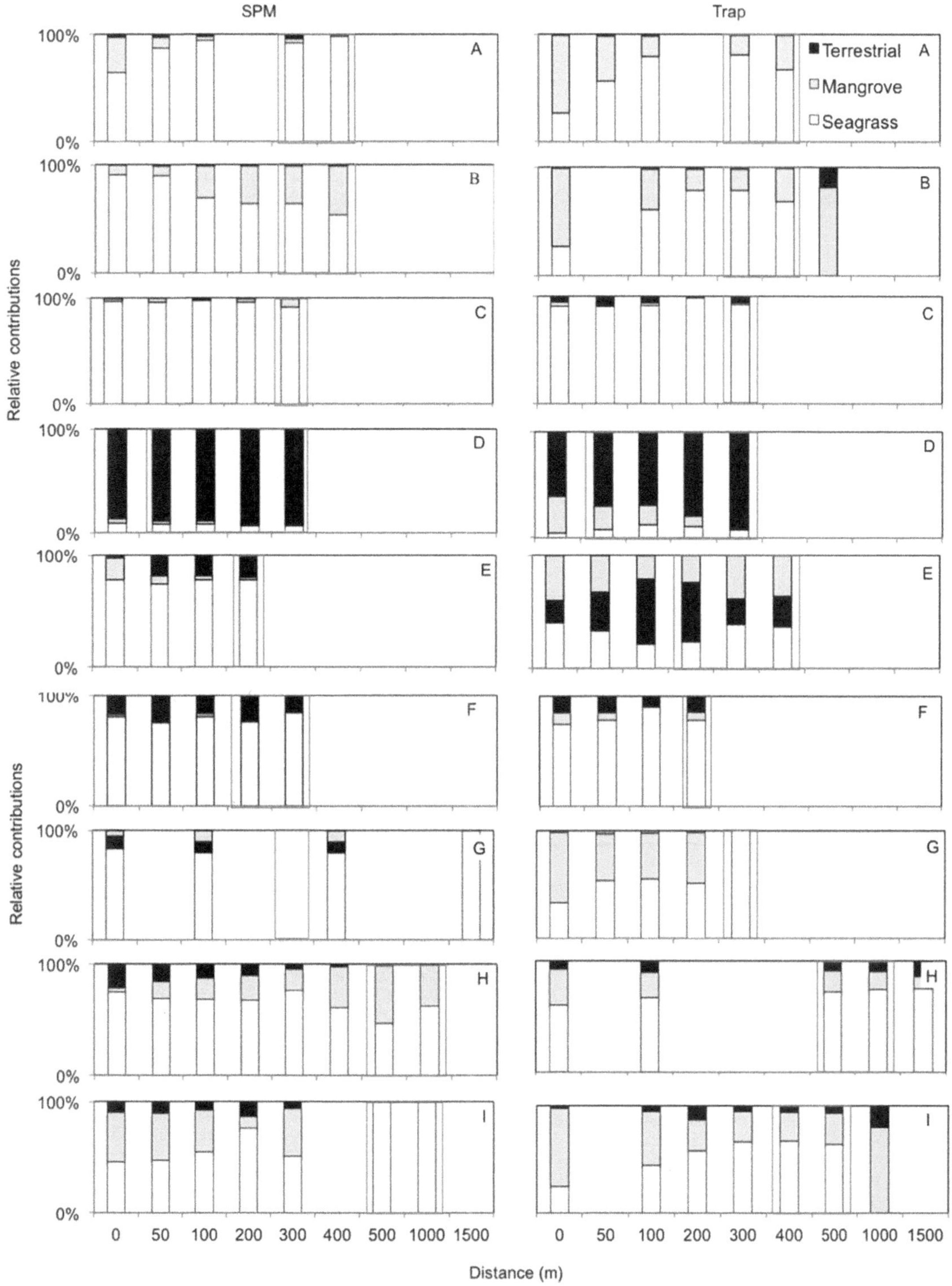

Figure 4. Dimensionless index indicating the relative contribution of terrestrial, mangrove and seagrass sources for suspended particulate matter (SPM) and sediment trap samples (trap), plotted in a stacked column chart. Distance (m) represents the distance along the length of the transect from the land (0 m) to the ocean for each site. SPM and trap are sample type, SPM is suspended particulate matter taken from the water column. Whilst trap represents sediment samples captured on the sediment floor over a 48 hour period. Each chart represents a site location, see letter at top left right side of chart (Figure 1). White areas represent seagrass beds, grey and black areas symbolise mangrove forests and terrestrial sources respectively. Green boxes around columns indicate when samples were taken in a seagrass bed. This index was not calculated for POM derived from oceanic sources, because no representative surface area for the 'ocean' could be defined.

increase litter movement through seagrass beds [34]. Herbivores digesting the seagrass plants and transferring nutrients via faeces across the tropical coastal seascape [37–39] may be another undetected factor.

Approximately one quarter of the samples showed a higher seagrass contribution within the bed but otherwise, the seagrass contribution seemed unrelated to the distance from or within the seagrass bed (Figures 3 and 4). This indicates that seagrasses have a spatially extended influence beyond the physical boundaries of the bed. Although the mechanisms are still unknown of what controls seagrass contribution, our results clearly show that seagrasses are a important contributor of organic matter in the tropical seascape of Phangnga bay relative to their spatial coverage.

Future Perspective and Management Implications

The high oceanic contribution (Table 3) shows how important oceanic derived nutrients are to ecosystems and organisms in these locations. We reason that this may be because of the high tidal flushing in all of our sites allowing for a large exchange of water. This could have implications for pollutants from industries such as fish farms being easily flushed into seagrass beds or mangrove forests from this tidal exchange.

Terrestrial input had the smallest contribution to the OM (Figure 3 and Figure 4). The differences in urbanizations across the 9 sites showed no significant correlation with terrestrial derived OM. Additionally, mangrove forests were not found to be an important input for OM in the coastal zone, compared to oceanic or seagrass organic matter. Recent research has found that mangrove forests may be an effective POM trap and therefore a strong nutrient sink [34]. Mangrove forests are effective filters, allowing them to retain most nutrients fluxes from terrestrial sources. For this reason the integrity and a critical size of the forest should always be ensured. Although many mangroves have been reduced in aerial extent in the Phuket region of Thailand, our results indicate that, though impoverished, mangroves may still maintain important ecosystem services of retaining mangrove and terrestrial derived nutrients in this region. However, continued deforestation or non-sustainable use of the forests may jeopardize this function.

An important consideration is that changes in mangrove forest aerial extent or biomass could reduce buffering and therefore increase the flux of terrestrial-derived nutrients to the coastal waters. This in turn could cause nutrient enrichment or increased turbidity for seagrass beds, causing physiological problems for the seagrass plants that could reduce the area and biomass of this ecosystem [42]. We have shown that seagrass beds are a comparatively large source of POM for adjacent ecosystems. However, seagrass beds can become unstable, for example following eutrophication [41] or hurricane damage [15]. This instability could have implications for the productivity of the seagrass beds and thus, their role as a nutrient source for other ecosystems and organisms could be affected. These alterations to the integrity of the system could have consequences on their ecosystem function as a nutrient sink or buffer for excess nutrients for sensitive adjacent ecosystems such as coral reefs. Therefore, a key role in management of these areas is to ensure the health and physical/physiological structure of seagrass beds. Both factors (health and physical/physiological structure of beds) can be related to quantities of POM being trapped and outwelled [41,43].

This study provides further evidence that the existence of connective particulate fluxes, which occur between ecosystems in the tropical coastal seascape, do exist. These connections could have implications for strengthening management especially from an ecosystem-based perspective.

Author Contributions

Conceived and designed the experiments: LG Gillis: principle author, data collection and analyses. AD Ziegler: daily supervision of paper, D Van Oevelen: data analyses and feedback on paper, C. Cathalot: data collection, analyses and feedback on paper, JW Wolters: data collection and analyses, PMJ Herman: feedback on paper and idea, TJ Bouma: perceived idea, daily supervision. Performed the experiments: LGG CC JWW. Analyzed the data: LGG DVO CC JWW. Contributed reagents/materials/analysis tools: LGG DVO CC. Wrote the paper: LGG ADZ DVO CC PMJ TJB. Principal author: LGG. Data collection: LGG CC JWW. Daily supervision of paper: ADZ TJB. Feedback on paper: DVO CC. Feedback on paper and idea: PMJH. Perceived idea: TJB.

References

1. Barbier EB, Hacker SD, Kennedy C, Koch EW, Stier AC, et al. (2011) The value of estuarine and coastal ecosystem services. Ecol Monog 81(2): 169–193.
2. Gattuso JP, Frankignoulle M, Wollast R (1998) Carbon and carbonate metabolism in coastal aquatic ecosystems. Annu Rev Ecol Syst 29: 405–434.
3. Krusche AV, Martinelli LA, Victoria RL, Bernardes M, de Camargo PB, et al. (2002) Compoistion of particulate and dissolved organic matter in a disturbed watershed of souteast Brazil (Piracicaba River basin). Wat Res 36: 2743–2752.
4. Kuramoto T, Minagawa M (2001) Stable carbon and nitrogen isotopic characterization of organic matter in a mangrove eocystem on the southwestern coast of Thailand. 2001. J Oceanogr 57: 421–431.
5. Thimdee W, Deein G, Sangrungruang C, Nishioka J, Matsunaga K (2003) Sources and fate of organic matter in Khung Krabaen Bay (Thailand) as traced by δ13C and C/N atomic ratios. Wetlands 23(4): 739–738.
6. Meksumpun S, Meksumpun C, Hoshika A, Mishima Y, Tanimoto T (2005) Stable carbon and nitrogen isotope ratios of sediment in the gulf of Thailand: Evidence for understanding of marine environment. Cont Shelf Res 25: 1905–1915.
7. Odum EP (1968) A research challenge: evaluating the productivity of coastal and estuarine water. In Proceedings of the Second Sea grant Conference, University of Rhode Island, Kingston, USA: 63–64.
8. Lee SY (1995) Mangrove outwelling- A Review. Hydrobiologia 295(1–3): 203–212.
9. Terrados J, Duarte CM (2000) Experimental evidence of reduced particle resuspension within a seagrass (*Posidonia oceanica* L.) meadow. J Exp Mar Biol Ecol 243: 45–53.
10. Vonk JA, Middelburg JJ, Stapel J, Bouma TJ (2008) Dissolved organic nitrogen uptake by seagrasses. Limnol Oceanogr 53: 542–548
11. Fonseca MS, Cahalan JA (1992) A preliminary evaluation of wave attentuation for four species of seagrass. Est Coast Shelf Sci 35: 565–576.
12. Hendriks IE, Sintes T, Bouma TJ, Duarte CM (2008) Experimental assessment and modeling evaluation of the effects of the seagrass *Posidonia oceanica* on flow and particle trapping. Mar Ecol Prog Ser 356: 163–173.
13. Hemminga MA, Marba N, Stapel J (1999) Leaf nutrient resorption, leaf lifespan and the retention of nutrients in seagrass systems. Aquat Bot 65: 141–158.
14. Koch EW, Verduin JJ (2001) Measurements of physical parameters in seagrass habitats. In Short FT, Short CA, Coles RG (Eds.). Global Seagrass Research Methods. Amsterdam: Elsevier Science pp. 325–344.
15. Infantes E, Terrados J, Orfila A, Canellas B, Alvarez-Ellacuria A (2009) Wave energy and the upper depth limit distribution of *Posidonia oceanica*. Bot Mar 52: 419–427.
16. Wilkie L, O'Hare MT, Davidson I (2012) Particle trapping and retention by *Zostera noltii*: A flume and field study. Aquat Bot 102: 15–22.
17. Hemminga MA, Slim FJ, Kazungu J, Ganssen GM, Nieuwenhuize J, et al. (1994) Carbon outwelling from a mangrove forest with adjacent seagrass beds and coral reefs (Gazi Bay, Kenya). Mar Ecol Prog Ser 106(3): 291–301.
18. McConnachie JL, Petticrew EL (2006) Tracing organic matter sources in riverine suspended sediment: Implications for fine sediment transfers. Geomorphology 79: 13–26.
19. Schindler Wildhaber Y, Liechti R, Alwell C (2012) Organic matter dynamics and stable isotope signature as tracers of the sources of suspended sediment. Biogeosciences 9: doi: 10.5194/bg-9-1985-2012.
20. Bouillon S, Dehairs F, Velimirov B, Abril G, Vieira Borges A (2007) Dynamics of organic and inorganic carbon across contiguous mangrove and seagrass

systems (Gazi Bay, Kenya). J Geophys Res G02018, doi:10.1029/2006JG000325.

21. Limpsaichol P, Khotiattiwong S, Bussarwait N, Sojisuporn P (1998) Environmental factors influencing the health and productivity of Phang-nga Bay. In Nickerson D. J. (Ed.) 1992: Community-based fisheries management in the Phang-nga bay, Thailand. Proceedings of the National Workshop on Community-based Management. Phuket, Thailand (BOBP Report No 78): 85–120.
22. Woodroffe C (1992) Mangrove Sediments and Geomorphology. In: Robertson AI and DM Alongi (eds.) Trophical Mangrove Ecosystems. American Geophysical Union Washington, USA: 7–41.
23. Nieuwenhuize J, Maas EMY, Middelburg JJ (1994) Rapid analysis of organic carbon and nitrogen in particulate materials. Mar Chem 45: 217–244.
24. Phillips DL, Gregg JW (2001) Uncertainty in source partioning using stablis isotopes. Oecologia 127: 171–179.
25. Phillips DL, Koch PL (2002) Incorating concentration dependence in stable isotope mixing modes. Oecologia 120: 114–125.
26. R Core Team (2012) R: A language and environment for statistical computing. R Foundation for Statistical Computing, Vienna, Austria. ISBN 3-900051-07-0, Available: http://www.R-project.org/. Accessed 2014 Oct 1.
27. Van Oevelen D, Van den Meersche K, Meysman F, Soetaert K, Middelburg JJ, et al. (2010) Quantitative reconstruction of food webs using linear inverse models. Ecosystems 13: 32–45.
28. Cline JD, Kaplan IR (1975) Isotopic fractionation of dissolved nitrate during denitrification in the eastern tropical North Pacific Ocean. Mar Chem 3: 271–299.
29. Dittmar T, Lara RJ (2001) Driving forces behind nutrient and organic matter dynamics in a mangrove tidal creek in North Brazil. Est Coast Shelf Sci 52: 249–259.
30. Hunsinger GB, Mitra S, Findlay SEG, Fischer DT (2010) Wetland-driven shifts in suspended particulate organic matter compoistion of the Hudson River estuary, New York. Limnol Oceanogr 55(4): 1653–1667.
31. Evrard V, Kiswara W, Bouma TJ, Middelburg JJ (2005) Nutrient dynamics of seagrass ecosystems: N-15 evidence for the importance of particulate organic matter and root systems. Mar Ecol Prog Ser 295: 49–55.
32. Van Engeland T, Bouma TJ, Morries EP, Brun FG, Peralta G, et al. (2011) Potential uptake of dissolved organic matter by seagrasses and macroalage. Mar Ecol Prog Ser 427: 71–81.
33. Alongi DM (1990) Effect of mangrove detrital outwelling on nutrient regeneration and oxygen fluxes in coastal sediments of the central-greater-barrier-reef lagoon. Est Coast Shelf Sci 31(5): 581–598.
34. Gillis LG, Bouma TJ, Kiswara W, Ziegler AD, Herman PMJ (2014) Leaf transport in mimic mangrove forests and seagrass beds. Mar Ecol Prog Ser 498: 95–102.
35. Bouillon S, Borges AV, Castaneda-Moya E, Diele K, Dittmar T, et al. (2008) Mangrove production and carbon sinks: A revision of global budget estimates. Global Biogeochem cycles 22, doi: 10.1029/2007GB003052.
36. Lai S, Gillis LG, Mueller C, Bouma TB, Guest JR, et al. (2013) First experimental evidence of corals feeding on seagrass matter. Coral Reefs 32 (4): 1061–1064.
37. Krumme U (2009) Dial and Tidal Movements by Fish and Decapods Linking Tropical Coastal Ecosystems. In: Nagelkerken I (Ed.). Ecological Connectivity among Tropical Coastal Ecosystems, Springer Netherlands: 271–324.
38. Forward RB, Tankersley RA (2001) Selective tidal-stream transport of marine animals. Oceanogr Mar Biol 39: 305–353.
39. Kimirei IA, Nagelkerken I, Mgaya YD, Huijbers CM (2013) The Mangrove Nursery Paradigm Revisited: Otolith Stable Isotopes Support Nursery-to-Reef Movements by Indo-Pacific Fishes. PLOS ONE 8 (6): e66320.
40. Heck KL, Valentine JF (2006) Plant-herbivore interactions in seagrass meadows. J Exp Mar Biol 330: 420–436.
41. Burkholder JM, Tomasko DA, Touchette BW (2007) Seagrasses and eutrophication. J Exp Mar Biol Ecol 350(1–2): 46–72.
42. Todd PA, Ong XY, Chou LM (2010) Impacts of pollution on marine life in Southeast Asia. Biodiversity Conserv 19: 1063–1082.
43. Perez M, Invers O, Ruiz JM, Frederiksen MS, Holmer M (2007) Physiological responses of the seagrass *Posidonia oceanica* to elevated organic matter content in sediments: An experimental assessment. J Exp Mar Biol Ecol 344: 149–160.

9

In Situ Natural Product Discovery via an Artificial Marine Sponge

James J. La Clair[1]*, Steven T. Loveridge[2], Karen Tenney[2], Mark O'Neil-Johnson[3], Eli Chapman[4]*, Phillip Crews[2]*

1 Xenobe Research Institute, San Diego, California, United States of America, **2** Department of Chemistry and Biochemistry, University of California Santa Cruz, Santa Cruz, California, United States of America, **3** Lead Discovery and Rapid Structure Elucidation Group, Sequoia Sciences, Inc., St. Louis, Missouri, United States of America, **4** Department of Pharmacology and Toxicology, College of Pharmacy, University of Arizona, Tucson, Arizona, United States of America

Abstract

There is continuing international interest in exploring and developing the therapeutic potential of marine–derived small molecules. Balancing the strategies for ocean based sampling of source organisms versus the potential to endanger fragile ecosystems poses a substantial challenge. In order to mitigate such environmental impacts, we have developed a deployable artificial sponge. This report provides details on its design followed by evidence that it faithfully recapitulates traditional natural product collection protocols. Retrieving this artificial sponge from a tropical ecosystem after deployment for 320 hours afforded three actin–targeting jasplakinolide depsipeptides that had been discovered two decades earlier using traditional sponge specimen collection and isolation procedures. The successful outcome achieved here could reinvigorate marine natural products research, by producing new environmentally innocuous sources of natural products and providing a means to probe the true biosynthetic origins of complex marine–derived scaffolds.

Editor: Tilmann Harder, University of New South Wales, Australia

Funding: This work was supported by funding from the NIH, R01CA047135 (P. C.), program funds from the Xenobe Research Institute (J. J. L.), as well as startup funds from the University of Arizona (E. C). The funders had no role in study design, data collection and analysis, decision to publish, or preparation of the manuscript.

Competing Interests: The authors have the following interests: Mark O'Neil-Johnson conducted the NMR studies included in this program. He is currently employed at Sequoia Biosciences. Sequoia Biosciences does not have commercial efforts underway on any part of this program including the development of the devices, compounds, or any related intellectual property within this program. Additionally, there has been no intellectual property filed on the inventions included within this manuscript. DeepSea Power and Light provided suggestions/guidance for this study. This does not alter the authors' adherence to all the PLOS ONE policies on sharing data and materials, as detailed online in the guide for authors.

* Email: i@xenobe.org (JJL); pcrews@ucsc.edu (PC); chapman@pharmacy.arizona.edu (EC)

Introduction

Seminal reviews note that marine ecosystems serve as outstanding sources of molecular structures with activities in antitumor, anti–inflammatory, analgesic, anti–allergy, antiviral, and immunomodulatory screens [1–4]. However, recent advances in genomic, proteomic, and microbiological techniques, plus insights from biosynthetic analyses, intimate that many of the key marine–derived therapeutic leads isolated to date such as patellamide [5], the didemnins [6–8], the bengamides [9], psymberin [10], or the manzamines [11,12] may arise from symbiotic microbes and not from the collected organism. In an increasingly appreciated view, while marine–derived small molecules continue to constitute a critical resource in drug discovery, contemporary methods are capable of accessing only a minute fraction of marine diversity. Unfortunately, the majority of marine microbial species have been recalcitrant to cultivation, possibly due to their reliance on complex symbioses [13]. Amazingly, current estimates indicate that over 99% of microbes from environmental samples cannot be cultured under laboratory conditions [14].

While marine–derived small molecules with new modes of action (MOA) are being identified annually [15], the rate of discovery is rather low. A typical natural product program obtains samples by expeditions, where organisms are gathered. The resulting specimens are then shipped back to the laboratory where they are either used as sources for natural product isolation or microbial culturing to obtain compounds. While these methods have produced a number of leads and clinical entities, the procedure is far from being efficient; often many small molecules and microbes are lost [16]. Another drawback is associated with the possible negative environmental impact associated with the large–scale collection of high priority invertebrates including soft corals, sponges, and tunicates [17]. These deficiencies combined with difficulties in correlating leads with clinically relevant MOA [15] and adequate pharmacological properties [16] pose the largest roadblocks to clinical translation.

In an effort to circumnavigate such difficulties, we began a campaign to develop an artificial sponge, allowing the first stages of the experimental work to be conducted at sea in an environmentally benign fashion. A guiding principle, inspired in part by prior marine biomimetic robots [18] and deployable isolation systems [19], involved designing a mechanical device that mimics the currently appreciated action of a marine sponge. This device could then serve as a host for microbial communities and as a biological medium for accumulation of secondary metabolites. A proof–of–concept demonstration is presented herein involving the successful use of this artificial sponge to isolate actin–targeting jasplakinolide depsipeptides [20–22], a family of natural products that were first identified from sponge hosts collected in the same geographic region as the deployment sites for our artificial sponge (Table 1). It should also be noted that these molecules possess remarkable similarity, as shown in Figure 1, to chondramide D [22], isolated from a terrestrial myxobacterium and the miuraenamides [23], produced by a marine–derived myxobacterium.

Materials and Methods

General

Unless otherwise noted, all reagents and chemical compounds were purchased from Fisher Scientific and used without further purification. NMR samples were dissolved in 13 μL of 100% CD_3OD (Cambridge Isotope Laboratories), vortexed, and added to a 1 mm tube (Bruker Biospin). ^1H NMR and ^1H–^1H gCOSY NMR spectra were collected on a 600 MHz Avance spectrometer equipped with a 1.7 TCI MicroCryoProbe (Bruker Biospin). FID files were processed using MestReNova version 8.1 (Mestrelab Research) and were referenced relative to the residual solvent peaks. Devices were launched at multiple locations in the Pacific Ocean and the South China Sea at 10–50 m depth. Permits were not required for this deployment. No animal, plant or living matter was collected or damaged during the study nor was biological material collected. The materials were not deployed within protected ecosystems or marine reserves or parks.

Artificial Sponge

Each artificial sponge was housed within a 2″ ID clear rigid schedule 80 PVC pipe (Harvel) with one of the ends fitted with a cap containing ports for seawater entry, seawater exit, and connections to power. The core of each device was a hollow–fiber bioreactor [24] (C2011, FiberCell Systems), which served to mimic the sponge mesohyl by providing a location to host microbial growth. This bioreactor contained two ports: one for inoculation and a second to pass media during incubation. Seawater was passed through the artificial sponge during both inoculation and incubation using a NF5 RP.51DC–M diaphragm pump (KNF Neuberger), whose flow rate was regulated to pump at 1.2±0.3 L/h through a self–built regulator. Power was provided by an external 6V SeaBattery (DeepSea Power and Light) or a self–built equivalent capable of deployment up to 100 m depth. During incubation, seawater leaving the hollow–fiber bioreactor was passed through three 10 cm ID sep–pak cartridges (ePlastics) containing 50±10 g Amberlite XAD–18 (Dow Chemical Co.) resin/pak, which served as the vehicle for natural product concentration. Components were connected using thick–walled Tygon tubing. Particle filters (OP–6620–14, Chemglass) and 2 μM Millex–AP microbial filters (SLA05010, Millipore) were placed on the inlets and outlets of the artificial sponge to prevent clogging and contamination. A schematic representation of the artificial sponge has been provided in Figure 2 and images of the components are provided in Figure S8 in File S1. The artificial sponges were evaluated during 11 separate deployments (Table 1). Each deployment involved three stages of application (inoculation, incubation, and harvesting) as outlined in Figure 3.

Inoculation of the Artificial Sponge

Each deployment began by inoculating the hollow–fiber bioreactor, as depicted by Step 1 in Figure 3. Sites were selected that contained multiple species of sponges. Through dive teams, each device was placed such that seawater could be pumped into the inoculation port using a 110 mL disposable funnel containing a polyethylene frit (OP–6602–14, ChemGlass) that was positioned within 1 m of live sponges (typically multiple species of sponges). This funnel served to prevent sediment and particulate matter from entering the artificial sponge. The outlet was capped with a 2 μm 50 mm Millex–AP filter (SLA05010, Millipore) to prevent backflow. After pumping for 5–7 h (Table 1), the inoculation port was closed and the apparatus was moved to a region proximal the sponges used for inoculation (typically within 100 m).

Incubation of the Artificial Sponge

Once inoculated, each device was incubated for several days proximal to its site of inoculation (Step 2, Figure 3). Each artificial sponge was held in place by anchoring with conventional diving

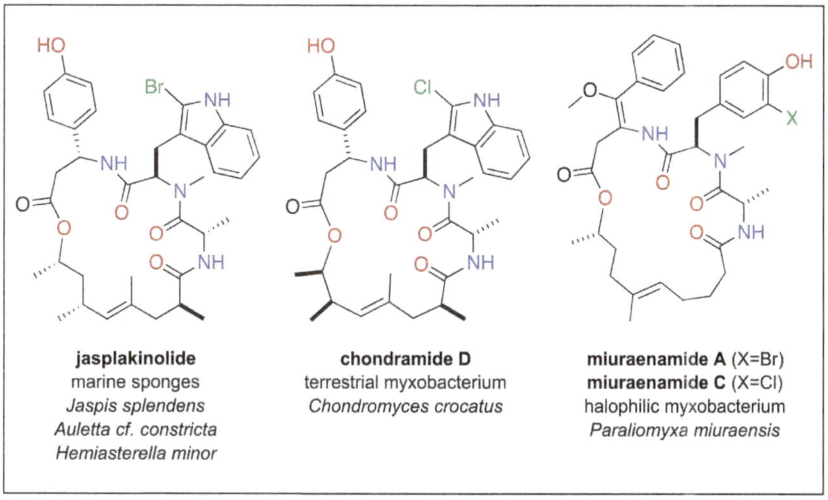

Figure 1. A sponge based hypothesis. A side–by–side comparison of the three natural products shown here illustrates an example of parallel biosynthetic pathways that operated in disparate organisms including marine sponges and myxobacteria. These compounds, arising from the fusion of a triketide with unusual tripeptide moieties, represent the types of biosynthetic products targeted in this study owing to their parallel biogenetic and potential microbial origins. Each compound was previously discovered from the indicated source organisms, and all were subsequently shown to be F–actin stabilizers.

Table 1. Deployments.

Date	Extract	Location (longitude, latitude)	Depth	T_I (h)	T_C (d)
05/2011	SES-001	+32°40′11.01″, −117°14′47.04″	35′	5	12
05/2011	SES-002	+32°40′6.46″, −117°14′44.42″	45′	5	12
06/2011	SES-003	+34°0′22.61″, −119°23′28.09″	25′	6	7
06/2011	SES-004	+34°0′22.61″, −119°23′28.09″	35′	6	7
06/2011	SES-005	+34°0′16.85″, −119°25′20.57″	40′	7	7
06/2011	-	+34°0′16.85″, −119°25′20.57″	35′	7	-
06/2011	SES-007	+34°0′16.85″, −119°25′20.57″	25′	7	7
07/2011	SES-008	+1°45′4.28″, +110°30′2.84″	35′	5	14
07/2011	SES-009	+1°45′4.28″, +110°30′2.84″	45′	5	14
07/2011	SES-010	+1°45′4.28″, +110°30′2.84″	25′	5	14
07/2011	SES-011	+1°45′4.28″, +110°30′2.84″	30′	5	14

Date, extract name, location, depth, time of inoculation (T_I) and time of incubation (T_C) for the 11 deployments explored in this program. Three expeditions were conducted between the period of 05/02/2011 and 07/31/2011 at three locations. Deployment 6 was lost at sea.

weights and marked by a single buoy at the surface. During this stage, both the inlet and outlet of the artificial sponge were capped with a 2 μm 50 mm Millex-AP filter (SLA05010, Millipore). Routine diving was required to maintain the filters as well as to recharge the battery, the latter was readily achieved by replacement with a charged battery followed by recharging the spent

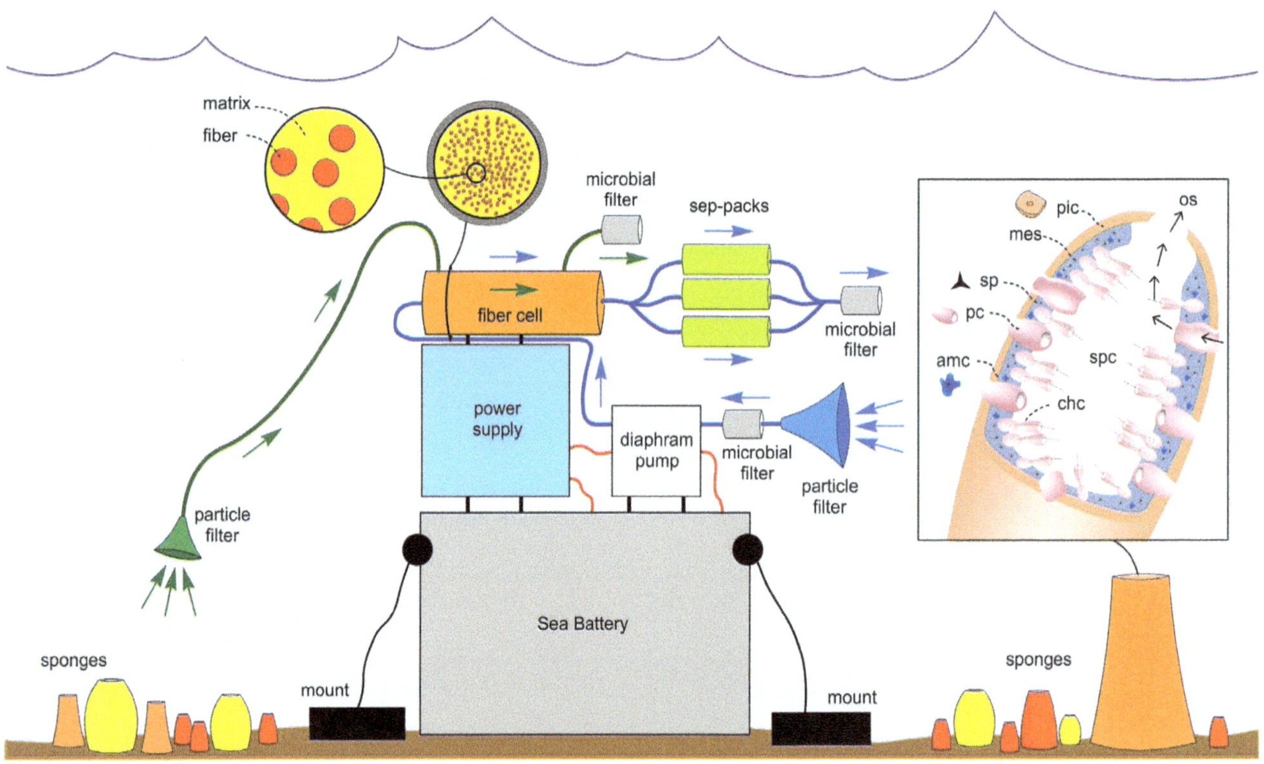

Figure 2. A schematic overview. The artificial sponge was mounted underwater proximal to sponges envisioned to be engaged in secondary metabolite biosynthesis, such as shown in Figure 1. Its components are: particle filters (disposable funnel containing a polyethylene frit, OP-6602-14, ChemGlass), microbial filters (2 μm pore, 50 mm OD, Millex-AP microbial filters, SLA05010, Millipore); a SeaBattery (DeepSea Power & Light); a power supply (self-built); a microdiaphram pump (NF5RP, KNF Neuberger); a hollow-fiber bioreactor (4300-C5011, FiberCell Systems); and parallel-bundle of sep-pak cartridges (ePlastics) containing Amberlite XAD-18 resin (Dow Chemicals). The hollow-fiber bioreactor can act as a bioreactor allowing microbial material to culture inside the artificial sponge. The sep-pak cartridges serve to collect materials either from the seawater or from the microbial content within the hollow-fiber bioreactor. Green arrows indicate flow direction during charging of the hollow-fiber bioreactor from the water column during the inoculation stage (step 1, Figure 3). Blue arrows depict the passage of seawater through the artificial sponge during the incubation stage (step 2, Figure 3). A generic depiction of the anatomy of a sponge is shown within the inset to illustrate the parallel engineered design, codes are: os = osculum, spc = spongocoel, chc = choanocytes, amc = amebocyte, pc = porocytes, sp = spicule, mes = mesohyl, pic = pinacocytes.

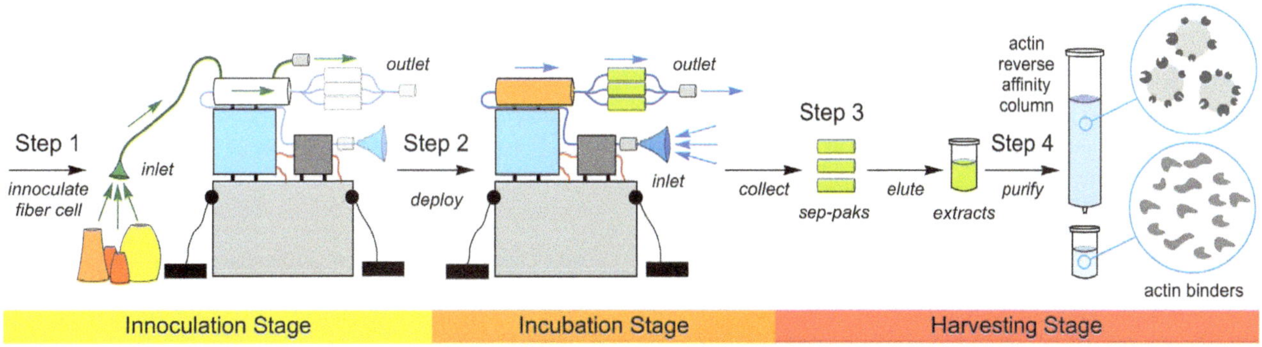

Figure 3. The capture and isolation of natural products through a three-stage, four-step procedure. The process began with inoculation of the hollow-fiber bioreactor (Step 1) followed by incubating the artificial sponge (Step 2) for 7–14 days (Table 1) proximal to the inoculation site. The systems were moved from their inoculation site to a second incubation site, devoid of sponges and fragile sea organisms, in an effort to prevent damage to marine specimens. After the completion, the sep-pak cartridges were harvested and eluted to provide a crude extract (Step 3). A total of 10 extracts were obtained (Table 1). Next, a reverse-affinity method (Step 4) was deployed to isolate natural products based on their ability to bind to actin. This provided a semi-pure fraction from each extract as illustrated in Figure 4 (code: AC-X-A) from extract SES-009 (Table 1). A detailed description of these steps has been provided in the Materials and Methods.

battery. Artificial sponges were operated for 7–14 days, as noted in Table 1.

Harvesting of the Artificial Sponge

After incubation, the artificial sponge was collected (Step 3, Figure 3). The sep-pak cartridges were removed immediately, washed with deionized H_2O (3×50 mL), capped, placed in zip lock bags, and stored at 4°C until extraction in a laboratory setting. Extracts were prepared by elution with 95% EtOH (5×50 mL). The fractions were pooled and dried by rotary evaporation in a 250 mL round bottom flask. The material was transferred with EtOH washing to a 20 mL scintillation vial and dried by rotary evaporation followed by lyophilization for 12–18 h, providing extracts SES–001 to SES–011 (Table 1) with 180–630 mg/extract.

Actin Polymerization Assay

We began by evaluating our extracts for their ability to regulate the polymerization of actin [25] using a commercially available, pyrene labeled assay (BK003, Cytoskeleton, Inc.). The assay was conducted as described in the manufacturer's guidelines. Briefly, purified rabbit muscle actin was modified to contain covalently linked pyrene at the Cys374 residue with 0.6 *N*–1–pyreneacetamide labels per actin monomer. Each assay was conducted with 20 µM pyrene–labeled actin in 5 mM Tris–HCl pH 8.0, 0.2 mM $CaCl_2$ and 0.1% DMSO used to solubilize the analyte. Analytes were incubated with the G–actin for 30 min and then polymerization was induced by the addition of 0.2 mM ATP. The assay was run in a final volume of 240 µL in a black 96 well plate (Costar) and monitored using an HTS700 plate reader (Perkin Elmer) with excitation at 360±10 nm and emission at 465±20 nm. Data were collected with a 2.8 s reading time every 34 s. The assay was run four times with three replicates, and each run was obtained with 99.5% of the data points occurring with ≤ 5% deviation.

Reverse-Affinity Purification

An aliquot of an Affigel 10 resin (2 mL) bearing 2.2±0.8 mg/mL of rabbit muscle actin (Worthington Biochemical) was incubated with 520 mg of the SES–009 extract in 5 mL of 50 mM phosphate buffered saline (PBS) buffer pH 8.0 containing 0.2 mM $CaCl_2$ and 5% DMSO. After shaking by inversion for 12 h at 4°C, the resin was collected via centrifugation on a C–1200 mini–centrifuge (VWR Scientific) for 2 min. The resin was then washed twice with 4 mL 50 mM PBS pH 8.0 containing 0.2 mM $CaCl_2$ at 4°C, and once with 4 mL of deionized H_2O. After each wash the resin was collected by mini–centrifugation. The resulting resin was washed with warm 95% EtOH (2×4 mL). The EtOH fractions were combined and dried via rotary evaporation. The final material was divided into triplicate samples (estimated to be <30 µL total volume) and two samples were used without further purification for NMR and LC–MS analyses, respectively.

LC-MS Analysis

Ultra high performance liquid chromatography coupled with time–of–flight mass spectrometry (UHPLC/TOF–MS) experiments were performed using an Agilent 1260 binary pump in low dwell volume mode, an Agilent column oven heated to 45°C, and an Agilent 6230 Time of Flight Mass Spectrometer with an electrospray ionization (ESI) source. Experimentally, 1 µL of the AC–X–A sample dissolved in 50% aq. MeOH was injected onto a 1.8 mm particle size, 50×2.3 mm I.D., Zorbax RRHT column. Each sample was subjected to a gradient from 10: 1 to 1: 10 H_2O: CH_3CN over 4 min with a 1.5 min hold at 1: 10 H_2O: CH_3CN before a 3 min re–equilibration. The flow rate was maintained at 0.8 mL/min. Formic acid, 200 µL/L, was added to both the H_2O and the CH_3CN. The TOF–MS was run in positive high–resolution 4 GHz mode with a detector range from 100 *m/z* to 1700 *m/z*. The ESI source was operated with a desolvation temperature of 350°C and a drying gas flow rate of 11 L/min. The fragmentor voltage was held at 135 V while the capillary voltage was ramped from 2500–3000 V.

Results

The goal to engage in a comprehensive evaluation of the novel artificial marine sponge required completion of several operations. The first step involved building and deploying the device (Figure 3) whose components were designed to mimic the generally understood properties of sponges [26]. This required that the apparatus be robust enough to withstand being placed, for an extended period of time, in the water column proximal an ecosystem rich with sponge biomass. After significant modifications, we settled on the overall design outlined in Figure 2. Ultimately this construct mimicked the general anatomy of a sponge (inset, Figure 2), and required the five primary components shown in Figure 2 and pictorially in Figure S8 in File S1. Briefly, the first module was comprised of a battery, power supply, and a microdiaphragm pump. This assemblage mirrored the function of the sponge aquiferous system (inhalant apertures or pores, inhalant canals, choanocyte chambers, exhalant canals, exhalant apertures or oscules) and provided a uniform flow of seawater at 1.2±0.3 L/h through the artificial sponge, a rate commonly observed in sponges [26]. The next module mimicked the sponge mesohyl, which provided a possible setting to facilitate symbiotic microbial growth. Creating this artificial mesohyl was accomplished by employing a hollow–fiber bioreactor. The idea here was to facilitate the possibility that chemically prolific microbial biota could be captured, maintained, and even flourish within such a matrix. Thus, it seemed essential to constantly bathe the bioreactor contents with nutrients from the water column via pumping seawater through the fibers. The final component, which served a role similar to that of sponge spherulous cells [27], consisted of a parallel bundle of three 10 cm ID sep–pak cartridges containing XAD–18 resin, and was the vehicle for natural product concentration.

Three expeditions were undertaken to test the artificial sponges, two in the Pacific Ocean and one in the South China Sea (Table 1). A total of 11 deployments were conducted using a four–step procedure (Figure 3). Using protocols that were developed by testing the systems in the laboratory (see Materials and Methods), we were able to obtain 10 extracts as given by SES–001 to SES–011, with SES–006 lost at sea (Table 1).

We turned to a commercial actin polymerization assay (see Materials and Methods) as a tool to guide compound isolation. We chose actin polymerization as our primary screen, given the proclivity of marine sponges to contain natural products that are potent modulators of cytoskeleton dynamics, such as the jasplakinolides, the latrunculins, and the swinholides [28–29]. To our delight, one of the extracts, coded as SES–009 (Table 1), obtained from placement of the artificial sponge at 45 m depth near Pulau Lakei, Sarawak, Malaysia, displayed an $IC_{50} \leq 50$ μM (IC_{50} value of 9.2±0.3 μM) in the actin assay.

We then combined this observation with a recently described reverse–affinity (RA) strategy developed in our laboratories [30] to isolate the active principles from extract SES–009 (Figure 3, Step 4). This reverse–affinity procedure (see Materials and Methods) delivered a miniscule sample (<30 μg) given the code AC–X–A (a code that was generated to identify this fraction), which was divided into triplicate samples.

We then turned to modern methods in MS and NMR analysis. LC–MS analysis (see Materials and Methods) of fraction AC–X–A resolved three major peaks in a relative ratio of 4:6:1, and each displayed a MS pseudo–molecular ion doublet (Figure 4A) indicating the presence of a bromine atom. The compound with $m/z = 709/711$ was reminiscent of jasplakinolide [20,21,31]. Further examination by accurate MS analysis and comparison with potential molecular formulas reported in the literature identified the trio (Figure 4B) as follows: jasplakinolide C (**2**) [27] $R_t = 2.34$ min. $m/z = 725.2560$ [MH^+] and molecular formula $C_{36}H_{45}BrN_4O_7$; jasplakinolide B (**3**) [31,32] $R_t = 2.47$ min. $m/z = 723.2401$ [MH^+] and molecular formula $C_{36}H_{43}BrN_4O_7$; and jasplakinolide (**1**) $R_t = 2.85$ min. $m/z = 709.2597$ [MH^+] and molecular formula $C_{36}H_{45}BrN_4O_6$.

Analytical data from NMR analyses, including ^1H NMR (Figure 5A) and ^1H–^1H gCOSY NMR spectra (Figure 5B and Figure S2 in File S1) were obtained prior to MS analysis using one of the triplicate samples. Effective data on this microsample could only be obtained using a high sensitivity 1.7 TCI MicroCryoProbe on a 600 MHz Avance (Bruker Biospin) spectrometer (see Figs. S1–S7). Based on the structural hypothesis derived by MS, it was possible to annotate ^1H NMR resonances for **2** (coded as "c", Figure 5) and **3** (coded as "b", Figure 5), while resonances for **1** were either hidden or too small in intensity to be identified. These assignments were additionally verified via side–by–side comparison of the ^1H NMR data of the mixture to that of natural jasplakinolide B (**3**) (Figure S1 in File S1). Inspection of the ^1H NMR resonances associated with Me–33, Me–32, and Me–29 (Figure S7 in File S1) indicated that the ratio of jasplakinolide B (**2**): jasplakinolide C (**3**) was 60:40, in agreement with the ratio derived by MS (Figure 4). Strikingly, this same mixture of three compounds, in a ratio of 18:1:1 (compounds **1**: **2**: **3**) was reported from *Jaspis splendens* collected near Vanuatu [32]. Furthermore, as shown in Figure 1, jasplakinolide and its congeners have been reported from three different sponge genera in different taxonomic classes [33]. In addition, the Crews laboratory and others have extensively investigated the structure activity relationships of this class against actin [31,34–36].

Discussion

One of the eleven deployed artificial sponges (SES–009, Table 1) provided an extract that inhibited actin polymerization. Given each extract was only tested in the actin assay, this is a quite reasonable rate of success. The other extracts likely contain molecules that will display bioactivity to other targets or pathways, and hence the success of this system is not accurately reflected by 1 out of 10. The bioactive fraction was then resolved by reverse–affinity purification. Eventually, a fraction (coded AC–X–A, Figure 3) was discovered to contain a mixture of three members of the jasplakinolide family of depsipeptides, compounds **1**–**3**. Since the discovery of the cyclodepsipeptide, jasplakinolide (**1**), from the sponge *Jaspis splendens* (order Astrophorida, family Ancorinidae) in 1986 by Crews [20] and the Ireland–Faulkner–Clardy consortium [21], sponges in the *Jaspis* (syn: *Dorypleres*) genus have received considerable attention as a source for structurally novel bioactive natural products. Since then, our laboratory has shown that **1** can be found in specimens of *Jaspis* sponges collected throughout the Indo–Pacific Ocean [22,31]. That we found **1**–**3** in our artificial sponge was in direct agreement with this observation. Importantly, this also represents a critical indicator of the power of such an approach to isolate natural products without having to upset the local ecosystem through invasive collection procedures.

It is relevant to briefly discuss the possible biosynthetic producers or origins of the jasplakinolide mixture isolated from the sep–pak cartridges of the artificial sponge. The only two logical natural sources of the compounds observed here are that: (a) metabolites observed here were those excreted from proximal sponges or microorganisms into the water column and then concentrated in the sep–pak cartridges during the inoculation and/or incubation stages, or (b) metabolites were produced from biosynthetic actions of microorganisms sequestered in the hollow–fiber bioreactor and subsequently collected in the sep–pak cartridges. For the moment, we favor the latter possibility but cannot rule out the former. A confounding issue is that it was impossible to precisely determine the yield of **1**–**3** obtained from the artificial sponges deployed in this study. Two of the triplicate samples of the AC–X–A fraction (<10 μg each), none of which could be accurately weighed, were used in the MS and NMR analyses. It can be assumed that the initial yield from the sep–pak cartridges was undoubtedly greater than the <30 μg obtained.

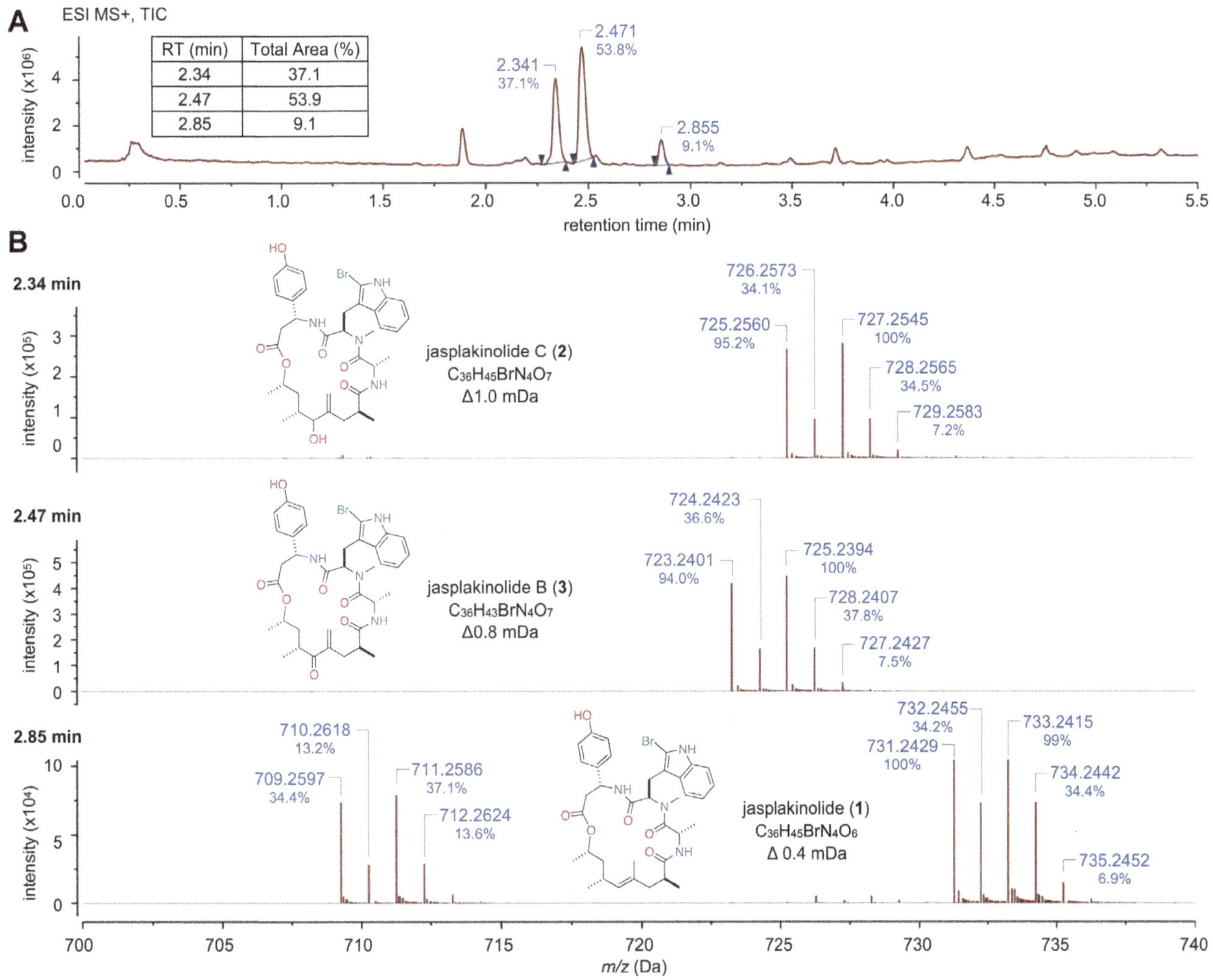

Figure 4. Accurate mass analysis on a <10 μg sample: The actin-bound biosynthetic products were isolated from a sample (code: AC-X-A) via the procedure shown in Figure 3, Step 4. The top panel shows, by Total Ion Chromatogram (TIC) areas, three compounds of interest observed in a ratio of 37:54:9 from this material. The three lower panels show AM-MS data for dereplication of compounds at: 2.34 min = jasplakinolide C (**2**), 2.47 min = jasplakinolide B (**3**), and 2.85 min = jasplakinolide (**1**), respectively.

During the incubation stage, the artificial sponge pumped an estimated 380 L of seawater (315 h at 1.2 L/h), at perfect efficiency the seawater column would have to contain ~0.2 nM of **1**–**3** for this material to be isolated from seawater alone. Alternatively, a microbial consortium would need to operate efficiently to generate the production of **1**–**3** at microgram scales observed in this study. At the present time, the identity of these microbial consortia from each of the bioreactors has not been analyzed. Further experiments are underway to deploy artificial sponges and conduct metagenomic and whole genome sequencing with the goal of unraveling the microbial consortia. This is predicted to be a powerful tool in relating microbiomes to chemical ecology [1]. In addition, these bioreactor data can be compared to the microbiomes of sponge species near the sites of deployment to gain further insight into the origin of these valuable secondary metabolites.

The results presented above delineate an alternative strategy to interrogate a coral reef environment for bioactive chemotypes. This work defines a first–generation deployable artificial marine sponge as a tool to isolate natural products at sea without disturbing the local, fragile ecosystem. We attribute the success realized here to developing an artificial sponge modeled on the anatomy and action of marine sponges. Historically such taxa have served as among the richest sources of natural products, possibly due to the presence of symbiotic colonies of bacteria [37]. However, pinpointing the true source of **1**–**3** concentrated in the sep–pak cartridges constitutes an important and challenging initiative for the future. While our artificial sponge was a first step, these results suggest that further design and optimization of similar marine mechanical artificial devices could provide an ecologically friendly means to access and potentially harvest marine microbes in their natural environment. Overall, this study shows that a mechanical mimic of a sponge provides a non–invasive tool to investigate marine bioorganic chemistry. In addition, with minor modifications the artificial sponges can be tailored to a specific class of molecules. This can be done by using sep–paks with pre–selected biological activity or chemical selectivity [38]. Further molecular diversity may also be obtained from the biological material growing in the fiber cell bioreactor (non–secreted microbial compounds). These artificial sponges also

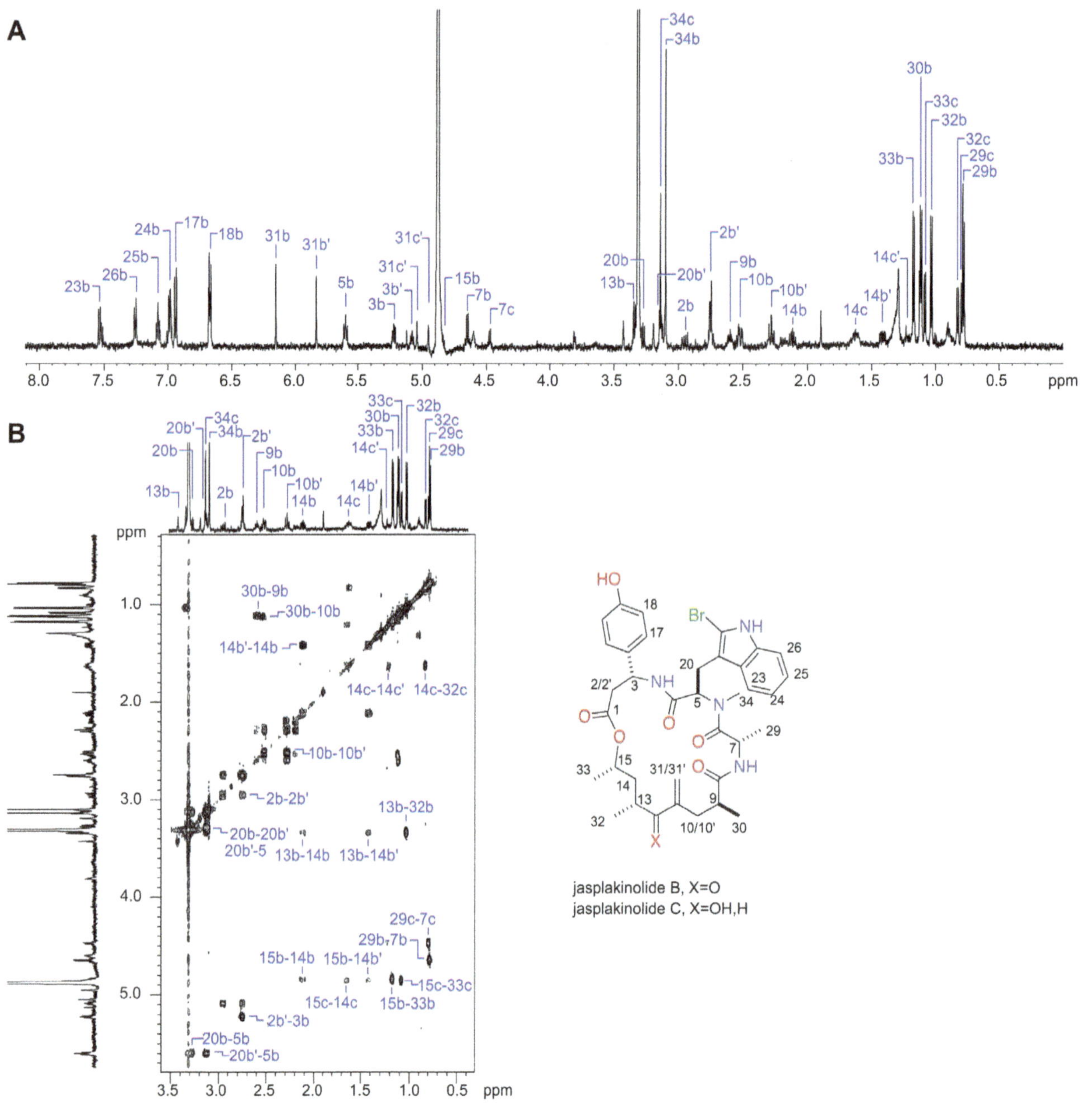

Figure 5. Final dereplication by NMR analysis: Sample AC-X-A (~10 μg) containing the mixture shown in Figure 4 was subjected to NMR determinations at 26°C in CD_3OD using a high sensitivity 1.7 mm TCI MicroCryoProbe on a 600 MHz Avance Spectrometer (Bruker Biospin). The annotations shown for the **(A)** ^{1}H and **(B)** ^{1}H–^{1}H gCOSY NMR spectra confirm the dereplication assignments proposed in Figure 4 for jasplakinolide B (**3**) (protons coded as "b") and jasplakinolide C (**2**) (protons coded as "c"), while resonances for jasplakinolde (**1**) could not be unambiguously observed. Ratios of **3** to **2** were estimated to be 60: 40 by peak areas shown in Figure S7 in File S1.

potentially offer a new approach to advance the mining of marine natural product resources and accessing the genomic and metabolomic potential of their associated marine microbes. As mentioned, we have already begun studies to relate microbiomes to chemical ecology using advanced sequencing procedures coupled with state-of-the-art analytical chemistry.

Supporting Information

File S1 This file contains Figure S1–Figure S8. Figure S1, ^{1}H NMR stacked plot of sample AC-X-A (bottom: 600 MHz, CD_3OD) showing jasplakinolide B (protons coded as "b") and jasplakinolide C (2) (protons coded as "c") versus natural jasplakinolide B (3) (top: 500 MHz, CDCl3). The jasplakinolide B spectrum was supplied by Prof. D'Aura [32]. This Figure corresponds to Figure 5 providing a direct comparison between the published ^{1}H-NMR spectrum of jasplakinolide B (1) (top) with the AC-X-A sample (bottom). Figure S2, Annotated gCOSY ^{1}H NMR spectrum of sample AC-X-A (600 MHz, CD_3OD) with assignments for jasplakinolide B (3) (protons coded as "b") and jasplakinolide C (2) (protons coded as "c"). This Figure corresponds to Figure 5 providing a copy of the ^{1}H–^{1}H gCOSY

NMR spectrum from the AC–X–A sample used to assign the protons showed in Figure 5. Figure S3, Annotated ^{1}H NMR expansion spectrum (0–2 ppm) of sample AC–X–A (600 MHz, CD3OD) with assignments for jasplakinolide B (3) (protons coded as "b") and jasplakinolide C (2) (protons coded as "c"). This Figure corresponds to Figure 5 offering an expansion of the spectrum from 0–2 ppm. This expansion offers increased resolution of the assigned peaks in the AC–X–A sample. Figure S4, Annotated ^{1}H NMR expansion spectrum (2–4 ppm) of sample AC–X–A (600 MHz, CD_3OD) with assignments for jasplakinolide B (3) (protons coded as "b") and jasplakinolide C (2) (protons coded as "c"). This Figure corresponds to Figure 5 offering an expansion of the spectrum from 2–4 ppm. This expansion offers increased resolution of the assigned peaks in the AC–X–A sample. Figure S5, Annotated ^{1}H NMR expansion spectrum (4–6 ppm) of sample AC–X–A (600 MHz, CD_3OD) with assignments for jasplakinolide B (3) (protons coded as "b") and jasplakinolide C (2) (protons coded as "c"). This Figure corresponds to Figure 5 offering an expansion of the spectrum from 4–6 ppm. This expansion offers increased resolution of the assigned peaks in the AC–X–A sample. Figure S6, Annotated ^{1}H NMR expansion spectrum (6–8 ppm) of sample AC–X–A (600 MHz, CD_3OD) with assignments for jasplakinolide B (3) (protons coded as "b") and jasplakinolide C (2) (protons coded as "c"). This Figure corresponds to Figure 5 offering an expansion of the spectrum from 6–8 ppm. This expansion offers increased resolution of the assigned peaks in the AC–X–A sample. Figure S7, Relative abundance of jasplakinolide B (3) and C (2) in sample AC–X–A (600 MHz, CD_3OD) as 60: 40, respectively. Annotated ^{1}H NMR expansion spectrum (0.7–1.3 ppm) with assignments for jasplakinolide B (3) (protons coded as "B") and jasplakinolide C (2) (protons coded as "C"). This Figure corresponds to Figure 5 indicating the relative abundance of jasplakinolides B (3) and C (2) in the AC–X–A sample. Figure S8, A close up image of a prototype XRI–S4 depicting the hollow fiber cell culture bioreactor and diaphragm pump. The inoculation and flow ports to the hollow fiber bioreactor are indicated, as well as the flow of seawater through the system (arrows). This Figure corresponds to Figure 2 providing a photograph of the diaphragm pump and fiber cell used within the system (XRI–S4 prototype) applied in these studies.
(PDF)

Acknowledgments

We thank J. Newton (U.C. San Diego), P. R. Jensen (Scripps Institution of Oceanography), Alison Murray (Desert Research Institute), Mark Butler (U. Melbourne), and Mark Olsson (DeepSea Power and Light) for suggestions and guidance.

Author Contributions

Conceived and designed the experiments: JJL EC PC. Performed the experiments: JJL STL KT MOJ EC PC. Analyzed the data: JJL STL KT MOJ EC PC. Contributed reagents/materials/analysis tools: JJL STL KT MOJ EC PC. Wrote the paper: JJL STL KT MOJ EC PC.

References

1. Hentschel U, Piel J, Degnan SM, Taylor MW (2012) Genomic insights into the marine sponge microbiome. Nat Rev Microbiol 10(9): 641–654. doi: 10.1038/nrmicro2839
2. Newman DJ, Cragg GM (2012) Natural products as sources of new drugs over the 30 years from 1981 to 2010. J Nat Prod 75(3): 311–335. doi: 10.1021/np200906s
3. Newman DJ, Cragg GM (2009) Natural product scaffolds as leads to drugs. Future Med Chem 1(8): 1415–1427. doi: 10.4155/fmc.09.113
4. Radjasa OK, Vaske YM, Navarro G, Vervoort HC, Tenney K, et al. (2011) Highlights of marine invertebrate–derived biosynthetic products: their biomedical potential and possible production by microbial associants. Bioorg Med Chem 19(22): 6658–6674. doi: 10.1016/j.bmc.2011.07.017.
5. Schmidt EW, Nelson JT, Rasko DA, Sudek S, Eisen JA, et al. (2005) Patellamide A and C biosynthesis by a microcin–like pathway in Prochloron didemni, the cyanobacterial symbiont of *Lissoclinum patella*. Proc Natl Acad Sci U S A 102(20): 7315–7320.
6. Xu Y, Kersten RD, Nam SJ, Lu L, Al–Suwailem AM, et al. (2012) Bacterial biosynthesis and maturation of the didemnin anti–cancer agents. J Am Chem Soc 134(20): 8625–8632. doi: 10.1021/ja301735a
7. Tsukimoto M, Nagaoka M, Shishido Y, Fujimoto J, Nishisaka F, et al. (2011) Bacterial production of the tunicate–derived antitumor cyclic depsipeptide didemnin B. J Nat Prod74(11): 2329–2331. doi: 10.1021/np200543z
8. Lee J, Currano JN, Carroll PJ, Joullié MM. (2012) Didemnins, tamandarins and related natural products. Nat Prod Rep 29(3): 404–424. doi: 10.1039/c2np00065b
9. Johnson TA, Sohn J, Vaske YM, White KN, Cohen TL, et al. (2012) Myxobacteria versus sponge–derived alkaloids: the bengamide family identified as potent immune modulating agents by scrutiny of LC–MS/ELSD libraries. Bioorg Med Chem 20(14): 4348–4355. doi: 10.1016/j.bmc.2012.05.043
10. Fisch KM, Gurgui C, Heycke N, van der Sar SA, Anderson SA, et al. (2009) Polyketide assembly lines of uncultivated sponge symbionts from structure–based gene targeting. Nat Chem Biol 5(7): 494–501. doi: 10.1038/nchembio.176.
11. Peraud O (2006) PhD Dissertation, Isolation and characterization of a sponge associated actinomycete that produces manzamines, Ph.D. Thesis, University of Maryland.
12. Hill RT, Hamann MT, Peraud O, Kasanah N (2005) Manzamine–producing actinomycetes US Patent 20050244938.
13. DeLong EF (1997) Marine microbial diversity: the tip of the iceberg. Trends Biotechnol 15(6): 203–207.
14. Sharma R, Ranjan R, Kapardar RK, Grover A (2005) Unculturable bacterial diversity: an untapped resource. Curr Sci 89: 72–77.
15. La Clair JJ (2010) Natural product mode of action (MOA) studies: a link between natural and synthetic worlds. Nat Prod Rep 27(7): 969–995. doi: 10.1039/b909989c
16. Butler MS (2008) Natural products to drugs: natural product–derived compounds in clinical trials. Nat Prod Rep 25(3): 475–516. doi: 10.1039/b514294f.
17. Li JW, Vederas JC (2009) Drug discovery and natural products: end of an era or an endless frontier? Science 325(5937): 161–165. doi: 10.1126/science.1168243
18. Ayers J, Witting J (2007) Biomimetic approaches to the control of underwater walking machines. Philos Trans A Math Phys Eng Sci 365(1850): 273–295.
19. Coll JC, Bowden BF, Tapiolas DM, Dunlap WC (1982) In situ isolation of allelochemicals released from soft corals (Coelenterata: *Octocorallia*): A totally submersible sampling apparatus. J Exp Mar Biol Ecol 60: 293–299.
20. Crews P, Manes LV, Boehler M (1986) Jasplakinolide, a cyclodepsipeptide from the marine sponge, *Jaspis* SP. Tetrahedron Lett 27(25): 2797–2800.
21. Zabriskie MT, Klocke JA, Ireland CM, Marcus AH, Molinski TF, et al. (1987) Jaspamide, a modified peptide from a Jaspis sponge, with insecticidal and antifungal activity J Am Chem Soc108(11): 3123–3124.
22. Robinson SJ, Morinaka BI, Amagata T, Tenney K, Bray WM, et al. (1010) New structures and bioactivity properties of jasplakinolide (jaspamide) analogues from marine sponges. J Med Chem 53(4): 1651–1661. doi: 10.1021/jm9013554
23. Iizuka T, Fudou R, Jojima Y, Ogawa S, Yamanaka S, et al. (2006) Miuraenamides A and B, novel antimicrobial cyclic depsipeptides from a new slightly halophilic myxobacterium: taxonomy, production, and biological properties. J Antibiot (Tokyo) 59(7): 385–391.
24. Park EY (2004) Recent progress in microbial cultivation techniques. Adv Biochem Eng Biotechnol 90: 1–30.
25. Doolittle LK, Rosen MK, Padrick SB (2013) Measurement and analysis of in vitro actin polymerization. Methods Mol Biol 1046: 273–293. doi: 10.1007/978–1–62703–538–5_16
26. Leys SP, Yahel G, Reidenbach MA, Tunnicliffe V, Shavit U, et al. (2011) The sponge pump: the role of current induced flow in the design of the sponge body plan. PLoS One 6(12): e27787. doi: 10.1371/journal.pone.0027787.
27. Thompson JE, Barrow KD, Faulkner DJ (1983) Localization of two brominated metabolites, aerothionin and homoaerothionin, in spherulous cells of the marine sponge *Aplysina fistularis*. Acta Zoologica 64: 199–210.
28. Allingham JS, Klenchin VA, Rayment I (2006) Actin–targeting natural products: structures, properties and mechanisms of action. Cell Mol Life Sci 63(18): 2119–2134.
29. Nagle A, Hur W, Gray NS (2006) Antimitotic agents of natural origin. Curr Drug Targets 7(3): 305–326.
30. Rodríguez AD, Lear MJ, La Clair JJ (2008) Identification of the binding of sceptrin to MreB via a bidirectional affinity protocol. J Am Chem Soc 130(23): 7256–7258. doi: 10.1021/ja7114019
31. Watts KR, Morinaka BI, Amagata T, Robinson SJ, et al. (2011) Biostructural features of additional jasplakinolide (jaspamide) analogues. J Nat Prod 74(3): 341–351. doi: 10.1021/np100721g

32. Zampella A, Giannini C, Debitus C, Roussakis C, D'Auria MV (1999) New jaspamide derivatives from the marine sponge *Jaspis splendans* collected in Vanuatu. J Nat Prod 62(2): 332–334.
33. Noro JC, Kalaitzis JA, Neilan BA (2012) Bioactive natural products from Papua New Guinea marine sponges. Chem Biodivers 9(10): 2077–2095. doi: 10.1002/cbdv.201100292.
34. Kunze B, Jansen R, Sasse F, Höfle G, Reichenbach H (1995) Chondramides A approximately D, new antifungal and cytostatic depsipeptides from *Chondromyces crocatus* (myxobacteria). Production, physico-chemical and biological properties. J Antibiot (Tokyo) 48(11): 1262–1266.
35. Tannert R, Milroy LG, Ellinger B, Hu TS, Arndt HD, et al. (2010) Synthesis and structure-activity correlation of natural-product inspired cyclodepsipeptides stabilizing F-actin. J Am Chem Soc 132(9): 3063–3077. doi: 10.1021/ja9095126.
36. Terracciano S, Bruno I, D'Amico E, Bifulco G, Zampella A, et al. (2008) Synthetic and pharmacological studies on new simplified analogues of the potent actin-targeting Jaspamide. Bioorg Med Chem 16(13): 6580–6588. doi: 10.1016/j.bmc.2008.05.019
37. Blunt J, Gerwick W, Fenical W, Proksch P, Menna M (2012) Taxonomy and natural products. In: Fattorusso E, Gerwick WH, Taglialatela-Scafati O, editors. Handbook of Marine Natural Products. Springer. pp. 3–54.
38. Schulte BA, De Nys R, Bakus GJ, Crews P, Eid C, et al. (1991) A modified allomone collecting apparatus. J Chem Ecol 17(7): 1327–1332. doi: 10.1007/BF00983765.

10

A New Cryptic Species of South American Freshwater Pufferfish of the Genus *Colomesus* (Tetraodontidae), Based on Both Morphology and DNA Data

Cesar R. L. Amaral[1,2]*, Paulo M. Brito[1], Dayse A. Silva[2], Elizeu F. Carvalho[2]

1 Department of Zoology, Universidade Estadual do Rio de Janeiro, Rio de Janeiro, Brazil, **2** Department of Ecology, Universidade Estadual do Rio de Janeiro, Rio de Janeiro, Brazil

Abstract

The Tetraodontidae are an Acantomorpha fish family with circumglobal distribution composed of 189 species grouped in 19 genera, occurring in seas, estuaries, and rivers between the tropical and temperate regions. Of these, the genus *Colomesus* is confined to South America, with what have been up to now considered only two species. *C. asellus* is spread over the entire Amazon, Tocantins-Araguaia drainages, and coastal environments from the Amazon mouth to Venezuela, and is the only freshwater puffers on that continent. *C. psittacus* is found in coastal marine and brackish water environments from Cuba to the northern coast of South America as far south as to Sergipe in Brazil. In the present contribution we used morphological data along with molecular systematics techniques to investigate the phylogeny and phylogeography of the freshwater pufferfishes of the genus *Colomesus*. The molecular part is based on a cytochrome C oxidase subunit I dataset constructed from both previously published and newly determined sequences, obtained from specimens collected from three distinct localities in South America. Our results from both molecular and morphological approaches enable us to identify and describe a new *Colomesus* species from the Tocantins River. We also discuss aspects of the historical biogeography and phylogeography of the South American freshwater pufferfishes, suggesting that it could be more recent than previously expected.

Editor: Dorothee Huchon, Tel-Aviv University, Israel

Funding: This study was supported by the Brazilian National Counsel of Technological and Scientific Development and Fundação Carlos Chagas Filho de Amparo à Pesquisa do Estado do Rio de Janeiro. The funders had no role in study design, data collection and analysis, decision to publish, or preparation of the manuscript.

Competing Interests: The authors have declared that no competing interests exist.

* E-mail: crlamaral@yahoo.com.br

Introduction

The Tetraodontidae is an Acantomorpha fish family with circumglobal distribution composed of 189 species in 19 genera, occurring in seas, estuaries, and rivers between the tropical and temperate regions [1]. They are mainly characterized by their typical four large dental plates; the ability to inflate their body in stressful situations; the presence of the neurotoxin Tetrodotoxin/Saxitoxin in its tissues, being responsible for numerous cases of fatal poisoning in many countries, including Brazil; and by having the smallest genome among vertebrates, therefore being considered as a model for the genome evolution of the group.

Among the Amazonian taxa exploited by the ornamental fish industry in South America are those of *Colomesus* [2], a genus confined to South America, with what is presently considered two species, *C. asellus* and *C. psittacus*. *C. asellus* [3] is spread in the entire Amazon, Tocantins-Araguaia drainages, and coastal environments from the Amazon mouth to Venezuela, being the only freshwater puffers on that continent. *C. psittacus* [4] is found in coastal marine and brackish water environments from Cuba and the northern coast of South America to Sergipe in Brazil.

Mainly located in tropical and subtropical regions all around the world, including the Amazon region, the ornamental fish industry is one of the largest transporters of live animals and plants with an annual trade volume estimated at U$15–25 billion [5–7], in a scenario where species identification problems, mainly related to border biosecurity are not rare.

The DNA barcode is a widely accepted tool for species determination mainly due to its enhanced attention on standardization and data validation [8], being a rapid and low cost method of identification [9]. The use of DNA barcoding techniques has been utilized in many taxa, including bacteria, birds, bivalves, butterflies, fishes, flies, macroalgae, mammals, spiders, sprigtails, and also for plants [10–27].

The DNA barcode technique for Metazoans uses a short (~650 bp) and standardized gene region from the mitochondrial 5′ region of the cytochrome C oxidase subunit I (COI) for a rapid and cost-effective animal identification. This has been demonstrated to be an effective fish identification tool in numerous situations, including consumer protection [28–30], fisheries management/conservation [31], border biosecurity in the ornamental fish trade [5], and in the identification of overlooked or cryptic species [32].

Here we used both morphological and molecular methodologies in an integrative taxonomical approach to investigate the diversity of the Amazonian freshwater pufferfishes of the genus *Colomesus* based on specimens collected from three distinct populations from both Brazil and Peru. Additionally, we describe a new *Colomesus*

Figure 1. Map of South America showing the northern hydrology and the localities where the specimens were collected (grey and green marks).

species from the Upper Tocantins drainage based on both morphological and molecular data.

Methods

Specimens of *Colomesus asellus* were collected from three distinct populations with about 2200 km of mean distance separating them. The collection localities were Ilha do Mosqueiro, Belém, Brazil; Upper Tocantins River - Porto Nacional, Tocantins, Brazil; and Nanay River - Iquitos, Peru (Figure 1).

Ethics Statement

No statement from an ethics committee was necessary, and the manuscript did not involve any endangered or protect species. All samples were extracted from dead specimens collected with appropriate permissions under authorization number 22512 issued by SISBIO/Instituto Chico Mendes de Conservação da Biodiversidade. We used the ice-slurry method for killing following [33] as they are tropical warm water species and the collected specimens are all smaller than 5 cm SL. All specimens were preserved in alcohol. The reported localities do not include protected areas.

Nomenclatural Acts

The electronic version of this document does not represent a published work according to the International Code of Zoological Nomenclature (ICZN), and hence the nomenclatural acts contained in the electronic version are not available under that Code from the electronic edition. Therefore, a separate edition of

Table 1. Taxonomic sampling and accession numbers.

Taxon	Accession No.	Taxon	Accession No.
Família Triodontidae			JQ841396
Triodon macropterus	AP009170		JQ840304
Família Diodontidae			GU225449
Diodon holocanthus	AP009177	*Sphoeroides annulatus*	GU440524
Chilomycterus reticulatus	AP009188	*Sphoeroides testudineus*	KC959927*
Família Tetraodontidae			KC959928*
Lagocephalus laevigatus	AP011934		GU225665
	JQ365394		GU225453
	JQ365392		GU225450
	JQ365395		JQ842706
	JQ365393		JQ843064
	KC959926*		JQ840306
Lagocephalus inermis	FJ434549		GU225664
	GU804920		JQ365576
Lagocephalus lunaris	DSFSG9111		JQ365575
Lagocephalus lagocephalus	AP011933		JQ365574
Lagocephalus guentheri	HQ149858		GU440524
	JF493722	*Colomesus psittacus*	KC959923*
	JF493724		KC959924*
Lagocephalus wheeleri	JF952772		KC959925*
	AP009538	*Colomesus asellus*	KC959904*
	FJ434551		KC959907*
Lagocephalus spadiceus	EU595163		KC959908*
	EU595161		KC959909*
Takifugu ocellatus	AP009536		KC959910*
Takifugu poecilonotus	AP009539		KC959911*
Takifugu snyderi	AP009531		KC959913*
Takifugu oblongus	AP009535		KC959914*
Takifugu pardalis	AP009528		KC959915*
Takifugu niphobles	AP009526		KC959916*
Takifugu porphyreus	AP009529	*Colomesus tocantinensis*	KC959905*
Tetraodon biocellatus	KC959929*		KC959906*
Tetraodon nigroviridis	KC959930*		KC959912*
Sphoeroides pachygaster	EU074597		KC959917*
	EU074598		KC959918*
	EU869843		KC959919*
	CSFOM07310		KC959920*
	JF494544		KC959921*
	JF494541		KC959922*
	EU869842		
	EU869839		
	AP006745		
Sphoeroides greeleyi	JQ365572		
Sphoeroides nephelus	JQ842698		
	JQ842695		
	JQ842699		
Sphoeroides spengleri	JQ842704		
	JQ842695		
	JQ842701		
	JQ841395		

(*)Sequences newly determined in this study.

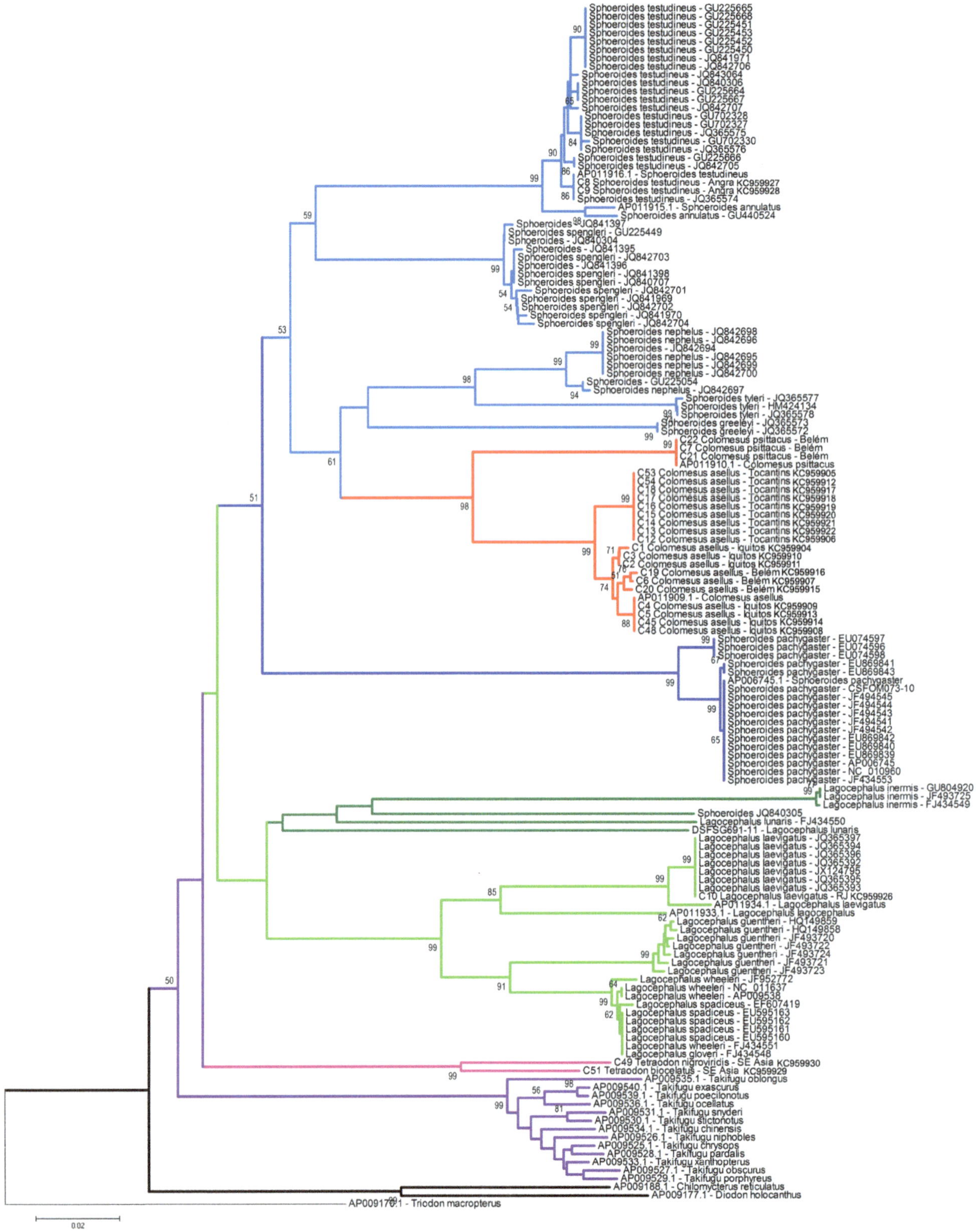

Figure 2. Neighbor-Joining tree based on the barcode region of the COI. The numbers near the branches represent bootstrap probabilities higher than 50%.

AP009170.1 - Triodon macropterus
AP009177.1 - Diodon holocanthus
AP009188.1 - Chilomycterus reticulatus
Sphoeroides testudineus - JQ842706
Sphoeroides testudineus - GU225668
Sphoeroides testudineus - GU225665
Sphoeroides testudineus - GU225451
Sphoeroides testudineus - GU225453
Sphoeroides testudineus - GU225452
Sphoeroides testudineus - JQ841971
Sphoeroides testudineus - GU225450
Sphoeroides testudineus - JQ842707
Sphoeroides testudineus - JQ843064
Sphoeroides testudineus - JQ840306
Sphoeroides testudineus - GU225667
Sphoeroides testudineus - GU225664
Sphoeroides testudineus - JQ365576
Sphoeroides testudineus - GU702330
Sphoeroides testudineus - GU702328
Sphoeroides testudineus - JQ365575
Sphoeroides testudineus - GU702327
Sphoeroides testudineus - JQ842705
Sphoeroides testudineus - GU225666
C8 Sphoeroides testudineus - Angra KC959927
AP011916.1 - Sphoeroides testudineus
Sphoeroides testudineus - JQ365574
C9 Sphoeroides testudineus - Angra KC959928
Sphoeroides annulatus - GU440524
AP011915.1 - Sphoeroides annulatus
Sphoeroides spengleri - JQ842702
Sphoeroides spengleri - JQ841969
Sphoeroides spengleri - JQ842704
Sphoeroides spengleri - JQ841970
Sphoeroides spengleri - JQ842701
Sphoeroides - JQ841395
Sphoeroides spengleri - JQ842703
Sphoeroides spengleri - JQ840707
Sphoeroides - JQ841396
Sphoeroides spengleri - JQ841398
Sphoeroides - JQ841397
Sphoeroides spengleri - GU225449
Sphoeroides - JQ840304
Sphoeroides nephelus - JQ842696
Sphoeroides nephelus - JQ842698
Sphoeroides - JQ842694
Sphoeroides nephelus - JQ842695
Sphoeroides nephelus - JQ842700
Sphoeroides nephelus - JQ842699
Sphoeroides nephelus - JQ842697
Sphoeroides - GU225054
Sphoeroides tyleri - JQ365577
Sphoeroides tyleri - JQ365578
Sphoeroides tyleri - HM424134
Sphoeroides greeleyi - JQ365572
Sphoeroides greeleyi - JQ365573
C7 Colomesus psittacus - Belém KC959924
C22 Colomesus psittacus - Belém KC959923
C21 Colomesus psittacus - Belém KC959925
AP011910.1 - Colomesus psittacus
C54 Colomesus asellus - Tocantins KC959912
C53 Colomesus asellus - Tocantins KC959905
C18 Colomesus asellus - Tocantins KC959917
C17 Colomesus asellus - Tocantins KC959918
C16 Colomesus asellus - Tocantins KC959919
C15 Colomesus asellus - Tocantins KC959920
C14 Colomesus asellus - Tocantins KC959921
C13 Colomesus asellus - Tocantins KC959922
C12 Colomesus asellus - Tocantins KC959906
C48 Colomesus asellus - Iquitos KC959908
C45 Colomesus asellus - Iquitos KC959914
C5 Colomesus asellus - Iquitos KC959913
AP011909.1 - Colomesus asellus
C4 Colomesus asellus - Iquitos KC959909
C6 Colomesus asellus - Belém KC959907
C19 Colomesus asellus - Belém KC959916
C20 Colomesus asellus - Belém KC959915
C1 Colomesus asellus - Iquitos KC959904
C3 Colomesus asellus - Iquitos KC959910
C2 Colomesus asellus - Iquitos KC959911
Sphoeroides pachygaster - EU074597
Sphoeroides pachygaster - EU074596
Sphoeroides pachygaster - EU074598
AP006745.1 - Sphoeroides pachygaster
Sphoeroides pachygaster - EU869841
Sphoeroides pachygaster - EU869843
Sphoeroides pachygaster - CSFOM073-10
Sphoeroides pachygaster - JF494545
Sphoeroides pachygaster - JF494544
Sphoeroides pachygaster - JF494543
Sphoeroides pachygaster - JF494541
Sphoeroides pachygaster - JF494542
Sphoeroides pachygaster - EU869842
Sphoeroides pachygaster - EU869840
Sphoeroides pachygaster - EU869839
Sphoeroides pachygaster - AP006745
Sphoeroides pachygaster - NC_010960
Sphoeroides pachygaster - JF434553
AP009536.1 - Takifugu ocellatus
AP009540.1 - Takifugu exascurus
AP009539.1 - Takifugu poecilonotus
AP009533.1 - Takifugu xanthopterus
AP009535.1 - Takifugu oblongus
AP009529.1 - Takifugu porphyreus
AP009527.1 - Takifugu obscurus
AP009525.1 - Takifugu chrysops
AP009528.1 - Takifugu pardalis
AP009534.1 - Takifugu chinensis
AP009526.1 - Takifugu niphobles
AP009531.1 - Takifugu snyderi
AP009530.1 - Takifugu stictonotus
C49 Tetraodon nigroviridis - SE Asia KC959930
C51 Tetraodon biocelatus - SE Asia KC959929
Lagocephalus inermis - FJ434549
Lagocephalus inermis - JF493725
Lagocephalus inermis - GU804920
Lagocephalus lunaris - FJ434550
DSFSG691-11 - Lagocephalus lunaris
Lagocephalus laevigatus - JQ365397
Lagocephalus laevigatus - JQ365394
Lagocephalus laevigatus - JQ365396
Lagocephalus laevigatus - JQ365392
Lagocephalus laevigatus - JX124795
Lagocephalus laevigatus - JQ365395
C10 Lagocephalus laevigatus - RJ KC959926
Lagocephalus laevigatus - JQ365393
AP011934.1 - Lagocephalus laevigatus
AP011933.1 - Lagocephalus lagocephalus
Lagocephalus guentheri - HQ149859
Lagocephalus guentheri - HQ149858
Lagocephalus guentheri - JF493720
Lagocephalus guentheri - JF493722
Lagocephalus guentheri - JF493723
Lagocephalus guentheri - JF493724
Lagocephalus guentheri - JF493721
Lagocephalus wheeleri - NC_011637
Lagocephalus wheeleri - AP009538
Lagocephalus wheeleri - JF952772
Lagocephalus spadiceus - EF607419
Lagocephalus gloveri - FJ434548
Lagocephalus spadiceus - EU595163
Lagocephalus spadiceus - EU595162
Lagocephalus spadiceus - EU595161
Lagocephalus spadiceus - EU595160
Lagocephalus wheeleri - FJ434551
0.05

Figure 3. Maximum-likelihood phylogeny based on the barcode region of the COI marker. The numbers near the branches represent the bootstrap probabilities.

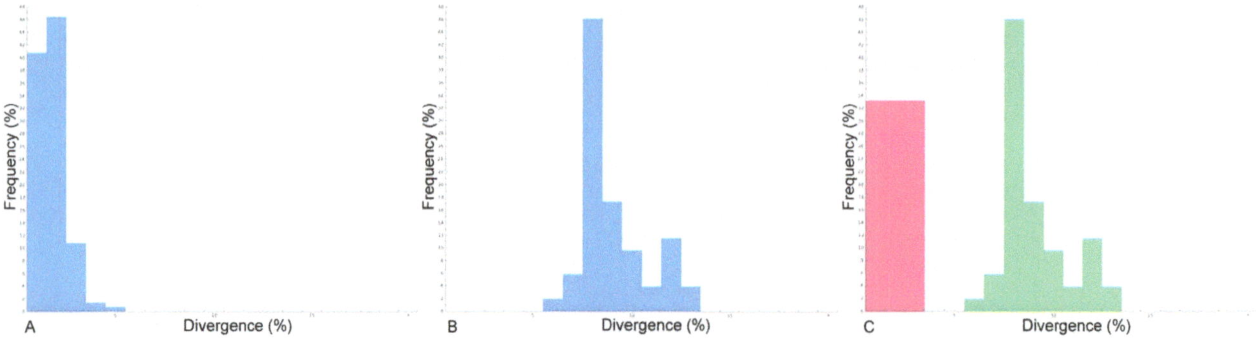

Figure 4. Distribution of K2P distances (%) for COI: A) within species; B) within genera; C, normalized distribution of K2P distance (%) within species. The analyses included the following taxa: *Tetraodon nigroviridis, Tetraodon biocellatus, Sphoeroides testudineus, Lagocephalus laevigatus, Colomesus asellus, Colomesus psittacus*, and the freshwater *Colomesus* from the Tocantins drainage.

this document was produced by a method that assures numerous identical and durable copies, and those copies were simultaneously obtainable (from the publication date noted on the first page of this article) for the purpose of providing a public and permanent scientific record, in accordance with Article 8.1 of the Code. The separate print-only edition is available on request from PLoS by sending a request to PLoS ONE, 1160 Battery Street, Suite 100, San Francisco, CA 94111, USA along with a check for $10 (to cover printing and postage) payable to "Public Library of Science".

In addition, this published work and the nomenclatural acts it contains have been registered in ZooBank, the proposed online registration system for the ICZN. The ZooBank LSIDs (Life Science Identifiers) can be resolved and the associated information viewed through any standard web browser by appending the LSID to the prefix "http://zoobank.org/". The LSID for this

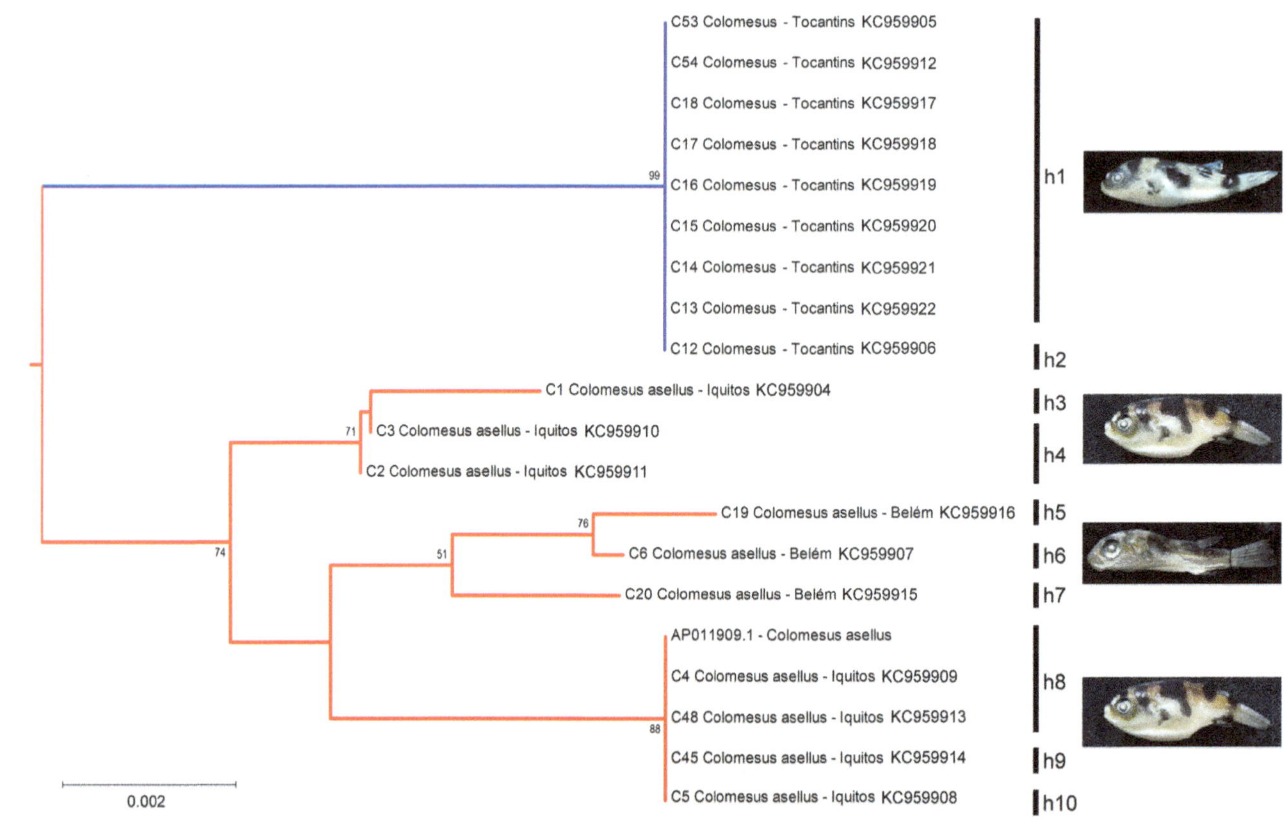

Figure 5. Neighbor-Joining phylogeny of the freshwater *Colomesus* and haplotype determination.

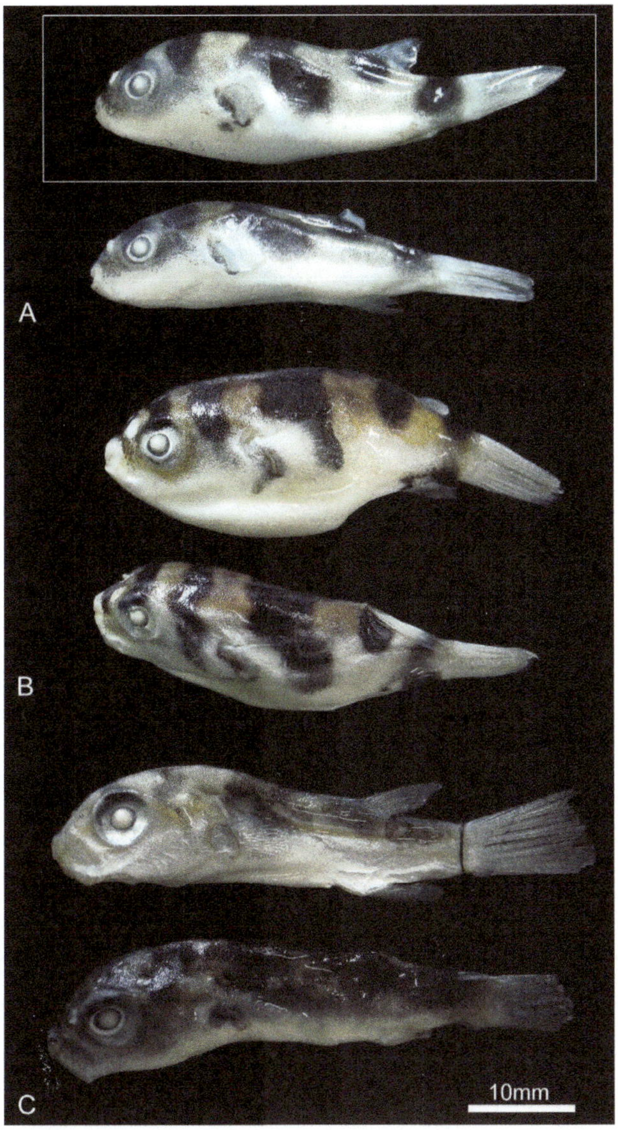

Figure 6. External morphology of the genus *Colomesus*. A) *Colomesus tocantinensis* **nov. sp.** – Tocantins (holotype PNT.UERJ.405 highlighted in white); B) *Colomesus asellus* – Iquitos; C) *Colomesus asellus* – Belém.

publication is: urn:lsid:zoobank.org:pub:033A323A-18F7-4788-8405-32D78BF65B13.

Morphological Analyses

Specimens from all three localities were cleared and stained following the methodology of [34].

Molecular Analyses

The molecular systematic analyses used newly determined sequences obtained from the mitochondrial barcode marker COI as well as previously published sequences obtained from the NCBI and BOLD databases.

For the newly determined sequences, a fragment of epaxial musculature was submitted to the standard protocol for DNA extraction and purification from the Qiagen QIAamp DNA FFPE Tissue kit. The fragments were amplified and sequenced using the primers VF2_t1 and FishR2_t1 [35–37]. All primers were appended with M13 tails on sequencing reactions. The PCR profile consisted of 2 min at 95°C, 35 cycles of 30 sec at 94°, 40 sec at 52°C, and 1 min at 72°C, with a final extension step for 10 min at 72°C. Sequencing reactions were performed with the use of the BigDye® Terminator v.3.1 Cycle Sequencing kit (Applied Biosystems, Inc.), with 25 cycles of 10 sec at 95°C, 5 sec at 50°C and 4 min at 60°C. Sequencing products were processed in an ABI 3500 capillary system (Applied Biosystems, Inc.).

Table 2. Morphometric and meristic data of the type series of *Colomesus tocantinensis* nov. sp.

Register	SL	HL	PR	DR	AR	CR	IOL
PNT.403	30.84	10.75	15	10	9	11	5.8
PNT.404	34.9	11.83	15	10	9	11	5.9
PNT.405*	29.62	10.37	16	10	9	11	5.47
PNT.395	30.35	11.25	15	10	9	11	6.35
PNT.396	29.28	11.1	16	9	9	11	6.78
PNT.397	29.59	10.79	16	10	9	11	5.48
PNT.398	29.46	10.6	15	10	9	11	5.77
PNT.399	30.66	11.99	16	10	9	11	5.55
PNT.400	32.92	11.75	15	10	9	11	6.03
PNT.401	27.02	9.61	16	10	9	11	5.12

SL, standard length; HL, head length; PR, pectoral fin rays; DR, dorsal fin rays; AR, anal fin rays; CR, caudal fin rays; IOL, interorbital length.
(*)Holotype.

The chromatograms were checked and aligned using the BioEdit 7.053 [38] software with its built-in ClustalW routine [39]. The alignment was visually inspected for accuracy and to minimize missing data. All the newly determined sequences are available at the BOLD database (http://www.barcodinglife.com) under the project acronym PUFER. The GenBank accession numbers for all newly determined and previously published sequences used in the present manuscript are summarized in Table 1. The dataset consisted of a 651 bp COI matrix, and we used the MEGA 5.06 software [40] to determinate the TN93+G+I as the most appropriate model of sequence evolution based on the Akaike criterion (AIC) [41].

The neighbor-joining (NJ) and maximum-likelihood (ML) trees that encompass the genera *Triodon*, *Diodon*, *Chilomycterus*, *Lagocephalus*, *Tetraodon*, *Takifugu*, *Sphoeroides*, and *Colomesus*, were constructed using the MEGA 5.06 software [40].

The neighbor-joining sequence divergences were calculated based on the Kimura Two Parameter (K2P) distance model [42] on BOLD workbench and MEGA 5.06 software [40]. The haplotype determination was carried with the use of the server FaBox (http://birc.au.dk/software/fabox/).

Results and Discussion

The neighbor-joining (NJ) and maximum-likelihood (ML) result trees are presented in Figures 2 and 3, respectively. The genus *Colomesus* was recovered as monophyletic inside the group formed by the sampled *Sphoeroides* species, in except for *Sphoeroides pachygaster*. *Lagocephalus* was recovered in a basal phylogenetic position in relation to *Sphoeroides* and *Colomesus*, therefore corroborating recent results such as those presented by [43–45].

Colomesus was recovered deep inside the group formed by the remaining *Sphoeroides* species, therefore suggesting *Sphoeroides* as paraphyletic, with *S. pachygaster* being recovered as basal in relation

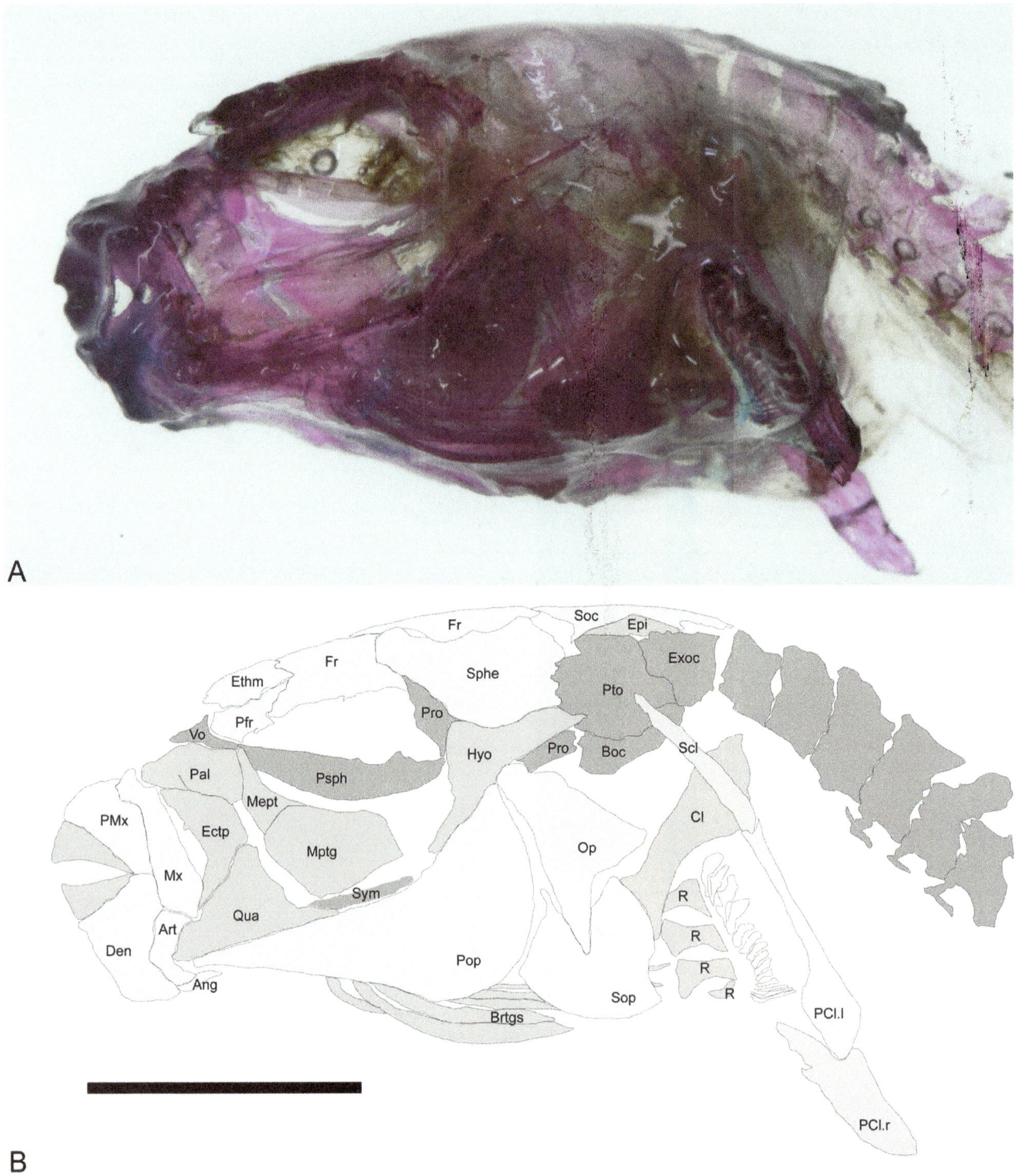

Figure 7. ***Colomesus tocantinensis*** **nov. sp. (PNT.UERJ.398).** A) left photograph of the head; B) anatomical interpretations. Abbreviations: Ang, angular; Art, articular; Boc, basioccipital; Brstgs, branchiostegals; Cl, cleithrum; Den, dentary; Epi, epiotic; Ethm, ethmoid; Exo, exoccipital; Fr, frontal; Hyo, hyomandibula; Ecptg, ectopterygoid; Mept, mesopterygoid; Mtptg, metapterygoid; Mx, Maxilla; Op, opercle; Pal, palatine; PCl.l/r, ventral postcleithrum left and right; Pfr, prefrontal; PMx, premaxilla; Pop, preopercle; Pro, prootic; Psph, parasphenoid; Pto, pterotic; Qua, quadrate; R, radials; Scl, supracleithrum; Soc, supraocciptal; Sop, subopercle; Sphe, sphenotic; Sym, sympletic; Vo, vomer. Scale bar equals 5 mm.

to all the remaining *Sphoeroides* species in all the analyses. Additionally, *Colomesus* was also recovered as the sister-taxa of the group formed by the species *Sphoeroides nephelus*, *S. tyleri*, and *S. greeleyi* in the NJ result, although it was recovered as the sister-taxa of *S. greeleyi* in the ML results.

In the same way, *Colomesus* is clearly distinguishable from the group formed by all the *Sphoeroides* species mainly by the banded

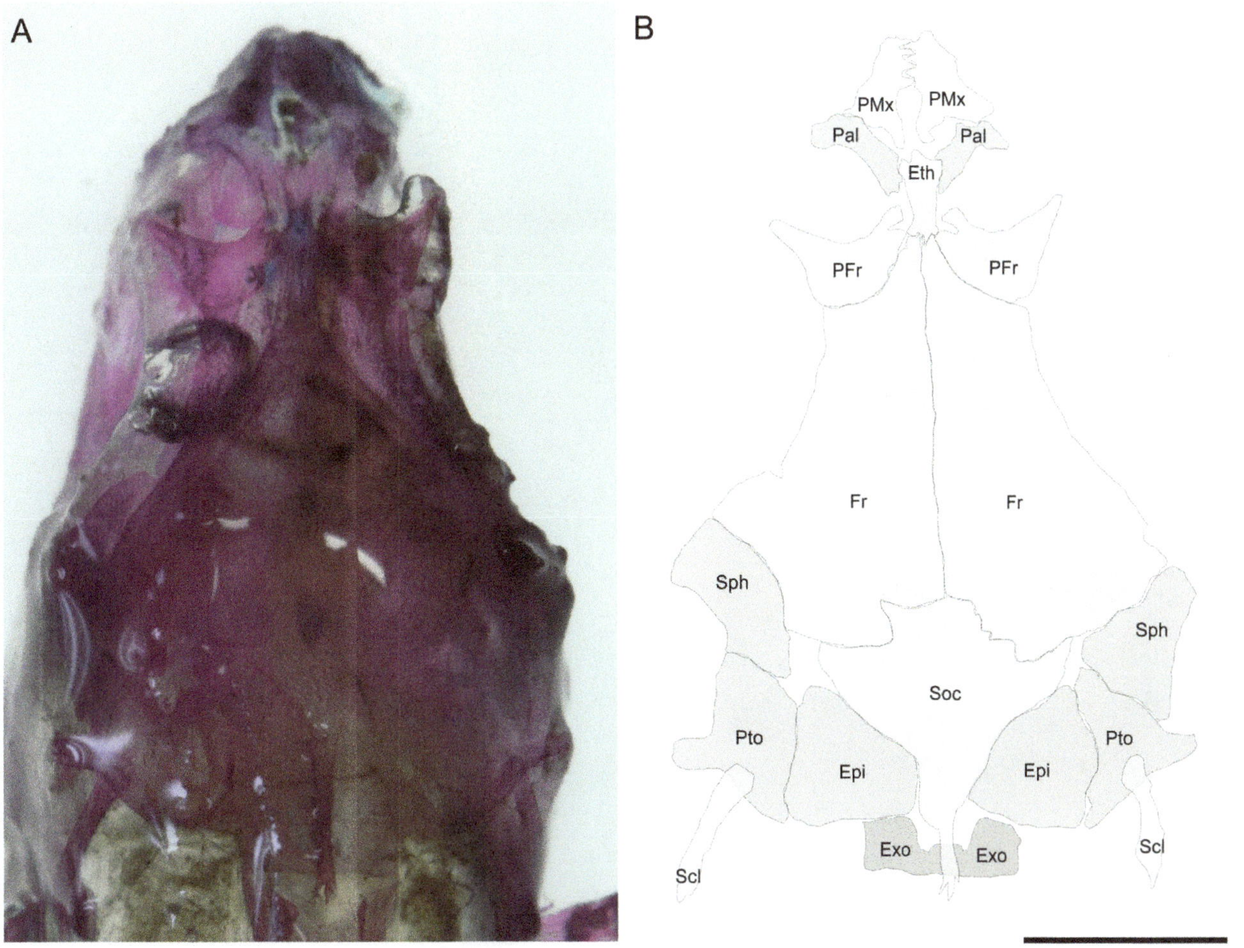

Figure 8. ***Colomesus tocantinensis*** **nov. sp. (PNT.UERJ.398).** A) top photograph of the head; B) anatomical interpretations. Abbreviations: Epi, epiotic; Ethm, ethmoid; Exo, exoccipital; Fr, frontal; Pal, palatine; Pfr, prefrontal; PMx, premaxilla; Pto, pterotic; Scl, supracleithrum; Soc, supraocciptal; Sphe, sphenotic. Scale bar equals 5 mm.

pigmentation pattern present in all the *Colomesus* species; the presence of two lateral lines, with the ventral line running the full length of the caudal peduncle; and the absence of an upraised horizontal ridge of skin ventrolaterally along the caudal peduncle. The color pattern was used by [46], along with pectoral fin ray counts, the presence of a dark bar underside of caudal peduncle, and the presence of dermal flaps across the chin, to distinguish between what at that time were considered to be the only two species of the genus, the marine/estuarine *C. psittacus*, and the freshwater *C. asellus*. The presence of a dark bar on the underside of the caudal peduncle is a prominent feature for specimens of *C. asellus* from Iquitos, but this bar is present or not in specimens from both Belém and Tocantins. Dermal flaps were observed in all specimens from both Iquitos and Belém, but such flaps were not observed in any of the examined specimens from the Tocantins drainage.

DNA Barcode and Deep Sequence Divergence

COI amplicons were obtained from all the specimens included in the analyses. The obtained sequences clearly identified both previous accepted *Colomesus* species (*C. asellus* and *C. psittacus*), therefore being in accordance with the previous morphological diagnose presented by [46].

The K2P divergence distances between congeneric species ranged from 5.557% to 12.394% with a mean distance of 8.546%, while the uncorrected K2P distance ranged from 0 to 4.472% within species. The mean K2P distance within the analyzed populations was 0.657% and the mean normalized distance within species is 1.079% (Figure 4).

Deep sequence divergence was observed regarding the freshwater *Colomesus* from the Tocantins drainage (Figure 5). The mean sequence divergence of the specimens from both Belém and Iquitos was estimated at 1.079%, while the Tocantins distances ranged from 1.955% to 3.063%, with a mean distance of 2.166%. The observed sequence divergence values together with the congruence observed from both molecular and morphological phylogenetic approaches used here suggest the existence of an overlooked species within the genus *Colomesus*.

A new *Colomesus* species from the Tocantins River, Brazil

Systematics. Tetraodontiformes *sensu* Tyler, 1980 [47]

Tetraodontidae *sensu* Santini & Tyler, 2003 [45]

Colomesus Gill, 1885 [2]

Colomesus tocantinensis **nov. sp.** urn:lsid:zoobank.org:act:9B8ACCB5-FF55-4514-901B-6366FB6EA307

Derivation of name. The specific epithet *tocantinensis* refers to the type locality, Porto Nacional, State of Tocantins, Brazil.

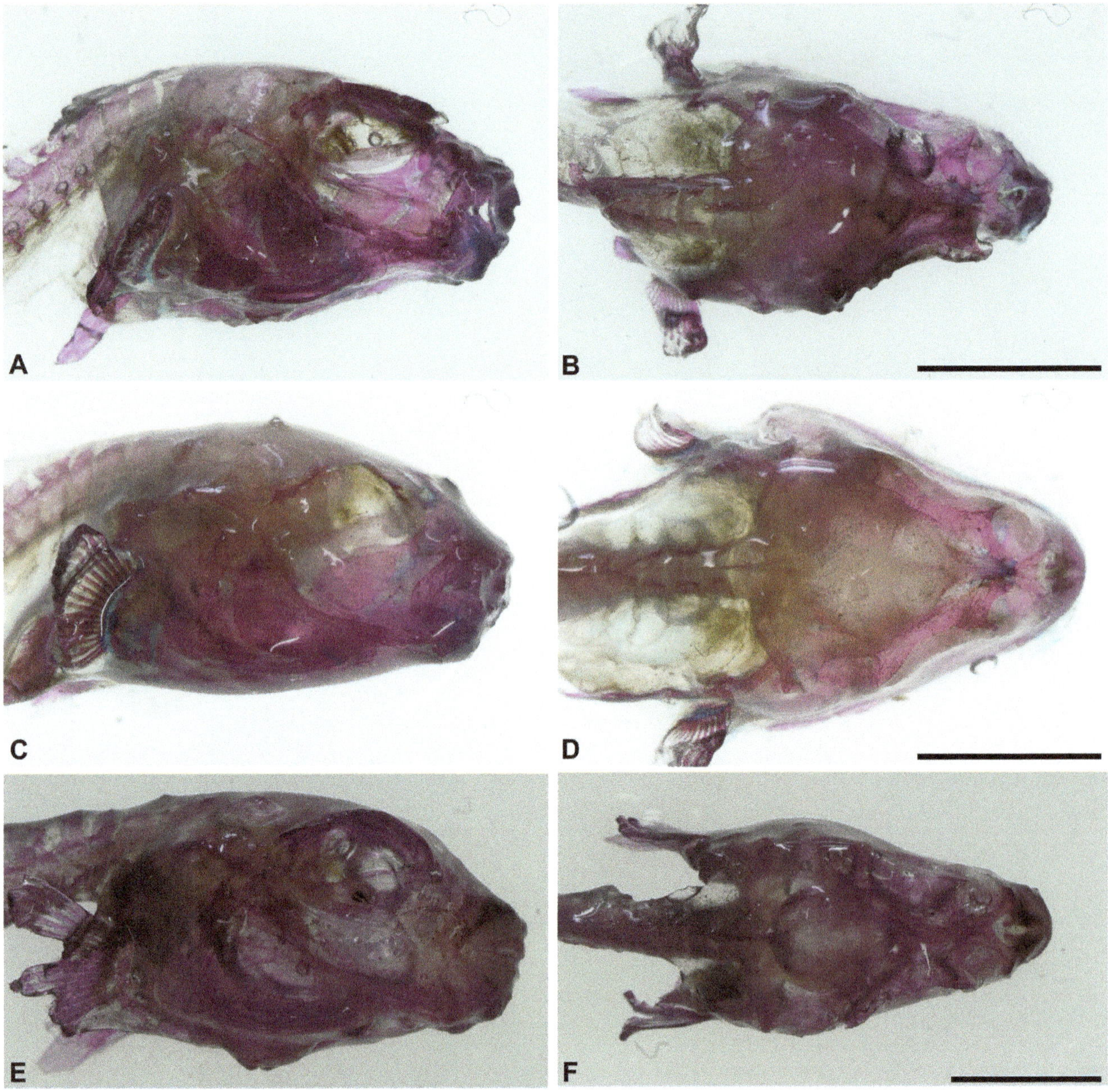

Figure 9. Right and top photographs from cleared-and-stained specimens of: A–B) *Colomesus tocantinensis* **nov. sp. – Tocantins (PNT.UERJ.398); C–D)** *Colomesus asellus* **– Iquitos (PNT.UERJ.470); E–F)** *Colomesus asellus* **– Belém (PNT.UERJ.386).**

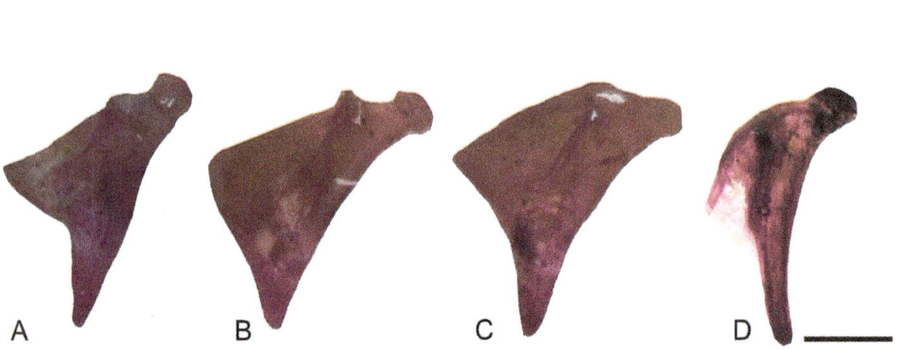

Figure 10. Isolated opercles from: A) *Colomesus tocantinensis* **nov. sp. – Tocantins (PNT.UERJ.398); B)** *Colomesus asellus* **– Iquitos (PNT.UERJ.470); C)** *Colomesus asellus* **– Belém (PNT.UERJ.386); D)** *Colomesus psittacus* **– Belém (PNT.UERJ.387).** Scale bar equals 1 mm.

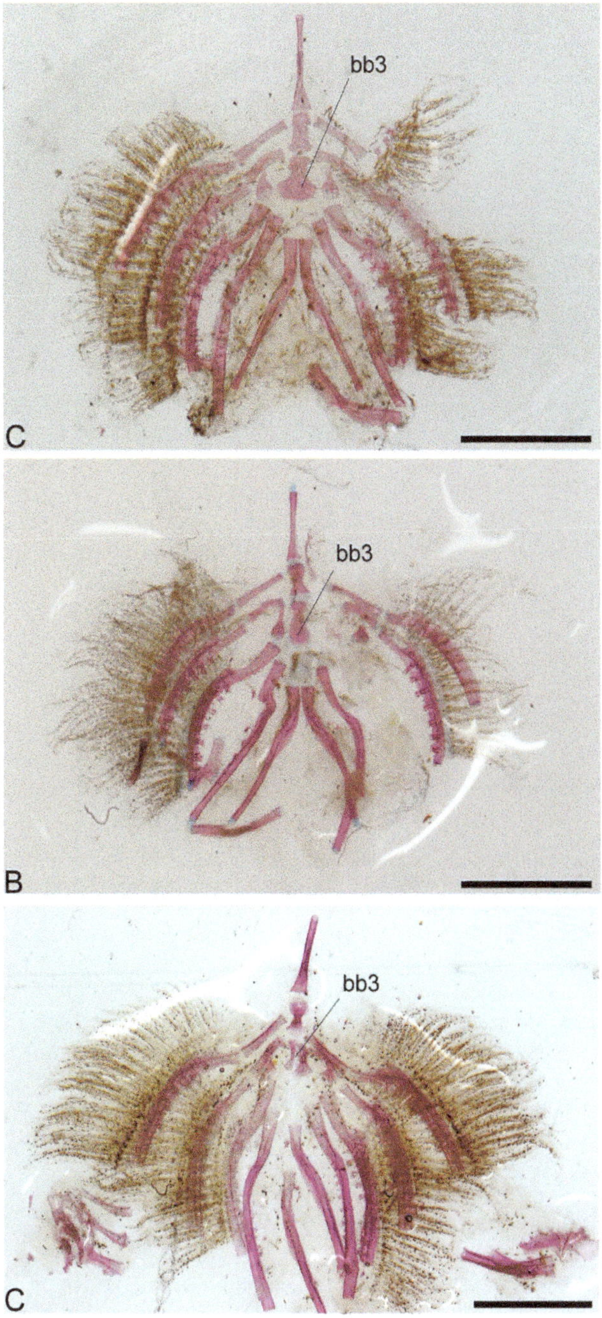

Figure 11. Isolated branchial apparatus from: A) ***Colomesus tocantinensis*** **nov. sp. - Tocantins (PNT.UERJ.404); B)** ***Colomesus asellus*** **- Iquitos (PNT.UERJ.470); C)** ***Colomesus asellus*** **- Belém (PNT.UERJ.386).** Scale bar equals 2.5 mm.

Holotype. PNT.UERJ.405 (Figure 6).

Paratypes. PNT.UERJ.396, PNT.UERJ.397, PNT.UERJ.398, PNT.UERJ.399, PNT.UERJ.400, PNT.UERJ.401, PNT.UERJ.402, PNT.UERJ.403, PNT.UERJ.404.

Type-locality. The specimens are from the Tocantins River near Porto Nacional, State of Tocantins, Brazil.

Diagnosis. *Colomesus* species diagnosed by six to seven basal pterygiophores and nine rays in the anal fin (*contra* ten to eleven in both *C. asellus* and *C. psittacus*); ten basal pterygiophores and rays in the dorsal fin (*contra* eleven for both *C. asellus* and *C. psittacus*); the absence of dermal flaps across the chin (*contra* its presence uniquely in *C. asellus*); a caudal peduncle with eight vertebrae; and an opercle with a posterior ventral border subdivided in a ventral and a posterior region, the herein called "inverted V" shape (*contra* the triangular opercle exhibited by both *C. asellus* and *C. psittacus*).

Description. The holotype (PNT.UERJ.405) is 29,62 mm SL (Figure 6), with 10,37 mm HL; the entire type-series ranges from 27.02 mm to 34.9 mm SL. The meristic and morphometric data of the type series is presented in Table 2. The extent of the dorsal and ventral lateral lines is similar to those found in *C. asellus*. The prickles extend along the dorsal, lateral, and ventral surfaces of the body, from the level of the eye to the origin of the dorsal fin.

The color pattern of *Colomesus tocantinensis* **nov. sp.** is essentially the same as that of *Colomesus asellus*, with five transverse dark bars across the dorsal region of the body. A dark blotch on the underside of the caudal peduncle, which is a state used by [46] to diagnose *Colomesus asellus*, *is* present or absent, being vestigial to unobservable or absent in several specimens. The interspaces between the dark bars are light yellow, with gradually decreasing pigmentation and becoming white in the ventral region (Figure 6). However, the light yellow to pale pattern presented by *C. tocantinensis* **nov. sp**. clearly contrasts with the gold-yellow pattern present in specimens from Iquitos and Belém.

The nasal sac is higher than that presented in the specimens of *C. asellus*. Two large lateral and anteromedial nostrils are present. They are similar to those found on *C. psittacus*, rather than the two small nostrils exhibited by *C. asellus*. The anterior surface of the nasal sac is smooth while the posterior surface of it is folded as in *C. psittacus*, exhibiting a "T-shaped" ridge with a relatively small dorsal flap. This flap seems much smaller than the one found on *C. asellus*, although more flexible when compared to *C. psittacus*.

The presence of dermal flaps across the chin is another character used by [46] to distinguish *C. asellus* from *C. psittacus*. No dermal flaps could be seen in the examined specimens from the Tocantins River, although they are always present in examined specimens from Iquitos and Belém.

The skull is partially similar to those found in *Colomesus asellus* described and figured by [46], although the frontals exhibit a wide posterior border and prominently participate in the orbital margin (Figures 7–9). The prefrontals are triangular and articulate medially with the ethmoid, which posteriorly articulates with the frontals and anteriorly with the palatines (Figure 8). The supraoccipital is roughly triangular and well developed, with an elongate posterior process which covers the first vertebrae (Figure 8). The sphenotics articulate postero-laterally with the frontals and, in the examined specimens, they neither contact nor closely approach the prefrontals. The lateral wing of the sphenotics is only partially developed (Figure 8). Posterior to the sphenotics, the pterotics (Figures 7 and 8) articulates posteriorly with the slender supracleithrum and medially with the epiotics, which articulates medially with the supraoccipital (Figure 9).

In lateral view, the skull is characterized by the wide preopercle with about 110 degrees between both horizontal and vertical rami (Figure 7), with the preopercular canal running along its anterior border, and by the opercle which is divided in two distinct regions, having ventral and posterior wings, the herein called "inverted V" shape, distinct from the condition found in all other examined specimens of *Colomesus* (Figure 10). The subopercle is sturdy, with two small dorsal processes.

The parasphenoid is elongate and does not exhibit any developed dorsal flange (Figures 7 and 9). The hyomandibula is roughly triangular and has a slender ventral region; its wide head articulates dorsally with the sphenotics, and its upper posterior edge with the anterior end of the opercle (Figure 7).

Figure 12. ***Colomesus tocantinensis*** **nov. sp. (PNT.UERJ.403).** A) left photograph of the unpaired fins and caudal endoskeleton; B) anatomical interpretations. Abbreviations: E, epural; H, dorsal hypural plate; H-H, ventral hypural plate fused with the ural centrum; Phy, parhypural; PU, pre-ural vertebrae; PbD, dorsal-fin basal pterigiophores; PbA, anal-fin basal pterigiophores. Scale bar equals 5 mm.

The palatine is wide and somewhat triangular, with a robust anterior process for the maxilla (Figure 7). The maxilla is robust, with an anterodorsal region articulating with the premaxilla and a posterior expanded region, medially concave for muscle insertion. The ectopterygoid articulates dorsally with the palatine and ventrally with the anterodorsal border of the quadrate. The metapterygoid is wide and composes almost the entire ventral orbital region (Figure 7). It articulates anteriorly with the mesopterygoid (Figure 7), and with the posterior end of the large and triangular quadrate (Figure 7). The quadrate exhibits a well-developed posteroventral spine articulating posteriorly with the slender symplectic (Figure 7), and anteriorly with the articular. The articular is "L" shaped and articulates anteriorly with the robust dentary and ventrally with the small angular (Figure 7).

Five branchiostegal rays (Figure 7) are present and the branchial apparatus is strikingly similar in all the examined specimens (Figure 11).

The pectoral girdle is robust and formed by a wide cleithrum, somewhat triangular and posteriorly expanded, articulating dorsally with the slender supracleithrum. The supracleithrum articulates ventrally with the two postcleithra; a slender dorsal postcleithrum, followed by the posteriorly expanded ventral postcleithrum (Figure 7). There are four radials and sixteen pectoral fin rays (Figure 7).

The axial skeleton has 19 vertebrae. The dorsal fin originates between vertebrae 7–8 and has ten basal pterygiophores and ten fin rays (Figures 12–14). The anal fin is located beneath the 9th vertebra and has six basal pterygiophores and nine fin rays.

The caudal skeleton (Figure 12) has a wide ural centrum formed by the preural centrum 1, the ural centrum, the ventral hypural plate, and the postero-dorsal expansion which articulates anteriorly with the almost triangular epural, and posteriorly with the dorsal hypural plate (Figure 12). Eleven caudal fin rays, five dorsal and six ventral, are present in all of the specimens, both the uppermost and the two lowermost rays are unbranched.

Phylogeography of the South American Freshwater Pufferfishes

Although the influence of marine incursions after the Miocene is still under debate, the Caribbean (or Miocene) marine incursion, via the Llanos Basin (Colombia-Venezuela), is well accepted based on both geological and paleontological evidence, suggesting that these incursions may have isolated marine taxa within the western South America freshwater environments [48–52]. This might be the case for the freshwater tetraodontids. As pointed by [53], this scenario predicts that the distribution of the marine sister groups of marine lineages should be related with the Caribbean or western Atlantic, the age of freshwater taxa should be coincident with marine incursions, and the biogeographic congruence should be observed among multiple unrelated taxa, conditions only partially filled by the genus *Colomesus*.

The timing of divergence between the brackish/marine *C. psittacus* and the freshwater *C. asellus* was recently discussed [45], based on a multiple loci approach including both nuclear and mitochondrial markers. The authors dated the split between 2,5-7My, therefore postdating the Miocene marine incursions usually

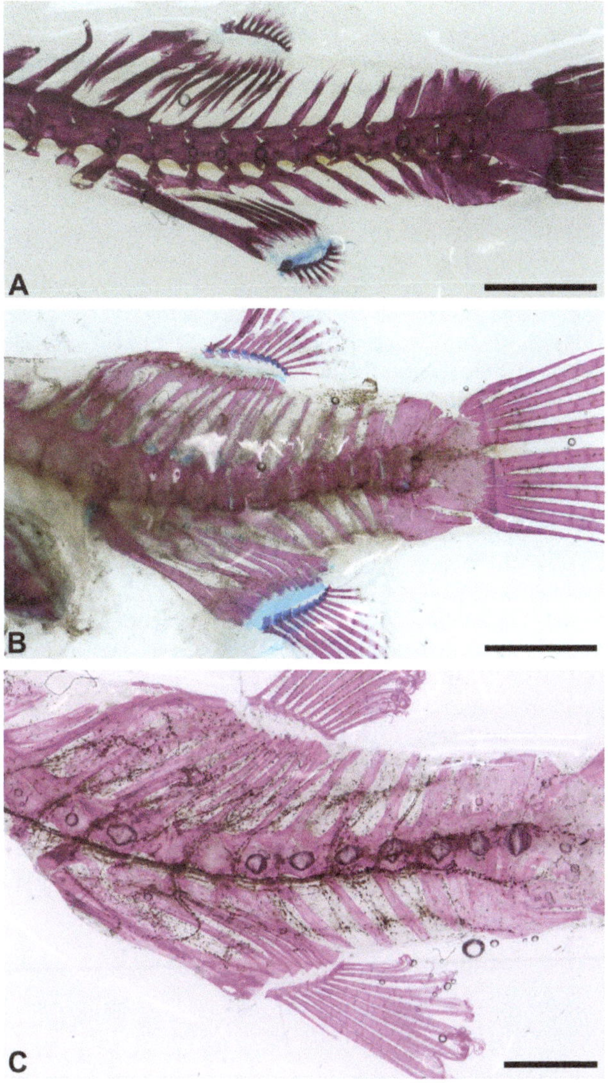

Figure 13. Left view photographs from cleared-and-stained specimens of: A) ***Colomesus tocantinensis*** **nov. sp. (PNT.UERJ.403); B)** ***Colomesus asellus*** **- Iquitos (PNT.UERJ.470); C)** ***Colomesus asellus*** **- Belém (PNT.UERJ.386).** Scale bar equals 5 mm.

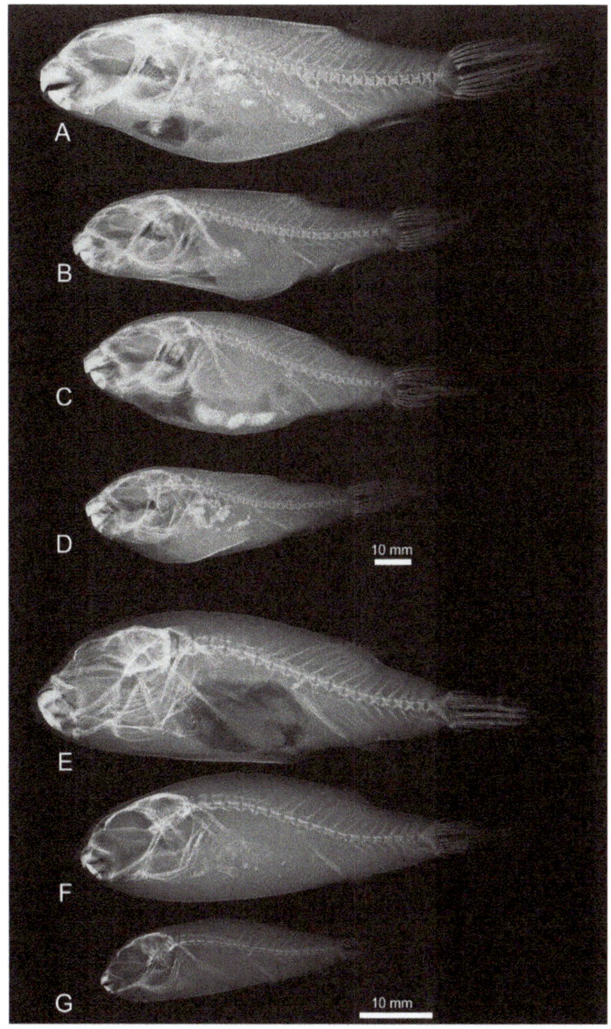

Figure 14. High-definition x-ray images of: A–D, ***Colomesus psittacus*** **USNM.393077; E–G,** ***Colomesus asellus*** **USNM.191569.** Scale bar equals 10 mm.

used to explain the presence of several marine groups within the western Amazon. In this sense, as observed by [45], the colonization carried by the tetraodontids in South America could be presumably related to the Pliocene global climate oscillations. Additionally, the basal split of the *Colomesus* from Tocantins and from Iquitos/Belém agrees with the general area cladogram of neotropical fishes presented by [54] in which the Xingu/ Tocantins-Araguaia group was recovered in a basal position in relation to the groups from the lowlands of Western and Eastern Amazon.

It was recently proposed [55], based on the distribution of characiforms, that recent marine incursions would have isolated fish populations in upland terrains or refuges, where lineage divergence is maximized, followed by dispersal episodes back to the lowlands. The "museum hypothesis" predicts that lowlands exhibit a higher number of species, but lower levels of endemism, than highlands, and that the upland refuges would represent areas of high endemism.

Looking on the molecular phylogeny of the serrasalmids *Pygocentrus* and *Serrasalmus*, [56] proposed a phylogenetic test which predicts that basal lineages in a phylogeny of widespread fishes would occur in highland areas, and lowland lineages would have originated only during the last 5 Ma. Additionally, [57] studying the genetics of *Symphysodon* cichlids, indicated the effects that marine incursions would have in population structure, stating that populations in upland terrains or refuges would exhibit reduced genetic variation, while populations in lowlands would represent multiple upland sources, therefore exhibiting a high level of genetic variation, and that populations in lowlands would show a demographic pattern of expansion.

Our results recovered the Upper Tocantins lineages in a basal position in relation to all the remaining specimens, with the sequences being collapsed in uniquely two haplotypes (Figure 5), the first one (h1), represented by eight sequences, and the second haplotype (h2), represented by a unique sequence. This suggests low genetic variation, at least among the studied sampling, and a history initially related with the eastern Amazon, followed by a subsequently expansion to the western South America.

The Tocantins-Araguaia Ichthyofauna

The Tocantins-Araguaia drainage is the fourth largest Brazilian drainage, draining part of the northern end of the Brazilian shield directly to the eastern end of the Amazon Basin. It exhibits a recent geomorphological history, within a still tectonically active sedimentary basin with recent subsidence episodes, which are related with the high load of sediments observed within the basin, leading the development of the Bananal Plain, in the lower part of the drainage, mainly during the Quaternary [58–59].

The Tocantins drainage, specially the Upper Tocantins River, is constantly regarded as an area of high endemism, with several fish taxa restricted to this area having been described, such as *Leporinus taeniofasciatus* (Anostomidae), *Sternarchorhynchus mesensis* (Apteronotidae), *Aspidoras albater*, *A. eurycephalus* (Callichthyidae), *Acestrocephalus maculosus*, *Astyanax unitaeniatus*, *Astyanacinus goyanensis*, *Creagrutus atrisignum*, *C. britskii*, *C. mucipu*, *C. saxatilis*, *Hyphessobrycon hamatus*, *Moenkhausia tergimaculata*, *Cetopsis caiapo*, *C. sarcodes* (Cetopsidae), *Characidium stigmosum* (Crenuchidae), *Pimelodella spelaea* (Heptapteridae), *Ancistrus aguaboensis*, *A. jataiensis*, *A. minutus*, *A. reisi*, *Corumbataia veadeiros*, *Hemiancistrus micrommatos*, *Hypostomus ericae*, *Gymnotocinclus anosteos*, *Lamontichthys avacanoeiro* (Loricariidae), *Apareiodon argenteus* and *A. cavalcante* (Parodontidae), *Cynolebias griseus*, *C. notatus*, *Rivulus planaltinus*, and *Simpsonichthys marginatus* (Rivulidae), the herein described *Colomesus tocantinensis* **nov. sp.** (Tetraodontidae), and *Ituglanis bambui* and *I. mambai* (Trichomycteridae).

In the same way, [60–62] pointed that the Tocantins River is the only river system related with the Amazon Basin, and draining shield areas which exhibit a considerable number of fish taxa known from the lowlands of the central and western Amazon. The faunal similarity between both basins is constantly related with the evolution of the Bananal Plain as a selective barrier for the lowland fauna and the upper part of the drainage, and the list of typical lowland ichthyofauna shared by the Tocantins-Araguaia and the Amazon drainages includes, as presented by [60], *Leporinus trifasciatus* (Anostomidae), *Arapaima gigas* (Arapaimatidae), *Auchenipterichthys coracoideus* (Auchenipteridae), *Mylossoma* spp., *Pygocentrus nattereri* (Characidae), *Cetopsis candiru*, *C. coecutiens* (Cetopsidae), *Cichla monoculus*, *C. pleiozona*, *C. kelberi* (Cichlidae), *Psectrogaster amazonica* (Curimatidae), *Thorachocharax stellatus* (Gasteropelecidae), *Osteoglossum bicirrhosum* (Osteoglossidae), *Pellona castelnaeana*, *Pristigaster cayana* (Pristigasteridae), and *Colomesus asellus* (Tetraodontidae). In this sense, even still under debate, it seems clear that the Tocantins-Araguaia drainage has a composite nature including both lowland and upland ichthyofauna, as argued by [62].

Conclusions

Based on a comprehensive analysis including both morphological and molecular methodologies using the cytochrome C oxidase I gene, we were able to discuss aspects of the phylogeny and phylogeography of the South American freshwater pufferfishes of the genus *Colomesus*.

Our molecular results based on the COI marker agrees with the recent results such as [43–45], and suggest that the genus *Sphoeroides* should be revised, mainly regarding the phylogenetic position recovered for the genus *Colomesus*, deeply nested within the *Sphoeroides* tree, and the basal position recovered for *S. pachygaster*. We plan further investigations along these lines to reconcile any conflicts between these molecular hypotheses presented herein and morphologically based interpretations [47] of the phylogeny of the taxa of *Colomesus*, *Sphoeroides*, and *Lagocephalus*.

The use of molecular systematic techniques together with morphological methodologies confirmed the identification of a new cryptic pufferfish species from the Upper Tocantins drainage, *Colomesus tocantinensis* **nov. sp.** Morphological features such as the color pattern, the absence of dermal flaps across the chin, the distinct 'inverted V' opercle shape, and the caudal peduncle morphology, all support the description of *Colomesus tocantinensis* **nov. sp.**, as a new pufferfish species from the South American freshwater drainages.

The timing of divergence between the marine/brackish species *Colomesus psittacus* and the freshwater group formed by *C. asellus* and *C. tocantinensis* **nov. sp.**, as recovered by [45], postdates the Miocene marine incursions usually used to explain the presence of tetraodontids within the Amazon freshwater environments. Therefore, it suggests that the freshwater colonization in South America, at least for the tetraodontids, could be more recent than previously expected. Additionally, together with the observed distribution of haplotypes, our results suggest that the history of tetraodontids into the Amazonian freshwater environments could be presumably related to the Pliocene global climate oscillations and its effects inside the eastern Amazon and subsequently to the western South America.

Finally, our results reinforce the Upper Tocantins drainage as an area of high endemism within the Tocantins-Araguaia drainage, although the composite nature of the entire drainage is unquestionable.

Acknowledgments

We would like to thank Dr. James C. Tyler (Smithsonian Institution, Washington) for his support and helpfull comments on the manuscript. We are also grateful to Dr. Francesco Santini (Universitá degli Studi di Torino), Dr. Dorothée Huchon (Tel Aviv University), and an anonymous reviewer for the valuable suggestions during the review of the manuscript, Dr. Leonor Gusmão and Dr. Antonio Amorim (Universidade do Porto) for the comments during the initial discussion of the results, Yuri Modesto (Universidade do Estado do Rio de Janeiro) for the specimens from the Tocantins drainage, Lúcio Paulo Machado and Diogo de Mayrink (Universidade do Estado do Rio de Janeiro) for the specimens from Iquitos; Ms. Sandra Raredon (Smithsonian Institution, Washington) for the x-rays of tetraodontids, Dr. Richard Pyle (Hawaii Biological Survey) for the LSID numbers, and Kleyton M. C. Severiano and Anna Carolina Chaves (Universidade do Estado do Rio de Janeiro) for the technical assistance.

Author Contributions

Conceived and designed the experiments: CRLA PMB DAS EFC. Performed the experiments: CRLA PMB DAS EFC. Analyzed the data: CRLA PMB DAS EFC. Contributed reagents/materials/analysis tools: CRLA PMB DAS EFC. Wrote the paper: CRLA.

References

1. Froese R, Pauly D (2012) *FishBase*. World Wide Web electronic publication. www.fishbase.org, 2012, version (02/2012).
2. Gill TN (1885) Synopsis of the Plectognath fishes. Proc United States Natl Mus 7: 411–427.
3. Müller J, Troschel FH (1849) Fische. In: Reisen in Britisch-Guiana in den Jahren, 1840–1844. Im Auftrag Sr. Mäjestat des Königs von Preussen ausgeführt von Richard Schomburgk. [Versuch einer Fauna und Flora von Britisch-Guiana.] 3. Berlin.
4. Bloch ME, Schneider JG (1801) Systema Ichthyologiae iconibus cx illustratum. Post obitum auctoris opus inchoatum absolvit, correxit, interpolavit Jo. Gottlob Schneider, Saxo. Berolini. Sumtibus Auctoris Impressum et Bibliopolio Sanderiano Commissum. M. E. Blochii, Systema Ichthyologiae.: i-lx +1–584, Pls. 1–110.
5. Collins RA, Armstrong KF, Meier R, Yi Y, Brown SDJ, et al. (2012) Barcoding and border Biosecurity: Identifying Cyprinid Fishes in the Aquarium Trade. PLoS ONE 7(1): e28381. doi:10.1371/journal.pone.0028381.

6. Padilla DK, Williams SL (2004) Beyond ballast water: aquarium and ornamental trades as sources of invasive species in aquatic ecosystems. Front Ecol Environ 2: 131–138.
7. Ploeg A, Bassleer G, Hensen R (2009) Biosecurity in the Ornamental Aquatic Industry. Maarssen: Orn Fish Intl. 148 p.
8. Mabragaña E, Díaz de Astarloa JM, Hanner R, Zhang J, González Castro M (2011) DNA Barcoding Identifies Argentine Fishes from Marine and Brackish Waters. PLoS ONE 6(12): e28655. doi:10.1371/journal.pone.0028655.
9. Golding GB, Hanner R, Hebert PDN (2009) Preface. Molecular Ecology Resources 9(Suppl. 1): iv–vi.
10. Hogg ID, Hebert PDN (2004) Biological identification of springtails (Collembola: Hexapoda) from the Canadian Arctic, using mitochondrial DNA barcodes. Can J Zoo 82: 749–754.
11. Barrett RDH, Hebert PDN (2005) Identifying spiders through DNA barcodes. Can J of Zoo 83: 481–491.
12. Hebert PDN, Cywinska A, Ball SL, deWaard JR (2003a) Biological identification through DNA barcodes. Proc Roy Soc London, Series B: Biol Sci 270: 313–321.
13. Hebert PDN, Ratnasingham S, deWaard JR (2003b) Barcoding animal life: cytochrome c oxidase subunit 1 divergences among closely related species. Proc Roy Soc London, Series B: Biol Sci (Suppl 1) 270: 96–99.
14. Janzen DH, Hajibabaei M, Burns JM (2005)Wedding biodiversity inventory of a large and complex Lepidoptera fauna with DNA barcoding. Phil Trans Roy Soc London, Series B, Biol Sci 1462: 1835–1846.
15. Hajibabaei M, Janzen DH, Burns JM, Hallwachs W, Hebert PDN (2006) DNA barcodes distinguish species of tropical Lepidoptera. Proc Natl Acad Sci U S A, 103: 968–971.
16. Lukhtanov VA, Sourakov A, Zakharov EV, Hebert PDN (2009) DNA barcoding Central Asian butterflies: increasing geographical dimension does not significantly reduce the success of species identification. Mol Ecol Res 9: 1302–1310.
17. Smith MA, Wood DM, Janzen DH, Hallwachs W, Hebert PDN (2007) DNA barcodes affirm that 16 species of apparently generalist tropical parasitoid flies (Diptera, Tachinidae) are not all generalists. Proc Natl Acad Sci U S A 104: 4967–4972.
18. Järnegren J, Schander C, Sneli JA, Ronningen V, Young CM (2007) Four genes, morphology and ecology: distinguishing a new species of Acesta (Mollusca; Bivalvia) from the Gulf of Mexico. Mar Biol 152: 43–55.
19. Ward RD, Zemlak TS, Innes BH, Last PR, Hebert PDN (2005) DNA barcoding Australia's fish species. Phil Trans Roy Soc London, Series B, Biol Sci 360: 1847–1857.
20. Hebert PDN, Stoeckle MY, Zemlak TS, Francis CM (2004) Identification of birds through DNA barcodes. PLoS Biol 2: 1657–1663.
21. Kerr KCR, Lijtmaer DA, Barreira AS, Hebert PDN, Tubaro PL (2009) Probing evolutionary patterns in neotropical birds through DNA barcodes. PloS ONE 4: e4379.
22. Clare EL, Lim BK, Engstrom MD, Eger JL, Hebert PDN (2007) DNA barcoding of Neotropical bats: species identification and discovery within Guyana. Mol Ecol Notes 7: 184–190.
23. Amaral AR, Sequeira M, Coelho MM (2007) A first approach to the usefulness of cytochrome c oxidase I barcodes in the identification of closely related delphinid cetacean species. Mar Fresh Res 58: 505–510.
24. Borisenko AB, Lim BK, Ivanova NV, Hanner RH, Hebert PDN (2008) DNA barcoding in surveys of small mammal communities: a field study in Suriname. Mol Ecol Res 8: 471–479.
25. Hollingsworth PM, Forrest LL, Spouge JL, Hajibabaei M, Ratnasingham S, et al. (2009) A DNA barcode for land plants. Proc Natl Acad Sci U S A 106: 12794–12797.
26. Saunders GW (2005) Applying DNA barcoding to red macroalgae: a preliminary appraisal holds promise for future applications. Phil Trans Roy Soc London, Series B, Biol Sci 360: 1879–1888.
27. Sogin ML, Morrison HG, Huber JA, Welch DM, Huse SM, et al. (2006) Microbial diversity in the deep sea and the underexplored 'rare biosphere'. Proc Natl Acad Sci U S A 103: 12115–12120.
28. Lowenstein JH, Amato G, Kolokotronis SO (2009) The real maccoyii: identifying tuna sushi with DNA barcodes - contrasting characteristic attributes and genetic distances. PLoS ONE 4: e7866.
29. Lowenstein JH, Burger J, Jeitner CW, Amato G, Kolokotronis SO, et al. (2010) DNA barcodes reveal species-specific mercury levels in tuna sushi that pose a health risk to consumers. Biol Lett 6: 692–695.
30. Cohen NJ, Deeds JR, Wong ES, Hanner RH, Yancy HF, et al. (2009) Public health response to puffer fish (tetrodotoxin) poisoning from mislabeled product. J Food Prot 72: 810–817.
31. Holmes BH, Steinke D, Ward RD (2009) Identification of shark and ray fins using DNA barcoding. Fish Res 95: 280–288.
32. Steinke D, Zemlak TS, Hebert PDN (2009) Barcoding Nemo: DNA-Based identifications for the Ornamental Fish Trade. PLoS ONE 4(7): e6300. doi:10.1371/journal.pone.0006300.
33. Blessing JJ, Marshall JC, Balcombe SR (2010) Humane killing of fishes for scientific research: a comparison of two methods. J Fish Biol. 76(10): 2571–2577. doi: 10.1111/j.1095-8649.2010.02633.x.
34. Song J, Parenti L (1995) Clearing and Staining Whole Fish Specimens for Simultaneous Demonstration of Bone, Cartilage, and Nerves. Copeia 1995(1): 114–118.
35. Palumbi SR (1996) Nucleic acids II: the polymerase chain reaction. In: Molecular Systematics, Hillis DM, Moritz C, Mable BK (Eds), Sinauer & Associates Inc., Sunderland, Massachusetts: 205–247.
36. Ward RD, Zemlak TS, Innes BH, Last PR, Hebert PDN (2005) DNA barcoding Australia's fish species. Phil Trans R Soc London. Series B, Biol Sci 360: 1847–1857.
37. Ivanova NV, Zemlak TS, Hanner RH, Hebert PDN (2007) Universal primer cocktails for fish DNA barcoding. Mol Ecol Notes 7: 544–548. doi: 10.1111/j.1471-8286.2007.01748.x.
38. Hall TA (1999) BioEdit: a user-friendly biological sequence alignment editor and analysis program for Windows 95/98/NT. Nucleic Acids Symposium Series 41: 95–98.
39. Thompson JD, Gibson TJ, Plewniak F, Jeanmougin F, Higgins DG (1994) The Clustal X windows interface: Flexible strategies for multiple sequence alignment aided by quality analysis tools. Nuc Acid Res 24: 4876–4882.
40. Tamura K, Peterson D, Peterson N, Stecher G, Nei M, et al. (2011) MEGA5: Molecular Evolutionary Genetics Analysis using Maximum Likelihood, Evolutionary Distance, and Maximum Parsimony Methods. Mol Biol Evol 28: 2731–2739.
41. Akaike H (1973). Information theory and an extension of the maximum likelihood principle. Proc. 2nd Inter. Symposium on Information Theory, Budapest: 267–281.
42. Kimura M (1980) A simple method for estimating evolutionary rate of base substitutions through comparative studies of nucleotide sequences. J Mol Evol 16: 111–120.
43. Yamanoue Y, Miya M, Doi H, Mabuchi K, Sakai H, et al. (2011) Multiple Invasions into Freshwater by Pufferfishes (Teleostei: Tetraodontidae): A Mitogenomic Perspective. PLoS ONE 6(2): e17410. doi:10.1371/journal.-pone.0017410.
44. Santini F, Tyler JC (2003) A phylogeny of the families of fossil and extant tetraodontiform fishes (Acanthomorpha, Tetraodontiformes), Upper Cretaceous to recent. Zoo J Linn Soc 139: 565–617.
45. Santini F, Nguyen MTT, Sorenson L, Waltzek TB, Lynch Alfaro JW, et al. (2013) Do habitat shifts drive diversification in teleost fishes? An example from the pufferfishes (Tetraodontidae). J Evol Biol: 1–16.
46. Tyler JC (1964) A diagnosis of the two species of South American puffer fishes (Tetraodontidae, Plectognathi) of the genus *Colomesus*. Proc Acad Natl Sci Philad 116: 119–148.
47. Tyler JC (1980) Osteology, phylogeny, and higher classification of the fishes of the order Plectognathi (Tetraodontiformes). National Oceanic and Atmospheric Administration technical reports, Natl Mar Fish Serv 434: 1–422.
48. Nuttall CP (1990) A review of the Tertiary non-marine molluscan faunas of the Pebasian and other inland basins of north-western South America. Bull Brit Mus Natl Hist, Geol 45: 165–371.
49. Webb SD (1995) Biological implications of the Middle Miocene Amazon seaway. Science 269: 361–362.
50. Lovejoy NR (1997) Stingrays, parasites, and Neotropical biogeography: A closer look at Brooks, et al's hypotheses concerning the origins of Neotropical freshwater rays (Potamotrygonidae). Sys Biol 46: 218–230.
51. Lovejoy NR, Bermingham E, Martin AP (1998) Marine incursion into South America. Nature 396: 421–422.
52. Lovejoy NR, Albert JS, Crampton WGR (2006) Miocene marine incursions and marine/freshwater transitions: evidence from Neotropical fishes. J Sou Am Ear Sci 21: 5–13.
53. Bloom DD, Lovejoy NR (2011) The Biogeography of Marine Incursions in South America. In: Albert JS, Reis RE (eds). Historical biogeography of neotropical freshwater fishes. 406p.
54. Albert JS, Carvalho TP (2011) Neogene Assembly of Modern Faunas. In: Albert JS, Reis RE (eds). Historical biogeography of neotropical freshwater fishes. 406p.
55. Hubert N, Renno J-F (2006) Historical biogeography of South American freshwater fishes. J Biog 33: 1414–1436.
56. Hubert N, Duponchelle F, Nuñez J, Garcia-Davila C, Paugy D, et al. (2007) Phylogeography of the piranha genera *Serrasalmus* and *Pygocentrus*: Implications for the diversification of the Neotropical ichthyofauna. Mol Ecol 16: 2115–2136.
57. Farias IP, Hrbek T (2008) Patterns of diversification in the discus fishes (*Symphysodon* spp. Cichlidae) of the Amazon basin. Mol Phyl Evol 49: 32–43.
58. Saadi A (1993) Neotectônica da plataforma brasileira: Esboço e interpretação preliminares. Geonomos 1: 1–15.
59. Saadi A, Bezerra FHR, Costa RD, Igreja HLS, Franzinelli E (2005) Neotectônica da Plataforma Brasileira. In: *Quaternário do Brasil*, edited by Souza CRG, Suguio K, Oliveira AMS, Oliveira PE, 211–234. Ribeirão Preto: Holos Editora.
60. Jégu M, Keith P (1999) Lower Oyapock River as northern limit for the Western Amazon fish fauna or only a stage in its northward progression. Compt Rend Biol Acad Sci 322: 1133–1143.
61. Hrbek T, Seckinger J, Meyer A (2007) A phylogenetic and biogeographic perspective on the evolution of poeciliid fishes. Mol Phyl Evol 43: 986–998.
62. Lima FCT, Ribeiro AC (2011) Continental-Scale Tectonic Controls of Biogeography and Ecology. In: Albert JS, Reis RE (eds). Historical biogeography of neotropical freshwater fishes. 406p.

Core Microbial Functional Activities in Ocean Environments Revealed by Global Metagenomic Profiling Analyses

Ari J. S. Ferreira[1], Rania Siam[1], João C. Setubal[2], Ahmed Moustafa[1], Ahmed Sayed[1], Felipe S. Chambergo[3], Adam S. Dawe[4], Mohamed A. Ghazy[1], Hazem Sharaf[1], Amged Ouf[1], Intikhab Alam[4], Alyaa M. Abdel-Haleem[1], Heikki Lehvaslaiho[4], Eman Ramadan[1], André Antunes[5], Ulrich Stingl[6], John A. C. Archer[4], Boris R. Jankovic[4], Mitchell Sogin[7], Vladimir B. Bajic[4], Hamza El-Dorry[1]*

1 Department of Biology and the Science and Technology Research Center, School of Sciences and Engineering, The American University in Cairo, Cairo, New Cairo, Egypt, **2** Departamento de Bioquímica, Instituto de Química, Universidade de São Paulo, São Paulo, Brazil, **3** Escola de Artes, Ciências e Humanidades, Universidade de São Paulo, São Paulo, Brazil, **4** Computational Bioscience Research Center, King Abdullah University of Science and Technology, Thuwal, Kingdom of Saudi Arabia, **5** Institute for Biotechnology and Bioengineering, Centre of Biological Engineering, University of Minho, Portugal, **6** Red Sea Research Center, King Abdullah University of Science and Technology, Thuwal, Kingdom of Saudi Arabia, **7** Josephine Bay Paul Center, Marine Biological Laboratory, Woods Hole, Massachusetts, United States of America

Abstract

Metagenomics-based functional profiling analysis is an effective means of gaining deeper insight into the composition of marine microbial populations and developing a better understanding of the interplay between the functional genome content of microbial communities and abiotic factors. Here we present a comprehensive analysis of 24 datasets covering surface and depth-related environments at 11 sites around the world's oceans. The complete datasets comprises approximately 12 million sequences, totaling 5,358 Mb. Based on profiling patterns of Clusters of Orthologous Groups (COGs) of proteins, a core set of reference photic and aphotic depth-related COGs, and a collection of COGs that are associated with extreme oxygen limitation were defined. Their inferred functions were utilized as indicators to characterize the distribution of light- and oxygen-related biological activities in marine environments. The results reveal that, while light level in the water column is a major determinant of phenotypic adaptation in marine microorganisms, oxygen concentration in the aphotic zone has a significant impact only in extremely hypoxic waters. Phylogenetic profiling of the reference photic/aphotic gene sets revealed a greater variety of source organisms in the aphotic zone, although the majority of individual photic and aphotic depth-related COGs are assigned to the same taxa across the different sites. This increase in phylogenetic and functional diversity of the core aphotic related COGs most probably reflects selection for the utilization of a broad range of alternate energy sources in the absence of light.

Editor: Fabiano Thompson, Universidade Federal do Rio de Janeiro, Brazil

Funding: This work was supported by King Abdullah University for Science and Technology Global Collaborative Partners (GCR) program. The funders had no role in study design, data collection and analysis, decision to publish, or preparation of the manuscript.

Competing Interests: Co-authors Ahmed Moustafa, John Archer and Vladimir Bajic are PLOS ONE Editorial Board members. This does not alter the authors' adherence to PLOS ONE Editorial policies and criteria.

* E-mail: dorry@aucegypt.edu

Introduction

Microbial communities dominate the oceans and seas [1] and are involved in fundamental processes in the global ecosystem, synthesizing half of the photosynthetic biomass and producing significant amounts of oxygen [2]. They also play pivotal roles in carbon, nitrogen, sulfur and phosphorus cycles [3,4,5,6]. The diverse capabilities of marine microbes are the result of selection and adaptation to environments in which light level, oxygen concentration, nutrient availability, salinity, temperature and hydrostatic pressure all vary widely [7,8,9,10,11,12,13,14,15,16,17].

Acquisition and loss of genetic information, and mutational variation of gene regulatory networks, are known to shape the lifestyles of microorganisms including those that reside in marine environments [18,19,20,21,22,23,24,25]. However, some basic questions regarding these processes in marine environments remain unresolved. What metabolic and physiological capabilities enable microbes to thrive in marine habitats? How do biotic and abiotic factors, including human activities and climate change, affect their capacity to adapt to a specific environment? Thanks to recent developments in molecular biology, remote sensing and deep-sea sampling, these issues can now be addressed [26,27,28].

Metagenomic sampling of marine microbial communities at various sites has contributed much to our understanding of their biochemical and physiological diversity [7,8,9,10,11,12,13,14,29,30,31]. However, a comprehensive comparative analysis of publicly accessible sequences from a wide range of depth-related locations in oceans, which is necessary for a better understanding of the distribution and dynamics of microorganisms in marine ecosystems, has been lacking. In this study we report such an analysis, based on 24 metagenomic datasets from known depths at 11 sites representing a wide range of marine settings. These comprise all publicly available data from studies that generated sufficient sequence coverage to permit broad comparative analysis.

Four of the datasets come from samples collected at different depths in the Atlantis II Basin in the Central Red Sea (ATII), and are reported here for the first time. In addition, we included in our analysis three collections of datasets, each representing four different depths in the water column, from ALOHA Station in the North Pacific Subtropical Gyre [32], BATS Station in the Sargasso Sea [32], and Station 3 off the Northern Chilean coast near Iquique (Iquique) [31]; four datasets derived from samples obtained at depths of 1,000 m in the Central Basin of the Sea of Marmara [8], 4,000 m at ALOHA [9], 6,000 m within the Puerto Rico Trench (PRT) [11], and 50 m in the Mediterranean off Alicante in Southeastern Spain (Med.) [10]. We also selected for inclusion in our global study just four sites from the vast array of near-surface datasets acquired by the Global Ocean Sampling (GOS) expedition [7,13]. The entire collection comprises approximately 12.0 million sequences encompassing 5,358 Mb.

As a benchmark for understanding the genomic changes underlying microbial adaptation to different light and oxygen regimes, we assembled a core set of reference depth-related photic and aphotic COGs and defined a group of functional activities associated with extreme hypoxia, based on functional profiling of all 24 datasets mentioned above. These were then utilized to characterize biological activities in two specific oceanographic settings in which the normal relationships between light, oxygen and depth are disrupted: the Humboldt Current System (HCS) - a major upwelling zone - and the salinity-based two-layer flow system in the Sea of Marmara. Phylogenetic assignment of core COGs indicated that the same microbial taxa dominated the photic zone across all the sites considered, while aphotic zones show greater diversification and differentiation.

Materials and Methods

Sample collection, DNA isolation, and Whole-genome shotgun pyrosequencing

Water column samples from Atlantis II Basin in the Red Sea (21°13' N, 37°58' E) were obtained during Leg2 of the WHOI/KAUST/HCMR sampling cruise aboard the R/V Aegeo in April 2010. Permits to collect water samples from all the depths of Atlantis II basin in the Red Sea were issued by the Ministry of Defense, Kingdome of Saudi Arabia. All the field studies did not involve endangered or protected species. Detailed of samples collection and processing, environmental DNA isolation, whole-genome shotgun pyrosequencing, and establishing of ATII depth-related datasets for the four depths are presented in Methods S1.

Nucleotide sequence accession numbers

The ATII 454 metagenome has been deposited in the GenBank Sequence Read Archive with the following accession numbers: ATII 50 m: SRS598124; ATII 200 m: SRS598125; ATII 700 m: SRS598128; and ATII 1500 m: SRS598129.

ATII and publicly available datasets employed in comparative analysis

In addition to the four ATII depth-related datasets newly generated for this work, metagenomic datasets from ten other sites were obtained from public databases (fig. S1 and table S1 in file S1). Detailed information regarding the publically available datasets, 454 shotgun read simulation, sub-sampling, functional assignment using the eggNOG database, and analysis of differentially abundant COGs are presented in Methods S1.

Comparative analysis of datasets based on COG abundance

Because there were three depth-related metagenomic collections of datasets that could be used as reference (ALOHA, BATS and ATII), we elected to first screen for COGs that differ significantly in abundance among the various depths in each of the three water columns. This was achieved by means of Fisher's exact test ($p \leq 0.01$, FDR-corrected), using the exact Test function of the R bioconductor package edgeR [33], in order to identify common COGs that displayed significant differences in normalized abundance levels between at least two depths in one of the three water columns. 176 depth-related COGs were identified at this step and were employed for hierarchical clustering of all datasets (table S2 in file S1).

Determination of reference photic/aphotic global-core depth-related COGs

Datasets were initially classified as photic or aphotic based on publicly available PAR and oceanographic data. Subsequently the average normalized abundances of each COG were determined for the identified photic and aphotic groups of datasets, obtained by the hierarchical clustering of the 24 datasets based on the normalized abundances of the 176 depth-related COGs previously identified (table S3 in file S1).

Then, unequal variance conservative Student's t-test (Welch's test; $p \leq 0.05$, FDR-corrected) was employed to determine the COGs that presented statistically significant differences in the means of the two groups ($n = 12$ for each group). 82 COGs were found to vary significantly ($p \leq 0.0001$, FDR-corrected) in abundance between the photic and aphotic datasets and considered subsequently as the reference photic/aphotic global-core depth-related set of COGs. Out of those, 54 COGs were assigned as photic and 28 COGs were assigned as aphotic based on the means of the normalized abundance of each COGs between the photic and aphotic datasets (tables S4A and S4B in file S1).

Comparison of photic to aphotic depth-related COG abundance ratios in water columns

Using the reference photic/aphotic global-core depth-related set of COGs, the normalized abundance means of the reference 54 photic and 28 aphotic depth-related COGs were determined for each dataset. The ratio of photic to aphotic normalized abundance means of each dataset was then calculated and log2 transformed in order to be plotted. The average and mean standard deviations of the log2-transformed ratios were determined for photic and aphotic datasets from ATII, ALOHA and BATS to establish ratio ranges indicative of photic or aphotic genomic content.

Assessment of the effect of hypoxic environment on COG abundance

To evaluate the effect of low oxygen concentration on the genomic content of microbial communities, COGs that presented statistically significant changes in normalized abundance levels between datasets of the Iquique water column were initially identified following the same procedure performed for the selection of differentially abundant depth-related COGs in the three reference columns (ATII, ALOHA, BATS). Such analysis resulted in a list of 162 COGs from which the 82 global photic/

aphotic common-core depth-related COGs were subsequently excluded to avoid influence of sunlight intensity level on the identification of abundance changes specifically related to dissolved oxygen concentration (D.O.). Datasets were initially divided into two groups based on publicly available dissolved oxygen concentration (D.O.) data for the 24 datasets analyzed. Iquique datasets from 85 m, 110 m and 200 m depths (D.O.< 20 μmol.kg-1) were considered as the low-oxygen standard (n = 3) to which all other remaining datasets (n = 21) were compared. Fisher's exact test was again employed to determine the COGs that were statistically different in abundance across the Iquique water column. COGs whose abundances had been shown to vary in association with light level were omitted from this analysis. Wilcoxon's test was then used to find the COGs whose normalized abundances varied in association with oxygen concentration. 18 COGs (table S5 in file S1) were found to vary significantly ($p \leq 0.05$, FDR-corrected) in abundance between low and high oxygen datasets.

Taxonomic distribution of photic/aphotic depth-related COGs among sites.

The taxonomic distribution analysis was performed on selected datasets representing euphotic and aphotic zones of the different oceanic sites sampled. Datasets obtained from the shallowest samples of the reference water columns (ATII, ALOHA and BATS) plus the four GOS and the Mediterranean ones were used as representative of the euphotic zone, whilst datasets from the deepest samples of the reference water columns and those from Marmara, PRT and ALOHA (4,000 m) corresponded to the aphotic zone in the analysis. Reads from each dataset mapped to the photic/aphotic core related COGs were subjected to BLASTX similarity search against the NCBI nr database using a maximum e-value threshold of 1e-05 [34]. A taxonoxomic identifier was assigned to each individual read based on the best BLAST hit using NCBI taxdump (version of March, 2013). Taxonomic counts were then summarized at the phylum level per COG for each analyzed dataset (tables S6A and S6B in file S1) by means of bash/perl scripts developed in-house. Eukarya or unidentified taxa were discarded from the analysis. Heatmaps were generated from the summarized raw counts for the core photic and aphotic depth-related COGs using the publicly available software environment R version 2.11.1 (www.r-project.org, RColorBrewer and gplots libraries).

Results and Discussion

Defining the reference depth-related environments

The metagenomic datasets employed in this work were derived from samples obtained at different sites in the world's oceans. Detailed information on sites and sampling methods is presented in figure S1, and table S1 in file S1. In order to avoid biases in the comparison of samples due to different sequencing methods and coverage, simulated 454 sequences were generated from datasets made up of assembled contigs (ALOHA 4,000 and GOS datasets) prior to extraction from all analyzed datasets of random subsamples of the same size as the smallest complete collection (BATS 500 m; for details, see Methods S1).

It is important to emphasize that the datasets analyzed here include all those based on sampling at multiple depths in water columns at single sites: ALOHA [32], BATS [32], Iquique [31], and ATII (this work). The Iquique site [31] lies within the Humboldt Current System (HCS), the most productive marine ecosystem in the global oceans [35], which dramatically affects environmental conditions at this location [35]. The HC flows along the West coast of South America and drives a major upwelling system that brings cold, nutrient-rich water toward the surface of the Eastern Tropical South Pacific off Northern Chile. This promotes the growth of a dense planktonic community and supports a high level of primary production [35]. Among the most notable effects is the efficient depletion of dissolved oxygen, from >200 μmol.kg^{-1} at the surface to 3 to 4 μmol.kg^{-1} at depths between 110 and 200 m, by aerobic heterotrophs [31,36]. In addition, due to dense growth of phytoplankton and the presence of organic particulates, the sunlit zone at the Iquique site is rather shallow compared to the other sites (fig. 1B). We therefore considered the water columns at ALOHA, BATS and Atlantis II, but not that at Iquique, as reference depth-related environments.

Clustering COGs according to depth

To identify known biochemical functions represented in the 24 datasets, we subjected each one to sequence similarity analysis using the eggNOG v2.0 database [37], and calculated the normalized abundances of the identified clusters of orthologous groups of proteins (COGs) [38]. COGs that exhibited statistically significant differences in abundance within the three reference water columns were retrieved individually using Fisher's exact test ($p \leq 0.01$, and FDR-corrected; for details see Methods S1). This resulted in 649, 413, and 463 COGs for ATII, ALOHA and BATS, respectively. Of these, 176 COGs were present at all three sites (table S2 in file S1).

We first performed hierarchical clustering on the normalized abundance data for these 176 COGs in the three reference columns. The major feature common to the ATII, ALOHA and BATS profiles is the division of the datasets into two depth-related groups, representing the surface/near-surface zone, and a deeper zone (fig. S2, and table S2 in file S1).

Next, we proceeded to a hierarchical clustering of the same 176-COG set in all 24 datasets. As expected for a generalized depth-related pattern, members of the set represented in the Sea of Marmara (1,000 m), ALOHA (4,000 m) and the Puerto Rico Trench (PRT, 6,000 m) samples clustered with the deeper water group, while those occurring in the four GOS surface-water and the 50 m Mediterranean Sea datasets fell into the surface/near-surface group (fig. 1A, and table S3 in file S1). This depth-related division reflects metabolic and physiological processes specific to the microbial communities residing in photic and aphotic zones. Note that the dividing line coincides with the bottom of the photic zones in the three columns, as determined by their PAR (photosynthetically active radiation) values (fig. 1B). Interestingly, inclusion of the Iquique datasets in the clustering analysis preserves the depth-related profile division found at the other ten sites. However, as shown in fig. 1A, all four datasets generated from the Iquique samples (50 m, 85 m, 110 m and 200 m) clustered with the aphotic group. Nevertheless, the apparently anomalous position of the datasets related to the Iquique column in the profile of the 11 sites (fig. 1A) is entirely compatible with the environmental conditions at the site [31] which, as mentioned above, differ quite significantly from those at ALOHA [12], BATS [39] and ATII sites. The PAR value at Iquique drops below 1% of its surface value at a depth of less than 50 m [31] (fig. 1B). If one defines this level as the lower limit of the photic zone, it is not surprising that all Iquique datasets cluster with the aphotic group.

The photic/aphotic global-core depth-related COGs

In order to identify specific photic/aphotic global-core depth-related COGs we compared the normalized abundance of the photic with the aphotic groups of datasets defined in the dendrogram shown in fig. 1A. Of the 176 COGs used in the

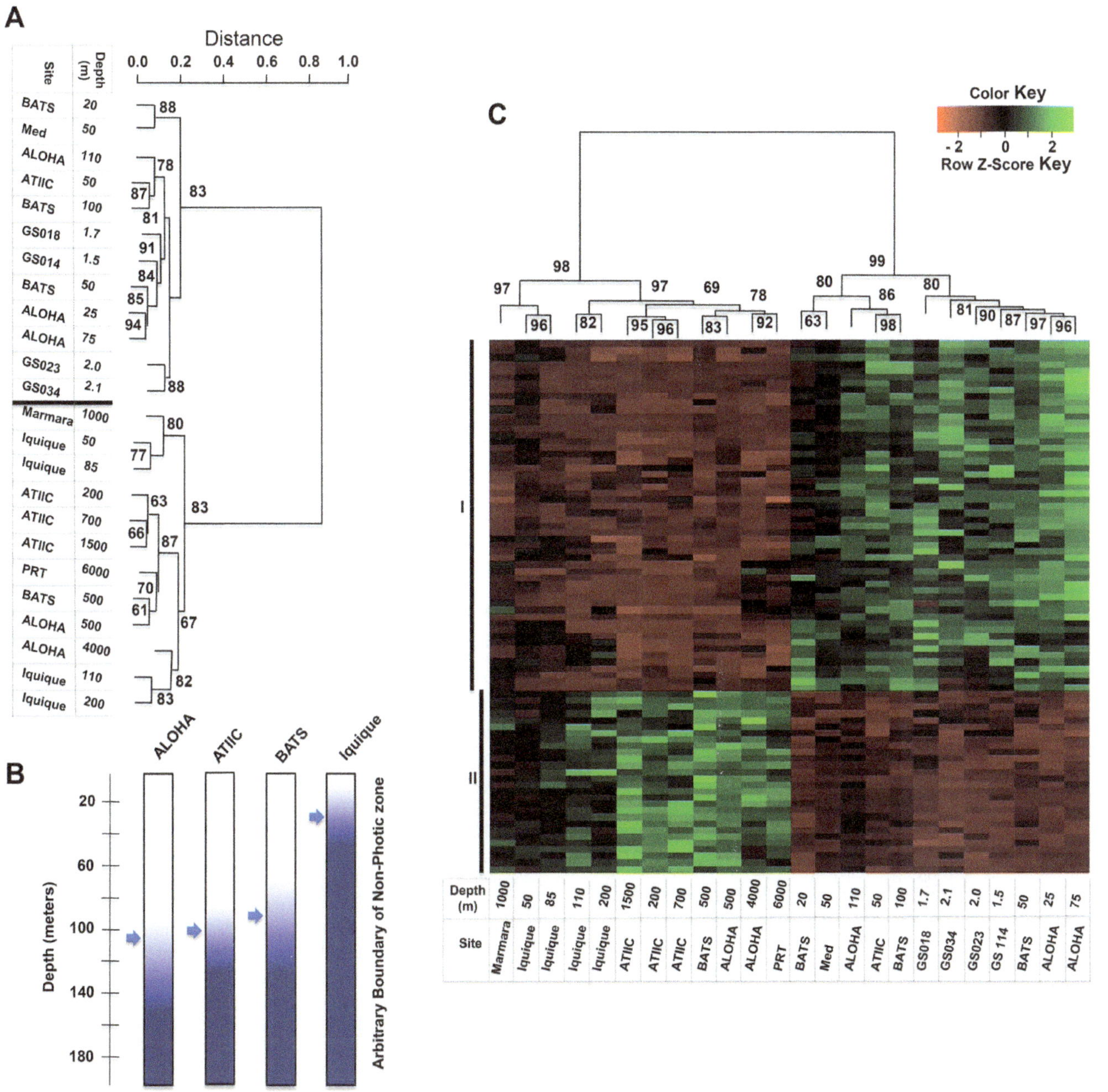

Figure 1. Clustering of the 176 depth-related COGs. (**A**) Hierarchical clustering of the 176 depth-related COGs in the 24 datasets. Clustering analysis is based on the normalized abundance profile of the 176 depth-related COGs that were shared by the three reference water columns (ATII, ALOHA and BATS) and significantly differed in abundance within at least one of them (details in Materials and Methods). The height indicates the relative distance between datasets. Bootstrap confidence values above 60 for the nodes are shown. The heatmap is shown in fig. S3A. (**B**) Location of the boundary between the photic and aphotic zones in each of the four water columns. The arrows indicate the depth at which PAR reaches 1% of the level at the surface. (**C**) Hierarchical clustering of the photic/aphotic global-core depth-related COGs. 82 COGs that showed statistically significant difference (Welch's test, FDR-corrected, $p \leq 1E-04$) in their normalized abundance between the photic and the aphotic groups of datasets were selected to establish a photic/aphotic global-core, depth-related reference set. 54 COGs had significantly higher abundance in the photic datasets (Group I, table S4A in file S1); contrary to the remaining 28 aphotic related COGs (Group II, table S4B in file S1). Bootstrap confidence values for the major nodes are shown. Heatmap coloring reflects the Z score of normalized abundances of each COG across all clustered datasets (details in Materials and Methods).

construction of this profile, a major subset of 82 COGs (19%) were found to show a statistically significant difference in abundance between photic and aphotic datasets ($p \leq 1e-04$ and FDR corrected; see details in Materials and Methods). Based on the mean normalized abundance for the whole set, 54 were considered photic related COGs and 28 were regarded as aphotic related COGs (fig. 1C, see also tables S4A and S4B in file S1). Bootstrapping evaluation indicated a high degree of confidence in the division of the photic and aphotic groups of the datasets presented in fig. 1A and 1C. This reflects the fact that functional adaptation to ambient light level is most probably one of the factors that strongly constrains the genetic variability of microorganisms in ocean environments. Thus, COGs related to photosynthesis, biosynthesis of light-harvesting pigments, assimilation of carbon dioxide by photosynthetic bacteria, as well as light-induced DNA repair and oxidative stress responses, were markedly more abundant in the photic than the aphotic datasets, as were COGs related to nitrogen fixation and phosphate metabolism. Conversely, in the aphotic datasets, COGs related to protein and amino acid catabolism, methane oxidation, sulfate assimilation and metabolism, selenocysteine metabolism and terpenoid biosynthesis were overrepresented, mirroring the exploitation of other energy sources in this zone.

It is important to note that previously detected differences in microbial gene abundances between the surface zone and the deep sea have been based on comparisons of limited metagenomic datasets [9,11,12]. In the present work all sufficiently extensive depth-related datasets available from different sites around the world's oceans (fig. S1, and table S1 in file S1) have been analyzed in a uniform fashion, allowing us to establish clearly defined sets of depth-related biological functions for the oceans as a whole. Such core functions shared between distinct sites around the global ocean can thus be used as a reliable diagnostic tool to analyze the distribution of light-related biological activities in sites that present atypical environmental conditions, as described below.

Abundance ratio of photic to aphotic depth-related COGs in the major upwelling zone of the Humboldt Current System and the salinity-based two-layer flow system in the Sea of Marmara

The environmental differences between the water columns at ALOHA, BATS and ATII relative to Iquique provide an excellent opportunity to explore the interplay between environmental factors - such as nutrient availability, light intensity, oxygen concentration - and the genomic changes that underlie microbial adaptation. When we examined the abundance ratio of the photic to the aphotic related COGs in all four water columns, we found, as expected, a well-defined correlation with the photic/aphotic boundary in the cases of ALOHA, BATS and ATII (fig. 2A) – high ratios of photic to aphotic related COG abundance in the photic samples and low ratios in aphotic ones. However, in the Iquique column the aphotic:photic ratio is merely 1.3 for the 50 m sample and 1.8 for the 85 m sample, whilst the mean ratio for the aphotic datasets of the reference columns is 3.8 (fig. 2A). The fact that microbial communities in this zone are genetically equipped for photosynthesis strongly suggests that the 50 m and 85 m levels are part of the mixed-water layer at this site.

To investigate this issue further, we inspected the ratios of photic to aphotic related COGs in all 24 datasets in relation to light levels. We selected one of the most striking positive functional associations with sunlight intensity as an example (correlation coefficient 0.89), the deoxyribodipyrimidine photolyase gene (COG0415), the product of which repairs DNA damage caused by exposure to ultraviolet light [40]. Figure 2B shows the abundance of this gene in each of the 24 datasets. It is overwhelmingly abundant in microbial communities that reside at all depth levels that clustered as photic in fig. 1A, whereas the gene is rarely found in those that reside in aphotic zones. Again, Iquique is an exception. Here, although the data from all four depths clustered with the aphotic group of datasets (fig. 1A), the abundance of the photolyase gene at 50 m and 85 m is nevertheless relatively high, while in the 110 m and 200 m datasets this drops to values comparable to those seen in aphotic zones (fig. 2B). Since the 50 m and 85 m levels at Iquique are most probably part of the mixed layer, microorganisms are presumably shuttled to the surface, where they are exposed to sunlight. Therefore, members of these communities are subjected to selective pressure to acquire and/or maintain biological functions that are crucial for survival in a sunlit environment. So, although its PAR value (fig. 1B) indicates that the 50 m layer at Iquique is already aphotic, the relative abundance of photic related COGs (fig. 2A) including photolyase (fig. 2B) argues that it belongs to a facultatively photic zone.

The data for the Sea of Marmara come from a depth of 1 km, which is unquestionably within the aphotic zone. The microbial community at this site was therefore expected to have low levels of the photolyase gene, comparable to those found in other aphotic zones. Paradoxically, however, the deep Marmara sample shows a relatively elevated abundance of this UV-protective gene (fig. 2B). Moreover, the ratio of aphotic to photic related COG abundances for the sample from the Sea of Marmara was found to be only 1.5, similar to that for the 50 m level in the Iquique column and much lower than the mean value for the aphotic datasets of the reference columns (fig. 2A). One possible explanation for this observation is provided by the hydrography and unusual circulation patterns in the Sea of Marmara [41], which receives an inflow of salty (around 3.9%) water from the Mediterranean Sea through the 65 m deep Strait of the Dardanelles. Owing to its density, this mainly euphotic water sinks to the bottom of the Marmara Basin [41]. Therefore, part of the microbial community in the 1-km deep aphotic Marmara environment is actually derived from the surface euphotic zone of the Mediterranean, and is continuously replenished. This would account for the unexpected presence of significant levels of photic genes at this site.

Effect of oxygen concentration on gene-content in marine microbial communities

Marine oxygen minimum zones (OMZs) occur naturally as a result of high oxygen consumption by heterotrophic microorganisms in nutrient-rich zones [35] like the upwelling system of the Humboldt Current (HC). They may also be induced by human activities such as fertilizer run-off and have a great environmental impact by creating zones dominated by microbial communities adapted to hypoxic conditions [42]. Although OMZs have important effects on the diversity, metabolism and physiology of microorganisms [43,44,45], the threshold oxygen concentration at which these begin to affect the content of microbial genomes in the ocean is not well defined. In order to clarify this issue, the three deepest Iquique datasets (85, 110 and 200 m) were chosen as reference hypoxic datasets because they represent the persistent OMZ present at this site [36] and comprise the datasets with the lowest oxygen concentrations among the 24 datasets studied in this work ($[O_2] = 11.0$, 4.0 and 3.2 $\mu mol.kg^{-1}$ at 85, 110 and 200 m, respectively). To avoid interference from any effects related to sunlight incidence, we removed members of the photic/aphotic global-core depth-related COGs from the list of 162 differentially abundant COGs that are statistically significant for the Iquique column. The normalized abundance medians of the remaining 149 COGs from these hypoxic Iquique samples were then compared with the corresponding values for the other 21 datasets (for details see Methods). As shown in fig. 3A (table S5 in file S1), 18 COGs in all were found to be significantly influenced by extreme oxygen limitation ($p <= 0.05$, FDR-corrected); 16 of these clustered in group I, and two in group II. Interestingly, approximately 43% of the COGs that are significantly more abundant in extremely hypoxic samples (Group I in fig. 3A) are related to nitrate metabolism.

Since the oxygen tension in the OMZ at the Iquique site is very low, the ability to use alternate final electron acceptors for the

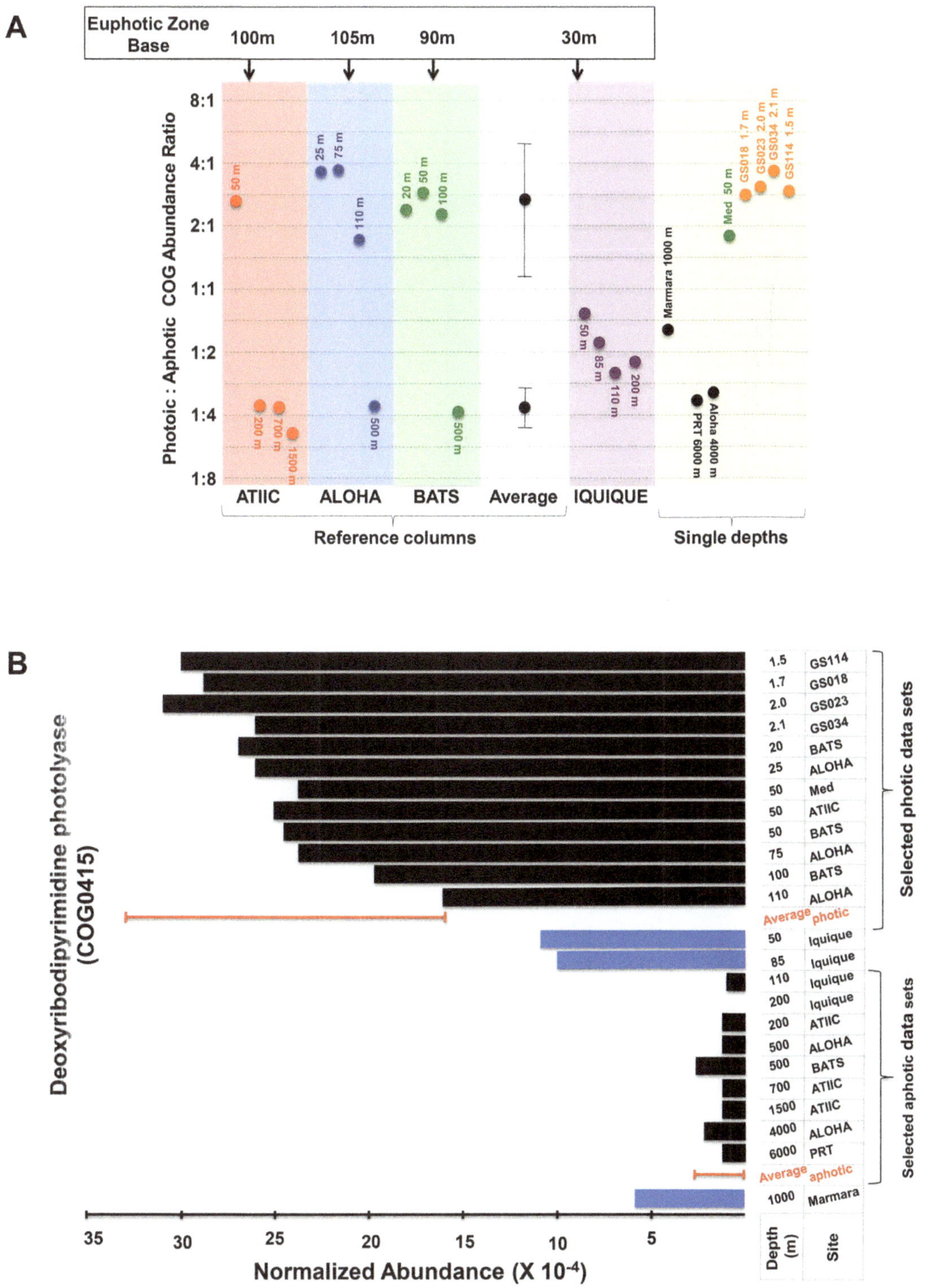

Figure 2. Abundance ratios of photic to aphotic related COGs. (**A**) Mean abundance ratios of photic to aphotic related COGs for datasets from the reference water columns, the Iquique water column, and from single-depth datasets. This analysis was initially determined using the 82-COG global-core, depth-related reference set, as described in Materials and Methods. Applying the ratios for the photic and aphotic datasets from the reference water columns, ratio ranges (mean±two SD) were established to indicate photic or aphotic functional genomic content enrichment. Mean abundance ratios for the Iquique water column and the single-depth datasets were calculated to verify their balance between photic and aphotic related biological functions. The arrows at the top of the diagram indicate the approximate depth of the boundary between photic and aphotic zones in each water column (fig. 1B). (**B**) Normalized abundance of the deoxyribodipyrimidine photolyase gene (COG0415; table S4A in file S1) in the 24 datasets obtained from the indicated depths. The average normalized abundance was calculated for the photic and for the aphotic groups of datasets. The bars represent two SD of the mean.

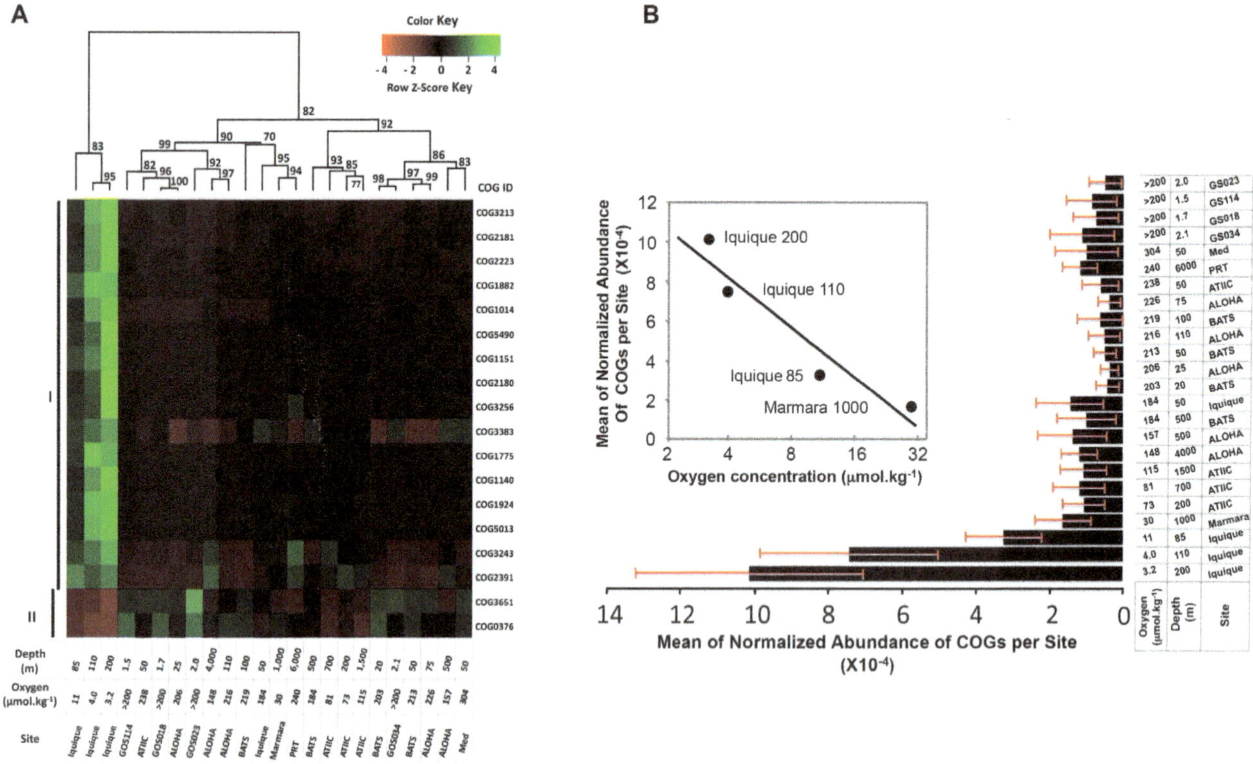

Figure 3. Effect of oxygen concentration on the functional genomic content of microbial communities represented in the 24 datasets. (A) Hierarchical clustering of COGs that significantly influenced by extreme oxygen limitation. Datasets analysis are based on the abundance profile of 18 COGs which significantly differed in abundance levels between an extremely hypoxic environment (Iquique 85-, 110 and 200 m deep, permanent OMZ) and environments with higher D.O. (Fisher's exact test, FDR-corrected, $p \leq 0.05$). Bootstrap confidence values for the nodes are shown. Heatmap coloring reflects the Z score of normalized abundances of each COG across all clustered datasets. Roman numbers on the left side of the figure present different groups of COGs as determined by abundance profile across the clustered datasets. (B) Mean abundances of the COGs in group I (figure 3A) as a function of the oxygen concentration. Inset shows the gradual increase in the abundance of these genes in datasets from Marmara 1000 m, Iquique 85 m, Iquique 110 m, and Iquique 200 m, as oxygen concentration drops from 30 μmol.kg^{-1} to 3.2 μmol.kg^{-1}.

replenishment of oxidative potential and energy production coupling becomes imperative for survival. This probably accounts for the fact that a number of the COGs overrepresented in this zone (COG2223, COG5013, COG1140, COG2181, COG2180, COG3256 and possibly COG3213) are related to nitrate reduction. The abundance of an enzyme involved in fermentation and specifically inactivated by oxygen (COG1882) was also significantly high. In addition, enzymes that are implicated in methanogenesis (COG1151) and pyruvate metabolism (COG1014) exhibited increased abundance in these datasets. Conversely, genes whose products require molecular oxygen for their activities - such as catalase (COG0376, fig. 3A, table S5 in file S1) - are poorly represented in this extremely hypoxic environment.

Analysis of the mean abundances of the COGs in group I as a function of the oxygen concentration reveals that these COGs are distinctively and overwhelmingly abundant in microbial communities from samples in which oxygen concentration was $\leq$ 11 μmol.kg^{-1} (fig. 3B). It is intriguing to note the gradual increase in the abundance of these genes in datasets as oxygen concentration drops from 30 μmol.kg^{-1} to 3.2 μmol.kg^{-1} (inset fig. 3B).

The fact that the relative abundances of genes in group I increases dramatically at oxygen concentrations of 11 μmol.kg^{-1} or less suggests that this level of oxygen represents the threshold at which selective effects on genome content in favor of anaerobic lifestyles set in.

Based on the content of the core sets of photic/aphotic depth-related COGs from all sites presented in fig. S3A, it appears that microbial communities living in photic zones share the same functions, regardless of site location. Thus, with respect to this zone, we do not see any specific branches clustering together that can be attributed to environmental conditions that are unique to any particular geographical location. The aphotic samples (based on PAR values), however, do display distinct branching profiles of genomic content that can be correlated with specific environmental conditions and oxygen concentrations. The three mixed-water layers, Iquique 50 and 85 m and Marmara 1 km (with aphotic:photic ratios of between 1.3 to 1.8; refer to fig. 2A), in which limitation of light and oxygen in these aphotic zones exert selective pressures, cluster together (fig. S3A, see also fig. S3B branch I), while the extremely hypoxic samples in the aphotic zone at Iquique (110 and 200 m) formed a separate cluster on the other side of the tree (fig. S3A, see also S3B branch II). The abundances of two COGs, photolyase (COG0415) and nitrate reductase (Alpha Subunit; COG5013), taken as representative of core photic and aphotic related COGs that contributed to the functional branching in these aphotic zone, are presented in fig. S3B. The results again argue that the concentration of oxygen at which selective pressure begins detectably to favor the acquisition of

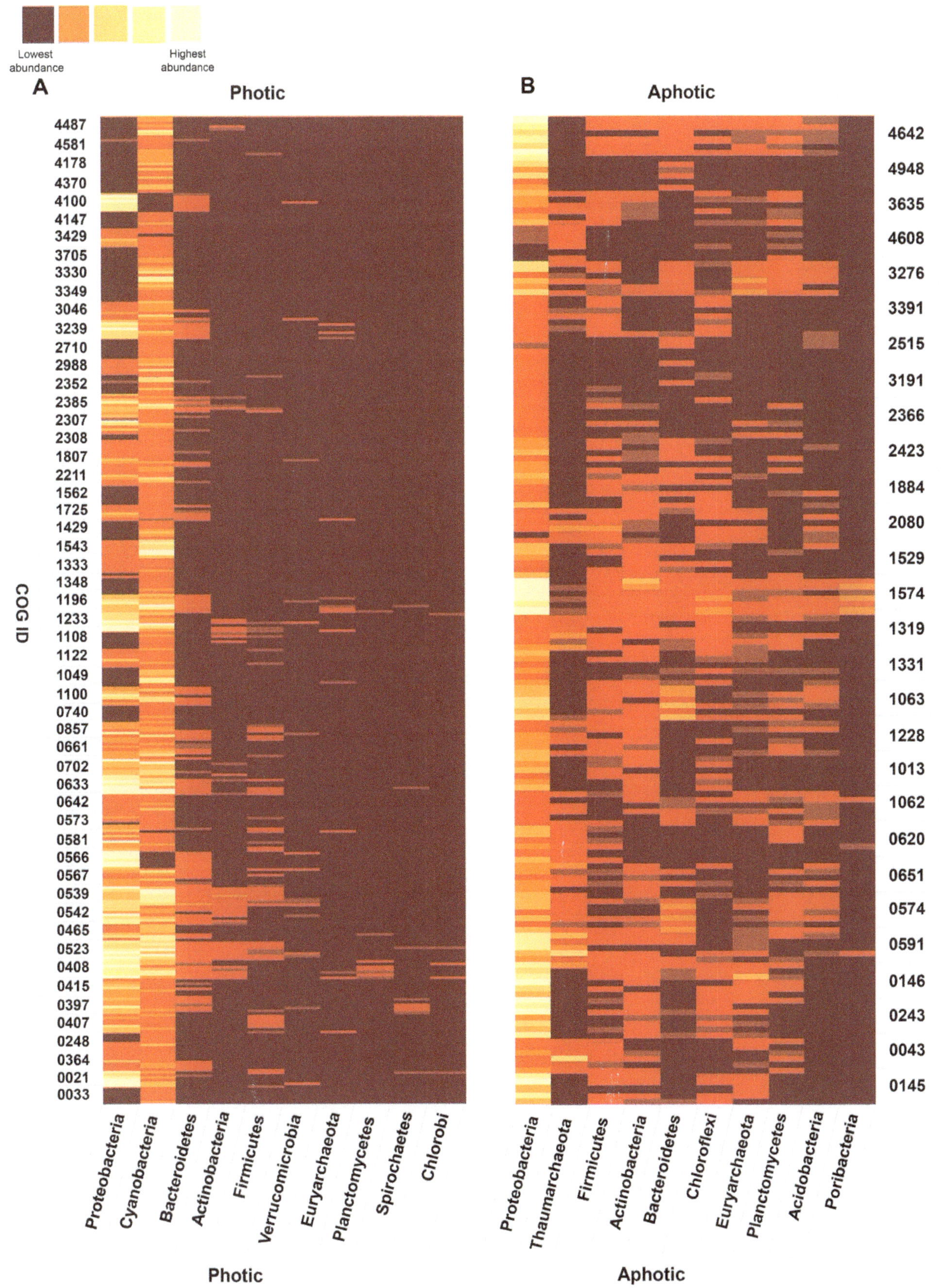
Lowest abundance
Highest abundance
A
Photic
B
Aphotic
COG ID
4487
4581
4178
4370
4100
4147
3429
3705
3330
3349
3046
3239
2710
2988
2352
2385
2307
2308
1807
2211
1562
1725
1429
1543
1333
1348
1196
1233
1108
1122
1049
1100
0740
0857
0661
0702
0633
0642
0573
0581
0566
0567
0539
0542
0465
0523
0408
0415
0397
0407
0248
0364
0021
0033
Proteobacteria
Cyanobacteria
Bacteroidetes
Actinobacteria
Firmicutes
Verrucomicrobia
Euryarchaeota
Planctomycetes
Spirochaetes
Chlorobi
Photic
4642
4948
3635
4608
3276
3391
2515
3191
2366
2423
1884
2080
1529
1574
1319
1331
1063
1228
1013
1062
0620
0651
0574
0591
0146
0243
0043
0145
COG ID
Proteobacteria
Thaumarchaeota
Firmicutes
Actinobacteria
Bacteroidetes
Chloroflexi
Euryarchaeota
Planctomycetes
Acidobacteria
Poribacteria
Aphotic

Figure 4. Taxonomical assignments of photic and aphotic depth-related COGs at the phylum level for the selected datasets. Reads assigned to COGs were compared to the NCBI nr database using BLASTX (E-value cutoff 1e-05) and the taxonomical identifiers of the best matches were retrieved. Subsequently, tables of relative frequency of taxa per COG per dataset were generated (tables S6A and S6B in file S1) and presented as heatmaps. The 10 phyla that have the highest average numbers of photic (**A**) and aphotic (**B**) COGs per site are presented. The taxa are sorted by the decreasing of their abundance. Datasets of the shallowest samples of the ALOHA, ATII and BATS water columns and single-depth datasets from GOS018, GOS023, GOS034, GOS114 and Mediterranean sites were used for the analysis of the photic related COGs, whereas the deepest datasets from the ALOHA, ATII and BATS water columns, and single-depth datasets from Marmara, PRT and ALOHA (4,000 m), were employed in the case of the aphotic related COGs. This analysis was based on a total of 36,938 sequences (24919 sequences are photic, and 12019 are aphotic).

adaptations for survival under hypoxic conditions lies below that of the Iquique 85 m sample (11 μmol.kg^{-1}), which is assigned to branch I.

Taxonomical distribution of photic and aphotic depth-related COGs

The establishment of collections of core photic and core aphotic depth-related COGs raises the question of how such functions are disseminated among phylogenetic clades present in the different environments covered by our analysis. To address this point, we generated taxonomic assignments, at the phylum level, for each core photic and aphotic depth-related COG based on the taxonomic identifiers of the respective reads (for details see Materials and Methods, and tables S6A and S6B in file S1). The results show that sequences related to core photic COGs derive from far fewer taxonomic clades than those of core aphotic ones. This reflects the predominance of certain taxa in most euphotic environments and the higher diversity of taxa found in deeper oceanic environments (fig. 4, and tables S6A and S6B in file S1). Moreover, the dominant taxon for any given COG is, with few exceptions, shared among all the sites analyzed.

The taxonomic distribution of the 54 core photic related COGs at the phylum level (fig. 4A, table S6A in file S1) reveals that the vast majority (91%) of them are most probably contributed by Proteobacteria (46%) and Cyanobacteria (45%). The taxonomic distribution of the 28 core aphotic related COGs at the phylum level is much more diverse than that observed for the photic functions. Although Proteobacteria are again predominant (50%), a total of ten phyla are represented at levels greater than 1%, compared to only five for the photic set (fig. 4B, table S6B in file S1).

The increased diversity of microbial communities in aphotic environments is perhaps the most striking result of the phylogenetic analysis of the core COGs. It can be attributed to adaptation to the wider variety (and lower abundance) of exploitable resources available in the aphotic zone, which contrasts with overwhelming importance of photosynthetic energy production in the photic zone. This interpretation is supported by the biochemical functions assigned to core photic/aphotic related COGs observed in our work.

In this study, using metagenomics-based functional profiling analysis, we have defined core sets of photic and aphotic depth-related biological activities that are shared by different sites distributed around the global oceans. These two sets of depth-related functional biological activities were used to provide insight into the interplay between light intensity - the most significant abiotic evolutionary force in the oceans - and the genetic and ecological differentiation of marine microbial communities. In addition, in conjunction with a set of functional biological activities related to extreme hypoxia, we provide a comprehensive picture of microbial biological activities in oceanographic settings in which light and oxygen imprint its effect in nominally aphotic zones. Moreover, our work highlights the importance of developing reference sets of functional biological activities that can be used as a tool for the diagnosis of the physiological and biochemical capabilities, and phylogenetic profiling, of marine microorganisms. This approach should be particularly helpful for investigations of the impact of anthropogenically induced environmental changes on marine ecosystems.

Supporting Information

Figure S1 Locations of the 11 sites and numbers of sequenced reads in the 24 datasets.

Figure S2 Hierarchical clustering of datasets from the three reference water columns (ATII, ALOHA, BATS). This analysis is based on the normalized abundance profile of 176 selected COGs that significantly differed in abundance within at least one of column (Fisher's exact test, FDR-corrected, $p \leq 0.01$) (table S2 in file S1). Heatmap coloring reflects the Z score of normalized abundances of each COG across clustered datasets. Roman numbers on the left side of the figure present different groups of COGs as determined by abundance profile across the clustered datasets.

Figure S3 Hierarchical clustering of the 176 depth-related COGs in the 24 datasets. (A) Heatmap of the datasets from 11 diverse marine sites (24 datasets). Refer to legend of figure 1A for details. Roman numbers on the left side of the figure present different groups of COGs as determined by abundance profile across the clustered datasets. The dendrogram of this figure is also presented in fig. 1A. (B) Distinct branching profiles of genomic content of the aphotic samples. The aphotic branch in figure S3A is presented together with the level of oxygen in each site and the normalized abundances of Deoxyribodipyrimidine photolyase (COG0415), and Nitrate reductase (Alpha Subunit; COG5013). The values of the normalized abundance of both COGs were obtained from table S4A and S5 in file S1.

File S1 Table S1, Sites locations, datasets, sampling and sequencing methods. Table S2, List of 176 depth-related COGs shared by the three reference columns, ATIIC, ALOHA and BATS. Fisher's exact test ($p \leq 0.01$, FDR corrected) was employed to identify COGs that presented a statistically significant difference in abundance between at least two depths of a reference column. COG cluster roman numbers refers to groups of COGs displaying similar normalized abundance profile. COGs are ordered as in figure S2. **Table S3, List of 176 depth-related COG set in all 24 datasets.** Fisher's exact test ($p \leq 0.01$, FDR corrected) was employed to identify the 176 depth-related COG set (table S2 in file S1) in all 24 datasets. Among these, the ones that were present in at least one dataset of each reference column were selected for subsequent comparative analysis of all datasets. COG cluster roman numbers refers to groups of COGs displaying similar abundance profile in the hierarchical clustering of all 24 datasets. COGs are ordered as in figures 1A and S3A. **Table S4, A. List of 54 photic global-**

core, depth-related COGs. Welch's test ($p \leq 1E-04$, FDR corrected) was employed to identify COGs that presented a statistically significant difference in the mean abundance between the photic and the aphotic group of datasets (see details in Methods). Photic related COGs are presented in the same order as shown in group I Fig. 1C. **B. List of 28 aphotic global-core depth-related COGs.** Welch's test ($p \leq 1E-04$, FDR corrected) was employed to identify COGs that presented a statistically significant difference in the mean abundance between the photic and the aphotic group of datasets (see details in Methods). Aphotic related COGs are presented in the same order as shown in group II Fig. 1C. **Table S5, List of 18 Oxygen Minimum Zone-related COGs.** Wilcoxon test ($p \leq 0.05$, FDR corrected) was employed to identify COGs that significantly differed in abundance between the highly hypoxic group of datasets (Iquique 85-, 110- and 200-m deep samples) and the group of remaining datasets with higher oxygen concentration (details in Methods). COGs are presented in the same order as that in the heatmap (fig. 3A).**Table S6. A. Taxonomic distribution of the photic COGs at the phylum level.** Phylogenetic analysis was performed based on the NCBI taxonomic identifiers of the best matches to the nr database. For each COG, the rawcounts of assigned taxa is presented for eight different photic datasets: the shallowest samples of the reference columns, and five singledepth samples (details in Methods). The 10 phyla that have the highest average numbers of photic COGs per site are presented in fig. 4A. **B. Taxonomic distribution of the aphotic COGs at the phylum level.** Phylogenetic analysis was performed based on the NCBI taxonomic identifiers of the best matches to the nr database. For each COG, the rawcounts of assigned taxa is presented for six different aphotic datasets: the deepest samples of the reference columns and three single-depth samples. The 10 phyla that have the highest average numbers of aphotic COGs per site are presented in Figure 4B.

Acknowledgments

We thank Drs. Amy S. Bower, Stephen A. Swift, and Abdulaziz Al-Suwailem for offering technical assistants and important information regarding sample collection. We are grateful to Dr. Paul Hardy for critical reading of the manuscript.

Author Contributions

Conceived and designed the experiments: AJSF RS JCS AM HD. Performed the experiments: AJSF RS JCS FSC ER AMA HD. Analyzed the data: AJSF JCS HS AMA FSC US VBB HD. Contributed reagents/materials/analysis tools: AS ASD MAG AO IA HL AA JACA BRJ MS VBB. Wrote the paper: AJSF JCS HD.

References

1. Whitman WB, Coleman DC, Wiebe WJ (1998) Prokaryotes: the unseen majority. Proc Natl Acad Sci U S A 95: 6578–6583.
2. Field CB, Behrenfeld MJ, Randerson JT, Falkowski P (1998) Primary production of the biosphere: integrating terrestrial and oceanic components. Science 281: 237–240.
3. Arrigo KR (2005) Marine microorganisms and global nutrient cycles. Nature 437: 349–355.
4. Schafer H, Myronova N, Boden R (2010) Microbial degradation of dimethylsulphide and related C1-sulphur compounds: organisms and pathways controlling fluxes of sulphur in the biosphere. J Exp Bot 61: 315–334.
5. Jorgensen BB, Boetius A (2007) Feast and famine—microbial life in the deep-sea bed. Nat Rev Microbiol 5: 770–781.
6. Azam F, Malfatti F (2007) Microbial structuring of marine ecosystems. Nat Rev Microbiol 5: 782–791.
7. Yooseph S, Nealson KH, Rusch DB, McCrow JP, Dupont CL, et al. (2010) Genomic and functional adaptation in surface ocean planktonic prokaryotes. Nature 468: 60–66.
8. Quaiser A, Zivanovic Y, Moreira D, Lopez-Garcia P (2011) Comparative metagenomics of bathypelagic plankton and bottom sediment from the Sea of Marmara. ISME J 5: 285–304.
9. Konstantinidis KT, Braff J, Karl DM, DeLong EF (2009) Comparative metagenomic analysis of a microbial community residing at a depth of 4,000 meters at station ALOHA in the North Pacific subtropical gyre. Appl Environ Microbiol 75: 5345–5355.
10. Ghai R, Martin-Cuadrado AB, Molto AG, Heredia IG, Cabrera R, et al. (2010) Metagenome of the Mediterranean deep chlorophyll maximum studied by direct and fosmid library 454 pyrosequencing. ISME J 4: 1154–1166.
11. Eloe EA, Fadrosh DW, Novotny M, Zeigler Allen L, Kim M, et al. (2011) Going deeper: metagenome of a hadopelagic microbial community. PLoS One 6: e20388.
12. DeLong EF, Preston CM, Mincer T, Rich V, Hallam SJ, et al. (2006) Community genomics among stratified microbial assemblages in the ocean's interior. Science 311: 496–503.
13. Rusch DB, Halpern AL, Sutton G, Heidelberg KB, Williamson S, et al. (2007) The Sorcerer II Global Ocean Sampling expedition: northwest Atlantic through eastern tropical Pacific. PLoS Biol 5: e77.
14. Venter JC, Remington K, Heidelberg JF, Halpern AL, Rusch D, et al. (2004) Environmental genome shotgun sequencing of the Sargasso Sea. Science 304: 66–74.
15. Lipson DA, Haggerty JM, Srinivas A, Raab TK, Sathe S, et al. (2013) Metagenomic insights into anaerobic metabolism along an Arctic peat soil profile. PLoS One 8: e64659.
16. Varin T, Lovejoy C, Jungblut AD, Vincent WF, Corbeil J (2012) Metagenomic analysis of stress genes in microbial mat communities from Antarctica and the High Arctic. Appl Environ Microbiol 78: 549–559.
17. Gregoracci GB, Nascimento JR, Cabral AS, Paranhos R, Valentin JL, et al. (2012) Structuring of bacterioplankton diversity in a large tropical bay. PLoS One 7: e31408.
18. Ochman H, Lawrence JG, Groisman EA (2000) Lateral gene transfer and the nature of bacterial innovation. Nature 405: 299–304.
19. Kazazian HH Jr (2004) Mobile elements: drivers of genome evolution. Science 303: 1626–1632.
20. Gogarten JP, Townsend JP (2005) Horizontal gene transfer, genome innovation and evolution. Nat Rev Microbiol 3: 679–687.
21. Sobecky PA, Hazen TH (2009) Horizontal gene transfer and mobile genetic elements in marine systems. Methods Mol Biol 532: 435–453.
22. Balaji S, Babu MM, Aravind L (2007) Interplay between network structures, regulatory modes and sensing mechanisms of transcription factors in the transcriptional regulatory network of E. coli. J Mol Biol 372: 1108–1122.
23. Baumbach J (2010) On the power and limits of evolutionary conservation—unraveling bacterial gene regulatory networks. Nucleic Acids Res 38: 7877–7884.
24. Crombach A, Hogeweg P (2008) Evolution of evolvability in gene regulatory networks. PLoS Comput Biol 4: e1000112.
25. Maslov S, Krishna S, Pang TY, Sneppen K (2009) Toolbox model of evolution of prokaryotic metabolic networks and their regulation. Proc Natl Acad Sci U S A 106: 9743–9748.
26. Lutz RA, Falkowski PG (2012) Ocean science. A dive to Challenger Deep. Science 336: 301–302.
27. Giovannoni SJ, Vergin KL (2012) Seasonality in ocean microbial communities. Science 335: 671–676.
28. Kroger S, Law RJ (2005) Sensing the sea. Trends Biotechnol 23: 250–256.
29. Karl DM (2007) Microbial oceanography: paradigms, processes and promise. Nat Rev Microbiol 5: 759–769.
30. DeLong EF, Karl DM (2005) Genomic perspectives in microbial oceanography. Nature 437: 336–342.
31. Stewart FJ, Ulloa O, DeLong EF (2012) Microbial metatranscriptomics in a permanent marine oxygen minimum zone. Environ Microbiol 14: 23–40.
32. Martinez A, Tyson GW, Delong EF (2010) Widespread known and novel phosphonate utilization pathways in marine bacteria revealed by functional screening and metagenomic analyses. Environ Microbiol 12: 222–238.
33. Robinson MD, McCarthy DJ, Smyth GK (2010) edgeR: a Bioconductor package for differential expression analysis of digital gene expression data. Bioinformatics 26: 139–140.
34. Altschul SF, Madden TL, Schaffer AA, Zhang J, Zhang Z, et al. (1997) Gapped BLAST and PSI-BLAST: a new generation of protein database search programs. Nucleic Acids Res 25: 3389–3402.
35. Daneri G, Dellarossa V, Quinones R, Jacob B, Montero P, et al. (2000) Primary production and community respiration in Humboldt Current System off Chile and associated oceanic areas. Marine Ecology Progress Series 197: 41–49.

36. Revsbech NR, Larsen LH, Gundersen J, Dalsgaard T, Ulloa O, et al. (2009) Determination of ultra-low oxygen concentrations in oxygen minimum zones by the STOX sensor. Limnology and Oceanography: Mehods 7: 371–381.
37. Muller J, Szklarczyk D, Julien P, Letunic I, Roth A, et al. (2010) eggNOG v2.0: extending the evolutionary genealogy of genes with enhanced non-supervised orthologous groups, species and functional annotations. Nucleic Acids Res 38: D190–195.
38. Tatusov RL, Galperin MY, Natale DA, Koonin EV (2000) The COG database: a tool for genome-scale analysis of protein functions and evolution. Nucleic Acids Res 28: 33–36.
39. Steinberg DK, Carlson CA, Bates NR, Johnson RJ, Michaels AF, et al. (2001) Overview of the US JGOFS Bermuda Atlantic Time-series Study (BATS): a decade-scale look at ocean biology and biogeochemistry. Deep-Sea Research II 48: 1405–1447.
40. Osburne MS, Holmbeck BM, Frias-Lopez J, Steen R, Huang K, et al. (2010) UV hyper-resistance in Prochlorococcus MED4 results from a single base pair deletion just upstream of an operon encoding nudix hydrolase and photolyase. Environ Microbiol 12: 1978–1988.
41. Besiktepe ST, Sur IH, Ozsoy E, Abdul Latif M, Oguz T, et al. (1994) The circulation and hydrography of the Marmara Sea. Prog Oceanogr 34: 285–334.
42. Diaz RJ, Rosenberg R (2008) Spreading dead zones and consequences for marine ecosystems. Science 321: 926–929.
43. Bryant JA, Stewart FJ, Eppley JM, DeLong EF (2012) Microbial community phylogenetic and trait diversity declines with depth in a marine oxygen minimum zone. Ecology 93: 1659–1673.
44. Lam P, Kuypers MM (2011) Microbial nitrogen cycling processes in oxygen minimum zones. Ann Rev Mar Sci 3: 317–345.
45. Ulloa O, Canfield DE, DeLong EF, Letelier RM, Stewart FJ (2012) Microbial oceanography of anoxic oxygen minimum zones. Proc Natl Acad Sci U S A 109: 15996–16003.

12

Rise of the Earliest Tetrapods: An Early Devonian Origin from Marine Environment

David George[1]*, Alain Blieck[2]

1 Department of Biotechnology, St. Peter's Engineering College, Chennai, Tamil Nadu, India, **2** Université Lille 1, Sciences de la Terre, FRE 3298 du CNRS Géosystèmes, Villeneuve d'Ascq, France

Abstract

Tetrapod fossil tracks are known from the Middle Devonian (Eifelian at ca. 397 million years ago - MYA), and their earliest bony remains from the Upper Devonian (Frasnian at 375–385 MYA). Tetrapods are now generally considered to have colonized land during the Carboniferous (i.e., after 359 MYA), which is considered to be one of the major events in the history of life. Our analysis on tetrapod evolution was performed using molecular data consisting of 13 proteins from 17 species and different paleontological data. The analysis on the molecular data was performed with the program TreeSAAP and the results were analyzed to see if they had implications on the paleontological data collected. The results have shown that tetrapods evolved from marine environments during times of higher oxygen levels. The change in environmental conditions played a major role in their evolution. According to our analysis this evolution occurred at about 397–416 MYA during the Early Devonian unlike previously thought. This idea is supported by various environmental factors such as sea levels and oxygen rate, and biotic factors such as biodiversity of arthropods and coral reefs. The molecular data also strongly supports lungfish as tetrapod's closest living relative.

Editor: Brock Fenton, University of Western Ontario, Canada

Funding: The authors have no support or funding to report.

Competing Interests: The authors have declared that no competing interests exist.

* E-mail: davetigerland@yahoo.com

Introduction

Terrestrialization may be defined as the series of processes (adaptation) that makes an aquatic organism capable of living and sustaining itself on land. It is usually considered as one of the most important events in the evolutionary history of life on Earth. It has occurred several times, that is among primitive organisms (bacteria, protists, fungi…), plants, invertebrates and vertebrates. Here we focus on vertebrate terrestrialization also known as evolution of fish to tetrapods. Lobe-finned fishes (Sarcopterygii) were a highly successful group during the Devonian, between ca. 416 and 359 million years ago (MYA). According to the most recent discoveries and ideas, terrestrialization of vertebrates has occurred in two steps: 1) the first tetrapods diverged from sarcopterygians during the Frasnian (about 375–385 MYA) or earlier in aquatic environments [1,2,3]; 2) this was followed by adaptation to terrestrial life much later in the earliest Carboniferous, about 345–359 MYA. Today only three groups of sarcopterygians survive, namely Tetrapoda, Dipnoi (lungfishes), and Actinistia (coelacanths). Tetrapods include about 21,100 extant species and a much greater number of extinct species; only six species of lungfishes and two species of coelacanths exist to-day, but both groups were much more abundant and diverse in the Devonian. It is generally accepted that elpistostegids, a group of extinct sarcopterygian fishes, are the closest relatives (sister-group) of tetrapods [4]. The early vertebrate expert community very often follows the idea originally proposed by A. S. Romer (e.g. [5]) that early tetrapod life occurred in freshwater under the "Drying Pond" scenario where tetrapods evolved from lobe-finned fishes driven onto the land by drought. A modern version of this scenario is that tetrapods evolved from the elpistostegids, probably in brackish to freshwater environments, in response to the modification of their environment [6]. This scenario has however been strongly contradicted as early as the 1950s (refs in [5]) and replaced by the idea that the transition from fish to tetrapod occurred in marine to land/sea transitional environments (tidal, intertidal or lagoonal zones) [7,8,9,10,11]. Devonian tetrapods and elpistostegids have been found in a wide range of geographic localities, including the Old Red Sandstone Continent, North China, and East Gondwana. This wide range could be due to marine tolerance. The first tetrapods, like their immediate piscine sister taxa, were capable of marine dispersal, thus explaining the widespread global distribution achieved in the Frasnian [12,13]. The recent discovery of tetrapod tracks from Poland [11] also suggests that the earliest evolution of tetrapods could have taken place in marine to land/sea transitional environments. Study of such relationships between living organisms and environmental conditions at global scale is generally known as geobiology (biosphere-geosphere interactions).

Attempts were made for a long time to determine whether or not sister relationships in the phylogenetic tree of vertebrates exist between tetrapods and lungfishes, tetrapods and coelacanths, or lungfishes and coelacanths. The mitochondrial [14,15], ribosomal [16], and nuclear [17,18] encoded sequences have been collected with the specific goal of resolving the relationships among living sarcopterygians, but the available molecular data has not provided complete resolution to whether lungfishes are the closest living relatives to tetrapods, or coelacanths are the closest, or the third case where both coelacanths and lungfishes could be equally related to tetrapods. The proteins used in this analysis on dipnoans, coelacanths and basal tetrapods (amphibians) are the 13 proteins synthesized by the mitochondrial genome. Modifications in the mitochondrial protein coding genes which are involved in oxidative phosphorylation (a process in cell metabolism by which respiratory enzymes in the mitochondria synthesize ATP, which is used to produce energy) can directly influence metabolic performance of an organism. Because of the importance of this biochemical pathway, evaluating selective pressures acting on mtDNA (mitochondrial DNA) proteins could provide key insight into the adaptive evolution of the mtDNA genome. One important goal of the present paper is to compare molecular to paleontological data in order to improve our view of early steps of tetrapod evolution, and place this process into a geobiological approach.

Materials and Methods

The 13 proteins synthesized by the mitochondrial genome of 17 species (see Table S1) were obtained from the NCBI (National Centre for Biotechnology Information). Only 17 sequences were used to avoid noise (convergent evolution) in the genes used, since amphibians seem to have a higher rate of nucleotide substitution while coelacanths and Australian lungfish have a lower rate. This will cause many problems when trying to compute any phylogenetic analysis. All the 12 species of tetrapods used were amphibians (in which 6 were salamanders). It was made sure that representatives from all the 3 orders of Amphibia namely Caudata (salamanders, newts, etc.), Anura (frogs, toads), and Gymnophiona (caecilians) were taken in the analysis. Since there is still debate about the relationship between coelacanths, lungfishes and tetrapods, an exact tree cannot be used for the analysis by TreeSAAP (Selection on Amino Acid Properties using phylogenetic Trees) [19]. Hence the analysis was performed by computing the 12 tetrapod species separately, 3 lungfishes and 2 coelacanths. Physicochemical amino acid changes among residues in mitochondrial protein coding genes were identified by the algorithm implemented in TreeSAAP, which compares the observed distribution of physicochemical changes inferred from a phylogenetic tree with an expected distribution based on the assumption of completely random amino acid replacement expected under the condition of selective neutrality. TreeSAAP also helps us to find positive or negative selection in the given sequences (Positive selection indicates that amino acid replacements are being preferred by natural selection, whereas negative selection means they are less frequent than expected by chance and are influenced by negative or purifying selection). This is done by computing the influence of amino acid properties in the given sequences. The positive selection was calculated by taking two different considerations. In the first consideration the values (called z-scores) of the individual amino acid sites were analyzed, and in the second consideration the entire protein sequence values was analyzed. For the calculation of positive selection when the entire protein was taken into analysis, the total sum value of all the individual amino acid sites needs to be calculated, this included the positive selection and the negative selection values of the individual amino acid sites. For example assume a protein has 4 amino acids. Assume the individual amino acid site values are 2, −2, −6, 4. Hence one amino acid site has been influenced by positive selection (any value above 3.09 was considered as positive selection, this value is most commonly used for this program). Now to calculate the positive selection when the whole protein sequence is taken to consideration add all the values $(2+(-2)+(-6)+4=-2)$. Since $-2<3.09$ this protein has not undergone positive selection (the detailed z-scores given by TreeSAAP are available on request to the corresponding author). Hence different results are possible when the entire protein sequence and individual amino acid sites are analyzed. Out of the 31 amino acid properties available in the software, only 20 were used in the analysis (see Text S1). This was done to increase the accuracy in detecting protein adaptation and to prevent false indications of protein adaptation. TreeSAAP was implemented by grouping changes into categories from 1 to 8, 1 being the most conservative and 8 being the most radical. When positive selection is detected in lower, more conservative magnitude ranges (categories 1, 2, or 3), the amino acid properties are considered to be under a type of stabilizing selection (here defined as selection that tends to maintain the overall biochemistry of the protein, despite a rate of change that exceeds the rate expected under conditions of chance). Conversely, when positive selection is detected in greater, more radical magnitude ranges (categories 6, 7, or 8), the amino acid property or properties are considered to be under destabilizing selection (here defined as selection that results in radical structural or functional shifts in local regions of the protein). We make the assumption that positive-destabilizing selection represents the unambiguous signature of molecular adaptation because when radical changes are favored by selection, they result in local directional shifts in biochemical function, structure, or both. For such changes to be favored by selection (i.e., for such changes to be more abundant than expected by chance), they must instill an increased level of survival and/or reproductive success in the individuals who possess and propagate them (refer [20] for more details). In this study we choose to focus on amino acid property changes of categories 6, 7, and 8 because they unambiguously indicate a significant change in the protein (See Table S2).

The GEOCARBSULF model [21] was used to know the atmospheric oxygen levels during the Devonian. It is a combination of earlier GEOCARB models for CO_2 and the isotope mass balance model for O_2. GEOCARBSULF is a computer model that takes account of the major factors thought to influence atmospheric O_2 and CO_2. These models account for "forcings," which are processes that affect the levels of these gases. Principal forcings for O_2 are burial of organic matter and pyrite (FeS_2) in sediments, their weathering on the continents, and rates of metamorphic and volcanic degassing of reduced carbon and sulfur-containing volcanic gases, such as sulfur dioxide.

Results

Adaptive Evolution Results

Amino acid properties with signals of strong positive selection accumulated at a rate roughly equivalent to the mutation rate of the gene itself (i.e. mutation rate of ATPase (Adenosine Triphosphate Synthase)>ND (Nicotinamide adenine dinucleotide dehydrogenase)>CYTB (Cytochrome b)>COX (Cytochrome c Oxidase)). This correlation is more pronounced for the protein-coding genes with higher mutation rates, such as ATPase and ND, as is most apparent in analyses of the variation. The best correlation between overall mutation rate and number of sites with radical amino acid changes was observed for NDs, while the existence of several outliers for ATPase slightly reduced the strength of the correlation. The biochemical complexity of the oxidative phosphorylation processes precludes a clear discussion on the functional implications of the amino acid properties that are under selection. The amino acid properties under positive destabilizing selection when the entire protein sequence was taken into consideration for analysis are Solvent accessible reduction ratio, Thermodynamic transfer hydrophobicity. This feature was observed only for the amphibians. No positive destabilizing selection was observed when the entire protein sequence of lungfishes and coelacanths were taken into consideration for analysis. But positive destabilizing selection was observed when individual amino acid sites were taken into consideration. Protein sequences in all the three groups namely amphibians, coelacanths and lungfishes (see Table 1) showed positive selection. Individual amino acid sites influenced by positive destabilizing selection were more similar between lungfishes and amphibians (439 similar sites from 13 proteins) than between coelacanths and amphibians (98 similar sites from 13 proteins). The least similarity was found between lungfishes and coelacanths (16 similar sites from 13 proteins). Please note that the values in Table 1 are NOT to be totaled since the table only lists the major properties, and also a same amino acid site might have been affected by more than one amino acid property, hence if values in Table 1 are added it will give an incorrect higher value (Refer Table S2).

Geobiological Data

Most recent interpretations about the origin of tetrapods and their Devonian representatives lead to conclude that they originated before the Middle Devonian, and probably in the Early Devonian [1,11,22]. It is here that the trackway found in the courtyard of Glenisla Homestead, in the Grampians Mountains, western Victoria, Australia [22] takes all its importance [1,23]. Because of its age (see here below under 'Discussion') it does indeed bring a physical argument for an Early Devonian origin of tetrapods. So, higher levels of atmospheric oxygen from the GEOCARBSULF model and the revised model [21] that were detected in the Early Devonian at ca. 397–416 MYA seem to coincide with the "elpistostegid-tetrapod changeover" (*sensu* [11]). Interestingly there seems to be an increase in terrestrial arthropod orders, autotrophic reef biodiversity, marine invertebrate size and genera during the same time [24,25,26,27]. Although coincidence is not necessarily evidence of correlation, it is suggested here that these events were indeed related. Another important finding from the analysis is the re-confirmation of the earliest Carboniferous Romer's Gap as a low oxygen interval [24] although the revised GEOCARBSULF model [21] shows a higher oxygen level than the previous model (Fig. 1). Compared to the Early Devonian genus-level biodiversity of marine invertebrates of about 585 genera, the Early Carboniferous one is below 400 genera [25] which fits with a lower oxygen interval although further analysis would provide better insights and improved clarity to the problem.

An important point to stress is the influence of the fossil record (likelihood of preservation, differences in paleoenvironments, abundance of field prospectionss, etc.) that certainly has an influence on the observed fossil diversity. Additionally it appears that many paleontologists (including the present senior author) had wrong impressions concerning the fossil diversity through time. A single example is given here, viz. the number of genera of reef builders that is higher in the late Early Devonian than in the rest of the Devonian (Fig. 1). It is indeed often said and tought that the most important period of coral reef development has been the Givetian-Frasnian (late Middle to early Late Devonian) time slice when huge reef systems, compared to the present day Australian Great Barrier, were developed in, e.g., Canadian Arctic or Western Australia [28,29,30]. However, most recent global evaluations of reef diversity in the Devonian show that this is not the case, the highest mean reef thickness and reef diversity being reached in the late Early Devonian [27,31]. Such reappraisal for a single group of organisms, if generalized to all terrestrial and aquatic Paleozoic taxa, will certainly give a very different picture from the classical one depicted by, e.g., the 'Sepkoski Curve' ([32], and later critical reevaluations such as, e.g., [25,33]). This very interesting topic is, however, out of the scope of the present paper.

Table 1. Total number of similar amino acid sites between the different groups of sarcopterygians which have been affected by positive-destabilizing during cladogenesis.

	A-L	A-C	L-C	A-L-C
pK'	117	26	6	0
R_a	157	19	7	0
a_n	2	17	0	17
H_p	14	9	0	0
H	16	0	1	3
K°	62	2	3	2
F	64	13	0	5
H_t	27	2	0	0
P	15	16	0	3

Abbreviations: A- Amphibians, L- Lungfishes, C- Coelacanths; (pK') - Equilibrium constant (ionization of COOH), (R_a) - Solvent accessible reduction ratio, (a_n) - Power to be at the N-terminal, (H_p) - Surrounding hydrophobicity, (H) - Hydropathy, (K°) - Compressibility, (F) - Mean r.m.s. fluctuation displacement, (H_t) - Thermodynamic transfer hydrophobicity, (P) – Polarity. The table shows clearly that the total number of amino acid sites commonly affected by positive selection is much higher between amphibians and lungfishes (A-L) when compared to the other groupings. Hence we come to a conclusion that amphibians and lungfishes are very closely related. Also pK' and R_a have affected the genes to a maximum extent unlike other properties. Please note that only the major amino acid properties are listed in the table (Refer to Table S2 for all the details).

Discussion

The major amino acid properties affecting the similar regions in the genes of amphibians, lungfishes, and coelacanths (see Table S2) are Equilibrium constant (ionization of COOH), Surrounding hydrophobicity, Power to be at the N-terminal, Solvent accessible reduction ratio, Hydropathy, Compressibility, Mean r.m.s. fluctuation displacement, Thermodynamic transfer hydrophobicity, Polarity. All the above properties were detected as influencing in a positive destabilizing selective direction. Since positive destabilizing selection indicates significant change in the protein, only such changes were taken in account for the analysis. When radical changes are favored by selection, they result in local directional shifts in biochemical function, structure, or both. Increase in Equilibrium constant (ionization of COOH) could influence the efficiency of a protein, interestingly this property would reduce the radical oxygen species production which would increase the longevity of the species since it is generally considered now that increase in radical oxygen species is the main reason for aging [34]. Surrounding hydrophobicity refers to the tendency for the region around the amino acid site in question to interact with water. This is important in our case since the proteins used here are transmembrane proteins. It is similar to hydropathy. The proteins becoming more hydrophobic could mean that many residues get buried inside making the protein more compact. Compressibility is a very important property since it influences the stability of the protein which shows that they have become more stable. Solvent accessible reduction ratio is the property that has mostly affected the proteins. The increase in this value suggests that proteins could have become bulkier and allowing more space for active site formation. The adaptive evolution data shows that

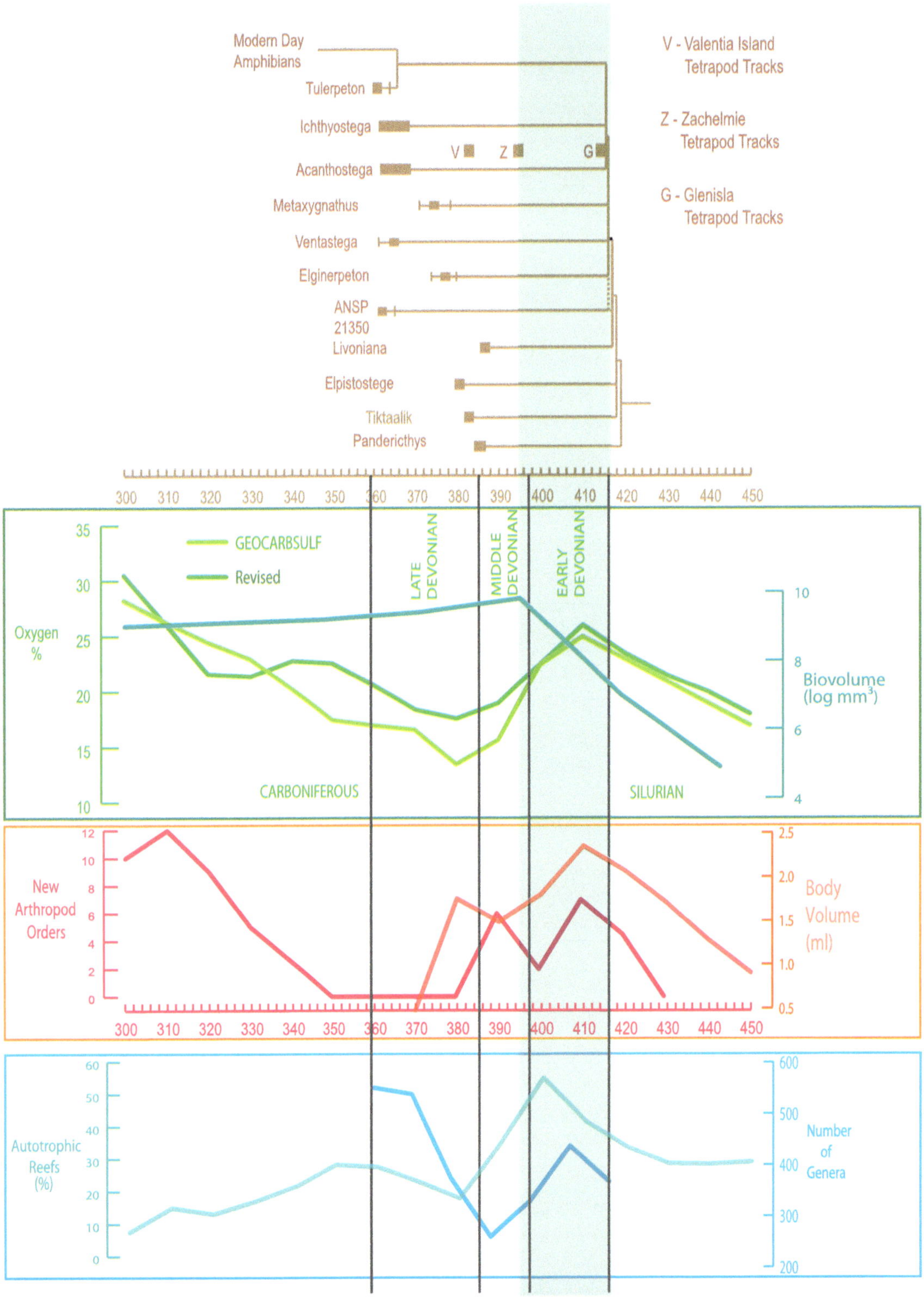

Figure 1. Genus-level biodiversity and phylogenetic relationships of elpistostegid fish and earliest tetrapods, as compared to abiotic and biotic features of Devonian environments. Due to the choosen phylogenetic scheme (after [11]), and the fact that we take into consideration the Glenisla trace fossils from Australia, ghost ranges of basal taxa (elpistostegids, from *Panderichthys* to *Livoniana*) and tetrapods (from ANSP 21350 to the crown group *'Tulerpeton* + modern amphibians') are increased in a significant amount. We use oxygen levels predicted by GEOCARBSULF [21], evolution of arthropod orders [24], evolution of autotrophic reefs [27], body volume of marine invertebrates [26], and genus-level diversity of marine invertebrates [25]. It must be noted that in the highlighted zone of the diagram, the arthropods concerned with are three clades of terrestrial arthropods (myriapods, arachnids, hexapods). Hence the image gives a view of changes in terrestrial and marine species, but giving stress about the changes in marine environment since this is where the tetrapods evolved. The Zachelmie tracks [11] are quite close to the highlighted region and the Glenisla tracks [22] find a satisfactory position amongst the controversy in our image.

Equilibrium constant (ionization of COOH) is the property that has influenced the genes to the second highest extent (Table 1), it drives a more product driven reaction in tetrapod mitochondrial proteins which is why we find a higher value of it affecting the genome.

The levels of oxygen predicted as per GEOCARBSULF [21] has already been studied in relation to tetrapods and arthropods [24]. It is interesting to note that 9 myriapod clades, 4 arachnid clades, and 3 hexapod clades have evolved about 397–416 MYA [24], which seems to confirm the presence of higher levels of oxygen as predicted. Also the increase in distribution of autotrophic reefs [27] could be indicating better formation of reefs in relation to higher levels of oxygen. The diversification of vascular plants [35] and the expansion of high energy predators, including large predatory fish [36,37], both events of major biological significance, occurred during the same period. An analysis with the use of isotopic composition and concentration of molybdenum in sedimentary rocks [38], and a review of maximum size of organisms through geological time [39] also indicate an increase in oxygen levels in relation to a strong increase in chordate maximum length around 400 MYA. Dahl et al. 's [38] study says that this event could have been the greatest oxygenation (rise in atmospheric oxygen level) event in Earth history.

As concerned with the Glenisla trackway of Australia [22] (G on Fig. 1), it has parallel tracks like some of the Zachelmie tracks of Poland [11]. For some authors (e.g. [40]), the tetrapod interpretation of the Glenisla trackway is very doubtful due to the lack of symmetry of the trackway and the absence of clear alternation in its supposed manus and pes tracks; nevertheless we added it in our Figure 1 because of the possibility that it is of a tetrapod. Even without including the Glenisla trackway in our figure, the radiation of early tetrapods is probably within the Early Devonian after the presently oldest known remains [11] (Z on Fig. 1). Including the Glenisla trackway extends the earliest occurrence of tetrapods near to the base of the Devonian. This has impact on the shape of the elpistostegid-tetrapod cladogram when drawn in front of the geologic time-scale (after [11]), by increasing the ghost range of elpistostegids (a ghost range is an interval of geological time where a fossil lineage *should* exist, but for which there is no direct evidence) down to the Silurian/Devonian boundary at ca. 416 MYA (Fig. 1). The abundance of oxygen during the Early Devonian could have led to the "elpistostegid-tetrapod changeover". The higher oxygen levels in the marine environments would have helped the tetrapods to evolve into larger organisms. This could mean that the "changeover" occurred during the arthropod terrestrialization unlike previously thought [24]. The higher oxygen levels suggest that earliest tetrapods never needed to breathe oxygen from the air. This feature might have evolved later when the oxygen levels were lower during the Late Devonian and Early Carboniferous (Fig. 1). In the Silurian, vertebrates (fishes) were generally smaller than in the Early Devonian when larger sizes were developed by both agnathans and jawed fishes, including predatory placoderms and sarcopterygians. The increase in biodiversity (number of genera, Fig. 1) and body size of marine invertebrates suggests almost surely that oxygen levels in the atmosphere and the marine environments did increase during the Early Devonian [26,27,38]. So, our hypothesis concerning tetrapods, even if it is highly speculative, does fit more global results on terrestrial and marine biodiversity in general, and in the Devonian in particular.

Another global event occurred in the Early Devonian, that is a relative lowering of sea levels that began in the late Silurian through the late Early Devonian [41]. But the higher oxygen levels means that even shallow marine regions were well oxygenated. These regions could be suitable regions for evolution of tetrapods where "walking" on the bottom or in water would prove useful (see recent results on "walking" chondrichthyans, e.g., [42,43]); these results let thinking that "walking has evolved many times among different lineages of benthic fishes" (in [43]). Walking is energetically less expensive than swimming and walking is used by thorny skates to capture live prey [43]. Such indications provide more possible reasons for evolution of tetrapods in a shallow marine environment. Generally larger organisms have higher metabolic rates [44] and the increase in efficiency of the proteins could be because the evolution of such species with larger mass can only occur during times of higher oxygen levels.

It may happen in the future that, after revision of the paleobiological databases, the Early Devonian biodiversification event has been as important as the Great Ordovician Biodiversification Event (GOBE: [45,46]) for Paleozoic life. This is supported by the interpretation of Klug et al. [47] who speak of the Devonian Nekton Revolution (DNR) for the re-organization of marine food webs in the Devonian. The nekton revolution of vertebrates did occur earlier in the Silurian [48], but the elpistostegid-tetrapod transition would participate of the early phase of the DNR and the Predation Revolution of vertebrates [48] at a time of high oxygen level. However, we must keep in mind that such scenarios linking global environmental physical factors (such as the atmospheric and oceanic oxygen rate) with development of life on Earth are a pure practice of uniformitarianism in Earth sciences. "Hypotheses linking evolutionary phenomena to atmospheric oxygen levels can be frustratingly difficult to disprove" [49], and "the fundamentally nonuniformitarian nature of Paleozoic and Proterozoic marine ecology must be taken into account" [50].

Conclusions

Here we suggest that the co-occurrence of a series of bio-events and physical properties of the oceans on Earth during the Early Devonian is not merely a coincidence, but reflects a global rearrangement of the biosphere. An increase in oxygen is likely to have occurred during the Early Devonian. It would have triggered the emergence of tetrapods in shallow marine environments, where "walking" on the bottom or in the water would have given them advantage in terms of energetic expenses and predation over other fishes. These shallow marine environments could have also proved as ideal regions for the growth of young tetrapods since they could have had fewer predators. We also conclude from the molecular data that lungfishes are much closer to tetrapods than to coelacanths, a result that is not in contradiction with most morphology-based phylogenetic analyses, although it would be hard to pin point and show that these changes shown in the phylogenetic analysis did occur only in the Early Devonian, it is very much a possibility that some of the changes did occur during the Early Devonian. Scenarios such as the ones described above [38,39,45,46,47] are attractive and represent possible solutions to the relation of global environmental factors and the development of life on Earth. This conclusion is applicable to most, if not all, geobiological scenarios through Earth's history [48].

Supporting Information

Text S1 References and definitions for the 20 physicochemical [amino acid] properties in TreeSAAP.

Table S1 Species used in the molecular adaptation study.

Table S2 Similar amino acid sites between the different groups of sarcopterygians which have been affected by positive-destabilizing during cladogenesis.

Acknowledgments

DG would like to thank Kate Jackson (Whitman College, Walla Walla, WA, USA) for inspiring him to do this study. He thanks Jebakumar, Pavan Kumar Singh and Senthil Kumar for the constant motivation; Yuichiro Suzuki (Wellesley College, Wellesley, MA, USA) and Antony Caesar (St Peter's College, Chennai, India) for providing him with key references and comments; Vinoth Kumar, Valivittan, David McClellan, Per Ahlberg, Alice Clement, Matt Friedman, Michael Coates, Mikolaj Zapalski, Michael Joachimski, John Speakman and David Wake for generously providing their papers and/or giving helpful comments to make this paper. He would also like to thank Hans-Peter Schultze (University of Kansas, Lawrence, KS, USA) for providing references and comments; Philip M. Novack-Gottshall (University of West Georgia, Carrollton, Georgia, USA) and Arnaud Bignon (Université Lille 1, Villeneuve d'Ascq, France) for reviewing an earlier version of the paper; and both referees of *PLoS ONE* for their precise and detailed reviews.

Author Contributions

Conceived and designed the experiments: DG. Performed the experiments: DG. Analyzed the data: DG AB. Wrote the paper: DG AB.

References

1. Young GC (2006) Biostratigraphic and biogeographic context for tetrapod origins during the Devonian: Australian evidence, in Proceedings of CAVEPS 2005, Reed L, Bourne S, Megirian D, Prideaux G, Young G, Wright A, eds. Alcheringa, Spec Iss 1: 409–428.
2. Blieck A, Clément G, Blom H, Lelièvre H, Luksevics E, et al. (2007) The biostratigraphical and palaeogeographical framework of the earliest diversification of tetrapods (Late Devonian), in Devonian Events and Correlations, Becker RT, Kirchgasser WT, eds. Geol Soc London Spec Publ 278: 219–235.
3. Clack JA (2007) Devonian climate change, breathing, and the origin of the tetrapod stem group. Integr Compar Biol 47: 510–523.
4. Daeschler EB, Shubin NE, Jenkins FA (2006) A Devonian tetrapod-like fish and the evolution of the tetrapod body plan. Nature 440: 757–763.
5. Romer AS (1958) Tetrapod limbs and early tetrapod life. Evolution 12: 365–369.
6. Clack JA (2002) Gaining Ground: The Origin and Early Evolution of Tetrapods. Indiana University Press, Bloomington and Indianapolis. 369 p.
7. Schultze HP (1997) Umweltbedingungen beim Übergang von Fisch zu Tetrapode [Paleoenvironment at the transition from fish to tetrapod]. Sitzber Ges Naturforsch Freunde Berlin, (NF) 36: 59–77. [In German, with English abstract].
8. Schultze HP (1998) The Fossil Record of the Intertidal Zone, in Intertidal Fishes - Life in Two Worlds, Horn MH, Martin KLM, Chotkowski MA, eds. (Academic Press, San Diego and London). pp 373–392.
9. Lebedev OA (2004) A new tetrapod Jakubsonia livnensis from the Early Famennian (Devonian) of Russia and palaeoecological remarks on the Late Devonian tetrapod habitats. Acta Universitatis Latviensis, Earth and Environment Sciences 679: 79–98.
10. Laurin M, Meunier FJ, Germain D, Lemoine M (2007) A microanatomical and histological study of the paired fin skeleton of the Devonian sarcopterygian Eusthenopteron foordi. Jl Paleont 81(1): 143–153.
11. Niedzwiedzki G, Szrek P, Narkiewicz K, Narkiewicz M, Ahlberg PE (2010) Tetrapod trackways from the early Middle Devonian period of Poland. Nature 463: 43–48.
12. Clack JA (2006) The emergence of early tetrapods. Palaeogeogr Palaeoclimat Palaeoecol 232: 167–189.
13. Long JA, Young GC, Holland T, Senden TJ, Fitzgerald EMG (2006) An exceptional Devonian fish from Australia sheds light on tetrapod origins. Nature 444: 199–202.
14. Meyer A, Dolven SI (1992) Molecules, fossils and the origin of tetrapods. Jl Mol Evol 35: 102–113.
15. Hedges SB, Hass CA, Maxson LR (1993) Relations of fish and tetrapods. Nature 363: 501–502.
16. Zardoya R, Meyer A (1996) Evolutionary relationships of the coelacanth, lungfishes, and tetrapods based on the 28S ribosomal RNA gene. Proc Natl Acad Sci USA 93: 5449–5454.
17. Brinkmann H, Venkatesh B, Brenner S, Meyer A (2004) Nuclear protein-coding genes support lungfish and not the coelacanth as the closest living relatives of land vertebrates. Proc Natl Acad Sci USA 101: 4900–4905.
18. Takezaki N, Figueroa F, Zaleska-Rutczynska Z, Takahata N, Klein J (2004) The phylogenetic relationship of tetrapod, coelacanth and lungfish revealed by the sequences of forty-four nuclear genes. Mol Biol Evol 21: 1512–1524.
19. Woolley S, Johnson J, Smith MJ, Crandall KA, McClellan DA (2003) Tree SAAP: Selection on Amino Acid Properties using phylogenetic trees. Bioinformatics 19: 671–672.
20. McClellan DA, Palfreyman EJ, Smith MJ, Moss JL, Christensen RG, et al. (2005) Physicochemical Evolution and Molecular Adaptation of the Cetacean and Artiodactyl Cytochrome b Proteins. Mol Biol Evol 22: 437–455.
21. Berner RA (2009) Phanerozoic atmospheric oxygen; new results using the GEOCARBSULF model. Amer Jl Sci 309(7): 603–606.
22. Warren A, Jupp R, Bolton B (1986) Earliest tetrapod trackway. Alcheringa 10: 183–186.
23. Blieck A, Clément G, Streel M (2010) The biostratigraphical distribution of earliest tetrapods (Late Devonian) – a revised version with comments on biodiversification, in The terrestrialization process: modelling complex interactions at the biosphere-geosphere interface, eds Vecoli M, Clément G, Meyer-Berthaud B. Geol Soc London Spec Publ 339: 129–138.
24. Ward P, Labandeira C, Laurin M, Berner RA (2006) Confirmation of Romer's gap as a low oxygen interval constraining the timing of initial arthropod and vertebrate terrestrialization. Proc Natl Acad Sci USA 103: 16818–16822.
25. Alroy J, Aberhan M, Bottjer DJ, Foote M, Fürsich FT, et al. (2008) Phanerozoic trends in the global diversity of marine invertebrates. Science 321: 97–100.
26. Novack-Gottshall PM (2008) Ecosystem-wide body size trends in Cambrian-Devonian marine invertebrate lineages. Paleobiology 34: 210–228.
27. Joachimski MM, Breisig S, Buggisch W, Talent JA, Mawson R, et al. (2009) Devonian climate and reef evolution: Insights from oxygen isotopes in apatite. Earth Planet Sci Lett 284: 599–609.
28. Copper P (1994) Ancient reef ecosystem expansion and collapse. Coral Reefs 13: 3–11.
29. Copper P (2002) Reef development at the Frasnian/Famennian mass extinction boundary. Paleogeogr Palaeoclimat Palaeoecol 181: 27–65.
30. Kiessling W, Flügel E, Golonka J (1999) Paleoreef maps: evaluation of a comprehensive database on Phanerozoic reefs. Bull Am Assoc Petrol Geol 83: 1552–1587.
31. Flügel E, Kiessling W (2002) Patterns of Phanerozoic Reef Crises, in Phanerozoic Reef Patterns. SEPM (Soc Sedim Geol) Spec Publ 72: 691–733.
32. Sepkoski Jr. JJ (1981) A Factor Analytic Description of the Phanerozoic Marine Fossil Record. Paleobiology 7(1): 36–53.
33. McGowan AJ, Smith AB (2008) Are global Phanerozoic marine diversity curves truly global? A study of the relationship between regional rock records and global Phanerozoic marine diversity. Paleobiology 34(1): 80–103.
34. Beckstead WA, Ebbert MT, Rowe MJ, McClellan DA (2009) Evolutionary pressure on mitochondrial cytochrome b is consistent with a role of CytbI7T affecting longevity during caloric restriction. PLoS ONE 4: e5836.
35. Gensel PG (2008) The Earliest Land Plants. Annu Rev Ecol Evol Syst 39: 459–477.
36. Bambach RK (1999) Energetics in the global marine fauna: A connection between terrestrial diversification and change in the marine biosphere. Geobios 32(2): 131–144.
37. Bambach RK (2002) Supporting predators: Changes in the global ecosystem inferred from changes in predator diversity, in The Fossil Record of Predation, eds Kowalewski M, Kelley PH, Dodson P. Paleont Soc Papers 8: 319–351.
38. Dahl TW, Hammarlund EU, Anbar AD, Bond DPG, Gill BC, et al. (2010) Devonian rise in atmospheric oxygen correlated to the radiations of terrestrial plants and large predatory fish. Proc Natl Acad Sci USA 107: 17911–17915.
39. Payne JL, Boyer AG, Brown JH, Finnegan S, Kowalewski M, et al. (2009) Two-phase increase in the maximum size of life over 3.5 billion years reflects biological innovation and environmental opportunity. Proc Natl Acad Sci USA 106(1): 24–27.
40. Clack JA (1997) Devonian tetrapod trackways and trackmakers; a review of the fossils and footprints. Palaeogeogr Palaeoclimat Palaeoecol 130: 227–250.
41. Haq BU, Schutter SR (2008) A chronology of Paleozoic sea-level changes. Science 322: 64–68.
42. Pridmore PA (1995) Submerged walking in the epaulette shark Hemiscyllium ocellatum (Hemiscyllidae) and its implications for locomotion in rhipidistian fishes and early tetrapods. Zoology: Analysis of Complex Systems 98: 278–297.
43. Lucifora LO, Vassallo AI (2002) Walking in skates (Chondrichthyes, Rajidae): anatomy, behaviour and analogies to tetrapod locomotion. Biol Jl Linn Soc 77: 35–41.
44. da Fonseca RR, Johnson WE, O'Brien SJ, Ramos MJ, Antunes A (2008) The adaptive evolution of the mammalian mitochondrial genome. BMC Genomics 9: 119.
45. Webby BD, Paris F, Droser ML, Percival IG, eds (2004) The Great Ordovician Biodiversification Event. "Critical Moments and Perspectives in Earth History and Paleobiology", Columbia University Press, New York. 484 p.
46. Servais T, Owen AW, Harper DAT, Kröger BR, Munnecke A (2010) The Great Ordovician Biodiversification Event (GOBE): the palaeoecological dimension. Palaeogeogr Palaeoclimat Palaeoecol 294(3–4): 99–119.

47. Klug C, Kröger B, Kiessling W, Mullins GL, Servais T, Fryda J, Korn D, Turner S (2010) The Devonian nekton revolution. Lethaia 43: 465–477.
48. Blieck A (2011) The André Dumont medallist lecture: From adaptive radiations to biotic crises in Palaeozoic vertebrates: a geobiological approach. Geologica Belgica 14(3–4): 203–227.
49. Butterfield NJ (2009) Oxygen, animals and oceanic ventilation: an alternative view. Geobiology 7: 1–7.
50. Butterfield NJ (2011) Was the Devonian radiation of large predatory fish a consequence of rising atmospheric oxygen concentration? Proc Natl Acad Sci USA 108(9): E28.

13

Capturing Ecosystem Services, Stakeholders' Preferences and Trade-Offs in Coastal Aquaculture Decisions: A Bayesian Belief Network Application

Laetitia Helene Marie Schmitt[1], Cecile Brugere[2]*

1 Centre for Health Economics/Department of Economics and Related Studies, University of York, Heslington, York, United Kingdom, **2** Stockholm Environment Institute, University of York, Heslington, York, United Kingdom

Abstract

Aquaculture activities are embedded in complex social-ecological systems. However, aquaculture development decisions have tended to be driven by revenue generation, failing to account for interactions with the environment and the full value of the benefits derived from services provided by local ecosystems. Trade-offs resulting from changes in ecosystem services provision and associated impacts on livelihoods are also often overlooked. This paper proposes an innovative application of Bayesian belief networks - influence diagrams - as a decision support system for mediating trade-offs arising from the development of shrimp aquaculture in Thailand. Senior experts were consulted (n = 12) and primary farm data on the economics of shrimp farming (n = 20) were collected alongside secondary information on ecosystem services, in order to construct and populate the network. Trade-offs were quantitatively assessed through the generation of a probabilistic impact matrix. This matrix captures nonlinearity and uncertainty and describes the relative performance and impacts of shrimp farming management scenarios on local livelihoods. It also incorporates export revenues and provision and value of ecosystem services such as coastal protection and biodiversity. This research shows that Bayesian belief modeling can support complex decision-making on pathways for sustainable coastal aquaculture development and thus contributes to the debate on the role of aquaculture in social-ecological resilience and economic development.

Editor: Jose Luis Balcazar, Catalan Institute for Water Research (ICRA), Spain

Funding: This research was supported by the Marie Curie Intra European Fellowship project ECOLIVA "Sustainable ECOsystem services and LIVelihoods through Aquaculture development" (Project No: PIEF-GA-2009- 235835, 2010–11) funded under the European Community 7th Framework Programme. The funders had no role in study design, data collection and analysis, decision to publish, or preparation of the manuscript.

Competing Interests: The authors have declared that no competing interests exist.

* E-mail: cecile.brugere@sei-international.org

Introduction

The shrimp aquaculture industry of Thailand, which accounted for 15.4% of the global aquaculture production of shrimps in 2009 [1], is a major contributor to the country's economy, bringing in over US$ 2 billion of foreign revenues annually [2] and employing more than one million people [3]. The majority of the industry (98%) is based on the production of the exotic Pacific white-leg shrimp (*Litopenaeus vannamei*), which has replaced the indigenous Giant black tiger shrimp (*Penaeus monodon*) since the early 2000s [4].

The introduction of intensive farming techniques in the 1980s led to Thailand becoming a world-leading shrimp producer. However, this shift was associated with water pollution and acidity build-up in shrimp ponds. In conjunction with poor husbandry methods, this resulted in the onset of catastrophic viral diseases and a collapse in production [5,6], prompting farmers to abandon their ponds and dig new ones further into mangrove-forested areas. Mangroves swamps were considered of low economic value as the ecosystem services they deliver (e.g. pollutant sink, storm protection, provision of wood and fish [7,8]) were not taken into account. As a result, Thailand's mangrove forests were halved between 1961 and 1993, although the extent to which shrimp farming alone was responsible for the loss of mangrove cover in the country remains debatable [5].

The Ecosystem Approach to Aquaculture (EAA) has recently emerged in response to calls for increased sustainability in the aquaculture sector and recognition that aquaculture is often embedded in sensitive social-ecological systems [9]. This approach advocates the consideration of biophysical and human dimensions of ecosystems in order to achieve the dual aims of sustainable production and human wellbeing [10]. At the heart of the EAA lies the notion of social-ecological resilience, defined as "the capacity to maintain integrity when responding to external changes and feedbacks" [11]. Decision-making in the context of coastal social-ecological systems requires considering several different issues: i) interconnectivity, i.e. the fact that "disruption in one system is likely to cause disruption in the other" [10], (ii) complexity and uncertainty, as coastal ecosystem responses are often non-linear [12] and the minimum level of ecosystem structure needed to maintain a constant flow of services is unknown [13], and (iii) stakeholder conflicts stemming from competing uses over coastal resources and institutional failures [14].

However, a gap in the methodology exists as these issues are not simultaneously considered in decision-making or when the EAA is implemented in coastal areas. This calls for a decision support system that can holistically assess and quantify the trade-offs arising from coastal aquaculture management decisions. A prior review of the range and uses of decision support tools currently available (e.g. extended cost-benefit analysis, multi-criteria decision analysis) showed that Bayesian Belief Networks (BBN) – also called influence diagrams – stand out for their ability to handle uncertainty, complex non-linear relationships as well as qualitative and quantitative data of an ecological and economic nature. BBNs also have the advantage of remaining straightforward enough to allow clear communication of complex decision outcomes to policy makers [15].

BBNs have been applied to natural resource management to model human pressures on diverse ecosystems [16–18], to assist in the assessment of the social, ecological and economic dimensions of coastal water and catchment management [19–21] and to evaluate and compare fisheries management plans [22]. Applications of BBNs to aquaculture have been limited to modeling the impacts of water management options on the simultaneous production of rice, fish, crab and shrimp in Viet Nam [23] and to quantifying the human drivers behind the adoption of sea cucumber aquaculture as an alternative livelihood strategy in Tanzania [15].

The purpose of the present study, and the novelty of this research, is to broaden the use of Bayesian Belief Networks by incorporating indicators of social-ecological resilience and refining their application to aquaculture. The specific objective of this work is to show how such a decision support tool can help mediate complex and multiple trade-offs in the context of shrimp aquaculture development in Thailand through: (i) the articulation of available knowledge on the impacts of coastal aquaculture on social, ecological and economic systems via a network structure, and (ii) the generation of a probabilistic impact matrix that highlights trade-offs – or "social conflicts over interest and values" [24] – and uncertainties in management decision outcomes. The underlying goal is to provide policy-makers with a full picture of the potential consequences of their aquaculture development decisions on specific characteristics of the system. Section 2 of this paper presents the methodology used to develop a BBN applied to shrimp aquaculture. Section 3 shows modeling results for six scenarios of aquaculture management and assesses their implications in terms of trade-offs between ecosystem services provision, local livelihoods, financial profits and economic development. Finally, section 4 discusses the findings and the role of BBN as a decision-support system for sustainable aquaculture development, in line with the EAA.

Materials and Methods

1. Study area and data collection

In Thailand, the shrimp farming sector is characterized by the prevalence of small production units (<2 ha), which coexist alongside a minority of very large farms [25]. As of 2007, more than 90% of shrimp farms were intensive, i.e. with high stocking densities of juvenile shrimp in ponds [26]. Production predominantly takes place in semi-closed systems characterized by minimal exchange of water with the outside environment to limit the introduction of viruses and pathogens and the release of pollution [27–29]. However, pond water and sludge are still partially discharged in adjacent waterways, notably at the time of harvest. In addition, *L. vannamei* shrimps have been reported in the vicinity of ponds, despite measures aimed at preventing their accidental release in the environment [30]. Fully closed systems also exist but their requirements of a water reservoir, sedimentation pond and continuous aeration make them capital-intensive operations [28]. Whilst shrimp aquaculture has increasingly taken place in supra-tidal areas, i.e. behind the mangrove fringe (S. Funge-Smith, personal communication, 2011), a large number of shrimp farms are still suspected to be operating in inter-tidal areas due to lack of law enforcement and legislation loopholes [5,31].

Nevertheless, awareness of Better Management Practices (BMPs), framed by the International Principles for Responsible Shrimp Farming [32], has been growing. BMPs aspire to reduce the impact of shrimp farming on the environment and improve the livelihoods of farmers by implementing measures such as storage ponds and increasing efficiency in production by optimizing the shrimp feeding process [33]. In Thailand, the implementation of BMPs has been promoted, in particular among small producers (e.g. [34]), but their adoption remains limited in comparison to other countries such as India and Viet Nam [35,36].

At the other end of the production spectrum is aquasilviculture, an extensive system that promotes harmonious aquatic production alongside mangrove forestation [37,38]. In the context of the present research, aquasilviculture is defined as a mangrove-shrimp system where 70% of the pond surface is planted with mangroves and 30% is dedicated to shrimp production. Although aquasilviculture remains marginal in Thailand in comparison to other Southeast Asian countries, it has enabled farmers to return to traditional low input-low output shrimp farming after crop failures [39]. It is also recognised that, under the right conditions, it can be used as a suitable management strategy to rehabilitate disused shrimp ponds, enhance the provision of mangrove ecosystem services, and provide a complementary source of income to farmers [40,41].

The characteristics of the development of shrimp farming in Thailand, its history and impacts have been extensively documented and general information and data on the sector were collected from the literature. In line with the EAA, Figure 1 captures the multiple interactions of the sector with the environment (including services provided by coastal ecosystems), the economy and livelihoods. Four groups of key stakeholders have been identified (underlined in Figure 1): (i) farmers themselves, organized in associations and seeking income from their activity, (ii) the Government of Thailand, seeking export revenues and employment through the entire shrimp value chain, (iii) members of coastal communities who can simultaneously benefit (e.g. employment) and suffer (e.g. loss of mangrove-dependent livelihoods) from the conversion of mangroves to ponds, and (iv) future generations, representing the stakes of long-term sustainable coastal development.

The bulk of the literature on the economics of shrimp farming relates to *P. monodon* and not *L. vannamei*, which is associated with different yields, costs and profits [4]. To fill this gap and provide up-to-date data on exploitation costs and revenues for *L. vannamei*, onsite surveys were implemented. Ten intensive shrimp farmers were sampled in Surat Thani, Surat Thani Province (sample stratified by farm size: 3 very small farms (<2 ha), 3 small farms (2.1–5 ha), 3 medium farms (5–50 ha) and 1 large farm (>50 ha)). Ten smaller-scale intensive farmers were also sampled in Samroyiot, Prachuap Kiri Khan Province (Figure 2). Of the Samroyiot farmers, six were implementing Better Management Practices (BMPs) and also provided economic and production data. These six farmers were gathered in a "cluster" and working together towards the implementation of BMPs.

Ethical clearance of the research was provided by the European Commission under Framework Programme 7 (PIEF-GA-2009-235835) as a condition of the award. Questionnaires were designed and tested to ensure an optimal balance between cultural sensitivity and achievement of the study objectives. Particular attention was given to gender and to the sensitiveness of the questions asked when interviewing women, minority or vulnerable

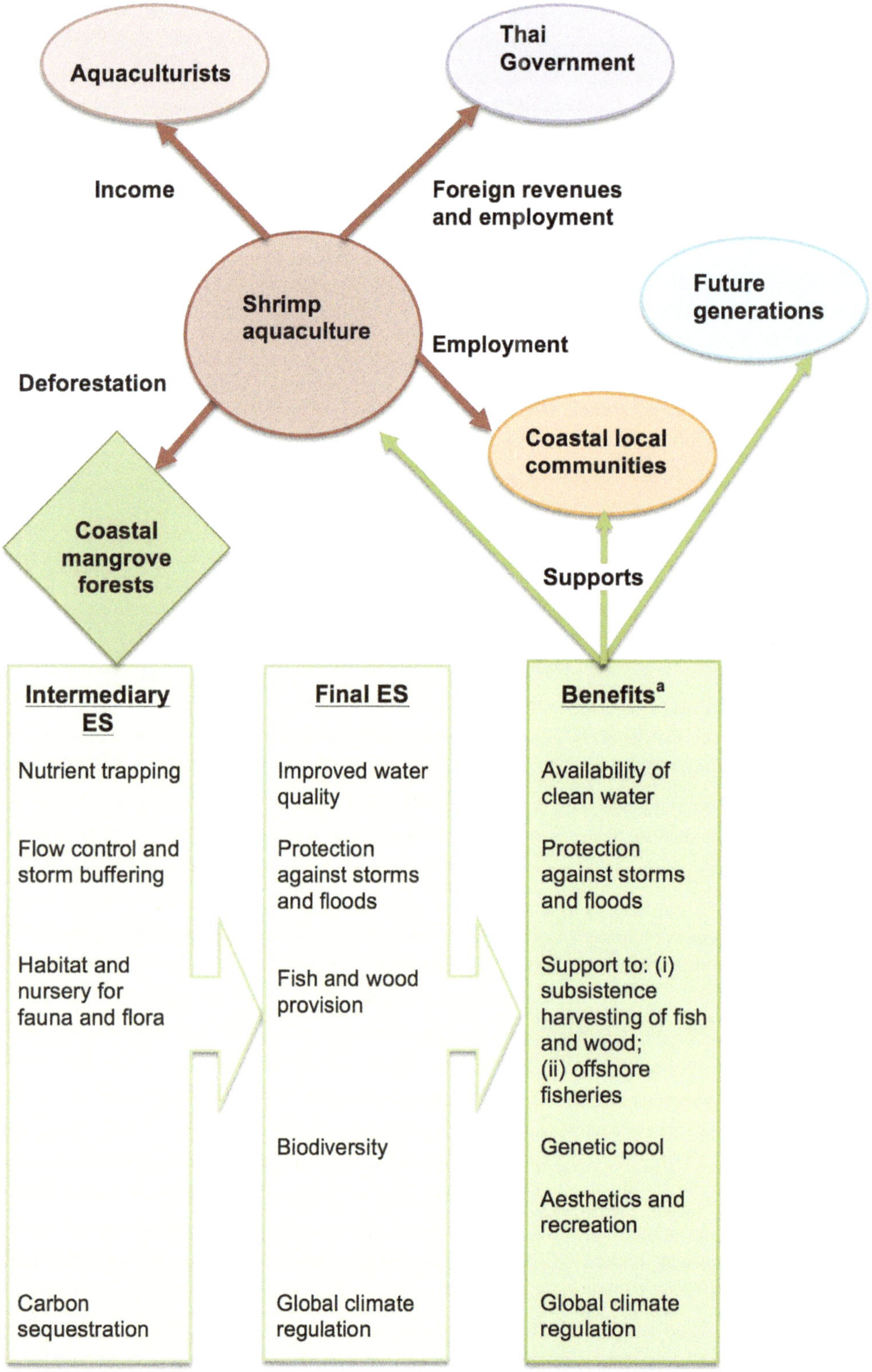

Figure 1. Aquaculture's main impacts on the environment, the economy and livelihoods and related user-conflicts. ES = Ecosystem Services. [a] Benefits should be distinguished from final ES since they are often a product of final ES and human inputs (e.g. fishing gear) and economic valuation of ecosystem should apply to ecosystem benefits only [13,54,55]. Figure based on information from [40] and [8], relying on the ecosystem services classification by [13,55].

Figure 2. Locations of onsite surveys in Thailand. 1: Surat Thani (Surat Thani Province) representing conventional shrimp farming. 2: Samroyiot (Prachap Kiri Khan Province) representing small-scale shrimp farming.

people (e.g. elders, disabled, illiterate, poor), in order to ensure the free expression of their views. Informed verbal consent was provided by each interviewee in light of guarantees of confidentiality, anonymity, privacy, data protection and the possibility to withdraw from the survey at any time. Ethical training was provided to all partners involved in data collection. Monitoring and evaluation of participation and of the research process occurred according to agreed codes of conduct and standards of research practice, including anonymizing the questionnaire recording forms and destroying these at the end of the study.

2. BBN development

2.1. BBN definition. A BBN is a directed acyclic graph, which parameters are treated as random variables and modeled via nodes. The typology of nodes used in BBN modeling is provided in Table 1. Nodes are interconnected in "parent-child" (i.e. predictive) relationships. A "child" node has a conditional probability distribution (CPD) for each possible combination of states of its "parent" nodes and the width of the distributions indicates the level of uncertainty in the cause-effect relation [42]. Unlike hierarchical Bayesian networks, which rely extensively on model simulations, the CPDs that populate BBN chance nodes are generally obtained analytically, with expert opinion used to overcome data gaps [43]. Scenario analysis allows the updating of chance nodes' CPDs into posterior probability distributions (PDs) via Bayes' rule of joint probabilities [44].

In addition to explicitly handling uncertainty via the notion of conditional probabilistic dependence between variables [45], a BBN can incorporate both qualitative and quantitative information by connecting a decision node with chance nodes and utility nodes (Figure 3). Finally, thanks to its network structure, a BBN provides a visual representation of the causal relationships that underpin complex systems to which management decisions are applied, as well as straightforward probabilistic information on uncertain outcomes of management actions [46,47].

2.2. Network construction. Information on the impacts of shrimp farming collected from the literature was incorporated in a network via chance nodes, whereby cause-effect relationships were made explicit. Leaf nodes, which are the nodes of focus for our scenario analysis, were specifically chosen to represent the diversity of interests of the four groups of stakeholders previously identified, so that user-conflicts could be made explicit in output results. The BBN was developed via the software GeNIe from the Decision Systems Laboratory of the University of Pittsburgh, USA (http://dsl.sis.pitt.edu). This software is freely available and has the advantage over other free commonly used software of not limiting the network size.

Following Marcot et al.'s guidance [48], the number of parent nodes per child node was limited to three so that the number of CPDs of child nodes would remain within reasonable limits. As a consequence, three modeling choices were made:

(i) The BBN was built at the shrimp farm level and thus did not model the whole shrimp value chain. Nevertheless, employment generated by the sector was indirectly dealt with via the node "long-term contribution to the country's shrimp exports".

(ii) The issue of abandoned ponds was addressed in the model via the node "long term contribution to the country's shrimp exports" (Thai government's perspective) and the computation of the net present value (NPV) of the profit earned by aquaculturists.

(iii) The modeling of coastal communities' resilience, which was inspired by Ashley et al.'s livelihoods framework [49], focused solely on natural, human and financial capital, which were the most straightforward to handle in the present analysis. Modeling of these three sources of capital was made via the nodes "fish/wood for locals", "health of locals" and "potential for income diversification" respectively.

The modeling of mangroves' ecosystem services called upon a value judgment on how the value of these services would be best communicated. Mangroves' ecosystem services were all modeled qualitatively via chance nodes. However, coastal protection was also modeled quantitatively in monetary units (utility node) as it is estimated to be mangroves' most economically valuable service [12].

2.3. Case study and management scenarios. A hypothetical case study representing the range of pond management possibilities was defined to show how the constructed BBN could support complex decision-making on coastal aquaculture development pathways. This case study used six mutually exclusive aquaculture and land management scenarios (Table 2), underpinned by a set of assumptions defined in Table 3.

2.4. Expert involvement. A group of 42 senior experts in coastal aquaculture from academia, research institutions and the industry were contacted by email to validate the network and the case study and to elicit the CPDs that populate chance nodes. They were sent conditional probability tables to fill for each

Table 1. Typology of nodes in BBN modeling, according to their modeling role and position in the network.

Modeling role	
Chance node	Node modeling a random variable over discrete states and defined by a joint conditional probability distribution
Utility node	Node populated with utility values that express preferences over outcomes
Decision node	Node modeling choices that can be made by the decision maker, comprising one state for each scenario
Position in network	
Parent node	Node predicting one (or more) child node(s)
Child node	Node linked to one (or more) parent node(s) via a predictive relationship. A child node has a conditional probability distribution for each combination of state of its parent nodes
Root node	A node with no parent, defined by a probability distribution
Leaf node	A node with no child

chance node, alongside extensive explanations about the constructed BBN, the case study and the management scenarios considered. Of these experts, twelve contributed to the fine-tuning of the network variables and causal relationships, and to the refinement of the assumptions underpinning the case study. CPDs were elicited by four of the twelve experts through a number of iterations and an in-depth dialogue.

2.5. BBN parameterisation. *Chance nodes.* Because biophysical modeling was outside the scope of this research and due to lack of available datasets, all the CPDs of the model's chance nodes were elicited by experts. As the BBN comprises a large number of chance nodes (24), all chance nodes were attributed the states of "high", "medium" and "low" to facilitate the CPD elicitation exercise. Although applying such a restriction on the states of variables clearly represents a simplification of reality, it ultimately enables information on trade-offs to be presented in an easily understandable form to decision-makers. Normalized average values of elicited CPDs were then calculated.

Utility node for coastal protection. Barbier at al. found that mangroves' level of wave attenuation was a quadratic function of habitat size, which allowed computation of a set of estimates for the monetary value of coastal protection services [12]. In order to link these estimates to the qualitative level of coastal protection quality modeled in the BBN, each coastal protection level ("high", "middle", "low") was arbitrarily associated to a non-linear change in wave height and the monetary value estimates were rescaled to rai units. This procedure, which enabled coastal protection values to be integrated into the model, is summarized in Figure 4.

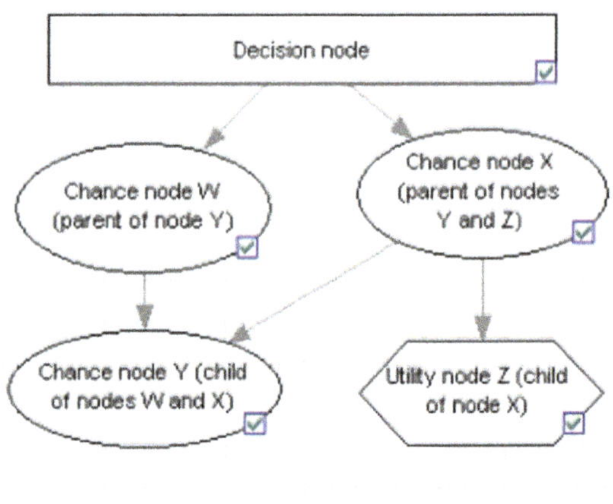

Figure 3. Convention adopted to graphically represent each type of node in a Bayesian Belief Network.

Utility nodes for aquaculturists' annual profit. The computation of the profit earned by the group of aquaculturists encompassed only the costs and revenues expected to be substantially impacted by the type of farm management. Therefore, the term "profit" here is an indicator of the financial performance of each management option, and not of the accounting profit earned. Annual profit was expressed in US$/rai so that it could be easily rescaled to any number of farms. Table S1, in supporting information, details how economic and production data from onsite surveys in Surat Thani and Samroyiot were integrated in the profit computations, in combination with secondary data and expert opinion.

For each management scenario$s$, aquaculturists' annual profit was computed in GeNIe, via several utility nodes, based on formulas (1) and (2). The net present value (NPV) was computed in Excel according to formula (3).

$$R_s = Q_s(1 - CL_s)P_s \tag{1}$$

$$\pi_s = R_s - VC_s \tag{2}$$

$$NPV(\pi_s) = -I_s + \sum_{t=1}^{T} \frac{\pi_{s,t}}{(1+r)^t} \tag{3}$$

R_s is aquaculturists' annual gross revenue (US$/rai, 1 rai = 0.16 ha); Q_s is the total annual shrimp production (kg/rai); CL_s is the annual crop loss (%); P_s is the shrimp farmgate price (US$/kg); π_s is the annual profit (US$/rai); VC_s is the annual production variable cost (US$/rai); I_s is the investment cost (US$/rai); T is the farm lifespan (years) and r is the discount rate.

Results

1. Network and impact matrix

The BBN in Figure 5 provides a graphical representation of the knowledge and belief of causal relationships between social, ecological and economic impacts associated with land and farm management scenarios relative to shrimp farming. Modeling output for the five leaf nodes (aquaculturists' annual profit; long-term contribution to the country's shrimp exports; value of coastal

Table 2. Mutually exclusive aquaculture and land management scenarios used in BBN modeling.

Scenario name	*Description*
BAU - Business as usual	Conventional intensive farming as currently practised
BMP	Introduce Better Management Practices
Restore forest+closed-system	Fully restore the existing farm as a mangrove forest and build a closed-system in the supra-tidal zone, behind the mangrove fringe
Replant 20%	Replant mangroves on 20% of the farm pond area
Replant 40%	Replant mangroves on 40% of the farm pond area
Aquasilviculture	Replant mangroves on 70% of the pond area and integrate the culture of mangroves with low-density shrimp aquaculture

Notes:
- The last three scenarios combine intertidal land conversion with mangroves replanting and conservation and were designed to capture non-linearity in mangroves provision of ecosystem services [12].
- The "aquasilviculture" and "restore forest+closed-system" scenarios were associated with a 100% probability of "high" level of coastal protection, while "BAU" and "BMP" scenarios were associated with a 100% probability of "low" level of coastal protection.

protection; biodiversity; resilience of local community), which were specifically chosen to reflect trade-offs and user-conflicts, are presented in the form of a probabilistic impact matrix (Figure 6). The scenario "restore forest+closed-system" scored the highest on each criterion modeled via the leaf nodes and thus represents the best-case scenario.

2. Interpretation and sensitivity analyses

2.1 Profit. Whilst levels of farm production drove the difference in annual profit between management scenarios, production loss due to mangrove replanting in ponds was partially compensated by a lower risk of shrimp loss from disease outbreak or typhoons and a higher likelihood of securing a price premium via production certification. For instance, the annual profit under scenarios "replant 20%" and "replant 40%" was lower than under "BAU" by only 15% and 32% respectively. This was due to a lower expected crop loss (10.3% and 8% of annual crop respectively, vs. 12.6% for "BAU") and a greater likelihood of certification (12.7% and 19.7% respectively, vs. 8% for "BAU"). The better financial performance of scenario "BMP" against "BAU" stemmed from lower feeding costs following optimization of the shrimp feeding process.

Sensitivity to assumptions underpinning the NPV computations was tested. As a result of compounded discounting, sensitivity to discount rate was especially acute for scenarios "aquasilviculture" and "restore forest+closed-system", since these management systems were assumed to be operating over the longest periods (50 years and 30 years respectively). Conversely, sensitivity to farm lifespan was stronger for scenarios "BAU" and "BMP" since they were associated with a shorter lifespan (10 and 13 years respectively). Under the assumption that all systems would operate for only 10 years, scenario "restore forest+closed-system" was no longer associated with the greatest NPV of profits due to the high investment cost incurred in year 0 (US$ 4,249/rai).

2.2 Coastal protection value. The larger the replanted pond surface was, the higher the expected value of coastal protection. Following assumptions pertaining to the probability of coastal protection levels associated with each management scenario (see Table 2), the BBN encompassed uncertainty about coastal protection quality only for the scenarios which involved replanting 20% and 40% of the pond surface. Economic values of coastal protection were underpinned by the assumed association between qualitative levels of protection and quantitative provision of wave-attenuation service (see Figure 3). Sensitivity analysis showed that the stronger the assumption of non-linearity, the more spread out across management scenarios coastal protection values were.

2.3 Long-term contribution to the country's exports. The performance of scenarios "aquasilviculture" and

Table 3. Assumptions behind the BBN modeling case study.

Assumption	*Description*
A1	Small-scale intensive shrimp farms relying on semi-closed systems are operating as a group and cultivate and export the non-native species *L. vannamei* (using specific pathogen free broodstock), thus generating cumulative impacts on the environment
A2	Farms are located in Thailand's intertidal area (i.e. formerly forested by mangroves), and have recently started operating
A3	The farm owner fully undertakes the management option of his/her choice at t = 0[(1)]
A4	The rate of survival of replanted mangroves is high and replanted mangroves provide their ecosystem service straightaway[(1) (2)]
A5	All production, including from aquasilviculture, is exported[(3)]
A6	To the exception of aquasilviculture, stocking densities remain identical among the different management scenarios

Notes:
[(1)]A3 and A4 were motivated by the difficulty to integrate time dynamics in BBNs.
[(2)]Mangroves are expected to deliver their services after 3 years of normal growth (J. Primavera, personal communication, 2011).
[(3)]While the quality and size of shrimps may greatly determine whether shrimps are exported or not, this assumption reflects the fact that 85% of the Thai shrimp production is exported [3].

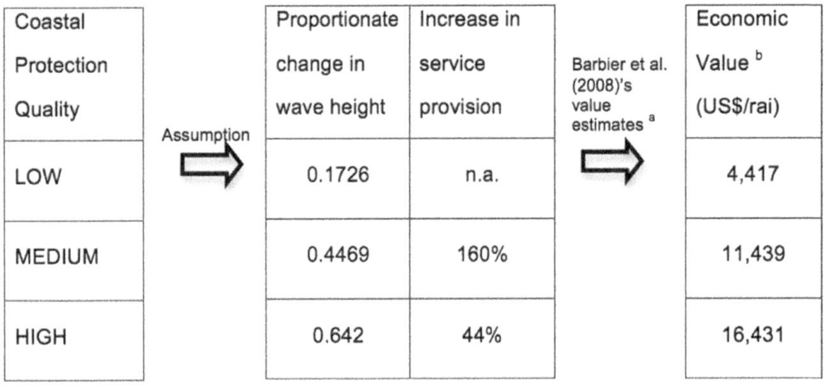

Figure 4. Parametrisation of the utility node "Value of coastal protection". [a] Estimates computed by [12] based on the expected damage cost method combined with a quadratic wave attenuation function and the assumption that each km^2 of mangroves represents a mangrove forest area of 100 m inshore along a 10 km coastline. Estimates were converted in (US$/rai) with 1 rai = 0.0016 km^2. [b] Net present values computed over a 20-year time-horizon with a 10% discount rate.

"restore forest+closed-system" on the long-term contribution to the country's exports differed sharply from the "BAU" scenario. It was expected to be high with a 62% probability under scenario "restore forest+closed-system" (vs. 37% under "BAU"), while it was expected to be low with a 69% probability under scenario "aquasilviculture" (vs. 28% under "BAU"). By contrast, in the other scenarios, the level of contribution to exports did not differ substantially from the "BAU" scenario as the decrease in production due to replanting was partially offset by an increase in farm lifespan (lower bioaccumulation of pollutants). For example, the likelihood of high contribution to exports under scenario "replant 40%" was only 4 percentage points lower than under "BAU".

2.4 Biodiversity. Whilst biodiversity levels ("high", "medium", "low") were not specifically defined here, they were meant to reflect the degree of variety of living organisms and genetic variability encapsulated in the definition of biodiversity adopted in Article 2 of the Convention on Biological Diversity in 1993. The likelihood of a high level of biodiversity was shown to increase from 7% under "BAU" to 8% under "BMP", 17% under "replant 20%", 25% under "replant 40%", 58% under "aquasilviculture" and 75% under "restore forest+closed-system". Interestingly, the likelihood of high biodiversity increased by 50% between scenarios "replant 20%" and "replant 40%" and by 130% between scenarios "replant 40%" and "aquasilviculture" (i.e. replant 70%), highlighting non-linearity in ecosystem service provision. Sensitivity analysis consisted of evaluating the separate impacts on biodiversity levels under each sensitivity scenario of: (i) a high risk of spread of diseases to native species; (ii) high water quality and (iii) low pollution from other activities, against the base case simulation (run 1) where the root node "pollution from other activities" was populated with a uniform PD. For these sensitivity analyses, the nodes "risk of spread of diseases to native species", "surface water quality" and "pollution from other activities" were separately controlled, in three successive runs, to be with a 100% probability in a given state of interest (e.g. "low" or "high"). Results, presented in Figure 7, showed that water quality, itself partially driven by levels of external pollution, is expected to have a greater influence on biodiversity levels than the risk of spread of disease to native species.

2.5 Resilience of local community. Scenarios "restore forest+closed-system" and "aquasilviculture" stood out as very likely to bring high coastal community resilience, with a 93% and 82% probability respectively, vs. 27% under "BAU". By contrast, the spread of posterior PDs under replanting scenarios showed much greater uncertainty in the expected level of resilience. Sensitivity analysis focused on the node "locals' subsistence capacity" as it strongly influenced the node "resilience of local community" (see Figure 5). Each parent node of the node "locals' subsistence capacity" (i.e. nodes "quantity of wood for locals", "quantity of fish for locals" and "local employment") was separately controlled, in three successive runs, to be in the state "low" with a 100% probability. Results against run 1, presented in Figure 8, showed that uncertainty in the level of restoration of productive services was the greatest contributor to overall uncertainty in subsistence capacity and thus, of resilience. Additionally, under the replanting and aquasilviculture scenarios, the likelihood of high subsistence capacity was found to decrease more under the "low wood" sensitivity run than under the "low fish" sensitivity run. However, when the mangrove forest was restored to its original state (scenario "restore forest+closed-system"), fish and wood provision were found to have an equal influence on locals' subsistence capacity.

Discussion

1. Highlights and contextualisation of the modeling results

Firstly, economic results aim solely to provide an indication of the difference in scenarios' financial performance and of the economic impact of improving coastal protection by replanting mangroves. As such, they should not be used for predictive purposes.

Secondly, the differential performance of scenarios in terms of contribution to Thailand's shrimp exports underlines substantial differences in production capacity and reliability between systems. Indeed, the productive capacity of aquasilviculture is small compared to closed systems, which are themselves more reliable than conventional systems ("BAU") due to better disease risk management. Additionally, results suggest that, in the long run, the loss in production capacity from mangroves replanting in ponds is at least partially offset by the increase in farm lifespan due to a lower bioaccumulation of pollutants.

Thirdly, the non-linearity in biodiversity provision highlighted in the results is of particular relevance for the identification of an optimal level of mangrove replanting. The impact of management

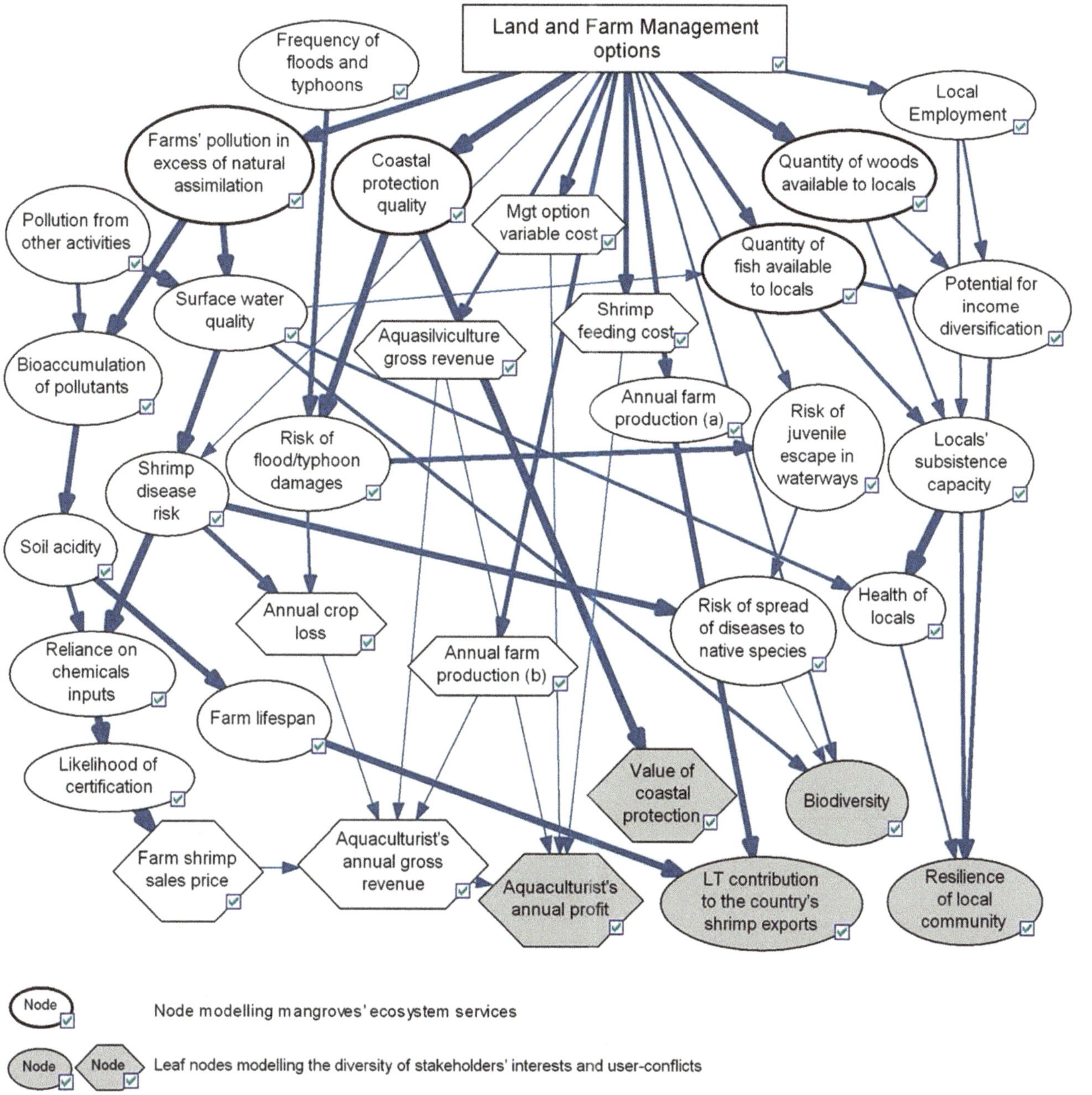

Figure 5. Bayesian belief network and strength of influence between variables. The width of the arrows linking nodes depicts the strength of influence between parameters ciphered by the conditional probability distributions. [(a)] Chance node defining farm production qualitatively over the states "high", "medium" and low". [(b)] Utility node defining farm production quantitatively (kg/rai).

scenarios on biodiversity is, however, expected to be more complex than our results suggest and should be validated by field data.

Fourthly, systems where farming is mixed with mangroves may differ in the way they control organisms entering and exiting the shrimp pond (M. Troell, personal communication, 2011). Therefore, when considering local community resilience, uncertainty about the level of restoration of productive services is likely to be larger for the provision of nursery habitat and fish seed than for wood. The greater uncertainty in the restoration of fish provision following mangrove replanting may explain why wood availability appears as a stronger contributor to locals' subsistence capacity than fish provision. This hypothesis should, however, be tested empirically.

2. Role of this BBN as a decision support tool for sustainable coastal resource use

The BBN presented here provided a holistic representation of the main user conflicts and trade-offs associated with various forms of shrimp farming development. Although not a predictive tool, this BBN is a means of articulating existing knowledge and beliefs about the multiple interactions of shrimp farming with the environment, the economy and livelihoods. It also enables to

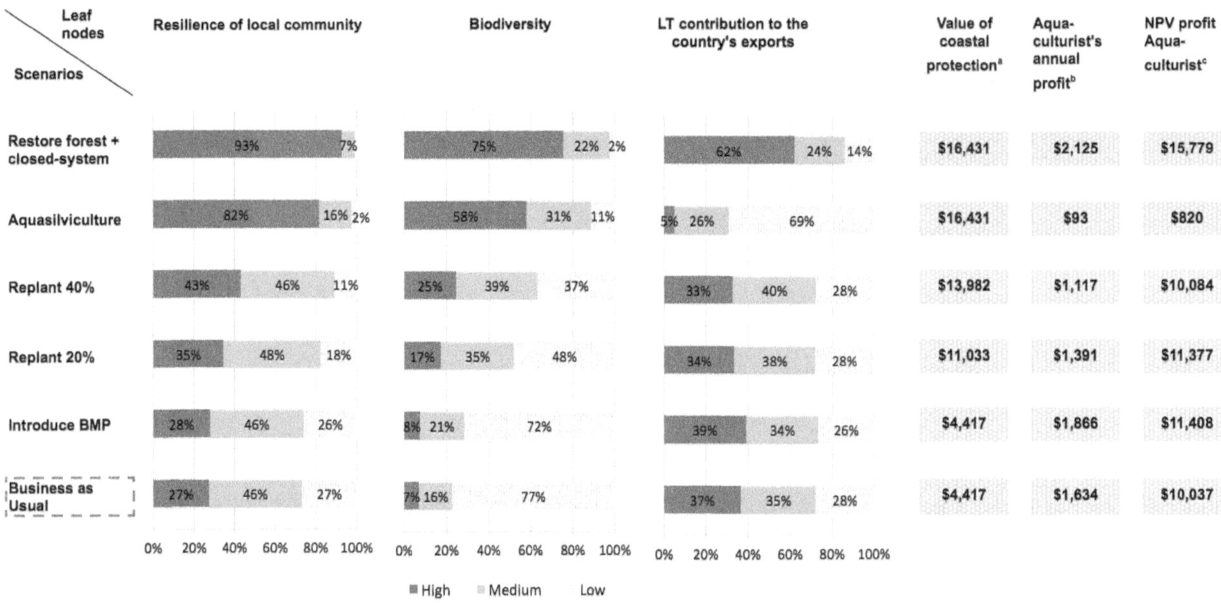

Figure 6. Impact matrix. Following belief propagation, posterior probability distributions and expected utility values for each leaf node were obtained for each management scenario. The matrix summarizes the performance of the five management scenarios against the BAU scenario on the criteria modeled via the leaf nodes. [a] NPV values (US$/rai) computed over a 20 year-time horizon, using a 10% discount rate. [b] Encompasses only exploitation costs and revenues impacted upon by the management option. [c] NPV values (US$/rai) computed using a 10% discount rate and based on estimates of farm lifespan (see Table S1). 1 ha = 6.25 rai.

quantitatively compare the trade-offs associated with a range of aquaculture development strategies. In a context of renewed calls for the improved governance of the sector, especially in countries where the social-ecological resilience of coastal areas has been eroded in the past, this approach could help policy-makers understand the potential consequences of their decisions and increase the transparency of their policy choices by prompting them to explicitly value pre-defined sustainability criteria.

Sustainable land and farm management practices should be in line with the principles of the EAA and adaptable to local

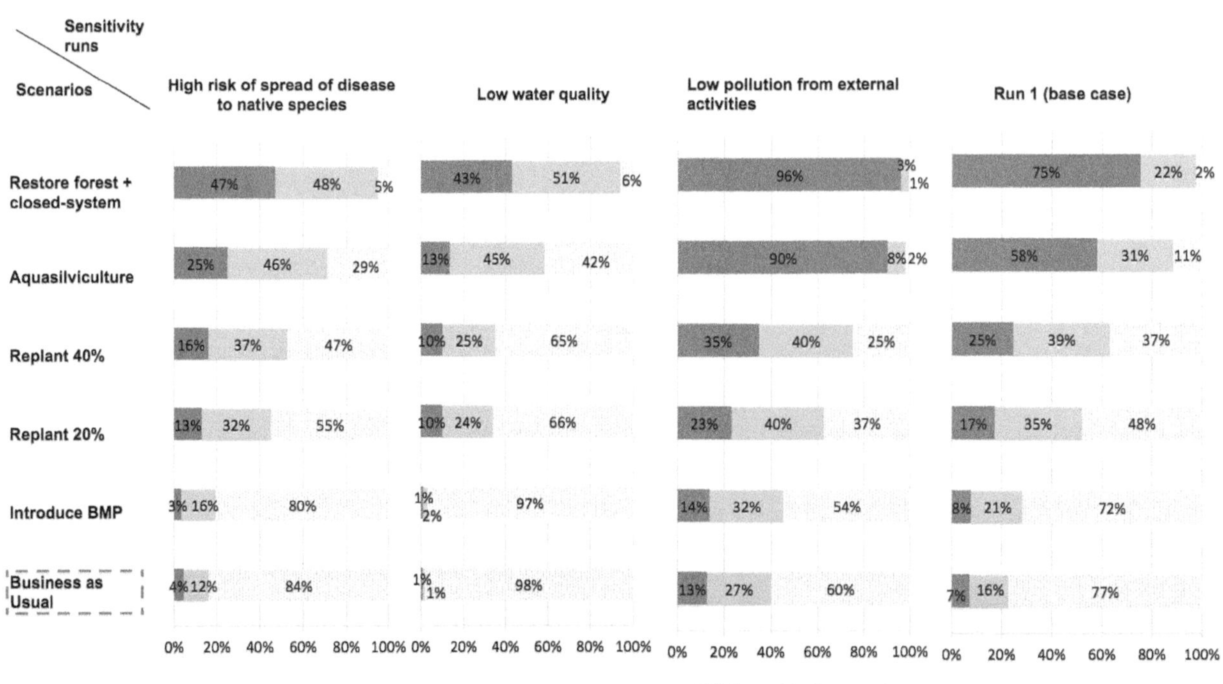

Figure 7. Sensitivity analyses for the node "Biodiversity". Assessment of the predictive influence of the variables "risk of spread of diseases to native species", "surface water quality" and "pollution from other activities" on the posterior probability distribution of the node "biodiversity".

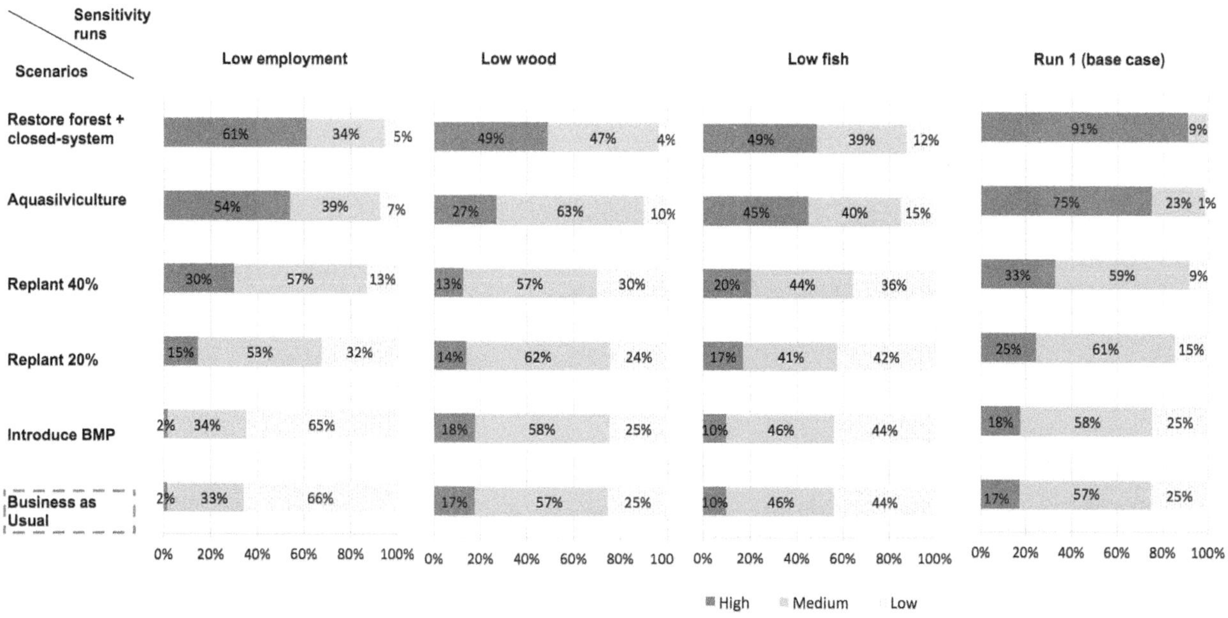

Figure 8. Sensitivity analyses for the node "Locals' subsistence capacity." Assessment of the predictive influence of the variables "quantity of wood for locals", "quantity of fish for locals" and "local employment" on the posterior probability distribution of the node "locals' subsistence capacity".

constraints as part of adaptive management [10]. Consequently, the objective of Thai policy-makers should not be to identify the best aquaculture scenario to promote to all coastal areas, but to define an appropriate diversified portfolio of environmentally-friendly and socially-acceptable practices, where aquasilviculture and closed-systems are only the extremes of a spectrum of possible options [11]. By making explicit how different alternatives may mediate the major trade-offs associated with shrimp farming, the BBN developed here can support the construction of such a diversified portfolio of management practices and inform policy choices.

Furthermore, since conventional intensive shrimp farmers have traditionally focussed on short-term profitability instead of sustainability [39], economic incentives are expected to play a key role in the successful diffusion and adoption of more sustainable management practices among aquaculturists. Potential governmental levers include the improvement of farmers' access to certified markets that drive a price premium, the set-up of microfinance schemes (e.g. to help fund capital-intensive closed-systems), and the implementation of schemes for payments for ecosystem services. Regarding the latter, and subject to property rights and farmers' acceptance, modeling findings on the estimation of the difference in financial performance between the "BAU" scenario and alternatives involving the restoration of mangroves ecosystem services could help in defining a level of compensation to provide to aquaculturists. However, as yields, costs and revenues can vary widely between sites [50,51], our data on aquaculturists' profits would need to be complemented by other data on costs and revenues to generate reliable estimates of the impacts of management scenarios on aquaculturists' profits at local, regional or national scales [50,51].

Although adjustments would be required, the model developed in this paper could be replicated for capturing trade-offs between aquaculture development and environmental and livelihood protection objectives in other countries of Southeast Asia (particularly the Philippines and Indonesia) and Latin America (e.g. Mexico, Honduras, Venezuela), where shrimp farming has displaced mangroves. The issue of scale, however, should be at the core of potential replications of the model. Local land and farm management practices should also be encompassed in wider integrated coastal assessments since "mangroves destruction goes beyond the shrimp industry alone" [52]. Additionally, in areas where shrimp pond abandonment has been a widespread phenomenon following disease outbreaks [5], the model could provide a basis for modeling the rehabilitation of disused shrimp ponds and restoration of mangrove ecosystem services. To transfer the application of the BBN to such cases, the extent to which the ecosystem has been altered, e.g. acidity levels, tidal hydrology and soil alteration, and the objective of ecosystem services restoration, e.g. coastline protection, supporting community livelihoods through restoration of coastal fisheries, or aquasilviculture development, should be incorporated [53]. Pond rehabilitation options for other commercial purposes such as salt production and coconut plantations could also be considered.

3. Suggestions for methodological improvements

Given the size of the network and for practical reasons, our elicitation exercise left what constituted "high", "middle" and "low" levels for each variable open to experts' interpretation. Ideally, the discrete states of all variables should have been characterised by a wider consultation with them.

Furthermore, interactions with experts at each stage of the network development should be complemented with broader consultations with key stakeholders, such as aquaculturists, local communities, environmental organisations and government representatives. Not only this is likely to strengthen the robustness of the model [23], it should also ease the implementation of policy measures stemming from the modeling findings [46].

Finally, although it was decided to focus only on ecosystem services that had a substantial economic value or that could be considered in terms of their qualitative contribution to local communities' resilience, the BBN could be further refined by incorporating other ecosystem services such as cultural services and carbon sequestration. The integration of profits from the latter, potentially secured via payments for ecosystem services, could enable to more accurately model the financial performance associated with each land management scenario.

Conclusion

This paper aimed to develop a Bayesian belief network (BBN) as a decision support system for mediating trade-offs between economic development, protection of natural ecosystems and coastal livelihoods, piloted in the case of the Thai coastal shrimp aquaculture. Modeling insights consisted of identifying for each land and aquaculture management scenario: (i) the expected magnitude of trade-offs due to user-conflicts and (ii) the level of uncertainty surrounding scenarios' performance on criteria reflecting stakeholders' diverse interests. Further analyses enabled quantitative measurement of the sensitivity of the model outputs to pre-defined assumptions (e.g. farm lifespan), input values (e.g. percentage of crop loss, pollution from external activities) and conditional probabilistic dependencies between the network's variables.

Whilst the BBN was developed for coastal shrimp farming in Thailand, suggestions were provided on how to apply this decision tool to other coastal aquaculture contexts. The presently developed BBN can therefore support the implementation of the Ecosystem Approach for Aquaculture in three ways: (i) by articulating available knowledge and beliefs on aquaculture's multiple interactions with the environment, the economy and livelihoods, (ii) by promoting comprehensiveness, explicit handling of uncertainty and transparency in the valuation of pre-defined sustainability criteria and (iii) by supporting innovative policy measures. Examples of such measures include the design of a diversified portfolio of sustainable farm management practices and of schemes of payments for ecosystem services. Finally, from a wider perspective, this research underlines the potential of BBNs to help frame the sustainable development of productive industries that interfere with the provision of ecosystem services.

Acknowledgments

The authors are grateful to project partners in Thailand (Coastal Resources Institute (CORIN) – Asia, and Seafresh Industry Public Co. Ltd) and to all the experts consulted during the study. They would like to acknowledge in particular the contributions of Dr. Dominique Gautier (Aquastar Ltd, UK); Dr. Malcolm Beveridge (WorldFish Center, Africa); Dr. Stuart Bunting (Essex University, UK) and Dr. Neil Ridler (University of Brunswick, Canada) in the probability elicitation exercise. The authors are also grateful to Drs. Murray Rudd and Richard Cookson at the University of York, UK, for their useful comments on an earlier draft, and to Sarah West for her proofreading of the manuscript.

Author Contributions

Conceived and designed the experiments: LS CB. Performed the experiments: LS. Analyzed the data: LS CB. Contributed reagents/materials/analysis tools: CB. Wrote the paper: LS CB.

References

1. FAO (2011) Aquaculture production (quantities and values) 1950–2009. Fishstats. Rome, Italy, Food and Agriculture Organization.
2. Thailand Customs Department (2007) Statistics on Shrimp Import and Export (in Thai). http:\\www.customs.go.th (Accessed 24 September 2013)
3. Fisheries Department (2008) Thailand experience and opportunities for aquaculture certification: Thai quality shrimp for international markets. Bangkok, Thailand: Department of Fisheries, Government of Thailand.
4. Wyban J (2007) Thailand's shrimp revolution. AQUA Culture Asia Pacific Magazine: 16–18.
5. Huitric M, Folke C, Kautsky N (2002) Development and government policies of the shrimp farming industry in Thailand in relation to mangrove ecosystems. Ecol Econ 40: 441–455.
6. Szuster B (2006) Coastal shrimp farming in Thailand: searching for sustainability. In: Hoanh CT, Tuong TP, Gowing JW, Hardy B, editors. Environment and Livelihoods in Tropical Coastal Zones. Oxford, UK: CAB International, pp. 86–98.
7. Sathirathai S (1998) Economic valuation of mangroves and the roles of local communities in the conservation of natural resources: case study of Surat Thani, South of Thailand. Singapore: Environment and Economics Program for South East Asia, International Development Research Centre.
8. Brander LM, Florax R, Vermaat JE (2006) The empirics of wetland valuation: A comprehensive summary and a meta-analysis of the literature. Environmental & Resource Economics 33: 223–250.
9. Soto D, Aguilar-Manjarrez J, Brugere C, Angel D, Bailey C, et al. (2008) Applying an ecosystem-based approach to aquaculture: principles, scales and some management measures. In: Soto D, Aguilar-Manjarrez J, Hishamunda N, editors. Building an ecosystem approach to aquaculture. FAO/Universitat de les Illes Balears Expert Workshop, 7–11 May 2007, Palma de Mallorca, Spain. FAO Fisheries and Aquaculture Proceedings. No. 14. Rome, Italy: Food and Agriculture Organization, pp. 15–35.
10. Bailey C. (2008) Human dimension of an ecosystem approach to aquaculture. In: Soto D, Aguilar-Manjarrez J, Hishamunda N, editors. Building an ecosystem approach to aquaculture. FAO/Universitat de les Illes Balears Expert Workshop, 7–11 May 2007, Palma de Mallorca, Spain. FAO Fisheries and Aquaculture Proceedings. No. 14. Rome, Italy: Food and Agriculture Organization, pp. 37–42.
11. Bush SR, van Zwieten PAM, Visser L, van Dijk H, Bosma R, et al. (2010) Scenarios for Resilient Shrimp Aquaculture in Tropical Coastal Areas. Ecology and Society 15.
12. Barbier EB, Koch EW, Silliman BR, Hacker SD, Wolanski E, et al. (2008) Coastal ecosystem-based management with nonlinear ecological functions and values. Science 319: 321–323.
13. Fisher B, Turner K, Zylstra M, Brouwer R, de Groot R, et al. (2008) Ecosystems services and economic theory: integration for policy-relevant research. Ecol Appl 18: 2050–2067.
14. Brugere C (2006) Can integrated coastal management solve agriculture-fisheries-aquaculture conflicts at the land-water interface? A perspective from New Institutional Economics. In: Hoanh CT, Tuong TP, Gowing JW, Hardy B, editors. Environment and Livelihoods in Tropical Coastal Zones. Oxford, UK: CAB International, pp. 258–273.
15. Slater MJ, Mgaya YD, Mill AC, Rushton SP, Stead SM (2013) Effect of social and economic drivers on choosing aquaculture as a coastal livelihood. Ocean Coast Manag 73: 22–30.
16. Borsuk ME, Stow CA, Reckhow KH (2004) A Bayesian network of eutrophication models for synthesis, prediction, and uncertainty analysis. Ecol Modell 173: 219–239.
17. Marcot BG, Holthausen RS, Raphael MG, Rowland MM, Wisdom MJ (2001) Using Bayesian belief networks to evaluate fish and wildlife population viability under land management alternatives from an environmental impact statement. For Ecol Manage 153: 29–42.
18. McNay RS, Marcot BG, Brumovsky V, Ellis R (2006) A Bayesian approach to evaluating habitat for woodland caribou in north-central British Columbia. Can J For Res 36: 3117–3133.
19. Sadoddin A, Kelly R, Letcher RA, Jakeman T, Croke B, et al. (2012) A bayesian decision support system model: dryland salinity management application. The International Journal of Environmental Resources Research 1: 1–17.
20. Kragt ME, Bennett J (2009) Integrated hydro-economic modelling: challenges and experiences in an Australian catchment. Canberra, Australia: Environmental Economics Research Hub, Crawford School of Economics and Government.
21. Ticehurst JL, Newham LTH, Rissik D, Letcher RA, Jakeman AJ (2007) A Bayesian network approach for assessing the sustainability of coastal lakes in New South Wales, Australia. Environ Model Softw 22: 1129–1139.
22. Levontin P, Kulmala S, Haapasaari P, Kuikka S (2011) Integration of biological, economic, and sociological knowledge by Bayesian belief networks: the interdisciplinary evaluation of potential management plans for Baltic salmon. ICES J Mar Sci 68: 632–638.
23. Baran E, Jantunen T, Cheng P (2006) Developing a consultative Bayesian model for integrated management of aquatic resources: an inland coastal zone case study. In: Hoanh CT, Tuong TP, Gowing JW, Hardy B, editors. Environment

and Livelihoods in Tropical Coastal Zones. Oxford, UK: CAB International, pp. 206–218.
24. Martinez-Alier J (2001) Ecological conflicts and valuation: mangroves versus shrimps in the late 1990s. Environ Plann C-Gov Policy 19: 713–728.
25. Davy FB, De Silva SS (2011) Success Stories in Asian Aquaculture. Ottawa, Canada: Springer.
26. Loawapong A (2010) Background report fisheries value chain: Kingdom of Thailand. Under project "A value-chain analysis of international fish trade and food security with an impact assessment of the small scale sector". Bangkok, Thailand: Department of Fisheries (DoF), Government of Thailand.
27. (ADB) Asian Development Bank, (NACA) Network of Aquaculture Centres in Asia-Pacific (1997) Shrimp and carp aquaculture sustainability. Beijing, China, October 1995. Bangkok, Thailand: NACA.
28. Kongkeo H (1997) Comparison of intensive shrimp farming systems in Indonesia, Philippines, Taiwan and Thailand. Aquaculture Research 28: 789–796.
29. Funge-Smith SJ, Aeron-Thomas M (1995) The economic factors and risks influencing the sustainability of Thai intensive shrimp farms. Stirling, UK: Institute of Aquaculture, University of Stirling.
30. Senanan W, Tangkrock-Olan N, Panutrakul S, Barnette P, Wongwiwatanawut C, et al. (2007) The presence of the Pacific whiteleg shrimp (Litopenaeus vannamei, Boone, 1931) in the wild in Thailand. J Shellfish Res 26: 1187–1192.
31. Hishamunda N, Ridler NB, Bueno P, Yap WG (2009) Commercial aquaculture in Southeast Asia: Some policy lessons. Food Policy 34: 102–107.
32. FAO NACA, UNEP WB (2006) International principles for responsible shrimp farming. Bangkok, Thailand: NACA.
33. NACA (2011) Network of Aquaculture Centers in Asia-Pacific. Bangkok, Thailand. http://www.enaca.org
34. NACA, WWF (2010) Supporting and demonstrating small-scale shrimp farmer group to access international and market through certification schemes. Bangkok, Thailand: NACA.
35. Umesh NR (2007) Development and adoption of BMPs by self-help farmer groups. Aquaculture Asia Magazine XII: 8–10.
36. Umesh NR, Chandra Mohan AB, Ravibadu G, Padiyar PA, Phillips MJ, et al. (2010) Shrimp farmers in India: Empowering small-scale farmers through a cluster-based approach. In: De Silva SS, Davy FB, editors. Success Stories in Asian Aquaculture. Ottawa, Canada: Springer, pp. 41–66.
37. Binh C, Phillips M, Demaine H (1997) Integrated shrimp-mangrove farming systems in the Mekong delta of Vietnam. Aquaculture Research 28: 599–610.
38. Primavera J (2000) Integrated mangrove-aquaculture systems in Asia. Integrated Coastal Zone Management, Autumn ed: 121–130.
39. Stevenson NJ (1997) Disused shrimp ponds: options for redevelopment of mangrove. Coast Manage 25: 423–425.
40. Troell M (2009) Integrated marine and brackishwater aquaculture in tropical regions; research, implementation and prospects. In: Soto S, editor. Integrated mariculture: a global review. FAO Fisheries and Aquaculture Technical Paper. No. 529. Rome, Italy: Food and Agriculture Organization, pp. 47–131.
41. Bunting SW, Bosma RH, van Zwieten PA, Sidik A (2013) Bioeconomic modeling of shrimp aquaculture strategies for the mahakam delta, Indonesia. Aquaculture Economics and Management 17: 51–70.
42. Varis O, Kuikka S (1999) Learning Bayesian decision analysis by doing: lessons from environmental and natural resources management. Ecol Model 119: 117–195.
43. Uusitalo L (2007) Advantages and challenges of Bayesian networks in environmental modelling. Ecol Model 203: 312–318.
44. Farmani R, Henriksen HJ, Savic D (2009) An evolutionary Bayesian belief network methodology for optimum management of groundwater contamination. Environ Model Softw 24: 303–310.
45. Castelletti A, Soncini-Sessa R (2007) Bayesian networks in water resource modelling and management. Environ Model Softw 22: 1073–1074.
46. Cain J (2001) Planning improvements in natural resources management. Wallingford, Oxon, UK: Centre for Ecology and Hydrology.
47. McCann RK, Marcot BG, Ellis R (2006) Bayesian belief networks: applications in ecology and natural resource management. Can J For Res 36: 3053–3062.
48. Marcot BG, Steventon JD, Sutherland GD, McCann RK (2006) Guidelines for developing and updating Bayesian belief networks applied to ecological modeling and conservation. Can J For Res 36: 3063–3074.
49. Ashley C, Carney D, Britain G (1999) Sustainable livelihoods: Lessons from early experience. London, UK: Department for International Development.
50. Kasai C, Nitiratsuwan T, Baba O, Kurokura H (2005) Incentive for shifts in water management systems by shrimp culturists in southern Thailand. Fish Sci 71: 791–798.
51. Raux P, Bailly D (2001) Literature review on world shrimp farming. Brussels, Belgium: European Commission Framework Programme 5.
52. Neiland AE, Soley N, Varley JB, Whitmarsh DJ (2001) Shrimp aquaculture: economic perspectives for policy development. Marine Policy 25: 265–279.
53. Lewis RR (2001) Mangrove restoration - costs and benefits of successful ecological restoration. University Sains, Penang, Malaysia, 4–8 April 2001. Stockholm, Sweden: Beijer Institute of Ecological Economics.
54. Boyd J, Banzhaf S (2007) What are ecosystem services? The need for standardized environmental accounting units. Ecol Econ 63: 616–626.
55. Fisher B, Turner RK, Morling P (2009) Defining and classifying ecosystem services for decision making. Ecol Econ 68: 643–653.

14

An Ecosystem-Based Approach to Assess the Status of a Mediterranean Ecosystem, the *Posidonia oceanica* Seagrass Meadow

Sébastien Personnic[1]*, Charles F. Boudouresque[1], Patrick Astruch[2], Enric Ballesteros[3], Sylvain Blouet[4], Denise Bellan-Santini[5], Patrick Bonhomme[2], Delphine Thibault-Botha[1], Eric Feunteun[6], Mireille Harmelin-Vivien[1], Gérard Pergent[7], Christine Pergent-Martini[7], Jérémy Pastor[8], Jean-Christophe Poggiale[1], Florent Renaud[5], Thierry Thibaut[1], Sandrine Ruitton[1]

1 Aix-Marseille University, Mediterranean Institute of Oceanography (MIO), Université de Toulon, CNRS/INSU, IRD, UM 110, Marseille, France, **2** GIS Posidonie, Pytheas Institute, Aix-Marseille University, Marseille, France, **3** Centre d'Estudis Avançats de Blanes - CSIC, Blanes, Spain, **4** Aire marine protégée de la côte Agathoise, site natura 2000, Agde, France, **5** Aix-Marseille University, Institut Méditerranéen de Biodiversité et d'Ecologie (IMBE), UMR 7263, Station Marine d'Endoume, Marseille, France, **6** Museum National d'Histoire Naturelle, UMR 7208, Station Marine de Dinard, France, **7** Equipe Ecosystèmes Littoraux, FRES 3041, University of Corsica, Corte, France, **8** Université de Perpignan, Via Domitia, Centre de Formation et de Recherche sur les Environnements Méditerranéens, UMR 5110, Perpignan, France

Abstract

Biotic indices, which reflect the quality of the environment, are widely used in the marine realm. Sometimes, key species or ecosystem engineers are selected for this purpose. This is the case of the Mediterranean seagrass *Posidonia oceanica*, widely used as a biological quality element in the context of the European Union Water Framework Directive (WFD). The good quality of a water body and the apparent health of a species, whether or not an ecosystem engineer such as *P. oceanica*, is not always indicative of the good structure and functioning of the whole ecosystem. A key point of the recent Marine Strategy Framework Directive (MSFD) is the ecosystem-based approach. Here, on the basis of a simplified conceptual model of the *P. oceanica* ecosystem, we have proposed an ecosystem-based index of the quality of its functioning, compliant with the MSFD requirements. This index (EBQI) is based upon a set of representative functional compartments, the weighting of these compartments and the assessment of the quality of each compartment by comparison of a supposed baseline. The index well discriminated 17 sites in the north-western Mediterranean (French Riviera, Provence, Corsica, Catalonia and Balearic Islands) covering a wide range of human pressure levels. The strong points of the EBQI are that it is easy to implement, non-destructive, relatively robust, according to the selection of the compartments and to their weighting, and associated with confidence indices that indicate possible weakness and biases and therefore the need for further field data acquisition.

Editor: Arga Chandrashekar Anil, CSIR- National institute of oceanography, India

Funding: This study has been funded by the French Centre National de la Recherche Scientifique (CNRS), the French Réseau des Stations et Observatoires Marins (RESOMAR), the Muséum National d'Histoire Naturelle (MNHN, Paris), the French Agence de l'eau Rhône Méditerranée Corse through the French part of the European MSFD project (in French: Directive Cadre Stratégie pour le Milieu Marin, DCSMM). The funders had no role in study design, data collection and analysis, decision to publish, or preparation of the manuscript.

Competing Interests: The authors have declared that no competing interests exist.

* E-mail: sebastien.personnic@univ-amu.fr

Introduction

Human activities can deeply alter the environment, species composition and functioning of ecosystems. These alterations can be tracked by the use of biotic indices, i.e. species or groups of species whose function, population, or status reflect the environmental quality. Thus, biotic indices are monitored for changes in presence and abundance. The occurrence of an organism in a specific environment indicates that overall, its biological requirements are satisfied, whereas its disappearance suggests a change in the environment. Species are also monitored for changes in biochemistry, physiology or behaviour induced by environmental conditions. Biotic indices are used in terrestrial, freshwater and marine habitats, because they enable the quality of an environment to be characterized in an integrated way [1–7].

Biotic indices are widely used in the marine realm to **(i)** assess the quality of a water body, **(ii)** assess processes such as currents, sedimentation and climate under natural and anthropogenic forcing, and **(iii)** monitor the status of species or communities of interest, either emblematic species, indicators of some ecosystemic processes or indicators of pollution. Sometimes, key species and ecosystem engineers [8–10] are selected for this purpose. This is the case of the seagrass *Posidonia oceanica* [11–17].

In the European Union (EU), the so called Habitats Directive of 1992 (92/43/ECC) listed habitats and species that are used to designate areas ('Natura 2000 sites'), where they are strictly protected. While the Habitats Directive also considered the marine realm, more recently, the EU Marine Strategy Framework Directive (MSFD: 2008/56/EC) established a framework for conservation in the field of marine environmental policy. The

MSFD is considered to be the environmental pillar of the Integrated Maritime Policy adopted in 2010 by the European Commission (2010/477/EU). This directive established eleven criteria, based on the descriptors set out in the MSFD, to determine 'good environmental status': **(i)** Biological diversity is maintained, **(ii)** Introduced species are at levels that do not adversely alter the ecosystems, **(iii)** Populations of all exploited fish and shellfish are within safely biological limits, **(iv)** All elements of the food webs occur at levels capable of ensuring the long-term abundance of the species, **(v)** Human-induced eutrophication is at a minimum, **(vi)** Sea-floor integrity is at a level that ensures that the structure and functions of the ecosystems are safeguarded, **(vii)** Permanent alteration of hydrographical conditions does not adversely affect marine ecosystems, **(viii)** Concentration of contaminants are at levels not giving rise to pollution effects, **(ix)** Contaminants in fish and other seafood do not exceed levels established by Community legislation, **(x)** Properties and quantities of marine litter do not cause harm to the coastal and marine environment, and **(xi)** Introduction of energy is at levels that do not adversely affect the marine environment. The EU MSFD established a framework within which Member States agreed to take the appropriate measures to achieve or maintain good environmental status in the marine realm by the year 2020 at the latest.

A key point of the MSFD is the ecosystem-based approach. Most previous strategies only dealing with Biological Quality Elements (BQE: species and/or communities) used in the European Union Water Framework Directive (WFD) were not indicative of ecosystem status, but merely indicative of the environmental status of water bodies [18–22,17,23] for a review of the approaches for classifying and assessing quality of benthic habitats. As an example of the possible shortcomings that may arise from the WFD descriptors, a seagrass meadow characterized by normal leaf growth, shoot density and the absence of mechanical injuries would be ranked as 'good', even if deprived of some basic compartments of the ecosystem.

According to the MSFD, 'good environmental status' means that the marine environment is at a level that allows uses and activities by current and future generations, i.e. the structure, functions and processes of the constituent marine ecosystems, together with the associated physiographic, geographic, geological and climatic factors, allow those ecosystems to function fully and to maintain their resilience to human induced environmental change. Overall, marine species and marine habitats are protected, human-induced decline of biodiversity is prevented and diverse biological components function in balance.

While the ecosystem-based approach constitutes the background and the guidelines of the MSFD, the pristine state of an ecosystem is not clearly defined. Obviously, such a notion is very complex and prone to anthropocentrism; reference conditions, as observed in areas distant from known human impact, may constitute a more realistic notion. Here we try to define: **(i)** how the status of an ecosystem can be measured and how much its current status differs from 'reference conditions', and **(ii)** which parameters we need to monitor to ascertain these targets. The *Posidonia oceanica* seagrass meadow was chosen for this attempt because: **(i)** it is widely present in almost the whole of the European Mediterranean; **(ii)** it is the only marine ecosystem considered as 'priority habitat' by the EU Habitats Directive; **(iii)** its functioning is relatively well known [11,24–30]; and **(iv)** like many seagrass ecosystems in the world ocean, *P. oceanica* meadows have been impacted or lost under the influence of direct and indirect effects of human activities and are therefore regarded as threatened [31–33]. Similar ecosystem-based approaches have been attempted for fisheries (e.g. [34–38,40,41]) and for the management and conservation of ecosystem services (e.g. [34,41,42]).

Materials and Methods

The conceptual model

Posidonia oceanica (L.) Delile is a seagrass species (Magnoliophyta, kingdom Archaeplastida) endemic to the Mediterranean Sea [43], which dwells in the sublittoral zone, from mean sea level down to 30 to 40 m depth, depending upon water transparency [27,28,32]. Due to the length, up to 120 cm, and density of the leaves, the seagrass canopy decreases water movement and traps sediments [44–47]. Rhizomes resist burial by vertical growth, so that the sea bottom slowly rises. Within the sediment, the deeper parts of the rhizomes, attached leaf sheaths and roots die, but their decay is extremely slow, so that they can persist for millennia [24,27]. The terrace constituted by live and dead intertwined rhizomes, together with the sediment, which fills the interstices, is named 'matte' [27,48,49]. When *P. oceanica* dies, the matte persists (hereafter 'dead matte'), since the decay of the rhizomes proceeds extremely slowly [44,50]. *P. oceanica* is the engineer of an ecosystem of major importance in the Mediterranean Sea [27,28]. A conceptual model of the functioning of the *P. oceanica* ecosystem has been proposed [11,27,28]. Here, we use an updated version of this conceptual model (Fig. 1; C.F. Boudouresque, unpublished).

The simplified conceptual model of the functioning of the *P. oceanica* ecosystem in the north-western Mediterranean Sea used in the present study (Fig. 1) encompasses the following compartments (boxes); these compartments are listed hereafter and detailed later on:

- *Posidonia oceanica* roots and rhizomes (box 1).
- Multicellular Photosynthetic Organisms (MPOs) epibiotic on *P. oceanica* rhizomes.
- *P. oceanica* leaves (box 2).
- MPO leaf epibiota (box 3).
- Dissolved Organic Carbon (DOC).
- Pelagic microbial loop.
- Filter- and suspension-feeder leaf epibiota (box 4).
- Filter- and suspension-feeder benthic epibiota on *P. oceanica* rhizomes: the bivalve *Pinna nobilis* (box 5) and other species, e.g. bryozoans and ascidians (box 6).
- The litter detritus (essentially dead *P. oceanica* leaves and some broken rhizomes) (box 7).
- Detritus feeders 1, e.g. Amphipoda, Isopoda (crustaceans) and *Psammechinus microtuberculatus* (sea urchin).
- Detritus feeders 2 and 3, e.g. Amphipoda, Isopoda (crustaceans) and *Holothuria* spp. (sea cucumber) (box 8).
- Decomposers, namely *Bacteria*, *Archaea*, Fungi and heterotrophic stramenopiles (BAFHS) such as Labyrinthulomycota and Oomycota.
- The benthic microbial loop.
- The matte endofauna, e.g. annelids and mollusks.
- Herbivores 1, e.g. *Sarpa salpa* (teleost), *Paracentrotus lividus* (sea urchin), *Idotea* spp. and *Pisa* spp. (crustaceans) (box 9).
- Herbivores 2, e.g. Amphipoda (crustaceans), *Jujubinus* spp. and *Rissoa* spp. (gastropods).
- Predatory teleosts (e.g. *Diplodus* spp., *Sparus aurata*, *Labrus* spp. and *Symphodus* spp.), cephalopods and seastars (e.g. *Marthasterias glacialis*) (box 10).

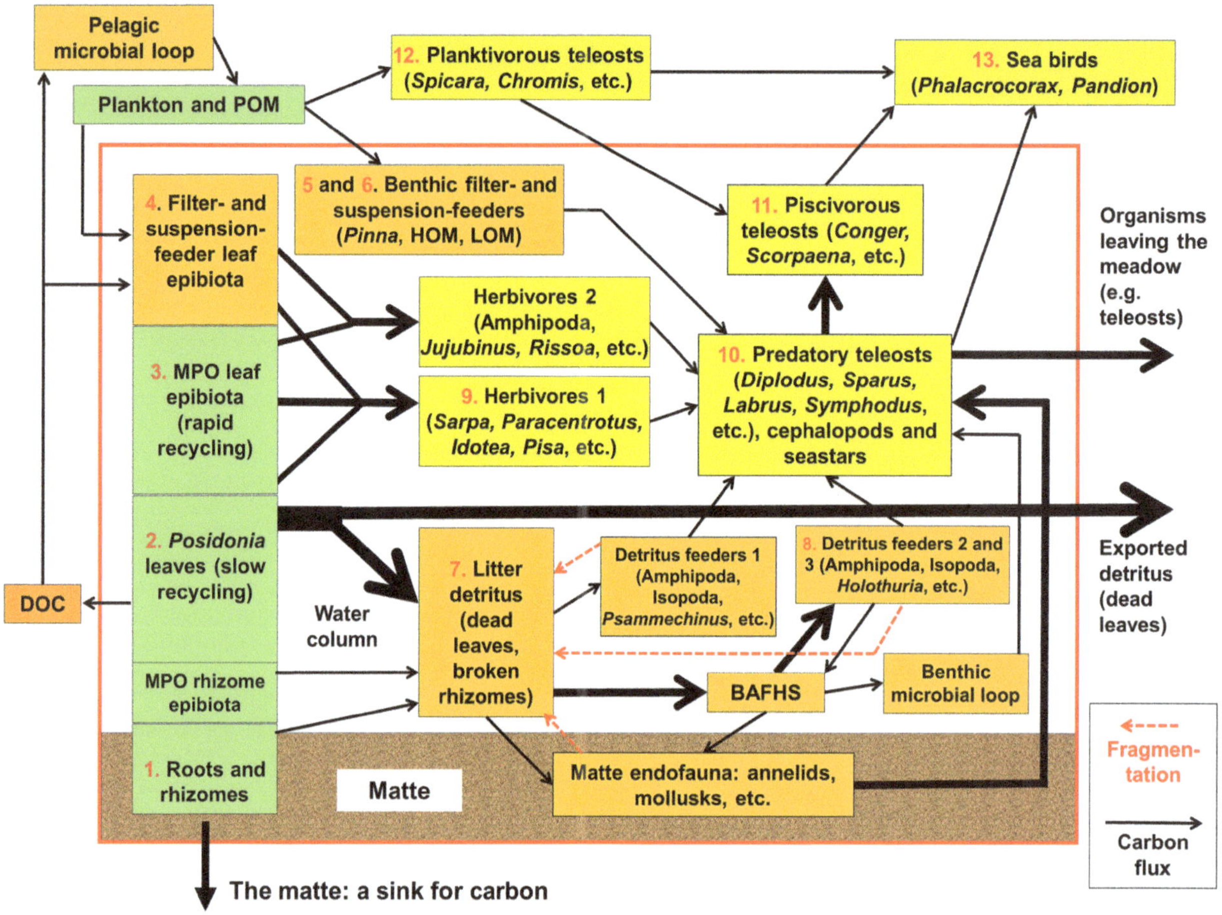

Figure 1. Conceptual model of the functioning of *Posidonia oceanica* seagrass ecosystem. For functional compartments and box numbers, see text. Primary producers are in green; filter-feeders, suspension-feeders, litter, detritus feeders, Dissolved Organic Carbon (DOC) and microbial loops are in orange; predators (including herbivores) are in yellow. The width of the arrows roughly represents the importance of the carbon flow. The proper *P. oceanica* ecosystem is included within the red rectangle. MPO: Multicellular Photosynthetic Organisms. POM: Particulate Organic Matter. From C.F. Boudouresque, unpublished.

- Piscivorous teleosts, e.g. *Conger conger*, *Scorpaena* spp. and *Serranus* spp. (box 11).
- Planktivorous teleosts of the water column, e.g. *Spicara* spp. and *Chromis chromis* (box 12).
- Sea birds, e.g. *Phalacrocorax aristotelis* ssp. *desmarestii* and *Pandion haliaetus* (box 13).
- Plankton (photosynthetic plankton and zooplankton) and non-living Particulate Organic Matter (POM).

Considered functional compartments (boxes)

Whenever possible, non-destructive methods were chosen to measure the parameters of the status of the functional compartments, as suggested by [51]. In the absence of further indications, measures are performed at a depth of ca. 15 meters (10 to 20 m), which is then considered as representative of the whole depth range [14,52]. The season of the sampling is indicated only for compartments which present seasonal variability. In the ranking of the status of the selected compartments (boxes), two cases were encountered; **(i)** a steady trend of the parameter, from very good to a low quality state; **(ii)** an upward/downward slope parameter, when the very good state corresponds to intermediate values.

***Posidonia* roots and rhizomes.** (box 1, Fig. 1). This compartment was estimated by the growth rate of vertical (orthotropic) rhizomes. The matte compartment results in carbon sequestration within the matte, which acts as a carbon sink [27,44,53,54] and was measured by means of lepidochronology [55,56]. Lepidochronology describes the annual cycle of leaf formation. A cycle includes a suite of dead leaf bases (called dead sheaths or scales) of increasing then decreasing thickness. An average of 7.5 leaves (generally between 6 and 9) are produced every year [57–60]. We considered that high and low rhizome growth rates are indicative of over sediment input or of deficit in sediment, respectively (Table 1). The highest growth rate of an orthotropic rhizome is 7 cm year^{-1}; higher sedimentation rates result in the death of the buried leaf bundle [44,61]. In contrast, deficit in sediment results in bare, non-sediment protected rhizomes, which are hence very vulnerable to storms, anchoring and trawling [27]. The nature of the substratum, namely meadows settled on rock, sand or matte, only slightly influences the growth rate of rhizomes [62] and no major differences were found at the community level [63] so that it was not necessary to adapt the scale. Thirty random *in situ* measures (growth of the rhizome corresponding to the last 8 dead leaf bases) are recommended. The obtained value was multiplied by 1.5 in order to take into account

Table 1. Relative weighting of each functional compartment (boxes; see Fig. 1 for box number) and ranges of each parameter defined for each grade of ecosystem status.

Box number	Functional compartment	Weighting (W)	Parameter	4	3	2	1	0
1	Roots and rhizomes ('rhizomes')	3	Growth of orthotropic rhizomes (mm a^{-1})	9 to 19	20 to 40	3 to 8	>40	<3
2	*Posidonia oceanica* leaves ('leaves')	5	- Density (shoots m^{-2})	≥490	489 to 370	369 to 250	249 to 130	<130
			- Cover (%)	>80	80 to 61	60 to 41	40 to 20	<20
3–4	MPOs, filter- and suspension-feeders leaf epibiota ('leaf epibiota')	4	Biomass (g DM $shoot^{-1}$) (only the 2 oldest - external - leaves)	0.3 to 0.7	0.1 to 0.2	0.8 to 1.5	<0.1	>1.5
5	Benthic filter-feeder: *Pinna nobilis* (bivalve) ('Pinna')	2	Density (individuals 100 m^{-2})	>3.0	3.0 to 1.1	1.0 to 0.6	0.5 to 0.1	<0.1
6	Other benthic filter- and suspension-feeders ('HOM LOM')	2						
	- HOM		- Density (m^{-2})	<0.1	0.1 to 0.9	1.0 to 1.9	2.0 to 5.0	>5.0
	- LOM		- Density (m^{-2})	>10.0	10.0 to 5.1	5.0 to 1.1	1.0 to 0.1	<0.1
7	Litter detritus: dead leaves and broken rhizomes ('litter')	2	g DM m^{-2}	>350	350-251	250-151	150-51	≤50
8	Detritus-feeders 2 and 3 (*Holothuria* spp.) ('Holothuria')	2	Individuals 10 m^{-2}	1.0 to 4.9	0.2 to 0.9	5.0 to 24.9	<0.2	≥25.0
9	Herbivores 1 ('herbivores')	5	- Density of *Paracentrotus lividus* (individuals m^{-2})	1.0 to 4.9	0.1 to 0.9	5.0 to 9.9	<0.1	≥10
			- Grazing index (% leaves)	30 to 59%	5 to 29%	60 to 95%	<5%	>95%
10	Predatory teleosts, cephalopods and seastars ('predators')	5	kg teleosts WM 100 m^{-2}	>1.5	1.5 to 1.1	1.0 to 0.6	0.5 to 0.3	<0.3
11	Piscivorous teleosts ('piscivores')	5	kg teleosts WM 100 m^{-2}	>1.0	1.0 to 0.6	0.5 to 0.3	0.2 to 0.1	<0.1
12	Planktivorous teleosts ('planktivores')	3						
	- Zooplankton feeders		- kg teleosts WM 100 m^{-2}	>3.0	3.0 to 1.6	1.5 to 0.8	0.7 to 0.3	<0.3
	- Omnivores		- kg teleosts WM 100 m^{-2}	>3.0	3.0 to 1.6	1.5 to 0.8	0.7 to 0.3	<0.3
9–12	All teleosts ('SRDI')	3	Specific Relative Diversity Index (SRDI)	>10	10 to 8	7 to 5	4 to 3	<3
13	Sea birds	1						
	- *Phalacrocorax* spp.		- Distance to the nearest nesting site (km)	<4	4 to 7	8 to 12	13 to 17	>17
	- *Pandion haliaetus*		- Distance to the nearest nesting site (km)	<4	4 to 7	8 to 12	13 to 17	>17

Between inverted commas: abbreviated name of the functional compartment in tables 4 and 7. DM: dry mass, including calcium carbonate. HOM: indicators of high level of organic matter. LOM: indicators of low level of organic matter. MPO: Multicellular Photosynthetic Organisms. POM: Particulate Organic Matter. WM: wet mass. Status: 4 (very good) through 0 (very low).

the fact that the rhizome continues to grow slowly during the following two years (Gérard Pergent, unpublished data).

***Posidonia* leaves.** (box 2, Fig. 1). A shoot is a rhizome tip with a bundle of living leaves. Shoot density is correlated with annual leaf primary production at local scale (patch) [64,65]. Primary production is a basic parameter for the functioning of the *P. oceanica* ecosystem. It was estimated by the number of shoots per square meter, measured within a small square frame (0.16 m^2 [66,67]) with at least 20 random replicates [5]. At a less local scale, the cover rate of the meadow is rarely 100%: it is broken by more or less extensive patches of either sand or dead matte, which reduce the overall cover. Cover rate was estimated by visual observation *via* a see-through plastic slide [68,69] or *via* vertical photographs [67,70]. Thirty random measures are recommended. Low cover rate is believed to characterize a poor condition of the meadow [5]. Some types of *P. oceanica* meadow, e.g. the hill meadow and the striped meadow, exhibit a low cover rate while in pristine state; these types of meadows do not, however, occur in the study area [27]. The final *P. oceanica* leaf index, for a given site, was the arithmetical mean between the density index, and the cover index.

MPOs, filter- and suspension-feeder leaf epibiota. (boxes 3 and 4, Fig. 1). *P. oceanica* leaf epibiota, both primary producers (MPOs and diatoms) and animals, share the same habitat (the leaf surface) and are usually co-consumed by the same species. In addition, some consumers eat simultaneously epibiota and the supporting leaf [68,71]. For these reasons, epibiota will be considered here as a single compartment. The colonisation of leaves by epibiota is a function of leaf age, the youngest leaves, in the center of the shoot, being less colonized, while the oldest, external leaves are the most colonized; in addition, leaf tips are more colonized than the lower parts of the leaves [72]. Leaf epibiota cover is believed to provide information on water quality, especially nutrient concentration in seawater [5,73,74]; however, it also reflects the herbivore pressure, epibiota biomass decreasing when macrograzer abundance increases [26]. By convention, the epibiota biomass was estimated on the two oldest (external) leaves, in July, on 30 randomly localized shoots (Table 1).

Filter- and suspension-feeder epibiota on *P. oceanica* rhizomes. (boxes 5 and 6, Fig. 1). A number of benthic filter- and suspension-feeders dwell on *P. oceanica* rhizomes, sometimes within the matte. They belong to bryozoans, hydroids, sponges, annelids (e.g. *Sabella spallanzani*), ascidians (e.g. *Halocynthia papillosa*, *Phallusia mammillata*, *P. fumigata*), gastropods and bivalves (such as the fan mussel *Pinna nobilis*) [75–79]. *Pinna nobilis* (box 5) density was estimated along 20 transects 10-m long and 1-m wide. Filter- and suspension-feeders other than *P. nobilis* (box 6) are indicators either of: **(i)** high level of organic matter in the water (hereafter HOM; e.g. *Sabella spallanzani*, *Phallusia mammillata*, *P. fumigata*, Didemnidae [80]); or **(ii)** low level of organic matter (hereafter LOM; e.g. bryozoans, sponges, *Halocynthia papillosa*, *Antedon mediterranea* [80]). HOM and LOM indicators are assessed within the same quadrats as the sea urchin *P. lividus* (see below, box 9). For colonial species, the number of colonies was taken into account, as suggested by [81]. For non-colonial species, the number of individuals was counted. Only individuals and colonies over 5 cm in diameter and/or height were considered. The final filter- and suspension-feeder epibiota (other than *P. nobilis*; box 6) on rhizomes index, for a given site, was estimated as the arithmetical mean between HOM and LOM indices.

Litter detritus. (box 7, Fig. 1). The litter detritus mass corresponds essentially to shed dead *P. oceanica* leaf blades and the epibiota they harbour, and some broken rhizomes. It therefore represents a kind of necromass [82]. Drift MPOs, exported from sublittoral reef habitats, can also occur within the litter. The litter detritus mass was estimated in July, within 5 randomly localized 0.1 m^2 quadrats; the litter was sucked up by an underwater vacuum cleaner. Litter detritus were dried at 50°C in an oven to constant weight.

Detritus-feeders 2 and 3. (box 8, Fig. 1). Detritus-feeders constitute a complex set of compartments. Here, the macro-detritus feeders *Holothuria* spp. were used as a proxy of detritus-feeders 2 and 3, as they are easy to count. Several species can be present, e.g. *H. polii* and *H. tubulosa* [83]. The abundance of *Holothuria* spp. was measured within the same quadrats as the sea urchin *P. lividus* (see below, box 9).

Herbivores 1. (box 9, Fig. 1). Macro-herbivores considered in this compartment were the sea urchin *Paracentrotus lividus* and the teleost *Sarpa salpa*, at 5 meters depth or, if the meadow is not present at this depth, at the upper limit of the meadow. They consume *P. oceanica* leaves, together with their epibiota, if present [68,71,84]. Other herbivores graze *P. oceanica* leaves, such as the spider crabs *Pisa* spp. and the isopod *Idotea hectica* [27,84], but their reduced size and habit make them more difficult to quantify and they were not considered. The abundance of *P. lividus* was assessed within 1-m^2 quadrats, with 30 replicates randomly localized. The census only considered individuals >3 cm (test diameter without spines), because small individuals can be hidden within the matte, generating bias in the census. A second proxy of the macro-herbivore pressure was the grazing index, i.e. the percentage of intermediate and adult leaves (*sensu* [85,86]) exhibiting bite scars due to *S. salpa* (all the intermediate and adult leaves of 30 shoots randomly localized). Bite scars were carefully distinguished from broken leaves, the latter being related to hydrodynamism. Most bite scars are due to the *S. salpa* browsing, and they are easy to distinguish from those due to other macro-herbivores [26,49]. The final macro-herbivore index was, for a given site, the arithmetical mean between the *Paracentrotus* index and the grazing index.

Predatory teleosts and cephalopods, piscivorous teleosts, planktivorous teleosts. (boxes 9 in part, 10, 11 and 12, Fig. 1). Teleost fishes associated with seagrass beds occupy different positions in and above the canopy during the day and at night, spend more or less time in this habitat depending on their life cycle, and naturally fluctuate in abundance and biomass with depth, season and years [87–90]. They participate actively in the functioning of the *P. oceanica* ecosystem (Fig. 1), but the perception of the composition and trophic structure of their assemblages largely depends on the methodology used [91]. Moreover, species richness, abundance and biomass of teleosts are favoured by the ecotones induced by the presence of rocky or sandy substrates in the middle of *Posidonia* beds. So these variations and methodological biases have to be taken into account to when measuring these compartments in order to assess the 'environmental status' of a site. These compartments were estimated *via* visual censuses at a standardized day time (10:00 to 16:00 UT) during the warm season (summer-autumn), preferentially in uniform beds (at least a long way from rocky substrates). All teleosts were counted within ten linear and 5-m wide transects, each census lasting 5 minutes. Around 50 m were covered so that each transect represents a surface area of nearly 250 m^2. Total length (at the nearest 2 cm) of individuals and the number of individual per species were noted. The Specific Relative Diversity Index (SRDI) is the mean number of species met with per transect. These data enable calculation of **(i)** predator biomass (predatory teleosts other than piscivorous and planktivorous; box 10), **(ii)** top predator biomass (piscivorous teleosts; box 11), **(iii)** planktivorous teleost biomass (box 12) and **(iv)** the Specific Relative Diversity Index (SRDI). The planktivorous teleost biomass is divided into 2 categories: the exclusive zooplankton feeders (*Chromis chromis*, *Spicara smaris*, *S. maena*, *Atherina* spp.) and the omnivorous feeders, which consume both zooplankton and POM (*Boops boops*, *Oblada melanura*). Some top predators (e.g. *Conger conger*, *Scorpaena* spp.) are active only during night time while *Serranus* spp. are active by day and thus more present in our visual counts. Some predators (e.g. *Symphodus rostratus*), though diurnal, are often hidden within the *P. oceanica* leaf canopy. As a result, their counts were underestimated to a greater or lesser extent. The parameter ranges within the status scale (Table 1) took into account these biases.

Sea birds. (box 13, Fig. 1). Most sea birds do not directly interact with the *P. oceanica* ecosystem, as they feed on offshore pelagic species, such as *Larus* spp. and *Puffinus* spp. The only exceptions are shags *Phalacrocorax* spp. and the osprey *Pandion haliaetus*. Shags can dive down to the benthic seagrass meadow; they mainly feed on pelagic planktivorous teleosts (e.g. *Spicara smaris*, *Chromis chromis*), but benthic teleosts (e.g. *Diplodus* spp., *Lithognathus mormyrus*, *Scorpaena notata*, *Serranus scriba* and *Symphodus mediterraneus*) account for 35% of the diet [92–94]. Osprey is an opportunistic fish-eating bird of prey. In Corsica, which harbours the only population within the study area, it mainly consumes mugilids (73% of the captures), together with *Diplodus* spp. (13%) and *Sarpa salpa* (11%) [95–97]. The sea bird compartment was estimated *via* the distance of the nearest *Phalacrocorax* spp. and *Pandion haliaetus*, respectively, nesting sites from the study site. As far as shags are concerned, the mean maximum foraging range is 16 km [95–97] (Table 1).

Some of the above-mentioned compartments correspond to inputs into the *P. oceanica* ecosystem from the pelagic habitat: plankton, POM and planktivorous teleosts. Outputs are also to be considered: **(i)** ca. 15–30% of the net primary production (NPP) corresponds to roots, rhizomes and dead sheaths buried and sequestrated within the matte [24,28,98,99]; **(ii)** ca. 6 to 50% of the NPP represents dead leaves that are exported as detritus towards beaches and adjacent habitats [28,31,98,100,101]; and **(iii)** a number of organisms (e.g. teleosts, crustaceans) leave the meadow, either temporarily, to feed in adjacent habitats, or permanently, to reach their adult quarters [87,102–104].

The Ecosystem-Based Quality Index (EBQI)

The rationale governing our ecosystem-based approach is trying to quantify and assess each compartment (box) of the conceptual model by means of a set of parameters, to balance their relative weighting and by using a simple algorithm to calculate a rank for the ecosystem status within a given area, matching the five Ecological Statuses of the Water Framework Directive (WFD), from bad to high. A great variety of parameters are available for the assessment of each compartment (box). Many of them are redundant. Others have been poorly used so that data are not available for most areas. For this reason a limited set of relevant parameters was selected for a restricted set of compartments.

Each parameter was assessed by means of a semi-quantitative scale (4 through 0), from very good (4) to very low (0). Calibration of the scale was based upon the available literature (e.g. [105]), including grey literature and expert judgment (the personal knowledge of the authors) based upon a Delphi process [106]. The highest grade (4) corresponds to the ecosystem status in the best-protected areas of well implemented MPAs, e.g. the Medes Islands reserve (Catalonia, Spain), the Port-Cros National Park (continental France), the Scandola reserve and the Bouches de Bonifacio reserve (Corsica, France).

Boxes were balanced, according to their relative weighting (W) in the ecosystem functioning, from 5 (highest weighting) to 1 (lowest weighting). The general principle in the ranking of the weighting of a box was that the boxes localized at the very base (bottom up control by primary producers), the herbivores (box 9) and the boxes localized at the very top (top down control via cascade effect) of the model were regarded as of major importance (with the exception of box 13), while intermediate ones were less weighted. Wasp-waist control has not been evidenced in that ecosystem [11,107]. The grade of each considered box was given by multiplying its status S (0 through 4) and weighting W (1 through 5), and therefore they are graded from 0 through 20. The grades of all considered boxes were added up, which gave the final grade of the ecosystem status (Ecosystem Based Quality Index, hereafter EBQI) at a given site. For practical purposes, the EBQI was converted to a scale from 0 to 10:

$$EBQI = \left[\sum_{i=1}^{13} (W_i \times S_i) / \sum_{i=1}^{13} (W_i \times S_{\max}) \right] \times 10$$

Where: W_i is the weighting of the box i, S_i the status of the box i, S_{max} the highest possible grade (= 4) for a box and i is the number of the box (1 through 13).

Five ecological status classes, from high to bad, according to the practice of the WFD, were delineated: bad ($EBQI<3.5$), poor ($3.5 \geq EBQI<4.5$), moderate ($4.5 \geq EBQI<6$), good ($6.0 \geq EBQI<7.5$) and high ($EBQI \geq 7.5$).

Since the box weightings were supported by partly subjective arguments (see above), we aimed to analyse the effect of the weighting choice on the EBQI and the ranking of the sites. In order to achieve our aim, we perturbed each weighting value and determined the new ranking obtained with the perturbed weightings. The perturbation on each weighting was obtained as follows. We first defined the maximum amplitude of the perturbation for all weightings and we then defined, for each weighting, a random perturbation, according to a uniform law between 0 and the maximum amplitude of perturbations. We then added or subtracted this perturbation term to or from the corresponding weighting. If the new weighting was less than 1, it was then set as equal to 1. If the new weighting was greater than 5, it was then set as equal to 5. We ended up with weightings between 1 and 5, close to the original ones if the maximum perturbation amplitude was low, and randomly selected otherwise. This has been calculated as follows: the perturbation method described previously was repeated 1,000 times and, for each site, we determined whether the rank of the site was the same as the initial one or if it had changed. An index of similarity was produced for each site, which was equal to 100% when the rank of the site was always unchanged after perturbation and 0% if it always changed.

For each compartment (box) status at each site, a Confidence Index (CI) was proposed (Table 2). The reason for the CI is **(i)** that data for one or several compartments may be missing or of poor quality in some sites, **(ii)** the reliability of available data may be different between compartments and sites, and **(iii)** it is worth drawing the attention of managers and scientists to those compartments that are poorly known and which merit further field studies. The grade of each considered compartment was given by its CI (0 through 4) and by its weighting (1 through 5), and therefore they are graded from 0 through 20. The grades of all considered compartments were added up, which gave the final mark of the CI at a given site. For practical purposes, the CI was converted to a scale from 0 to 4:

Table 2. Criteria to assess the Confidence Index (CI) of the status of a compartment (box).

CI	Criteria
4	Field data available, recent and suitable with the recommended methods
3	Field data recent, partially completed with expert judgment
2	No quantitative field data but recent expert judgment
1	No quantitative field data, but ancient expert judgment
0	No quantitative field data and no suitable expert judgment

$$CI_{EBQI} = \left[\sum_{i=1}^{13} (W_i \times CI_i) / \sum_{i=1}^{13} (W_i \times CI_{\max}) \right] \times 4$$

Where W_i is the weighting of the box i, CI_i the Confidence Index of the box i, CI_{max} the highest possible Confidence Index (= 4) for a compartment (box) and i is the number of the box (1 through 13).

In order to test the efficiency of the proposed method, it was applied to seventeen sites (Table 3) with a variety of available data (published, unpublished, expert judgment). A site is defined as a *P. oceanica* meadow, from its upper limit down to the lower limit, covering in the order of a dozen to several hundred hectares. The sites are localized in the north-western Mediterranean Sea, including continental France (French Riviera, Provence, French Catalonia), Corsica, Balearic Islands and Spanish Catalonia. This area is considered as homogenous and is consistent with the marine subregions, as defined in the MSFD. These sites also cover of a wide range of human pressures, from lesser impact, within well implemented Marine Protected Areas (MPAs), such as the Port-Cros National Park, to highly impacted areas due to different disturbance and/or stress sources (pollution, overfishing, fish farms, port facilities, anchoring and mooring; see table 3).

Results

EBQI assessment

The results of the EBQI application for assessing the status of the *P. oceanica* ecosystem functioning in the sites across the NW Mediterranean area (Table 3) are presented in Table 4. An example of calculation of the EBQI is given (Table 5). The content of Table 4 is based upon unpublished observations from the authors of the present article, together with some published data [108–111] and whenever necessary (CI<4) on expert judgment; when several co-authors of the present work were involved in the assessment of a site, the Delphi method [106] was used for assessing the status of each compartment (box). According to the EBQI, the 17 study sites were placed within five Ecological Status classes, from Bad to High:

- Bad (EBQI<3.5): Sitges.
- Poor (3.5≥EBQI<4.5): Niolon, Gulf of Giens and Porquerolles Island northern coast.
- Moderate (4.5≥EBQI<6): Villefranche Bay, Saint-Cyr Bay, Plateau des Chèvres (Marseilles), Prado Bay (Marseilles), Valincu Gulf, Tossa de Mar, Scandola (Elbu Bay) and Peyrefite Bay.

Table 3. Sites used to test the proposed ecosystem-based approach to assess the status of the *P. oceanica* ecosystem. MPA: Marine Protected Area. NTZ: No Take Zone.

Site	Region	Protection status	Pressure
Espardell	Balearic Islands (Spain)	MPA, Natura 2000	Artisanal fishery
Sitges	Spanish Catalonia	Natura 2000	Pollution, artisanal and recreational fishery, sedimentation
Tossa de Mar	Spanish Catalonia	Natura 2000	Artisanal and recreational fishery
Medes Islands	Spanish Catalonia	MPA, NTZ, Natura 2000	River mouth
Peyrefite Bay	French Catalonia	MPA[a]	Artisanal fishery, anchoring[b]
Niolon (Côte Bleue)	Provence (France)	MPA, Natura 2000	River mouth, artisanal fishery, trawling
Prado Bay, Marseilles	Provence (France)		Coastal development, artisanal fishery, nutrients
Plateau des chèvres, Marseilles	Provence (France)	MPA[c], Natura 2000	Artisanal fishery, sewage outfall,
Saint Cyr Bay	Provence (France)		Coastal development, artisanal fishery, sewage outfall
Gulf of Giens	Provence (France)		Sewage outfall, artisanal fishery, trawling
Porquerolles Island, northern coast	Provence (France)	MPA, Natura 2000	Trawling, artisanal fishery, anchoring
Porquerolles Island, southern coast	Provence (France)	MPA, Natura 2000	Trawling, artisanal fishery
Bagaud Pass, Port-Cros Island	Provence (France)	National Park, MPA, Natura 2000	Artisanal fishery, anchoring
Port-Cros Island, southern coast	Provence (France)	National Park, MPA, Natura 2000	Artisanal fishery
Villefranche-sur-Mer Bay	French Riviera		Coastal development, sewage outfall artisanal fishery, anchoring
Scandola, Elbu Bay	West Corsica (France)	MPA, Natura 2000	Artisanal fishery
Valincu Gulf	West Corsica (France)		Artisanal fishery

[a]This area is close to the Natural Marine Reserve of Cerbère-Banyuls. Since October 2011, this area has been included within a natural marine park ('Parc naturel marin du golfe du Lion').
[b]Since 2010, anchoring is banned.
[c]Since May, 2012, this area has been included within a National Park ('Parc national des Calanques'). However, it is unlikely that this new status would have already resulted in a perceptible effect.

Table 4. Ecosystem status (EBQI) and Confidence Index (CI) for each sampling site, and scale value and CI of each parameter used to obtain them.

Box	1 Rhizomes	2 Leaves	3–4 Leaf epibiota	5 Pinna	6 HOM LOM	7 Litter	8 Holothuria	9 Herbivores	10 Predators	11 Piscivores	12 Planktivores	9–12 (SRDI)	13 Sea birds	EBQI and (CI)
Espardell	4(0)	4(4)	3(3)	3(4)	3(0)	3(0)	3(0)	3(3)	1(4)	1(4)	1(4)	3(4)	2(3)	6.4(2.9)
Sitges	2(0)	0(4)	0(3)	0(3)	2(0)	2(0)	2(3)	3.5 (3.5)	0(3)	0(3)	1(3)	0(3)	0(3)	2.3(2.7)
Tossa de Mar	2(3)	3(4)	4(4)	0(3)	2(0)	2(0)	4(2)	3(2.5)	2(4)	0(4)	2(4)	3(4)	1(2)	5.6(3.2)
Medes Islands	2(3)	3.5(4)	4(4)	3(3)	2(0)	2(0)	2(0)	2.5(3)	4(2)	4(2)	3(2)	4(2)	2(2)	7.9(2.4)
Peyrefite Bay	2(0)	3.5(4)	2(0)	4(4)	2(0)	2(0)	2(0)	2(4)	3(3)	1(3)	1.5(3)	4(3)	0(2)	5.8(2.3)
Niolon	2(2)	2.5(4)	2(2)	0(2)	1.5(2)	1(0)	3(0)	2(1.5)	1(2)	0(0)	2(0)	2(2)	1(4)	3.9(1.7)
Prado Bay	2(4)	2.5(3)	2(2)	0(2)	2.5(2)	2(2)	3(2)	2.5 (1.5)	3(2)	1(2)	1.5(2)	3(2)	2(4)	5.3(2.3)
Plateau des Chèvres	2(4)	2.5(4)	4(3)	0(2)	1.5(2)	2(2)	3(2)	2.5 (1.5)	2(4)	1(4)	0.5(2)	2(4)	2(2)	5.0(3.1)
Saint-Cyr Bay	1(2)	3(3)	2(0)	1(0)	2(0)	2(2)	2(0)	2(0)	2(0)	2(0)	2(0)	2(0)	0.5(4)	4.9(0.7)
Gulf of Giens	3(4)	4(4)	2(1)	2(0)	2(0)	1(0)	3(2)	1.5(2)	1(4)	0(2)	1(2)	1(3)	0.5(4)	4.3(2.4)
Porquerolles Island northern coast	3(3)	2(4)	3(3)	2(3)	2(3)	0(2)	1(2)	1.5(1)	1(2)	1(2)	2(2)	2(2)	1(4)	4.3(2.4)
Porquerolles Island southern coast	3(3)	4(5)	4(3)	3(3)	3(3)	3(2)	2(2)	2(1)	2(2)	2(2)	3(2)	3(2)	1(4)	6.9(2.6)
Bagaud Pass, Port-Cros Island	4(2)	3(4)	2(4)	4(4)	3(2)	4(2)	4(2)	3(2.5)	3(2)	2(2)	3(2)	4(2)	1(4)	7.6(2.6)
Port-Cros Island southern coast	4(0)	4(4)	4(3)	4(2)	3(2)	4(0)	3(0)	3.5(0)	4(2)	4(0)	3(2)	4(2)	1.5(4)	9.3(1.6)
Villefranche-sur-Mer Bay	2 (1)	1.5 (4)	2(1)	1(2)	3(2)	0(0)	0(2)	2(1)	3(0)	2(1)	1.5(1)	4(0)	0(4)	4.8(1.3)
Scandola, Elbu Bay	4(4)	3(4)	1(3)	4(4)	3(2)	2(2)	2(2)	2(1.5)	2(2)	1(2)	1.5(2)	3(3)	4(4)	5.7(2.6)
Valincu Gulf	(4.4)	3(4)	2(3)	2(2)	2(0)	2(2)	3(2)	2(0)	2(0)	2(0)	2(0)	2(0)	1(0)	5.4(1.4)

Boxes 1 through 13 and SRDI. The status and the Confidence Index (in brackets; CI) is indicated for each box. EBQI ranges from 0 to 10, while the CI is graded from 0 to 4).

Table 5. Example of calculation of the EBQI at the site Espardell (Balearic Islands).

Box number	Functional compartment	Weighting (W)	Parameter	Status and mean status when 2 parameters (S)	CI	Status grade: W×S	CI grade: W×CI
1	Roots and rhizomes ('rhizomes')	3	Growth of orthotropic rhizomes (mm a^{-1})	4	0	12	0
2	*Posidonia oceanica* leaves ('leaves')	5	- Density (shoots m^{-2})	(4)	(4)	(20)	(20)
			- Cover (%)	(4)	(4)	(20)	(20)
				4	4	20	20
3–4	MPOs, filter- and suspension-feeders leaf epibiota ('leaf epibiota')	4	Biomass (g DM shoot^{-1})	3	3	12	12
5	Benthic filter-feeder: *Pinna nobilis* (bivalve) ('Pinna')	2	Density (individuals 100 m^{-2})	3	4	6	8
6	Other benthic filter- and suspension-feeders ('HOM LOM')	2					
	- HOM		- Density (m^{-2})	(3)	(0)	(6)	(0)
	- LOM		- Density (m^{-2})	(3)	(0)	(6)	(0)
				3	0	6	0
7	Litter detritus: dead leaves and broken rhizomes ('litter')	2	g DM m^{-2}	3	0	6	0
8	Detritus-feeders 2 and 3 (*Holothuria* spp.) ('Holothuria')	2	Individuals 10 m^{-2}	3	0	6	0
9	Herbivores 1 ('herbivores')	5	- Density of *Paracentrotus lividus* (individuals m^{-2})	(3)	(3)	(15)	(15)
			- Grazing index (% leaves)	(3)	(3)	(15)	(15)
				3	3	15	15
10	Predatory teleosts, cephalopods and seastars ('predators')	5	kg teleosts WM 100 m^{-2}	1	4	5	20
11	Piscivorous teleosts ('piscivores')	5	kg teleosts WM 100 m^{-2}	1	4	5	20
12	Planktivorous teleosts ('planktivores')	3					
	- Zooplankton feeders		- kg teleosts WM 100 m^{-2}	(1)	(4)	(3)	(12)
	- Omnivores		- kg teleosts WM 100 m^{-2}	(1)	(4)	(3)	(12)
				1	4	3	12
9–12	All teleosts ('SRDI')	3	Specific Relative Diversity Index (SRDI)	3	4	9	12
13	Sea birds	1					
	- *Phalacrocorax* spp.		- Distance to the nearest nesting site (km)	(4)	(3)	(4)	(3)
	- *Pandion haliaetus*		- Distance to the nearest nesting site (km)	0	(3)	0	(3)
				2	3	2	3
					Sum of the weighted status grades (left) and of the weighted CI grades (right) of the 13 boxes[a]	107	122
					EBQI[b] (left) and CI[c] (right) of the site	6.4	2.9

[a]The maximum value of the sum is 168 (see text).
[b]Maximum value: 10 (see text).
[c]Maximum value: 4 (see text).
CI: Confidence Index. DM: dry mass, including calcium carbonate. HOM: indicators of high level of organic matter. LOM: indicators of low level of organic matter. MPO: Multicellular Photosynthetic Organisms. POM: Particulate Organic Matter. WM: wet mass.

- Good (6.0≥EBQI<7.5): Espardell and Porquerolles Island southern coast.
- High (EBQI≥7.5): Bagaud Pass (Port-Cros Island), Medes Islands and Port-Cros Island (southern coast).

Redundancy of EBQI with already existing indices

Is the EBQI superfluous, i.e. redundant with already existing indices? A number of indices (EQR, Ecological Quality Ratio) based upon *P. oceanica* (the species itself, sometimes leaf epibiota,

Table 6. Comparison of EBQI with Ecological Quality Ratios (EQRs) based mainly upon *P. oceanica* (the organism itself) and aimed at establishing the ecological status of the seawater body.

Site	EBQI/10 (CI)	EQR/1	Type of index	Reference
Port-Cros Island, southern coast	9.3 (1.6)	0.802	PREI[a]	[14]
Medes Islands	7.9 (2.4)	0.752	POMI[b]	[12]
Scandola, Elbu Bay	5.7 (2.6)	0.802	BiPo[c]	[15]
Tossa de Mar	5.6 (3.2)	0.682	POMI	[12]
Valincu Gulf	5.4 (1.4)	0.386	PREI	[14]
		0.729	BiPo	[15]
Prado Bay, Marseilles	5.3 (2.3)	0.636	PREI	[14]
Plateau des chèvres, Marseilles	5.0 (3.1)	0.477	PREI	[14]
Saint Cyr Bay	4.9 (0.7)	0.682	PREI	[14]
Villefranche-sur-Mer Bay	4.8 (1.3)	0.280	PREI	[14]
Gulf of Giens	4.3 (2.4)	0.708	PREI	[14]
Porquerolles Island, northern coast	4.3 (2.4)	0.819	PREI	[14]
Niolon (Côte Bleue)	3.9 (1.7)	0.465	PREI	[14]
Sitges	2.3 (2.7)	0.238	POMI	[12]

[a]The metrics of PREI are: shoot density, shoot leaf surface area, ratio between epibiota biomass and leaf biomass, depth of the lower limit and type of this limit [14].
[b]The metrics of POMI are: shoot density, meadow cover, percentage of plagiotropic rhizomes, shoot leaf surface area, percentage of foliar necrosis, P, N and sucrose content in rhizomes, $\delta^{15}N$ and $\delta^{34}S$ isotopic ratio in rhizomes, N content in epiphytes, Cu, Pb and Zn content in rhizomes [12].
[c]The metrics of BiPo are: shoot density, shoot leaf surface area, lower depth limit and lower limit type [15].
PREI, POMI and BiPo indices are based upon distinct but similar metrics and range from 0 (lowest ecological status) to 1 (highest ecological status).

not the ecosystem) have been proposed and are currently routinely used, generally for the purpose of monitoring the ecological status of a water body, e.g. POMI [12], PREI [14] and BiPo [15]. They are based on different combinations of a number of parameters, e.g. shoot density, leaf surface area per shoot, depth of the lower limit of the meadow, percent cover of the dead matte and epibiota biomass. One or several of these indices are available from the literature for 13 out of the 17 study sites (Table 6). As these indices are more or less congruent [16], we have investigated the possible correlation between EBQI and the available EQR (POMI, BiPo or PREI) in these 13 sites, and between EBQI and PREI. There is no significant correlation (Spearman coefficient), either when all available EQRs are considered ($n = 13$, $r_s = 0.469$, p value = 0.11) or when only PREI data are concerned ($n = 9$, $r_s = -0.017$, p value = 0.97). This result is not unexpected: EQRs mainly assess the health of the seagrass, linked to e.g. the water quality, while EBQI measures the status of the ecosystem, linked not only to water quality but also to e.g. the overfishing. The contrasting ranks of Porquerolles Island (northern coast) through EBQI (poor) and EQR (high, first ranking, see Table 6), together with those of Valincu Gulf and Gulf of Giens, can be due to impacts other than the water quality, e.g. artisanal and recreational overfishing and anchoring of pleasure boats.

Redundancy of parameters to calculate the EBQI

Are some of the parameters used (functional compartments, boxes) redundant with other boxes or with the EBQI? Should such a redundancy exist, this could mean either that the box alone is sufficient to estimate the whole ecosystem's status, or that some boxes are useless. Obviously, the higher weighting is given to a box, the stronger is the probability of such a correlation. Some boxes (e.g. box 5 'Pinna' and box 10 'predators') are significantly correlated with others and/or with the EBQI (Table 7). However, the removal of the boxes correlated with others (boxes 5, 9, 11 and SRDI) results in changes (loss of accuracy?) in the EBQI: the five-class ranking (see below) of 53% of the study sites moves from one class to the next (result not presented). In addition, the removal of these possibly superfluous boxes would not save field time (data acquisition), since data for these boxes are acquired simultaneously with those of non-removed compartments.

Robustness of the EBQI

Is the EBQI robust, i.e. only slightly influenced by arbitrary choices in its conception and biases? Possible arbitrary choices are e.g. choice of the considered compartments (boxes) and the weighting (1 through 5) of the boxes. As far as the choice of the compartments taken into account is considered, it should be first emphasized that the choice is anything but arbitrary; boxes encompass the whole spectrum of an ecosystem functioning, from primary producers, herbivores, filter-feeders, suspension-feeders and detritus-feeders to top predators (Fig. 1). It is worth noting that the removal of up to 4 boxes (see above) only slightly changes the five-class ranking of the study sites: at the most, half of the sites move from one class to the next (e.g. good to moderate, poor to bad; results not presented). Finally, the changing of the weighting of the boxes (hereafter 'perturbation') alters the ranking of the sites; the changes increase with the importance of the perturbation (±1, ±2, fig. 2; ±3 and ±4, not presented). However, the changes due to weighting perturbation are relatively slight. In addition, an index of similarity was produced for each site, which is equal to 100% when the rank of the site is always unchanged after perturbation and 0% if it is always changed (Fig. 3). Some sites are very robust (e.g. Porquerolles Island northern coast and Bagaud Pass), other are more sensitive to the choice of weightings (e.g. Medes Islands and Tossa de Mar). The similarity ranges between 100 and 95% (perturbation ±1), 100 and 71% (±2), 100 and 57% (±3) and 100 and 12% (±4). The low values of the similarity index for high levels of perturbation of the weighting emphasize the fact that weighing the boxes is useful, despite the robustness of this parameter.

Table 7. Spearman's rank correlations between the status of each compartment (boxes 1 through 13 and SRDI), the other compartments (boxes) and the EBQI (see Table 5).

	1.Rhizomes	2. Leaves	3–4. Leaf epibiota	5. Pinna	6. HOM LOM	7. Litter	8. Holothuria	9. Herbivores	10. Predators	11. Piscivores	9–12 (SRDI)	12. Planktivores	13. Sea birds
1. Rhizomes	-												
2. Leaves	0.44	-											
3–4. Leaf epibiota	0.03	0.41	-										
5. Pinna	0.71	0.64	0.02	-									
6. HOM-LOM	0.57	0.33	−0.02	0.63	-								
7. Litter	0.44	0.55	0.29	0.48	0.51	-							
8. Holothuria	0.27	0.27	0.20	−0.09	−0.08	0.44	-						
9. Herbivores	0.07	0.03	0.20	−0.08	0.23	0.67	0.46	-					
10. Predators	−0.02	0.23	0.29	0.37	0.36	0.34	0.02	0.24	-				
11. Piscivores	0.22	0.30	0.33	0.51	0.46	0.42	−0.22	0.10	0.73	-			
12. Planktivores	0.18	0.28	0.38	0.34	0.24	0.34	0.05	0.06	0.48	0.62	-		
9–12 (SRDI)	0.16	0.32	0.30	0.56	0.59	0.41	−0.02	0.29	0.85	0.60	0.46	-	
13. Sea birds	0.37	0.22	0.37	0.12	0.19	0.36	0.30	0.26	0.20	0.18	0.08	0.19	-
EBQI	0.45	0.69	0.50	0.72	0.55	0.78	0.24	0.40	0.69	0.65	0.54	0.77	0.47

Bold italic characters: significant ($p<0.01$).

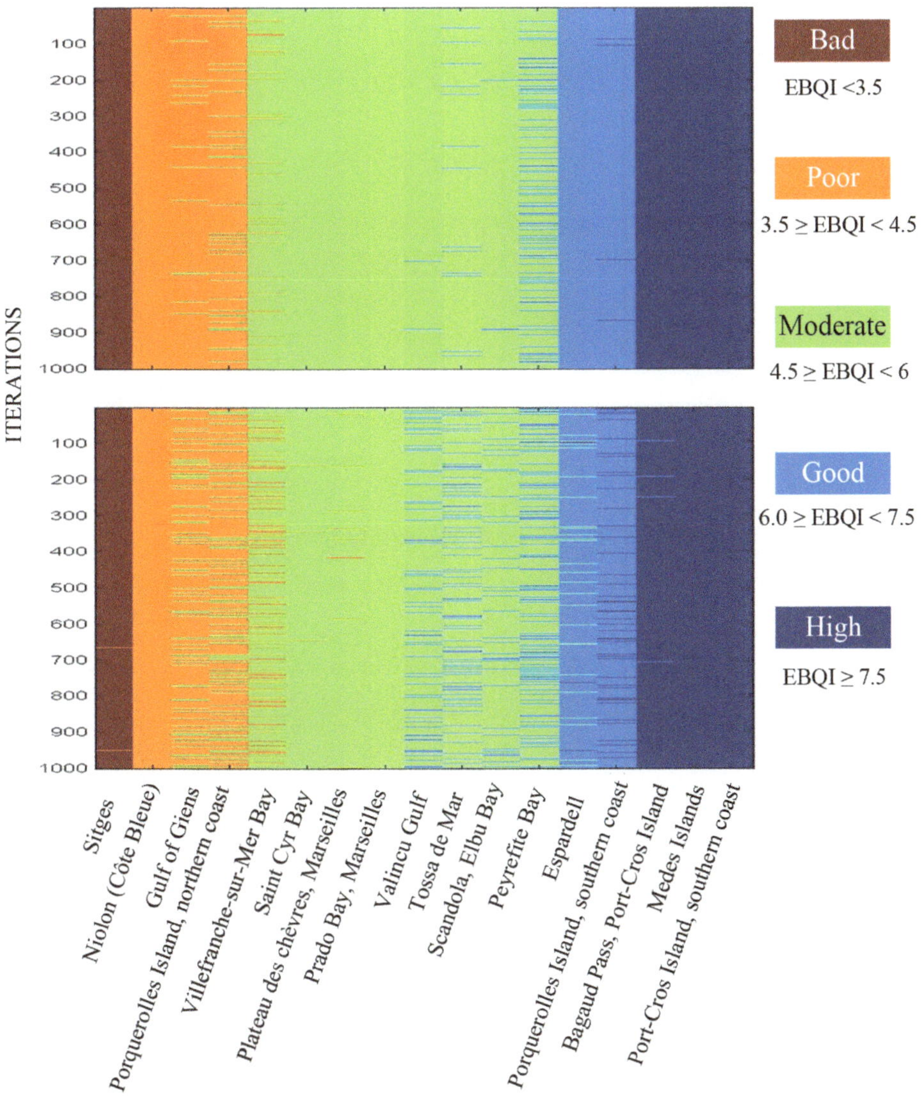

Figure 2. Robustness of the quality index (EBQI) with regard to the weighting of compartments (boxes). The 17 sites are arranged from left to right according to their growing EBQI (see Table 4) and ecological status (bad through high). Deep red = bad, orange = poor, green = moderate, light blue = good, deep blue = high. In order to test the effect of the weighting of the compartments (boxes) on the EBQI (robustness), weighting values have been randomly perturbed (above, ±1; below, ±2). 1000 iterations were performed. The change of the ecological status (bad through high) of a site, for a given iteration, is shown by the color of the new class within which it falls.

EBQI implementation

Is the EBQI excessively time-consuming or easy and rapid to implement? The acquisition of the parameters requires SCUBA diving field work. Considering that it is possible to work for around 1 hour at 15 m depth, we estimate that 6 dives involving 2 scientific divers are necessary to acquire all of the data for the assessment of one site (Table 8). For safety reasons, a diver can perform a maximum of two dives per day, so this will require 3 days of field work for a two man diving team. This field work requires expertise in seagrass and fish visual censuses. In addition, for a rapid and provisional assessment of the EBQI, already available data and expert judgment can be used (provided that the CI is specified).

Within the northern Mediterranean, the study sites are spread within three eco-regions, namely Corsica, Provence and French Riviera and Languedoc-Catalonia. Is the EBQI biased by regional environmental conditions, e.g. a higher mean temperature in Corsica and more water turbidity in Languedoc-Catalonia? The present study cannot answer this question as the sites were not randomly selected but chosen according to data availability. However, the EBQI range within each one of the three eco-regions does not exhibit obvious differences (Table 4), as far as the EBQI mean is concerned (~5.6 within each one of the three eco-regions).

Discussion

In the context of European Directives (Habitats Directive and WFD) and national regulations, a number of species and indices, based upon one or several species, have been designated to assess the quality of water bodies, the health status of emblematic species and habitats and to establish Marine Protected Areas (e.g. PMN [112], CI [113], EEI [2,3], SI [114], POMI [12], PREI [14], BiPo [15], PoSte [15], ZoNI [22] for European seagrasses

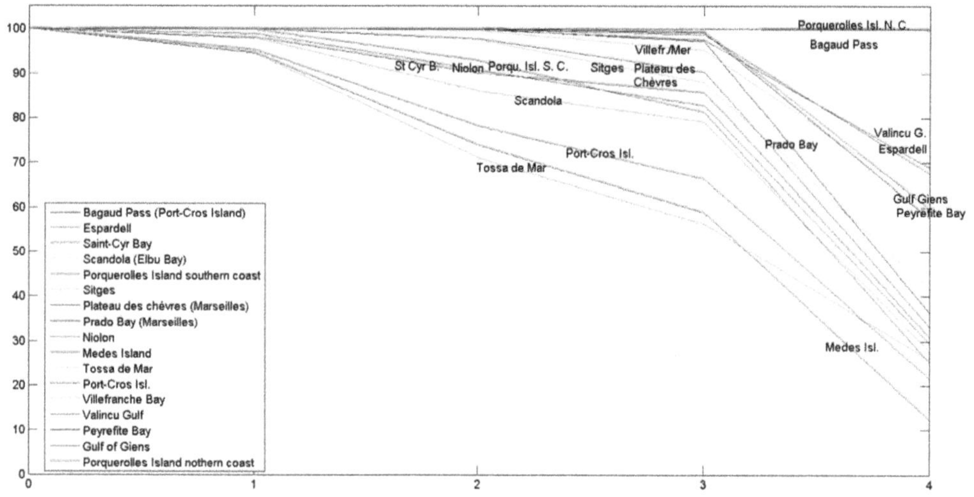

Figure 3. Robustness (index of similarity) of the quality index (EBQI) to the weighting of compartments (boxes). Percentage of times the ecological status of each site was unaltered by random perturbation of the weighting of the boxes (1,000 iterations) by ±1 through ±4. The similarity is equal to 100% when the class of the site is never changed after perturbation and 0% if it is always changed.

([17,51,115,116] for comprehensive reviews). These indices provide a valuable body of tools and information to coastal waters managers and make it possible to assess the status of a water body. However, the good quality of a water body and the apparent health of a species, whether emblematic or not, such as the seagrass *P. oceanica*, is not always indicative of the good structure and functioning of the whole ecosystem, a network of compartments, fluxes, functions, inputs and exportations. The most original contribution of the new MSFD European Directive is to provide an ecosystem-based approach to assess the ecological

Table 8. Estimation of time and diving effort needed for data acquisition within each box used in the EBQI.

Box	Proxy	Time	Dive organization
1 - *P. oceanica* roots and rhizomes	30 random *in situ* measures (growth of the rhizome corresponding to the last 8 leaf bases)	30 min	1 dive for 2 scientific divers
2 - *P. oceanica* leaves	20 random measures of shoot density in a square frame (0.16 m^2)	60 min	
	30 random measures of cover	10 min	
3 - MPO leaf epibiota	Sampling of the two oldest external leaves on 30 shoot randomly localized to estimate the epibiota biomass	10 min	
4 - Filter- and suspension-feeder leaf epibiota	Sampling of the two oldest external leaves on 30 shoot randomly localized to estimate the epibiota biomass	Same as box 3	
5 - Filter feeder benthic epibiota	Density of *Pinna nobilis* along 20 transects 10-m long and 1-m wide	120 min	1 dive for 2 scientists divers
7 - Litter detritus	Litter detritus mass collected in July, within 5 randomly localized 0.1-m^2 quadrats	90 min	1 dive for 2 scientist divers
6 - Other filter- and suspension-feeder benthic epibiota	Abundance of filter- and suspension-feeder benthic epibiota other than *Pinna nobilis* within 1-m^2 quadrats, with 30 replicates randomly localized	120 min	1 dive for 2 scientist divers
8 - Detritus feeders	Abundance of *Holothuria* spp. within 1-m^2 quadrats, with 30 replicates randomly localized	Same as box 6	
9 - Herbivores 1	Abundance of *Paracentrotus lividus* within 1-m^2 quadrats, with 30 replicates randomly localized at 5 m depth	60 min	1 dive for 2 scientist divers
	Percentage of intermediate and adult leaves exhibiting bite scars due to *S. salpa* (all the intermediate and adult leaves of 30 shoots randomly localized) at 5 m depth	30 min	
10 - Predatory teleosts	All teleosts counted within ten linear 50-m long and 5-m wide transects, each census lasting 5 minutes.	60 min	1 dive for 2 scientist divers
11 - Piscivorous teleosts	All teleosts counted within ten linear 50-m long and 5-m wide transects, each census lasting 5 minutes	Same as boxes 10 and 12	
12 - Planktivorous teleosts of the water column	All teleosts counted within ten linear 50-m long and 5-m wide transects, each census lasting 5 minutes	Same as boxes 10 and 11	
9–12 - SRDI	All teleosts counted within ten linear 50-m long and 5-m wide transects, each census lasting 5 minutes	Same as boxes 10, 11 and 12	

status of marine regions. The present approach constitutes a contribution towards this goal, and focuses on the *P. oceanica* seagrass meadow.

Here, we have developed, applied and tested an ecosystem-based index of the structure and functioning of the most emblematic and best-known Mediterranean coastal ecosystem, the *P. oceanica* meadow. This index (EBQI) is based upon a set of representative functional compartments, the weighting of these compartments and the assessment of each compartment quality by comparison of a supposed baseline based upon the available literature. The strong points of the EBQI are: **(i)** It is easy to implement, not too time-consuming and therefore relatively cheap; **(ii)** It is non-destructive, which is particularly important, dealing with protected species (e.g. *P. oceanica* and *Pinna nobilis*); sampling only concerns old leaves (not the shoot) and the litter, i.e. shed dead leaves; **(iii)** It is relatively robust, according to the selection of the compartments and to their weighting; **(iv)** It is associated with confidence indices CI (each compartment, and the overall mark) which indicate possible weakness and biases and therefore the need for further field data acquisition; **(v)** It can draw the attention of managers to sites whose CI is high and where routine monitoring can therefore be implemented, and conversely those with a low CI indicating a lack of knowledge.

The Integrated Coastal Zone Management (ICZM) contributes towards maximizing the benefits provided by the coastal zone and minimizing conflicts and the harmful effects of activities upon each other (e.g. [117]). ICZM, together with Ecosystem-Based Management (EBM; e.g. [118]) and, more generally, the implementation of European Directives, are in need of indices to assess ecological quality, either based upon a single or a few species, or ecosystem-based. The sites corresponding to well implemented MPAs, such as the Port-Cros National Park (Port-Cros Island southern coast, Bagaud Pass) and the Medes Islands marine reserve, not unexpectedly get the best marks, which is congruent with the overall excellent status of their habitats [108]. In contrast, sites localized in areas highly impacted by most human activities and waste, such as Sitges (South of Barcelona) and Niolon (close to the port of Marseilles and outfall of wastewater), obtain a very low mark. Though sites with obviously good or bad ecosystem status naturally obtain a good and bad, EBQI mark, respectively, some sites (e.g. Saint-Cyr Bay and Villefranche-sur-Mer Bay) do not obtain the mark they would seemingly have deserved through the apparent health of the seagrass itself. This confirms the usefulness of an index, EBQI, based upon the whole ecosystem rather than only upon the seagrass (sometimes also epibiota and other parameters). It is worth noting that the Confidence Index (CI) is overall relatively low. This is especially the case of Saint-Cyr Bay (CI = 0.7; table 4) and Villefranche-sur-Mer Bay (CI = 1.3), which could also account for the surprisingly low and not so low, EBQI, respectively, of these sites, according to literature data [119,120].

The weak points of the EBQI are: **(i)** The baselines used to assess the compartments, which may prove to be biased either by the poor knowledge or availability of totally non-human impacted areas, or by climate change which renders the supposed baseline obsolete; **(ii)** The probable need for testing and/or adapting the EBQI in areas distant from the study area -the north-western Mediterranean- and more generally to other seagrass ecosystems; (iii) The conceptual model which constitutes the basis for the present ecosystem-based approach is obviously oversimplified. For example, small-sized predatory 'invertebrates' were not taken into consideration. Some of them belong to the matte endofauna box, which encompasses detritus feeders, suspension-feeders and predators. Other are creeping or clinging organisms on the *P. oceanica* leaves. However, considering them would have made the model more complex and weighed down the assessment of ecosystem status. Moreover, other compartments are likely to provide redundant information, so that no improvement in accuracy is to be expected.

A number of attempts to review and compare the biotic indices using benthic (e.g. macrophytes) and pelagic marine, lagoon and estuarine organisms, have been performed [16,39,51,116,121]. For example, [116] analysed the strengths and weaknesses of 90 published indices. They identified several weaknesses: **(i)** problems of applicability due to practical and conceptual difficulties, affecting most of the indices related to ecosystem function, **(ii)** the failure of many indices using e.g. the taxonomic composition of the community to connect its condition with the stressors, and **(iii)**, as far as indices based upon the sub-individual level are concerned (e.g. multi-biomarkers), their poor strength to assess the ecological integrity of the habitat. They concluded that the most promising approach would be to aggregate indices with complementary strengths. In a sense, the EBQI could be considered as such an aggregative index. Although the goal of the EBQI is not to compete with the already available biotic indices utilizing the *P. oceanica* seagrass meadow for assessing the quality of a water body, but to assess the status of the proper seagrass ecosystem, the comparison of the EBQI with some other biotic indices will be of relevance. An attempt has been made here, using some published data; a more accurate comparison would require data acquisition at the same sites as those used for the EBQI, and is therefore beyond the scope of the present work. In addition, the correlation between the EBQI and anthropogenic gradients, including overfishing, will be useful, in order to assess which human impacts are of major relevance for the status of the ecosystem. This will constitute the next step.

Conclusions

Overall, the Ecosystem Based Quality Index (EBQI) is easy to implement, relatively robust and does not seem to be redundant with existing indices based upon the seagrass itself. In addition, the non-congruence of the EBQI, i.e. the quality of the ecosystem functioning, with the empirical idea one may have of the *P. oceanica* ecosystem at a given site, due e.g. to the health of the seagrass, the clearness of the water (and even the beauty of the landscape and the seascape) confirms the usefulness of the EBQI index, based upon the whole ecosystem rather than upon only the seagrass. Other anthropogenic impacts, e.g. overfishing, are putatively more important than the above mentioned parameters for the structure and functioning of the ecosystem. Finally, the EBQI that is proposed here for the *P. oceanica* ecosystem could constitute a model for similar indices designed for other ecosystems, such as coralligenous outcrops, underwater caves, soft bottoms and sublittoral reefs.

Acknowledgments

The authors are indebted to Fabien Morat for data concerning *Phalacrocorax aristotelis*, Philippe Lenfant, Boris Daniel (AAMP) and Pierre Boissery (AERMC) for contributions and valuable comments, Michael Paul for improvement of the English text and two anonymous reviewers for their highly valuable suggestions and improvement of the manuscript.

Author Contributions

Conceived and designed the experiments: CFB CPM DTB EB FR GP MHV SP SR. Performed the experiments: CPM EB GP JP PA PB SB SR TT. Analyzed the data: CFB JCP SP SR. Wrote the paper: DBS EB EF CFB JCP MHV SP SR.

References

1. Karr JR (1981) Assessment of biotic integrity using fish communities. Fisheries 6: 21–27.
2. Orfanidis S, Panayotidis P, Stamatis N (2001) Ecological evaluation of transitional and coastal waters: a marine benthic macrophytes-based model. Medit Mar Sci 2: 45–65.
3. Orfanidis S, Panayotidis P, Stamatis N (2003) An insight to the ecological evaluation index (EEI). Ecol Indic 3: 27–33.
4. White J, Irvine K (2003) The use of littoral mesohabitats and their macroinvertebrates in the assemblages of lakes. Aquat Conserv 13: 331–351.
5. Pergent-Martini C, Leoni V, Pasqualini V, Ardizzone GD, Balestri E, et al. (2005) Descriptors of *Posidonia oceanica* meadows: Use and application. Ecol Indic 5: 213–230.
6. Orfanidis S, Panayotidis P, Ugland KI (2011) Ecological Evaluation Index continuous formula (EEI-c) application: a step forward for functional groups, the formula and reference condition values. Medit Mar Sci 12: 199–231.
7. Hoare JM, Monks A, O'Donnel CFJ (2013) Do population indicators work? Investigating correlated responses of bird populations in relation to predator management. Ecol Indic 25: 23–34.
8. Jones CG, Lawton JH, Shachak M (1994) Organisms as ecosystem engineers. Oikos 69: 373–386.
9. Lawton JH (1994) What do species do in ecosystems? Oikos 71: 367–374.
10. Bond W (2001) Keystone species – Hunting the snark? Science 292: 63–64.
11. Boudouresque CF, Mayot N, Pergent G (2006) The outstanding traits of the functioning of the *Posidonia oceanica* seagrass ecosystem. Biol Mar Medit 13(4): 109–113.
12. Romero J, Martínez-Crego B, Alcoverro T, Pérez M (2007) A multivariate index based on the seagrass *Posidonia oceanica* (POMI) to assess ecological status of coastal waters under the water framework directive (WFD). Mar Poll Bull 55: 196–204.
13. Fernández-Torquemada Y, Díaz-Valdés M, Colilla F, Luna B, Sánchez-Lizaso JL, et al. (2008) Descriptors from *Posidonia oceanica* (L.) Delile meadows in coastal waters of Valencia, Spain, in the context of the EU Water Framework Directive. ICES J Mar Sci 65: 1492–1497.
14. Gobert S, Sartoretto S, Rico-Raimondino V, Andral B, Chery A, et al. (2009) Assessment of the ecological status of Mediterranean French coastal waters as required by the Water Framework Directive using the *Posidonia oceanica* Rapid Easy Index: PREI. Mar Poll Bull 58: 1727–1733.
15. Lopez y Royo C, Casazza G, Pergent-Martini C, Pergent G (2010) A biotic index using the seagrass *Posidonia oceanica* (BiPo), to evaluate ecological status of coastal waters. Ecol Indic 10: 380–389.
16. Lopez y Royo C, Pergent G, Alcoverro T, Buia MC, Casazza G, et al. (2011) The seagrass *Posidonia oceanica* as indicator of coastal water quality: experimental intercalibration of classification systems. Ecol Indic 11: 557–563.
17. Marbà N, Krause-Jensen D, Alcoverro T, Birk S, Pedersen A (2012) Diversity of European seagrass indicators: patterns within and accross regions. Hydrobiologia 704: 265–278.
18. Panayotidis P, Feretopoulos J, Montesanto B (1999) Benthic vegetation as an ecological quality descriptor in an Eastern Mediterranean coastal area (Kalloni Bay, Aegean Sea, Greece). Estuar Coast Shelf Sci 48: 205–214.
19. Ballesteros E, Torras X, Pinedo S, Garcia M, Mangialajo L, et al. (2007) A new methodology based on littoral community cartography by macroalgae for the implementation of the European Water Framework Directive. Mar Poll Bull 55: 172–180.
20. Guinda X, Juanes JA, Puente A, Revilla JA (2008) Comparison of two methods for quality assessment of macroalgae assemblages, under pollution types. Ecol Indic 8: 743–753.
21. Oliva S, Mascaró O, Llagostera I, Pérez M, Romero J (2012) Selection of metrics based on the seagrass Cymodocea nodosa and development of a biotic index (CYMOX) for assessing ecological status of coastal and transitional waters. Estuar Coast Shelf Sci 114: 7–17.
22. García-Marín P, Cabaço S, Hernández I, Vergara JJ, Silva J (2013) Multi-metric index based on the seagrass Zostera noltii (ZoNI) for ecological quality assessment of coastal and estuarine systems in SW Iberian Peninsula. Mar Poll Bull 68: 46–54.
23. Diaz R, Solan M, Valente R (2004) A review of approaches for classifying benthic habitats and evaluating habitat quality. J Environ Manage 73: 165–181.
24. Mateo MA, Romero J, Pérez M, Littler M, Littler D (1997) Dynamics of millenary organic deposits resulting from the growth of the Mediterranean seagrass *Posidonia oceanica*. Estuar Coast Shelf Sci 44: 103–110.
25. Alcoverro T, Manzanera M, Romero J (2001) Annual metabolic carbon balance of the seagrass *Posidonia oceanica*: the importance of carbohydrate reserves. Mar Ecol Prog Ser 211: 105–116.
26. Prado P, Tomas F, Alcoverro T, Romero J (2007) Extensive direct measurements of *Posidonia oceanica* defoliation confirm the importance of herbivory in temperate seagrass meadows. Mar Ecol Prog Ser 340: 63–71.
27. Boudouresque CF, Bernard G, Bonhomme P, Charbonnel E, Diviacco G, et al. (2012) Protection and conservation of *Posidonia oceanica* meadows. Marseilles: RAMOGE, RAC/SPA and Gis Posidonie publ. pp. 202 p.
28. Pergent G, Bazairi H, Bianchi CN, Boudouresque CF, Buia MC, et al. (2012) Mediterranean seagrass meadows: resilience and contribution to climate change mitigation. A short summary. Gland, Málaga: IUCN. 40 p.
29. Mazzuca S, Björk M, Beer S, Felisberto P, Gobert S, et al. (2013) Establishing research strategies, methodologies and technologies to link genomics and proteomics to seagrass productivity, community metabolism, and ecosystem carbon fluxes. Front Plant Sci 4(38): 1–19.
30. Sala E (2004) The past and present topology and structure of Mediterranean subtidal rocky-shore food webs. Ecosystems 7: 333–340.
31. Bellan-Santini D, Picard J (1984) Conséquences de la régression des herbiers de Posidonies pour le benthos animal des milieux méditerranéens. In: Boudouresque CF, Jeudy de Grissac A, Olivier J, editors. International Workshop on *Posidonia oceanica* beds. Marseilles: GIS Posidonie publ. pp. 423–429.
32. Boudouresque CF, Bernard G, Pergent G, Shili A, Verlaque M (2009) Regression of Mediterranean seagrasses caused by natural processes and anthropogenic disturbances and stress: a critical review. Botanica Marina 52: 395–418.
33. Coles R, Grech A, McKenzie L (2013) Seagrass under pressure. Seagrass-Watch 47: 2–6.
34. Ward T, Hegerl E (2003) Marine protected areas in ecosystem-based management of fisheries. Canberra: Commonweath of Australia publ. 66 p.
35. Pikitch EK, Santora C, Babcock EA, Bakun A, Bonfil R, et al. (2004) Ecosystem-based fishery management. Science 305: 346–347.
36. Turrell B (2004) The policy basis of the "Ecosystem Approach" to fisheries management. Norrköping, Sweden: EuroGOOS publ. 21: 1–28.
37. Rice JC (2005) Implementation of the Ecosystem Approach to Fishery Management – asynchronous co-evolution at the interface between science and policy. Mar Ecol Prog Ser 300: 265–270.
38. Tudela S, Short K (2005) Paradigm shifts, gaps, inertia, and political agendas in ecosystem-based fisheries management. Mar Ecol Prog Ser 300: 281–286.
39. Martínez-Crego B, Vergés A, Alcoverro T, Romero J (2008) Selection of multiple seagrass indicators for environmental biomonitoring. Mar Ecol Progr Ser 361: 93–109.
40. Christensen V, Maclean J (eds) (2011) Ecosystem approach to fisheries. A global perpective. Cambridge University Press. 342 p.
41. Kenchington E, Heino M, Nielsen EE (2003) Managing marine genetic diversity: time for action? ICES J Mar Sci 60: 1172–1176.
42. Rosenberg AA, McLeod KL (2005) Implementing ecosystem-based approaches to management for the conservation of ecosystem services. Mar Ecol Prog Ser 300: 270–274.
43. Ruíz JM, Boudouresque CF, Enríquez S (2009) Mediterranean seagrasses. Botanica Marina 52: 369–382.
44. Boudouresque CF, Jeudy de Grissac A (1983) L'herbier à *Posidonia oceanica* en Méditerranée: les interactions entre la plante et le sédiment. J Rech Océanogr 8: 99–122.
45. Gacia E, Granata TC, Duarte CM (1999) An approach to measurement of particule flux and sed-iment retention within seagrass (*Posidonia oceanica*) meadows. Aquat Bot 65: 255–268.
46. Sánchez-González JF, Sánchez-Rojas V, Memos CD (2011) Wave attenuation due to *Posidonia oceanica* meadows. J Hydraul Res 49: 503–514.
47. Infantes E, Orfila A, Simarro G, Terrados J, Luhar M, et al. (2012) Effect of a seagrass (*Posidonia oceanica*) meadow on wave propagation. Mar Ecol Prog Ser 456: 63–72.
48. Molinier R, Picard J (1952) Recherches sur les herbiers de Phanérogames marines du littoral méditerranéen français. Ann Inst Océanogr 27: 157–234.
49. Boudouresque CF, Meinesz A (1982) Découverte de l'herbier de posidonie. Hyères: Parc National de Port-Cros publ. iii + 79 p.
50. López-Sáez JA, López-Merino L, Mateo MÁ, Serrano Ó, Pérez-Díaz S, et al. (2009) Paleoecological potential of marine organic deposits of *Posidonia oceanica*: a case study in the NE Iberian Peninsula. Palaeogeogr Palaeoclimatol Palaeoecol 271: 215–224.
51. Montefalcone M (2009) Ecosystem health assessment using the seagrass *Posidonia oceanica*: a review. Ecological Indicators 9: 595–604.
52. Sartoretto S, Beaucour C (2011) Habitats particuliers de l'infralittoral : Herbier à *Posidonia oceanica* de la sous-région marine Méditerranée occidentale DCSMM/EI/MO. Ministère de l'Écologie, du Développement Durable, des Transports et du Logement DCSMM/EI/EE/MO/2.2.4.c. 14p.
53. Pergent-Martini C, Pergent G (1994) Lepidochronological analysis in the Mediterranean seagrass *Posidonia oceanica*: state of the art and future development. Oceanol Acta 17: 673–681.
54. Fourqurean JW, Duarte CM, Kennedy H, Marbà N, Holmer M, et al. (2012) Seagrass ecosystems as a globally significant carbon stock. Nat Geosci 5:505–509.
55. Boudouresque CF, Crouzet A, Pergent G (1983) Un nouvel outil au service de l'étude des herbiers à *Posidonia oceanica*: la lépidochronologie. Rapp PV Réun Commiss Internatl Explor Sci Médit 28(3): 111–112.
56. Pergent G, Boudouresque CF, Crouzet A, Meinesz A (1989) Cyclic changes along *Posidonia oceanica* rhizomes (lepidochronology): present state and perspectives. PSZN I Mar Ecol 10: 221–230.

57. Thélin I, Mossé A, Boudouresque CF, Lion R (1985) Le benthos littoral d'El Dabaa (Méditerranée, Égypte). II. L'herbier à *Posidonia oceanica*. Rapp PV Réun Commiss Internatl Explor Sci Médit 29(5): 247–248.
58. Pergent G, Pergent-Martini C (1991) Leaf renewal cycle and primary production of *Posidonia oceanica* in the bay of Lacco Ameno (Ischia, Italy) using lepidochronological analysis. Aquat Bot 42: 49–66.
59. Pergent G, Pergent-Martini C, Boudouresque CF (1995) Utilisation de l'herbier à *Posidonia oceanica* comme indicateur biologique de la qualité du milieu littoral en Méditerranée: état des connaissances. Mésogée 54: 3–27.
60. Martín MA, Sánchez Povedo M, Marhuenda E, Martínez JM, Sánchez Lizaso JL (1994) Estudio del estrato hipogeo de la pradera de *Posidonia oceanica* (L.) Delile de la reserva marina de Tabarca. In: Ramos Esplá A, editor. Trabajos de campo en la reserva marina de Tabarca (Alicante). Alicante: Universidad de Alicante publ. pp. 25–31.
61. Boudouresque CF, Jeudy de Grissac A, Meinesz A (1984) Relations entre la sédimentation et l'allongement des rhizomes orthotropes de *Posidonia oceanica* dans la baie d'Elbu (Corse). In: Boudouresque CF, Jeudy de Grissac A, Olivier J, editors. International Workshop on *Posidonia oceanica* Beds. Marseilles: GIS Posidonie publ. pp. 185–191.
62. Di Maida G, Tomasello A, Sciandra M, Pirrotta M, Milazzo M, et al. (2013) Effect of different substrata on rhizome growth, leaf biometry and shoot density of *Posidonia oceanica*. Mar Environ Res 87–88: 96–102.
63. Giovannetti E, Lasagna R, Montefalcone M, Bianchi CN, Albertelli G, et al. (2008) Inconsistent responses to substratum nature in *Posidonia oceanica* meadows: an integration through complexity levels? Chemistry and Ecology 24 (S1): 145–153.
64. Bay D (1984) A field study of the growth dynamics and productivity of *Posidonia oceanica* (L.) Delile in Calvi Bay, Corsica. Aquat Bot 20: 43–64.
65. Pergent-Martini C, Rico-Raimondino V, Pergent G (1994) Primary production of *Posidonia oceanica* in the Mediterranean basin. Mar Biol 120: 9–15.
66. Panayotidis P, Boudouresque CF, Marcot-Coqueugniot J (1981) Microstructure de l'herbier de *Posidonia oceanica* (Linnaeus) Delile. Botanica Marina: 115–124.
67. UNEP-MAP-RAC/SPA (2011) Draft guidelines for the standardization of mapping and monitoring methods of marine magnoliophyta in the Mediterranean. Working document, prepared by C Pergent-Martini for the tenth meeting of Focal Points for SPAs, Marseilles, France, 17–20 May 2011, UNEP(DEPI)/MED WG 359/9. Tunis: RAC/SPA Publ. 59 p.
68. Boudouresque CF, Verlaque M (2007) Ecology of *Paracentrotus lividus*. In: Lawrence J, editor. Edible sea urchins: biology and ecology, Second edition. Amsterdam: Elsevier. pp. 243–285.
69. Boudouresque CF, Bernard G, Bonhomme P, Charbonnel E, Le Diréach L, et al. (2007) Monitoring methods for *Posidonia oceanica* seagrass meadows in Provence and the French Riviera. Sci Rep Port-Cros Natl Park 22: 17–38.
70. Short FT, McKenzie LJ, Coles RG, Vidler KP (2002) SeagrassNet manual for scientific monitoring of seagrass habitat. Cairns: QDPI, QFS. 56 p.
71. Boudouresque CF, Verlaque M (2013) *Paracentrotus lividus*. In: Lawrence JM, editor. Sea urchins: Biology and ecology, Third Edition. Amsterdam: Elsevier. pp. 297–327.
72. van der Ben D (1971) Les épiphytes des feuilles de *Posidonia oceanica* Delile sur les côtes françaises de la Méditerranée. Mém Inst R Sci Nat Belg 168 : 1–101 + 6 tables.
73. Giovannetti E, Montefalcone M, Morri C, Bianchi CN, Albertelli G (2010) Early warning response of *Posidonia oceanica* epiphyte community to environmental alterations (Ligurian Sea, NW Mediterranean). Mar Poll Bull 60 (7): 1031–1039.
74. Mabrouk L, Hamza A, Ben Brahim M, Bradai MN (2013) Variability in the structure of epiphyte assemblages on leaves and rhizomes of *Posidonia oceanica* in relation to human disturbances in a seagrass meadow of Tunisia. Aquat Bot 108: 33–40.
75. Pérès JM, Picard J (1964) Nouveau manuel de bionomie benthique de la Mer Méditerranée. Rec Trav Stat Mar Endoume 31(47): 3–137.
76. Vicente N, De Gaulejac B (1993) *Pinna nobilis* L. (Mollusque Bivalve), indicateur biologique du littoral méditerranéen. In : Boudouresque CF, Avon M, Pergent-Martini C, editors. Qualité de milieu marin. Indicateurs biologiques et physico-chimiques. Marseilles: GIS Posidonie publ. pp. 141–150.
77. García-March JR, García-Carrascosa AM, Peña Cantero AL, Wang YG (2007) Population structure, mortality and growth of Pinna nobilis Linnaeus, 1758 (Mollusca, Bivalvia) at different depths in Moraira Bay (Alicante, Western Mediterranean). Mar Biol 150: 861–871.
78. García-March JR, Vicente N (2011) *Pinna nobilis* (Mollusca, Bivalvia) indicateur biologique de la qualité du littoral méditerranéen: densité des populations. In: Francour P, Bodilis P, Arceo H, editors. Actes du Congrès 2011 de la Société Zoologique de France. Nice: Université de Nice publ. pp. 80–81.
79. Urra J, Ramirez AM, Marina P, Salas C, Gofas S, et al. (2013) Highly diverse molluscan assemblages of *Posidonia oceanica* meadows in northwestern Alboran Sea (W Mediterranean): seasonal dynamics and environmental drivers. Estuar Coast Shelf Sci 117: 136–147.
80. Naranjo SA, Carball JL, Garcia-Gomez JC (1996) Effects of environmental stress on ascidian populations in Algeciras Bay (southern Spain). Possible marine bioindicators? Mar Ecol Progr Ser 144: 119–131.
81. Bianchi CN, Bedulli D, Morri C, Occhipinti Ambrogi A (1989) L'herbier de Posidonies: écosystème ou carrefour éco-éthologique? In: International Workshop on Posidonia Beds (CF . Boudouresque, A . Meinesz, E . Fresi & V . Gravez, eds). GIS Posidonie, Marseille 2: 257–272.
82. Romero J, Pergent G, Pergent-Martini C, Mateo M.A, Régnier C (1992). The detritic compartment in a *Posidonia oceanica* meadow: litter features, decomposition rates and mineral stocks. Mar Ecol PSZNI 13(1): 69–83.
83. Mezali K, Chekaba B, Zupo V, Asslah B (2003) Comportement alimentaire de cinq espèces d'holothuries aspidochirotes (*Holothuria*: Echinodermata) de la presqu'île de Sidi-Fredj (Algérie). Bull Soc Zool Fr 128: 49–62.
84. Verlaque M (1981) Preliminary data on some *Posidonia* feeders. Rapp PV Réun Commiss Internatl Explor Sci Méditerr 27(2): 201–202.
85. Giraud G (1979) Sur une méthode de mesure et de comptage des structures foliaires de *Posidonia oceanica* (Linnaeus) Delile. Bull Mus Hist Nat Marseille 39: 33–39.
86. Peirano A, Niccolai I, Mauro R, Bianchi CN (2001) Seasonal grazing and food preference of herbivores in a *Posidonia oceanica* meadow. Scientia Marina 65 (4): 367–374.
87. Francour P (1997) Fish assemblages of *Posidonia oceanica* beds at Port-Cros (France, NW Mediterranean): assessment of composition and long term fluctuations by visual census. PSZN I Mar Ecol 18: 157–173.
88. Bell JD, Harmelin-Vivien ML (1982) Fish fauna of French Mediterranean *Posidonia oceanica* seagrass meadows. 1. Community structure. Téthys 10: 337–347.
89. Harmelin-Vivien M (1982) Ichtyofaune des herbiers de posidonies du Parc national de Port-Cros: I. Composition et variations spatio-temporelles. Trav Sci Parc Natl Port-Cros 8: 69–92.
90. Francour P, Harmelin-Vivien M, Harmelin JG, Duclerc J (1995) Impact of *Caulerpa taxifolia* colonization on the littoral ichthyofauna of north-western Mediterranean Sea: preliminary results. Hydrobiologia 300/301: 345–353.
91. Harmelin-Vivien M, Francour P (1992) Trawling or visual censuses? Methodological bias in the assessment of fish populations in seagrass beds. PSZN I Mar Ecol 13: 41–51.
92. Guyot I (1990) Le cormoran huppé de Corse: biologie et interactions avec la pêche professionnelle. Trav Sci Parc Nat Rég Rés Nat Corse 28: 1–40.
93. Morat F (2007) Régime alimentaire de la population de cormoran huppé de Méditerranée (*Phalacrocorax aristotelis desmarestii*) de Riou. Marseilles: CEEP and Centre d'Océanologie de Marseille. 23 + 6 p.
94. Morat F, Mante A, Drunat E, Dabat J, Bonhomme P, et al. (2011) Diet of the Mediterranean European shag, *Phalacrocorax aristotelis desmaretii*, its ecological significance and interaction with local fisheries in the Riou Archipelago (Marseilles, France). Vie Milieu 61: 77–86.
95. Francour P, Thibault JC (1995) Le régime alimentaire du balbuzard pêcheur (*Pandion haliaetus*) dans la Réserve naturelle de Scandola. Trav Sci Parc Nat rég Rés Nat. Corse 53: 11–27.
96. Francour P, Thibault JC (1996) The diet of breeding osprey *Pandion haliaetus* on Corsica: exploitation of a coastal marine environment. Bird Study 43: 129–133.
97. Thibault JC, Bretagnolle V, Dominici JM (2001) Le balbuzard pêcheur en Corse. Ajaccio: Alain Piazzola. 184 p.
98. Pergent G, Romero J, Pergent-Martini C, Mateo MA, Boudouresque CF (1994) Primary production, stocks and fluxes in the Mediterranean seagrass *Posidonia oceanica*. Mar Ecol Prog Ser 106: 139–146.
99. Mateo MA, Romero J (1997) Detritus dynamics in the seagrass *Posidonia oceanica*: elements for an ecosystem carbon and nutrient budget. Mar Ecol Prog Ser 151: 43–53.
100. Pergent G, Rico-Raimondino V, Pergent-Martini C (1997) Fate of primary production in *Posidonia oceanica* meadows of the Mediterranean. Aquat Bot 59: 307–321.
101. Mateo MA, Sánchez-Lizaso JL, Romero J (2002) *Posidonia oceanica* 'banquettes': a preliminary assessment of the relevance for meadow carbon and nutrient budget. Estuar Coast Shelf Sci 56: 85–90.
102. Macquart-Moulin C (1968) Les amphipodes bentho-planctoniques du golfe de Marseille. Analyse des captures faites au cours de pêches planctoniques nocturnes régulières (années 1963–1964). Rec Trav Stat Mar Endoume 43(59): 311–332.
103. Harmelin-Vivien ML, Harmelin JG, Leboulleux V (1995) Microhabitat requirements for settlement of juvenile sparid fishes on Mediterranean rocky shores. Hydrobiologia 300–301: 309–320.
104. Jiménez S, Cano R, Bayle J, Ramos A, Sánchez Lizaso JL (1996) Las praderas de *Posidonia oceanica* (L.) Delile como zona de protección de juveniles de especies de interés comercial. Real Soc Esp Hist Nat, tomo extraordinario: 375–378.
105. UNEP-MAP-RAC/SPA (2011) Proposal of standard methods for inventorying and monitoring coralligenous and rhodoliths populations. Working document, prepared by G Pergent for the tenth meeting of Focal Points for SPAs, Marseilles, France, 17–20 May 2011, UNEP(DEPI)/MED WG 359/10. Tunis: RAC/SPA Publ. 23 p.
106. Dalkey N, Helmer O (1963) An experimental application of the Delphi method to the use of experts. Manage Sci 9: 458–467.
107. Ferrari B, Raventos N, Planes S (2008) Assessing effects of fishing prohibition on *Posidonia oceanica* seagrass meadows in the Marine Natural Reserve of Cerbère-Banyuls. Aquatic Botany 88: 295–302.
108. Astruch P, Boudouresque CF, Bonhomme D, Goujard A, Antonioli PA, et al. (2012) Mapping and state of conservation of benthic marine habitats and assemblages of Port-Cros national Park (Provence, France, northwestern Mediterranean Sea). Sci Rep Port-Cros Natl Park 26: 45–90.

109. Bazin N, Imbert M (2012) Le cormoran huppé de Méditerranée *Phalacrocorax aristotelis desmarestii.* Etat des connaissances et de la conservation actualisé des populations nicheuses des petites îles de Méditerranée. Marseilles: Initiative PIM. 18 p.
110. Romero J, Pérez M, Alcoverro T (2012) L'alguer de *Posidonia oceanica* de les illes Medes: més de trenta anys d'estudi. In: Quintana X, Hereu B, editors. El fons marí de les illes Medes i el Montgrí: quatre dècades de recerca per la conservatió. Toroella de Montrgrí: Càtedra d'Ecosistems Marins Mediterranis. pp. 81–100.
111. Rouanet E, Astruch P, Bonhomme D, Bonhomme P, Rogeau E, et al. (2012) Suivi de l'herbier de posidonie de la passe de Bagaud, impact de l'ancrage (Parc national de Port-Cros, Var, France). Marseilles: GIS Posidonie publ. 82 p.
112. Boudouresque CF, Charbonnel E, Meinesz A, Pergent G, Pergent-Martini C, et al. (2000) A monitoring network based on the seagrass *Posidonia oceanica* in the nortwestern Mediterranean Sea. Biol mar medit 7 (2): 328–331.
113. Moreno D, Aguilera PA, Castro H (2001) Assessment of the conservation status of seagrass (*Posidonia oceanica*) meadows: implications for monitoring strategy and the decision-making process. Biol Conservation 102: 325–332.
114. Montefalcone M, Albertelli G, Morri C, Bianchi CN (2007) Urban seagrass: status of *Posidonia oceanica* facing Genoa city waterfront (Italy) and implications for management. Mar Poll Bull 54: 206–213.
115. Van Hoey G, Borja A, Birchenough S, Buhl-Mortensen L, Degraer S, et al. (2010) The use of benthic indicators in Europe: from the Water Framework Directive to the Marine Strategy Framework Directive. Mar Poll Bull 60: 2187–2196.
116. Martínez-Crego B, Alcoverro T, Romero J (2010) Biotic indices for assessing the status of coastal waters: a review of strengths and weaknesses. J Environm Monit 12 (5): 1013–1028.
117. Pergent-Martini C, Pasqualini V, Ferrat L, Pergent G (2006) Ecological data in Integrated Coastal Zone Management: case study of *Posidonia oceanica* meadows along the Corsican coastline (Mediterranean Sea). Environ Manage 38: 889–895.
118. Parravicini V, Rovere A, Vassallo P, Micheli F, Montefalcone M, et al. (2012) Understanding relationships between conflicting human uses and coastal ecosystem status: a geospatial modelling approach. Ecol Indic 19: 253–263.
119. Meinesz A, Lefevre JR, Astier JM (1991) Impact of coastal development on the infralittoral zone along the southern Mediterranean shore of continental France. Mar Poll Bull 23: 343–347.
120. Leriche A, Pasqualini V, Boudouresque CF, Bernard G, Bonhomme P, et al. (2006) Spatial, temporal and structural variations of a *Posidonia oceanica* seagrass meadow facing human activities. Aquat Bot 84: 287–293.
121. Mascaró O, Alcoverro T, Dencheva K, Díez I, Gorostiaga JM, et al. (2013) Exploring the robustness of macrophyte-based classification methods to assess the ecological status of coastal and transitional ecosystems under the Water Framework Directive. Hydrobiologia 704: 279–291.

15

Vanishing Clams on an Iberian Beach: Local Consequences and Global Implications of Accelerating Loss of Shells to Tourism

Michał Kowalewski[1]*, Rosa Domènech[2], Jordi Martinell[2]

1 Florida Museum of Natural History, University of Florida, Gainesville, Florida, United States of America, 2 Institut de Recerca de la Biodiversitat and Departament d'Estratigrafia, Paleontologia i Geociències Marines, Facultat de Geologia, Universitat de Barcelona, Barcelona, Spain

Abstract

Multi-decadal increase in shell removal by tourists, a process that may accelerate degradation of natural habitats, was quantified via two series of monthly surveys, conducted thirty years apart (1978–1981 and 2008–2010) in one small embayment on the Mediterranean coast of the Iberian Peninsula. Over the last three decades, the local tourist arrivals have increased almost three-fold (2.74), while the area has remained unaffected by urban encroachment and commercial fisheries. During the same time interval the abundance of mollusk shells along the shoreline decreased by a comparable factor (2.62) and was significantly and negatively correlated with tourist arrivals ($r = -0.52$). The strength of the correlation increased when data were restricted to months with high tourist arrivals ($r = -0.72$). In contrast, the maximum monthly wave energy (an indirect proxy for changes in rate of onshore shell transport) was not significantly correlated with shell abundance ($r = 0.10$). Similarly, rank dominance of common species, drilling predation intensity, and body size-frequency distribution patterns have all remained stable over recent decades. A four-fold increase in global tourist arrivals over the last 30 years may have induced a comparable worldwide acceleration in shell removal from marine shorelines, resulting in multiple, currently unquantifiable, habitat changes such as increased beach erosion, changes in calcium carbonate recycling, and declines in diversity and abundance of organisms, which are dependent on shell availability.

Editor: Pauline Ross, University of Western Sydney, Australia

Funding: The study was supported by the research project CGL2010-15047 (Spanish Science and Innovation Ministry) and by the Jon A. and Beverly L. Thompson Foundation (University of Florida). The funders had no role in study design, data collection and analysis, decision to publish, or preparation of the manuscript.

Competing Interests: The authors have declared that no competing interests exist.

* E-mail: kowalewski@ufl.edu

Introduction

Local and global environmental impacts of tourism [1,2,3,4] are intensifying due to rapidly expanding human populations [5]. Marine shorelines are particularly vulnerable because of the massive residential shift toward coastlines [6] and the fact that seashores remain among the most alluring tourist destinations [7,8]. Tourists and single-day visitors *sensu* [9] affect beaches in diverse direct ways, including trampling, vehicle use, camping, shellfish harvesting, beach grooming, and many other factors [10,11,12]. However, long-term studies measuring human impacts on shoreline habitats are scant and fragmentary [10,13]. In particular, the removal of dead shell remains by tourists represents one of the most understudied and least understood processes associated with human activities along marine shorelines.

Rigorous assessments of shell removal by tourists are needed because skeletal materials left behind by dead organisms perform many important environmental and ecosystem services. In pristine coastal habitats shells can occur in great abundance [14,15] and serve multiple functions, from beach stabilization [16] to building materials for bird nests [17]. Also, shells may be intermittently submerged or transported into subtidal settings where they can provide various ecosystem services, including shelter for diverse endobiolithic algae [18], substrate for seagrass meadows [19], colonization sites for encrusting organisms [20], or armored protection for hermit crabs and fish [21,22,23]. Shell material is also continuously dissolved [24] in most coastal areas resulting in elemental recycling back into the global marine reservoir. The tourism-related removal of shell material from shorelines may thus have diverse environmental consequences.

Diverse processes can induce seasonal or decadal changes in shell abundance along shorelines and multiple hypotheses can be postulated *a priori* to explain variation in shell abundance through time. First, changes in local hydrodynamics may alter the magnitude of onshore transport of shell material [25,26]. This model predicts a significant correlation between proxy measures of onshore transport (e.g., wave energy) and shell abundance. Second, changes in ecosystem structure and population dynamics (especially spawning and mortality patterns) may induce variation in local shell accumulation rates, which may be partly controlled by input of newly dead specimens [27]. This model predicts seasonal or decadal-scale correlations between temporal ecosystem changes and trends in shell abundance and also implies possible changes in rank abundance of dominant species or other ecological proxies (e.g., frequency of predation events and per-species size-frequency distributions). Finally, as postulated here, changes in tourism activities and tourism-related beach management practices may be responsible for changes in shell abundance.

This last model predicts an inverse seasonal (or decadal) correlation between shell abundance and tourist arrivals.

In this study, local changes in shell abundance observed over the last 30 years were evaluated for one Mediterranean beach using two quantitative datasets from the same embayment, one collected more than 30 years ago and one collected in recent years. During a time interval of three years (7/1978 - 7/1981), quantitative surveys of shell material were conducted monthly along Llarga Beach Salou, Catalonia [28]. The same shoreline transects were resurveyed 30 years later (2008–2010). Because the sampled beach has not been altered considerably over the last 30 years, the resulting comparative data should be particularly suitable for testing the hypothesis that an increase in local tourism leads to the accelerated shell removal. Specifically, we evaluated the three models outlined above regarding changes in shell abundance in a local shoreline habitat. We tested for correlatives between the intensity of local tourism and shell abundance at two temporal scales: intra-annual/seasonal and multi-decadal. Within the same reference timeframe, we evaluated putative roles of other causative processes unrelated to tourism that could have led to shifts in shell abundance, either seasonally or through time, and explored global implications of the tourism-related removal of shells from coastal environments.

Our focus on removal of empty shells due to tourism in an area that has remained relatively unchanged differs in goals from studies targeting regions that experienced substantial ecosystem degradation through time [29]. Also, previous projects aimed at assessing the impact of human activities on live shellfish, including subsistence and recreational harvesting [30,31], collecting for curio trade [32], bait collecting [33], or inadvertent trampling of live organisms [33]. Instead, we focused here on activities that may be intuitively perceived as least harmful: the impact of beachcombing (leisurely collecting, inadvertent trampling, use of recreational vehicles, etc.) on empty seashells scattered along marine shorelines.

Figure 1. The study area. (a) A location map of the Catalonia Province; (b) A physiographic map of the Catalonia Province (non-copyrighted map, courtesy of Marine Geosciences Group, University of Barcelona); (c) A close-up map of the Llarga Beach (indicated with a blue arrow) and adjacent areas (image generated from an open access website: http://www.openstreetmap.org/); (d) an aerial photograph of Llarga Beach (flight # 10173, March 1, 1976; ©Cartographic Institute of Catalonia; reproduced with permission of the Cartographic Institute of Catalonia); (e) an aerial photograph of Llarga Beach (flight # 2011072000290012, February 18, 2011; ©Cartographic Institute of Catalonia; reproduced with permission of the Cartographic Institute of Catalonia).

Materials and Methods

Ethics Statement

The Llarga Beach is not included on the list of sites of natural interest protected by law and the endangered mollusk taxa (*Lithophaga lithophaga* (Linnaeus 1758) and *Pinna nobilis* Linnaeus 1758) have not been reported at the sampled locality. Consequently, the field study did not involve endangered or protected species. Live specimens were not collected in this study and permits were not required to collect modern shell material for scientific research in the study area [34]. Shell data used in this study have been archived as a PLoS One online-access appendix (Table S1).

Study Area

Llarga Beach (*Platja Llarga*) is a sandy beach ~600 m in length and ~25 m in width. It is bounded by a Jurassic limestone to the west and delineated by Pleistocene eolian sandstone outcrops in the central and eastern part of the beach (Fig. 1). None of those rock units contribute shells that could be confused with modern shell material. The fossil record of the Jurassic limestone in the area is represented by scarce belemnite guards, ammonites molds, and fragments of oyster shells. The Pleistocene sandstone contains small gastropods and bivalves readily distinguishable from modern materials based both on taxonomic and taphonomic (i.e., alterations due to fossilization processes) criteria [28].

The satellite images indicated that beach topography, its aerial extent, and vegetation have remained virtually unchanged (Fig. 1) over the 30 years that separated the two intervals of monthly surveys reported here. Also, the embayment has not been altered by direct urban encroachment, commercial activities, fisheries, and shellfish harvesting, which all concentrate in other areas of the Mediterranean coast of Spain. Shellfish fishery that could affect locally common bivalves such as *Chamelea gallina* (Linnaeus, 1758) and *Donax trunculus* (Linnaeus, 1758) has been intermittent and mostly concentrated in other areas located much farther south (90 km) or much further north (260 km) from the study area (Castelló, F., 2012, pers. comm).

Llarga Beach has not changed notably in terms of local weather and hydrodynamics (Fig. 2): neither average monthly temperature (Fig. 2A) nor wave height (Fig. 2B) changed substantially over those 30 years. In contrast, the area has seen a nearly three-fold increase in tourist arrivals across all months (see below for data sources). A median number of local tourist arrivals increased by a factor of 2.74 over the last 30 years, while relative monthly changes in tourist arrivals have remained virtually identical (Fig. 2C).

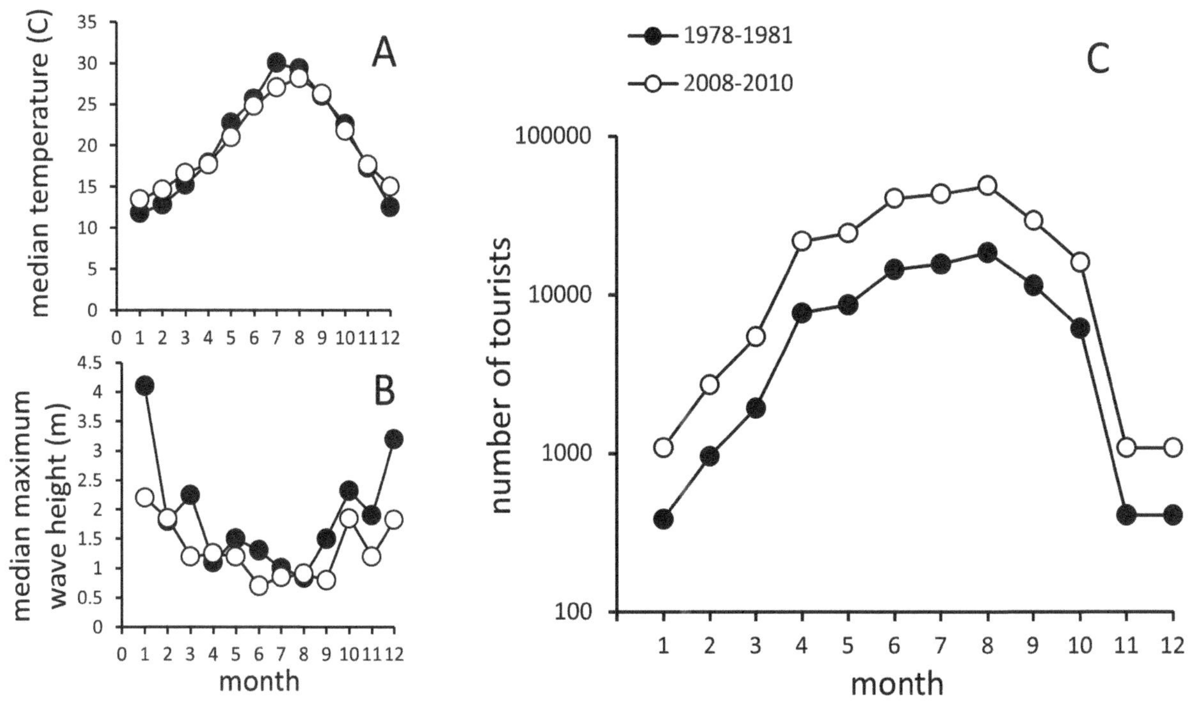

Figure 2. Local temperature, wave height and tourist arrival data for the studied time intervals. (a) Median monthly air temperature recorded at the Meteorological Station of Reus Airport (data provided by Servei Meteoròlogic de Catalunya); (b) Median maximum monthly wave height measured at a buoy (Simar-44 [2057048]) located proximally to the study area (publicly available data obtained from http://www.puertos.es; Puertos del Estado, Ministerio de Fomento); (c) A log-linear plot of monthly tourist arrivals for the 1978–1981 and 2008–2010 time intervals (see text for data sources).

In summary, over the last three decades, a small coastal area represented by Llarga Beach has experienced a considerable increase in tourist arrivals, but has remained relatively static otherwise and should provide a useful model system for assessing the localized impact of tourists on abundance of shells in coastal environments.

Shell assemblages

Primary data were derived by systematic transect sampling of mollusk material present on the surface of shoreface zone. Qualitative visual surveys of faunal composition were performed continuously throughout the field work. These qualitative assessments indicated consistently that three species of bivalves (*Chamelea gallina, Donax trunculus,* and *Donax semistriatus* Poli, 1795) dominated local beach assemblages throughout the sampled time interval. Consequently, their combined abundance should represent a robust proxy for abundance of all shell material along the shoreline. Only those three species have been targeted in quantitative analyses reported here. All other species observed during sampling were present intermittently, and only three were present occasionally in notable quantities: *Spisula subtruncata* (da Costa, 1778), *Mactra stultorum* (Linnaeus, 1758), and *Acanthocardia tuberculata* (Linnaeus, 1758). Other bivalve species observed occasionally included *Glycymeris insubrica* (Brocchi, 1714), *Chlamys varia* (Linnaeus, 1758), *Tellina planata* Linnaeus, 1758, *Scrobicularia plana* (Da Costa, 1778) and *Ensis ensis* (Linnaeus, 1758).

The three dominant species included *C. gallina,* which is restricted to the Mediterranean and Black Seas, *D. semistriatus,* which also appears on Atlantic coasts of the Iberian peninsula, and *D. trunculus,* which is known from littoral habitats from the Great Britain to Senegal. In the Mediterranean, all three species occur in shoreface sand communities (SFC *sensu* [35]). Both donacid species occur together exclusively in this habitat, whereas *C. gallina* is also present in subtidal sands down to 20 m depth (SFBC community *sensu* [35]). *C. gallina* is a mobile suspension feeder and the reproductive cycle of this commercially important species has been studied extensively in the Mediterranean [36,37]. *C. gallina* appears to reproduce primarily in the summer (or both spring and summer), with one or two peaks of gamete emission [36]. *D. trunculus* and *D. semistriatus* are detritivorous species. *D. trunculus*, the larger and more abundant of the two species, spawns from March-April to October in unpolluted areas of the southern Mediterranean littoral [38,39].

Both *C. gallina* and *D. trunculus* are a food source for carnivorous gastropods, especially naticid gastropods, as demonstrated by the ubiquitous presence of valves with drillholes characterized by a countersunk shelf (*Oichnus paraboloides* Bromley, 1981). Shells of the common drilling predator (*Natica hebraea* Martyn, (1784)) have been collected sporadically on Llarga Beach.

Sampling Methods

Monthly surveys were conducted during two time intervals: (1) 1978–1981 and (2) 2008–2010. During the first time interval the surveys were conducted every month from July 1978 through July 1981 (37 monthly transects total). Due to time and funding constraints, surveys conducted during the second time interval were more limited and included transects conducted monthly from August 2008 through October 2008 and from July 2009 through June 2010 (15 monthly transects total). For both time intervals, every calendar month was represented by at least one transect. Samples collected during those surveys were retained and are currently housed at the University of Barcelona.

In both time intervals, monthly surveys were carried early in the morning on weekdays near the end of each month along the 300×10 m (3000 m^2) shoreline transect with its endpoints defined by a carbonate cliff at the west end of the beach and a sandstone outcrop located centrally (Fig. 1). All transects were conducted along a 10 m wide swath most proximal to the waterline. Because of microtidal regime, daily tidal oscillations in the region do not exceed 10 cm and show a minimal annual variation [40]. Thus, tidal variation resulted in minimal lateral shifts from one survey to another, when using water line as a reference system.

Quantitative surveys focused on the three dominant species used as a proxy for the overall quantity of coarse biogenic material present on the beach (see above). The surveys were carried out by exhaustive sampling of all specimens found on the surface within the transect area. No subsurface material was counted or collected. Species identity and evidence of predation (i.e., presence/absence of drill holes) were recorded for each specimen. In addition, shell length was measured using digital calipers for specimens of *C. gallina* for each August transect available for each time interval. The resulting variables (relative abundance of dominant species, drilling frequency, and body size-frequency distributions) provided proxies of ecological processes for local benthic communities. This approach is justified here by growing evidence that shell assemblages can approximate local benthic ecosystems with high fidelity [41,42]. Even intertidal shell assemblages, including beach materials, can effectively capture information on local biodiversity [43,44].

Tourist arrivals were estimated using data available from town hall publications and unpublished counts (S. Antón 2013, University of Barcelona, pers. comm.) for lodging facilities directly adjacent to Llarga Beach. The arrival estimates were compiled to assess seasonal changes in tourism, including monthly estimates, seasonal estimates ("high season" represented by summer months and "low season" represented by non-summer months), and annual estimates. All local facilities existed throughout both of the studied time intervals. The only exception was the campground, which was active in 1978–1981 time interval but closed permanently before the second sampling time interval. The campground has been replaced by a private recreational area with a low density of occupation. The number of arrivals is likely an underestimate of the actual number of tourists visiting Llarga Beach because same-day visitors were not accounted for by the metric used here. However, estimates of tourist arrivals were used here only as a relative metric of changes in intensity of tourism across seasons and through time.

Analytical Methods

To test the three models postulated above, repeated sampling surveys of the same transect were conducted monthly across seasons for two time intervals ~30 years apart. This approach generated a series of shell abundance estimates that can be assessed for correlatives against parallel datasets representing potential controlling factors, including corresponding trends in tourist activities, local physical characteristics, and indirect ecological proxies. Whereas additional spatial control for the same time interval (ideally derived from comparable habitats inaccessible to tourists) would have strengthened the research design, such data were not available. Consequently, all causal interpretations proposed below are tentative and should be viewed as initial hypotheses that require further testing in other settings.

The data (Table S1) resulting from the repeated surveys were analyzed by time interval and seasonally within each time interval. Seasonal changes were estimated by averaging monthly or bimonthly data within each time interval (e.g., a bimonthly January-February estimate for the 2008–2010 time interval was computed as an arithmetic mean based on all available 2008–2010 January and February transects). Data were also grouped into non-tourist (September-May) and tourist (June-August) months. Bimonthly averages were used in some analyses to increase sample size because, for certain months, only one transect was available in a given time-interval (e.g., only one February transect for 2008–2010 time interval). The resulting time series were analyzed using raw values. In addition, first differencing was applied in order to detrend time series.

Non-parametric rank-based methods (Spearman rank correlation) and permutation tests were used as a primary tool in statistical testing. These methods were selected because some of the analyzed variables (e.g., wave energy, tourist arrivals) represented notably skewed distributions, sample sizes were small in some cases, and groups varied in sample size. The parametric approach (2-Way Unbalanced ANOVA) was applied in one case to evaluate differences in mean shell size for monthly and decadal data groups. This approach was deemed appropriate because groups had comparable dispersions (standard deviation ranged from 2.45 to 5.07 mm), all distributions were unimodal and did not display pronounced departures from normality (absolute values of skewness and kurtosis were <2 in all cases), and sample sizes were relatively large (see Results section below) further minimizing the effect of slight non-normality that may have affected the data. However, because of notable variation in sample size across groups, an unbalanced ANOVA was employed (GLM procedure, SAS).

A significance level of alpha = 0.05 was used for hypothesis testing and Bonferroni correction for multiple tests was applied when appropriate. As used here, significance values (p) denote probability of Type I Error and should not be misconstrued as indicating likelihood estimates in support of a given null hypotheses. SAS software, PAST freeware, and SAS/IML programming language were used in all statistical analyses.

Results

A comparison of 1978–1981 and 2008–2010 surveys indicated that shells at the shoreline of Llarga Beach (Fig. 3) were almost three times more abundant three decades ago: on average, 1506.5 specimens were recovered per transect in 1978–1981 compared to only 578.3 specimens in 2008–2010. The decline in shell abundance was comparable in magnitude for the tourist season (July–August) (70.1%) and for other months (60.0%). When 1978–1981 and 2008–2010 transect estimates were binned bimonthly (due to small number of transects for some of the winter months, monthly bins were inadequate), there was a remarkable congruence in seasonal changes in shell abundance with notable declines in shell quantity observed in the late spring, summer, and the late fall (Fig. 3B). One notable difference between the two time-intervals was a much steeper decline in average specimen abundance, which was observed only for the July-August months for the 2008–2010 sampling interval (Fig. 3B).

The maximum monthly wave energy [MMWE] for 1978–1981 and 2008–2010 was not correlated significantly with shell abundance [SA] (Spearman $r = 0.10$, $p = 0.46$) and the pattern persisted for first differences (month-to-month changes) [Δ_{SA}] ($r = -0.18$, $p = 0.21$). When MMWE values were downweighted by the length of a time lag between the last day in which wave height was at monthly maximum and the day of sampling [MMWE* = MMWE/#days], correlations remained non-significant: (SA: $r = 0.26$, $p = 0.06$ and Δ_{SA}: $r = -0.23$, $p = 0.10$, respectively). All correlation coefficients were low ($r < 0.3$), the first

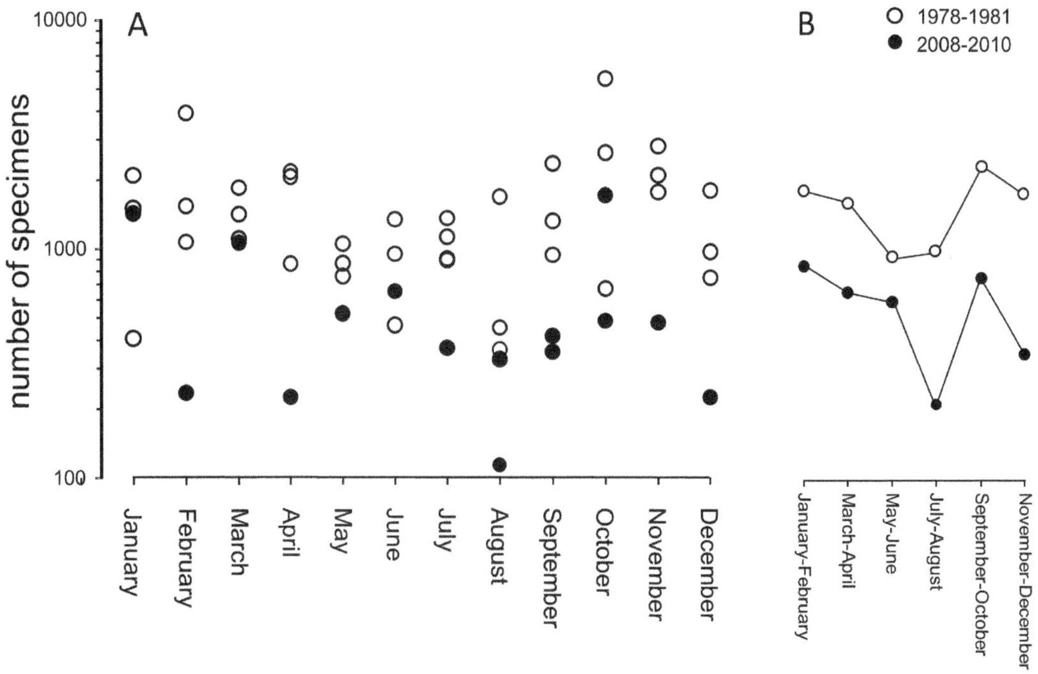

Figure 3. Seasonal changes in abundance of common bivalve species based on transects conducted along the shoreline of the Llarga Beach. (a) A bivariate log-linear plot of monthly changes in abundance of common bivalve species based on transects conducted along the shoreline of the Llarga Beach during 1978–1981 and 2008–2010 sampling intervals. Each point represents a single monthly transect; (b) A comparison of bimonthly changes in average abundance for 1978–1981 versus 2008–2010 sampling intervals.

differences [Δ_{SA}] yielded correlation coefficients with an opposite sign relative to the raw correlations [SA], and none of the tests was significant. When data were subdivided by season, time-interval, or both, correlation coefficients either remained low or were non-interpretable due to excessively small sample sizes.

No dramatic changes in indirect ecological proxies measurable from shell assemblages could be demonstrated across seasons or through time. Whereas the absolute abundance of specimens per transect decreased notably over the last 30 years (Fig. 4), the rank abundance of the three dominant species remained unchanged regardless of the season, with *C. gallina* being the dominant species and *D. semistriatus* the least abundant species (Fig. 4). Similarly, drilling frequency, a proxy for predator-prey interactions between carnivorous gastropods and bivalves, showed consistent patterns through time. *C. gallina* was drilled frequently in both time intervals (48.3% of drilled valves in 1978–1981 and 41.7% of drilled valves in 2008–2010, respectively), *D. semistriatus* was drilled infrequently (8.2% for 1978–1981 and 8.9% for 2008–2010), and *D. trunculus* was drilled rarely (1.3% for 1978–1981 and 4.3% for 2008–2010). Seasonal changes in drilling frequency also persisted through time. For both time intervals and for each species, drilling frequencies were notably lower during summer months. Finally, comparison of size frequency distributions for specimens of *C. gallina* obtained for August transects did not suggest any dramatic shifts in body size between the two time intervals (Fig. 5). The monthly median values for valve length varied in a comparable range for 1978–1980 August transects ($n = 826$, $M_{1978} = 8.11$, $n = 205$, $M_{1979} = 12.48$, $n = 452$, $M_{1980} = 12.30$) and 2008–2010 transects ($n = 259$, $M_{2008} = 10.69$, $n = 46$, $M_{2008} = 9.99$). Moreover, the variation in median shell size was more variable between the transects within the 1978–1980 time interval than across the two time intervals. The maximum difference in medians for 1978–1980 was 3.19 mm (M_{1980} vs. M_{1979}), whereas the maximum difference in medians observed between the two time intervals was 2.58 mm (M_{1978} vs. M_{2008}). When data were grouped by both, time interval and month, the two time intervals did not differ significantly from one another, once among-transect variation within time intervals was considered explicitly (Unbalanced 2-Way Anova, $df = 1$, $MS = 40.27$, $F = 2.85$, $p = 0.09$).

When shell abundance estimates for each transect was compared with corresponding local tourist arrivals, significant negative correlations were observed for monthly data ($n = 51$, $r = -0.52$, $p < 0.0001$) and for data averaged by month within each time interval ($n = 24$, $r = -0.49$, $p = 0.015$). Absolute values of correlation coefficients were relatively modest and the bivariate interrelation was not strong when examined visually for data averaged by month (Fig. 6A). However, the tourist arrivals were distinctly bimodal (note a gap along x-axis of Fig. 6A). When data were restricted to months when tourist arrivals had been high (>5000), a strong negative correlation was observed (Fig. 6B), whether analyzed for monthly data ($n = 32$, $r = -0.72$, $p < 0.0001$) or averaged by month within each time interval ($n = 15$, $r = -0.84$, $p = 0.0001$). For restricted data, tourist arrivals were a reasonable predictor of shell abundance ($r^2 = 0.64$, Reduced Major Axis Regression, $p = 0.0008$, Permutation Test). When data were analyzed separately for each time interval, correlation coefficients remained negative, although not always significant (as may be expected given a loss of statistical power associated with reduced sample size). Correlations between first differences (month-to-month change in shell abundance versus month-to-month change in tourist arrivals) were also negative, but less pronounced and non-significant in most cases. Because adjacent transect estimates were separated on average by 30 days, they did not represent directly adjacent time series datapoints with high potential for strong co-dependences. Consequently, detrending may be an overly conservative and potentially misleading corrective strategy.

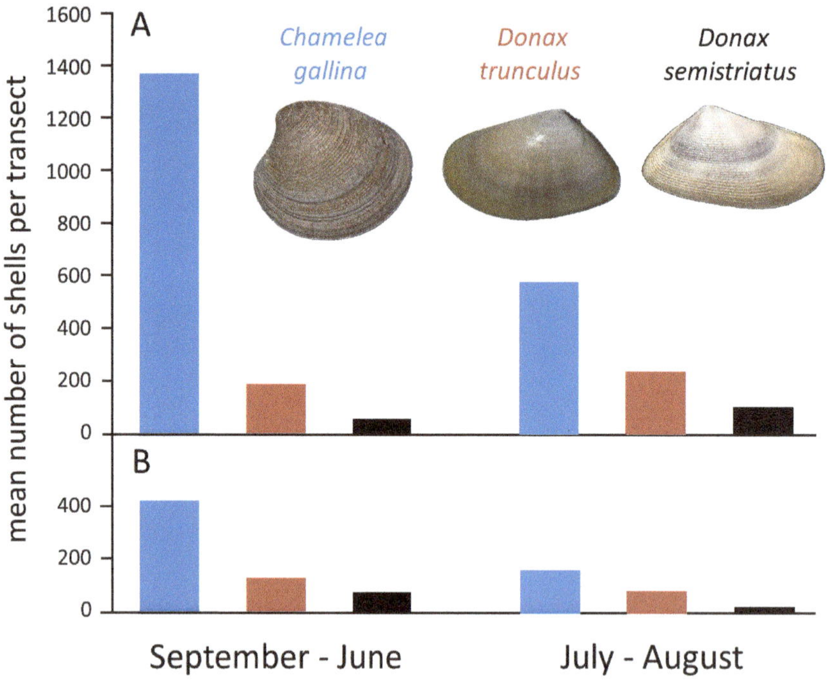

Figure 4. Rank abundance of shells of the three dominant species grouped by tourist seasons for the 1978–1981 and 2008–2010 sampling intervals. Values averaged across all transects conducted during a given time interval and season. (a) Data for 1978–1981 transects; (b) Data for 2008–2010 transects.

When data were grouped by tourist arrivals, the strength of correlation increased as a function of tourist arrivals (Fig. 7). The Spearman correlation coefficient reached the most negative value for data restricted to 28 months with the highest tourist arrivals and fluctuated around 0 for months with the lowest tourist arrivals (Fig. 7).

Discussion

Causes of Decline in Shell Abundance

The three models of shell decline evaluated in this study include trends in hydrodynamics resulting in changes in rates of onshore transport of shell material; ecological shifts in population dynamics and ecosystem characteristics resulting in time-variant shell input; and changes in tourism activities and tourism-related beach management practices resulting in varying rates of shell removal from shoreline areas. Whereas these models do not represent an exhaustive list of causal explanations, many other obvious factors such as changes in beach geomorphology, urban encroachment, or an increase in commercial fisheries are unlikely to apply. Other possible causal drivers are expected to be reflected in patterns evaluated by the three postulated models (e.g., an invasive species could alter rank order of dominant species, thus supporting ecological changes through time).

Seasonal changes in wave energy may result in differential delivery of shell material to the beach [25,45] and decadal-scale changes may consequently produce similar changes over longer time scales. Because wave energy has remained relatively unchanged over the last 30 years and shell abundance did not show any significant correlations with wave height regardless of data groupings and tests, it is unlikely that changes in local hydrodynamic played a notable role in the observed decline in shell abundance.

Population dynamics of the three studied species may have varied seasonally or through time, resulting in a variable input of new shell material to the beach. Because the initial shell input into death assemblages is expected to be primarily controlled by biological productivity [46,47], including spawning patterns, mortality rates, and related locally-controlled processes, seasonal and multi-decadal changes in shell input at Llarga Beach cannot be evaluated directly (bivalve population data for the study area were not available). However, over the last 30 years the same three species have dominated the local shell assemblages and have shown comparable seasonal trends. It is, thus, unlikely that the observed decline in shell abundance in the last three decades was driven by changes in population dynamics. Moreover, for each of the three species size frequency distributions remained comparable through time, suggesting temporal stability in mortality and recruitment patterns. Finally, patterns of drilling attacks by predatory snails (within and across prey species) did not change notably between the two time intervals. Thus, all indirect ecological proxies measurable from the sampled shell assemblages consistently suggested that no major changes took place in local ecosystems over the last 30 years.

A significant negative correlation was found between tourism and shell abundance consistently at several levels: (1) over the last 30 years, shell abundance declined almost three-fold (2.62), a value which is remarkably close to the concurrent increase in local tourist arrivals (2.74); (2) seasonally, the substantial increase in summer tourism was congruent with concurrent decline in shell abundance; and (3) monthly, negative correlation between shells and tourist arrivals was observed, especially for months for which local tourist arrivals are high. The increase in strength of the correlation with increase in tourist arrivals is particularly compelling because it suggests that when tourist activity was high shell abundance decreased. Conversely, when local tourist arrivals were low, the tourism-associated shell loss was less important than

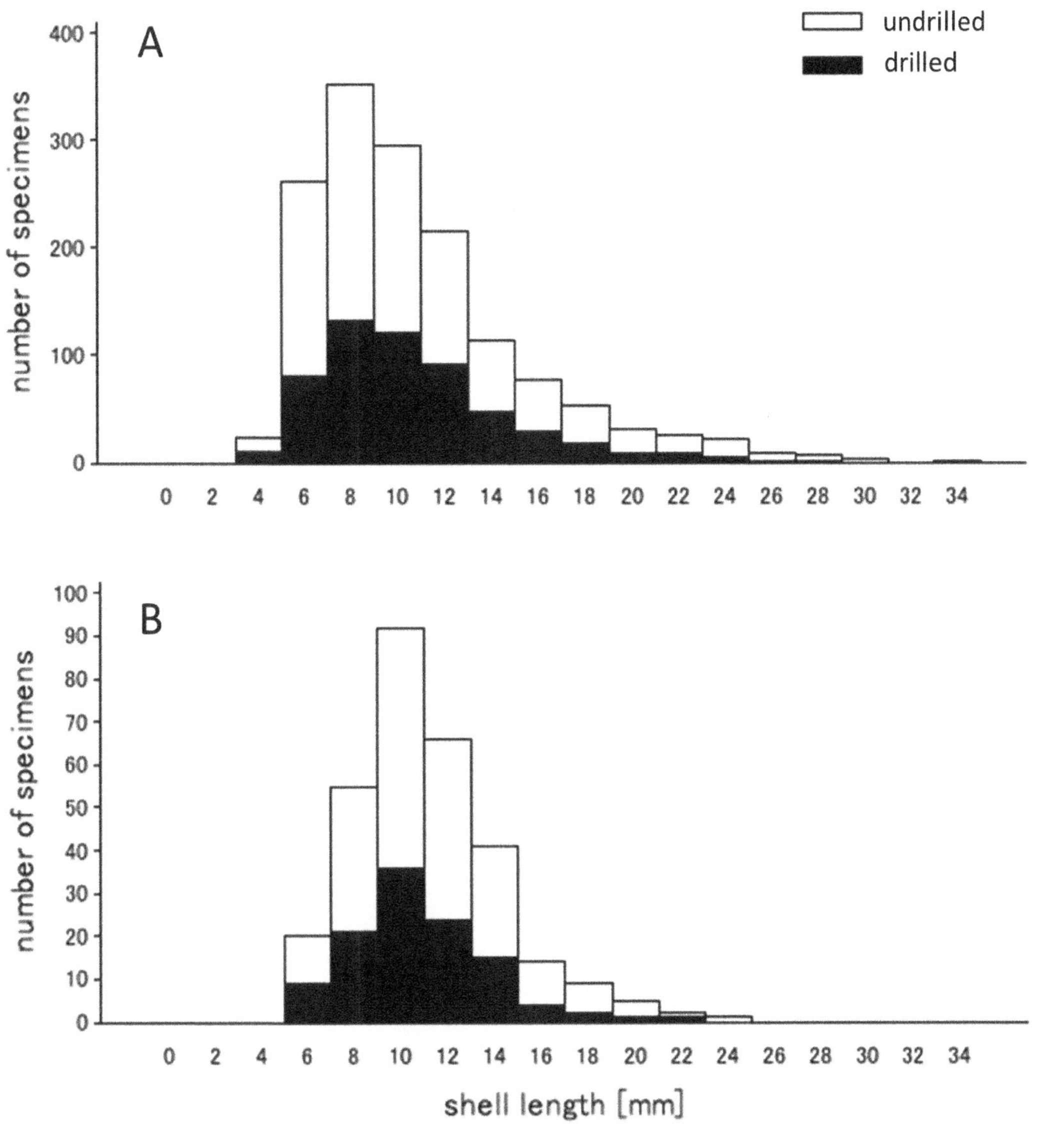

Figure 5. Size-frequency distributions of shells of *Chamelea gallina* from August transects. Drilled and undrilled specimens are differentiated. (a) Data for 1978–1981 time interval; (b) Data for 2008–2010 time interval.

other processes that may have contributed to variation in rates of shell accumulation and shell removal.

Active collecting versus other shell removal processes

The tourism-related processes that contribute to shell removal are certainly not limited to shell collecting, although collecting is an important activity along marine shorelines [48], as also demonstrated by hundreds of published shell guides aimed at leisurely and avocational shell collectors. However, other processes may be also important, including trampling [10,33], recreational clam harvesting [49], use of recreational vehicles [12], sand dredging for beach recovery [10], and other similar activities. In addition, beach maintenance, which can be expected to correlate with tourism intensity, often involves extensive cleaning and grooming of beach areas with heavy equipment [10,29,50,51]. This is the case for Llarga Beach, where maintenance activities, using tractors with rakes, occurred daily during summer months of 2008–2010, but not during the 1978–1981 sampling interval. These maintenance activities may also explain the anomalously high decline observed for the 2008–2010 July-August months when comparing with the 1978–1981 time interval (Fig. 3B).

The relative importance of these various tourism-related processes is impossible to evaluate for Llarga Beach due to lack of relevant data. In fact, we are not aware of any published assessments addressing this issue for dead shell material (there exist numerous studies on subtidal harvesting of live mollusks for curio trading [32,52] or impact of other tourism-related processes on beach and intertidal communities [10,11,12,50,51]). The results reported here suggest that an increase in tourism can trigger a major decrease in shell abundance along shorelines thus providing justification for future studies exploring specific processes that contribute to removal of shells from beaches by tourists.

As mentioned above, other human-related activities such as commercial fisheries are unlikely to have played a significant role in the study area. Although *C. gallina* and *D. trunculus* are commercial species, fisheries along this coast have been operating

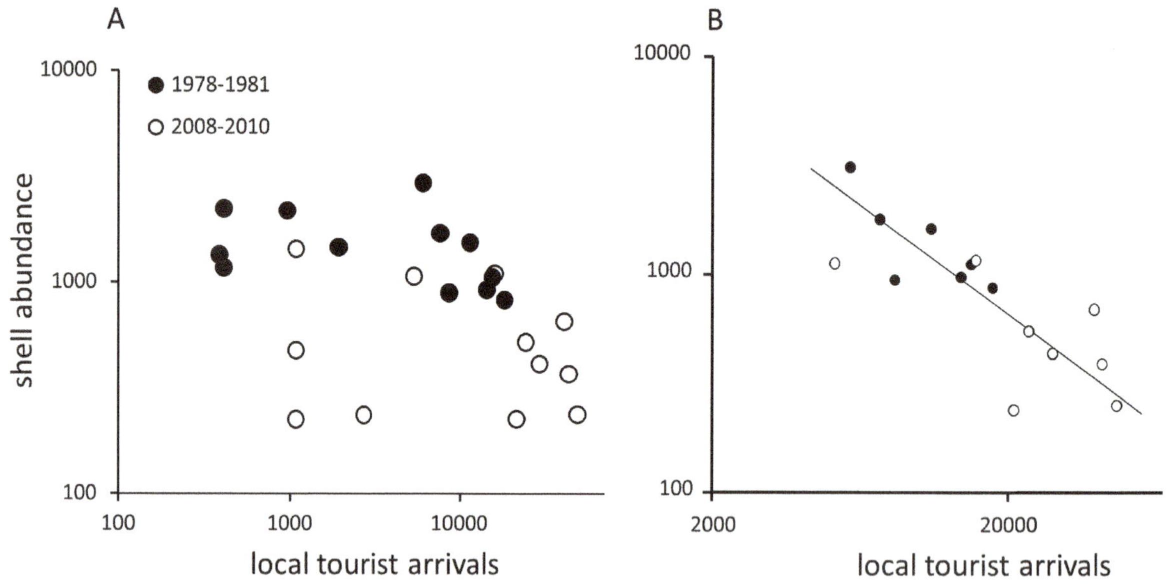

Figure 6. Log-log bivariate plots of tourist arrivals and shell abundance. Each data point represent average for the same months sampled in a given time interval (e.g., all July transects obtained in the 1978–1981 sampling period will be represented by one mean value. (a) All data; (b) Data restricted to months with tourist activity >5000 (a solid line represents a Reduced Major Axis Regression model).

over the last three decades in areas located far away from Llarga Beach (Castelló, F., 2012, pers. comm).

Local Shell Removal as a Proxy for Global Trends

Over the last 30 years, global tourism has grown four-fold, from ~250 million tourist arrivals in 1980 to ~billion arrivals in 2010 (UNWTO Tourism Highlights 2010 by the World Tourism Organization). The nearly three-fold increase in tourist arrivals at Llarga Beach is thus not unusual. Given this increase in intensity and diversity of tourism-related processes in coastal habitats [10,11,12], it is likely that shell abundance has decreased on many marine shorelines, paralleling shell losses due to curio trading [52,53]. In fact, some shell-rich countries that experience intense tourism have already recognized the severity of this problem and explicitly regulate not only the type, but also the quantity, of shell material that visitors are allowed to export out of the country (e.g.,

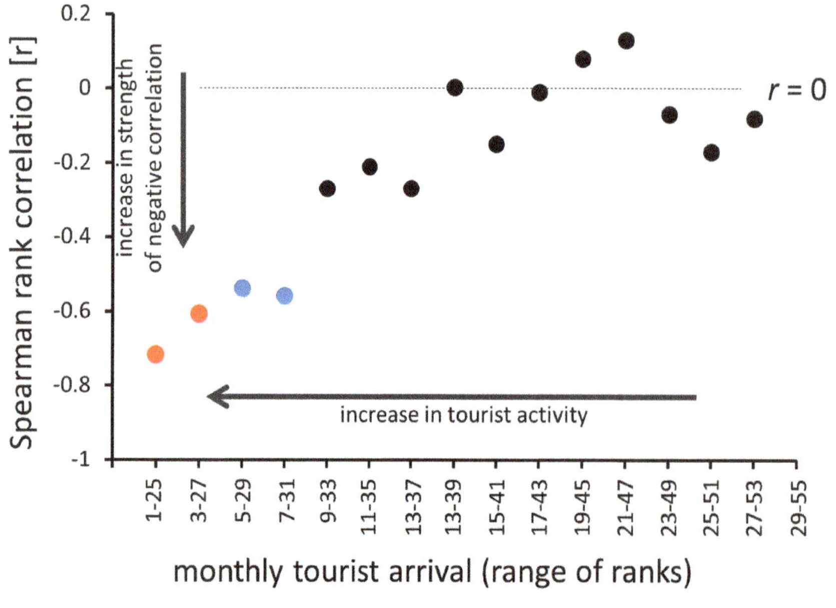

Figure 7. Changes in strength of rank correlation as a function of tourist activity. Monthly estimates are binned into 25-transect groups based on their rank in terms of tourist arrivals. The left endpoint along the x axis (ranks "1–25") represents the dataset restricted to 25 monthly transects with the highest tourist arrivals. The right endpoint (ranks "29–55") represent the 25 months with the smallest numbers of tourists. All restricted datasets are comparable in sample size ($n = \sim 25$ transects) although due to tied ranks, some rank ranges represent 24 or 26 monthly transects. Red symbols are significant at $p<0.001$ and blue symbols at $p<0.01$.

the Commonwealth of Bahamas). However, we lack assessments of impact of tourism on shelly organisms over multi-decadal time scales [10] largely because of the absence of methodologically comparative samples from past decades [29]. Estimates reported here may serve as a starting point toward developing empirical estimates for shell loss due to tourism.

It is likely that the nearly three-fold decline in shell abundance observed at Llarga beach is a conservative estimate compared to other beaches visited by tourists. While a popular tourist area, Llarga Beach is not a highest-tier tourist destination and, as noted above, the increase in tourist arrivals observed for Llarga Beach was well below global estimates for the same time interval. Llarga Beach provides shell material dominated by small, common bivalves that are neither attractive to professional shell collectors nor spectacular enough to attract attention of every casual beachcomber. Compared to other beaches, which may harbor esthetically appealing specimens, Llarga Beach is unlikely to trigger enthusiastic collecting by tourists or motivate intense subtidal harvesting of live fauna for curio trading. Finally, the negative effects of collecting live specimens for subsistence and recreational shellfish harvesting [10,31,54] have been minimal in the study area compared to many other marine shorelines.

Eco-Environmental Consequences

Removal of shell material from shorelines may trigger negative eco-environmental changes. Among others, such changes may include physical, chemical, and biological alterations.

For example, shell material scattered along the shorelines may stabilize sediments by forming pavements that hamper sediment movement [55], although eolian transport of fine fraction may be enhanced locally around coarse objects such as shells [56]. Selective removal of macroscopic shells may change the rates and patterns of sediment erosion along the shoreline.

Similarly, removal of shells potentially changes calcium carbonate budget. The carbonate skeletons of marine macro-organisms such as echinoderms and mollusks tend to be overlooked in modeling the global $CaCO_3$ cycles in the oceans [57], although their role may be significant [58,59]. The importance of mollusks in carbon cycle has been also acknowledged by listing them as a potentially important target for carbon sequestration [59]. As our understanding of the role of mollusks in the global elemental cycling continues to improve, we will also need to account for the removal of large quantities of mollusk shells from the natural cycle by tourism. If the removal of shells due to tourism along shorelines increased three-fold over the last three decades, as this study potentially implies, the tourists may be significantly altering the $CaCO_3$ cycling on our planet.

Removal or destruction of shells, which serve variety of ecosystem functions, may also negatively affect a broad spectrum of organisms. The obvious examples include use of shells for anchoring by seagrass [19], settling for diverse encrusting organisms [20,60,61], and dwelling and mining by endobiolithic algae [18] and bioeroding sponges [62,63]. Hermit crabs and fish also rely heavily on mollusk shells (mostly gastropods) and are often limited by lack of suitable shells [21,22,23]. Shells also serve as building materials for variety of coastal organisms, both terrestrial [17] and marine [64], and perform many other ecosystem services to a diverse array of organisms. It is fair to note here that the increase in shell abundance can also have negative effects. For example, beetles of the genus *Bledius* do not live in sand too rich in shells because of their digging behavior [65,66] and sediments with a high fraction of shell fragments may negatively affect intertidal clams [66,67].

Conclusions

By exploiting a unique opportunity for developing comparative data on shell abundance for one specific coastal area on the Iberian Peninsula, the study demonstrated that shell abundance declined substantially over the last three decades. Moreover, the shell abundance patterns showed strong negative correlations (over multiple time scales) with tourist arrivals suggesting that tourism and tourism-related activities may be a driving force behind the accelerating removal of shells from marine shorelines. The shell loss may have substantial ecological and environmental consequences, although a rigorous quantitative assessment is not yet possible.

There is a growing realization that the most recent fossil record, including surficial skeletal remains found on land and in the sea, may yield diverse data for assessing ecosystem changes [14,15,41,68,69]. This study suggests that shell accumulations found along shorelines may also provide quantifiable information about local consequences and global implications of seasonal and multi-decadal changes in tourism.

Supporting Information

Table S1 Raw data on shell abundance used in this study. Abbreviations: RD - number of right drilled valves; LD- number of left drilled valves; RUD - number of right non-drilled valves; LUD - number of left non-drilled valves; D - total number of drilled valves; UD - total number of non-drilled valves; R - total number of right valves; L - total number of left valves; Total - Total number of valves.

Acknowledgments

The authors thank Dr. F. Castelló (Barcelona University) for shell fishery information in the area, Dr. S. Antón (Rovira i Virgili University) for data on tourist occupation, Dr. A. Rius (Servei Meteorològic de Catalunya) for weather records, the Marine Geosciences Group (Barcelona University) and Institut Cartogràfic de Catalunya for providing images used in Fig. 1. We thank two anonymous reviewers and P. Ross (PLoS One Academic Editor) for numerous constructive criticisms that substantially improved the clarity and quality of this report.

Author Contributions

Conceived and designed the experiments: MK RD JM. Performed the experiments: MK RD JM. Analyzed the data: MK RD JM. Contributed reagents/materials/analysis tools: MK RD JM. Wrote the paper: MK RD JM.

References

1. Mathieson A, Wall G (1982) Tourism: economic, social and physical impacts. London: Longman. 208 p.
2. Hunter C, Green H (1995) Tourism and the environment: a sustainable relationship? Routledge: London. 256 p.
3. Wong PP (2004) Environmental impacts of tourism. In: Lew C, Hall M, Williams AM, editors. A companion to tourism. Malden: Blackwell. pp. 450–461.
4. Gössling S (2009) Global environmental consequences of tourism. Global Envir Change 12: 283–302.
5. Roberts L (2011) 9 Billion? Science 333: 540–543.
6. Roberts CM, Hawkins JP (1999) Extinction risk in the sea. TREE 14: 241–246.
7. Lencek L, Bosker G (1999) The beach: the history of paradise on Earth. Pimlico: London: 310 p.

8. Preston-Whyte R (2004) The beach as a liminal space. In: Lew C, Hall M, Williams AM, editors. A companion to tourism. Malden: Blackwell. pp. 349–359.
9. Smith SLJ (2004) The measurement of global tourism: old debates, new consensus, and continuing challenges. In: Lew C, Hall M, Williams AM, editors. A companion to tourism. Malden: Blackwell. pp. 25–35.
10. Defeo O, McLachlan A, Schoeman DS, Schlacher TA, Dugan J, et al. (2009) Threats to sandy beach ecosystems: a review. Estuar Coast Shelf Sci 81:1–12.
11. Schlacher TA, Dugan J, Schoeman DS, Lastra M, Jones A, et al. (2007) Sandy beaches at the brink. Divers Distrib 13: 556–560.
12. Davenport J, Davenport JL (2006) The impact of tourism and personal leisure transport on coastal environments: a review. Estuar Coast Shelf Sci 67: 280–292.
13. Gössling S, Kunkel T, Schumacher K, Zilger M (2004) Use of molluscs, fish, and other marine taxa by tourism in Zanzibar, Tanzania. Biodiver Conserv 13: 2623–2639.
14. Kowalewski M, Serrano GEA, Flessa KW, Goodfriend GA (2000) Dead delta's former productivity: two trillion shells at the mouth of the Colorado River. Geology 28: 1059–1062.
15. Kowalewski M (2009) The youngest fossil record and conservation biology: Holocene shells as eco-environmental recorders. In: Dietl GD, Flessa, KW, editors. Conservation paleobiology: using the past to manage for the future. Pal Soc Pap 15: 1–23.
16. Scyphers SB, Powers SP, Heck Jr KL, Byron D (2011) Oyster reef as natural breakwaters mitigate shoreline loss and facilitate fisheries. PlosOne 6(8): e22396. doi:10.1371/journal.pone.0022396.
17. Masio B, Handel SN, Power T (2011) Restoring beaches for Atlantic Coast piping plovers (*Charadius merodius*): a classification and regression tree analysis of nest-site selection. Restor Ecol 19: 194–203.
18. Pawlowska AM, Palinska KA, Piekarek-Jankowska H (2008) Colonisation and bioerosion of marine bivalve shells from the Baltic Sea by euendolithic cyanobacteria: an experimental study. In: Wisshak M, Tapanila L, editors. Current developments in bioerosion. Springer: Berlin. pp. 109–122.
19. Guo H, Pennings SC (2012) Post-mortem ecosystem engineering by oysters habitat for a rare marsh plant. Oecol 170: 789–798.
20. Zuschin M, Baal C (2007) Large gryphaeid oysters as habitats for numerous sclerobionts: a case study from the northern Red Sea. Facies 53: 319–327.
21. Kellog CW (1976) Gastropod shells: a potentially limiting resource for hermit crabs. J Exp Mar Biol Ecol 22: 101–111.
22. Wilson SK (2005) Discarded queen conch (*Strombus gigas*) shells as shelter sites for fish. Mar Biol 147: 179–188.
23. Laidre ME (2011) Ecological relations between hermit crabs and their shell-supplying gastropods: constrained consumers. J Exp Mar Biol Ecol 397: 65–70.
24. Davies DJ, Powell EN, Stanton Jr, RJ (1989) Taphonomic signature as a function of environmental process: shells and shell beds in a hurricane-influenced inlet on the Texas coast. Palaeogeog Palaeoclim Palaeoecol 72: 317–356.
25. Aigner T, Reineck HE (1983) Seasonal variations of wave-base on the shoreface of the barrier island Norderney, North Sea. Sencken Mar 15: 87–92.
26. Flessa KW (1998) Well-traveled cockles: shell transport during the Holocene transgression of the southern North Sea. Geology 26: 187–190.
27. Tomašových A, Fürsich FT, Olszewski TD (2009) Modeling shelliness and alteration in shell beds: variation in hardpart input and burial rates leads to opposing predictions. Paleob 32: 278–298.
28. Martinell J, de Porta J (1982) Observations on the molluscan thanatocoenoses from Platja Llarga (Salou, Spain). Malacol 22: 587–591.
29. Van der Sancia ET, Moolenbeek RG, Hoeksema BW (2009) Decline of the Jakarta Bay molluscan fauna linked to human impact. Mar Pollut Bull 59:101–107.
30. Kyle R, Pearson B, Fielding P, Robertson WD, Birnie SL (1997) Subsistence shellfish harvesting in the Maputuland Marine Reserve in Northern Kwazulu-Natal, South Africa: rocky shore organisms. Biol Conser 82: 183–192.
31. Jimenez H, Dumas P, Léopold M, Ferraris J (2011) Invertebrate harvesting on tropical urban areas: trends and impact on natural populations (New Caledonia, South Pacific). Fish Res 108: 195–204.
32. Dias TL, Neto NAL, Alves RR (2011) Molluscs in the marine curio and souvenir trade in NE Brazil: species composition and implications for their conservation and management. Biodiver. Conserv. 20: 2393–2405.
33. Murray SN, Denis TG, Kido JS, Smith JR (1999) Human visitation and the frequency and potential effects of collecting on rocky intertidal populations in southern California marine reserves. In: Murray SN, et al., editors. Visitation and collecting in intertidal marine reserves. CalCOFI Rep 40. pp. 100–106.
34. Spanish Government (2007) Spanish Law on Natural Heritage and Biodiversity(-Law 42/2007 [14712/2007]) on Natural Heritage and Biodiversity. Bolet Ofic Estado 299: 51275–51327.
35. Pères JM, Picard J (1964) Nouveau Manuel de Bionomie Benthique de la Mer Méditerranée. Rec Trav Station marit Endoume 47: 5–137.
36. Rodríguez de la Rúa A, Prado MA, Bruzón MA (2003) Estudio del ciclo reproductor de *Chamelea gallina* (L., 1758) (Mollusca: Bivalvia) en tres poblaciones del litoral andaluz. Bolet Inst Español Oceanogr 19: 57–63.
37. Dalgic G, Karayucel S, Okumus I (2009) Reproduction cycle of striped venus *Chamelea gallina* from the Black Sea Coast of Turkey. J Anim Veterin Adv 8: 2009–2012.
38. Tirado C, Salas C (1999) Reproduction of *Donax venustus* Poli 1795, *Donax semistriatus* Poli 1795 and intermediate morphotypes (Bivalvia: Donacidae) in the littoral of Málaga (Southern Spain). Mar Ecol 20: 111–130.
39. Tlili S, Métais I, Ayache N, Boussetta H, Mouneyrac C (2011) Is the reproduction of *Donax trunculus* affected by their sites of origin contrasted by their level of contamination? Chemosphere 84: 1362–1370.
40. Hofrichter R (2004) El Mar Mediterráneo. Fauna-Flora-Ecología. I parte general. Ediciones Omega: Barcelona. 592 p.
41. Kidwell SM (2007) Discordance between living and death assemblages as evidence for anthropogenic ecological change. Proc Nat Acad Sci 104: 17701–17706.
42. Kidwell SM (2008) Ecological fidelity of open marine molluscan death assemblages: effects of post-mortem transportation, shelf health, and taphonomic inertia. Lethaia 41: 199–217.
43. Warwick RM, Light J (2002) Death assemblages of molluscs on St Martin's Flats, Isles of Scilly: a surrogate for regional biodiversity? Biodivers Conserv 11: 99–112.
44. Warwick RM, Turk SM (2002) Predicting climate change, effects on marine biodiversity: comparison of recent and fossil molluscan death assemblages. J Mar Biol Assoc UK 82: 847–850.
45. Yanes Y, Kowalewski M, Romanek CS (2012). Seasonal variation in ecological and taphonomic processes recorded in shelly death assemblages. Palaios 27: 373–385.
46. Olszewski TD (2012) Remembrance of things past: modelling the relationship between species' abundances in living communities and death assemblages. Biol Let 8: 131–134.
47. Powell EN (1992) A model for death assemblage formation: Can sediment shelliness be explained? J Mar Res 50: 229–265.
48. Dance SP (1986) A history of shell collecting. EJ Brill: Leiden. 265 p.
49. Griffiths J, Dethier MN, Newsom A, Byers JE, Meyer JJ, et al. (2006) Invertebrate community responses to recreational clam digging. Mar Biol 149: 1489–1497.
50. Kinzelman JL, Whitman RL, Byappanahalli M, Jackson E, Bagley RC (2003) Evaluation of beach grooming techniques on *Escherichia coli* density in foreshore sand at North Beach, Racine, WI. Lake Reserv Manag 19: 349–354.
51. Fanini L, Martin CC, Scapini F (2005) Relationships between the dynamics of two *Talitrus saltator* populations and the impacts of activities linked to tourism. Oceanologia 47: 93–112.
52. Wells SM (1989) Impacts of the precious shell harvest and trade: conservation of rare or fragile resources. In: Caddy J, editor. Marine invertebrate fisheries: their assessment and management. Wiley, New York. pp.
53. Grey M, Blais MA, Vincent ACJ (2005) Magnitude and trends of marine fish curio imports to the USA. Oryx 39: 413–420.
54. McLachlan A, Dugan J, Defeo O, Ansell A, Hubbard D, et al. (1996) Beach clam fisheries. Oceanography and marine biology: an annual review 34: 163–232.
55. Carter RWG (1976) Formation, maintenance and geomorphological significance of an aeolian shell pavement. J Sed Res 46: 418–429.
56. van der Wal D (1998) The impact of the grain-size distribution of nourishment sand on aeolian sand transport. J Coast Res 14: 620–631.
57. Lebrato M, Iglesias-Rodríguez D, Feely RA, Greeley D, Jones DO, et al. (2010). Global contribution of echinoderms to the marine carbon cycle: $CaCO_3$ budget and benthic compartments. Ecol Monogr 80: 441–467.
58. Chauvaud L, Thompson JK, Cloern JE, Thouzeau G (2003) Clams as CO_2 generators: the *Potamocorbula amurensis* example in San Francisco Bay. Limn Oceanogr 48: 2086–2092.
59. Seung-Woo L, Seung-Bin P, Soon-Kwan J, Kyoung-Soo L, Si-Hyun L, et al. (2010) On carbon dioxide storage based on biomineralization strategies. Micron 41: 273–282.
60. Lescinsky HL, Edinger E, Risk M (2002) Mollusc shell encrustation and bioerosion rates in a modern epeiric sea: Taphonomy experiments in the Java Sea, Indonesia. Palaios 17: 171–191.
61. Rodland DL, Kowalewski M, Carroll M, Simões MG (2004) Colonization of a 'Lost World': Encrustation patterns in modern subtropical brachiopod assemblages. Palaios 19: 381–395.
62. Evans JW (1969) Borers in the shell of the sea scallop, *Placopecten magellanicus*. Amer. Zool. 9: 775–782.
63. Carver CE, Thériault I, Mallet AL (2010) Infection of cultured eastern oysters *Crassostrea virginica* by the boring sponge *Cliona celata*, with emphasis on sponge life history and mitigation strategies. J. Shellfish Res. 29: 905–915.
64. Poppe GT, Goto Y (1991) European Seashells (vol. 1). Verlag Christa Hemmen, 352 pp.
65. Den Hollander J, Van Etten JCA (1974) De oekologie van *Bledius arenarius* en *B. subniger* op het Oostvoornse strand (Coleoptera, Staphylinidae). Entomol Berich 34: 155–160.
66. Speybroeck J, Bonte D, Courtens W, Gheskiere T, Grootaert P, et al. (2006) Beach nourishment: an ecologically sound coastal defence alternative? A review. Aquatic Conserv: Mar Freshw Ecosyst 16: 419–435.
67. Peterson CH, Hickerson DHM, Johnson GG. (2000) Short-term consequences of nourishment and bulldozing on the dominant large invertebrates of a sandy beach. J Coast Res 16: 368–378.
68. Dietl GP, Flessa KW (2011) Conservation paleobiology: putting the dead to work. TREE 26: 30–37.
69. Terry RC (2010) The dead do not lie: using skeletal remains for rapid assessment of historical small-mammal community baselines. Proc Roy Soc B 277: 1193–1201.

16

Salmon Aquaculture and Antimicrobial Resistance in the Marine Environment

Alejandro H. Buschmann[1], Alexandra Tomova[2], Alejandra López[1], Miguel A. Maldonado[1], Luis A. Henríquez[1], Larisa Ivanova[2], Fred Moy[3], Henry P. Godfrey[3], Felipe C. Cabello[2]*

1 Centro i~mar, Universidad de Los Lagos, Puerto Montt, Chile, **2** Department of Microbiology and Immunology, New York Medical College, Valhalla, New York, United States of America, **3** Department of Pathology, New York Medical College, Valhalla, New York, United States of America

Abstract

Antimicrobials used in salmon aquaculture pass into the marine environment. This could have negative impacts on marine environmental biodiversity, and on terrestrial animal and human health as a result of selection for bacteria containing antimicrobial resistance genes. We therefore measured the numbers of culturable bacteria and antimicrobial-resistant bacteria in marine sediments in the Calbuco Archipelago, Chile, over 12-month period at a salmon aquaculture site approximately 20 m from a salmon farm and at a control site 8 km distant without observable aquaculture activities. Three antimicrobials extensively used in Chilean salmon aquaculture (oxytetracycline, oxolinic acid, and florfenicol) were studied. Although none of these antimicrobials was detected in sediments from either site, traces of flumequine, a fluoroquinolone antimicrobial also widely used in Chile, were present in sediments from both sites during this period. There were significant increases in bacterial numbers and antimicrobial-resistant fractions to oxytetracycline, oxolinic acid, and florfenicol in sediments from the aquaculture site compared to those from the control site. Interestingly, there were similar numbers of presumably plasmid-mediated resistance genes for oxytetracycline, oxolinic acid and florfenicol in unselected marine bacteria isolated from both aquaculture and control sites. These preliminary findings in one location may suggest that the current use of large amounts of antimicrobials in Chilean aquaculture has the potential to select for antimicrobial-resistant bacteria in marine sediments.

Editor: Martin Krkosek, University of Otago, New Zealand

Funding: This research was supported by a grant from the Lenfest Ocean Program/Pew Charitable Trusts to F.C.C. and A.H.B. The funders had no role in study design, data collection and analysis, decision to publish, or preparation of the manuscript.

Competing Interests: The authors have declared that no competing interests exist.

* E-mail: cabello@nymc.edu

Introduction

It is believed that aquaculture will constitute the source of over more than half of the seafood consumed in the world in coming years because of the collapse of natural fisheries [1]. However, this optimistic view needs to be tempered by increasing information suggesting such expansion may be unsustainable as aquaculture generates untoward effects such as habitat destruction, eutrophication and environmental contamination with chemicals and antimicrobials [2]. The therapeutic, growth-promoting and prophylactic use of antimicrobials was introduced into agricultural practice in the 1940s and became widespread in Europe and the United States [3–6]. Antimicrobial resistance in antimicrobial-fed animals was soon noted [7], concerns about the possibilities of transmission of this resistance to human pathogens followed shortly thereafter [6,8,9], and indeed, has been demonstrated to occur [10,11]. Voluntary and legislated bans on the use of antimicrobials as growth promoters in the member states of the European Union since the 1990s have been associated with a marked decrease in antimicrobial usage without negative impacts on productivity in fowl and swine [8,12,13].

Salmon aquaculture is an exponentially growing industry worldwide, particularly in two countries – Norway and Chile [14,15]. In Chile, this growth has been accompanied by major mortalities of salmon reared in net pens. These can reach 50% of production under some conditions with ensuing large economic losses [16,17]. This growth has triggered concerns regarding many environmental issues, particularly because large amounts of chemotherapeutics and antimicrobials in the feed readily pass into the marine environment and potentially alter bacterial biodiversity [2,18–22]. Because the use of vaccines to prevent bacterial diseases in fish is limited [16], this in turn has led to increased use of therapeutic and prophylactic antimicrobials [23–25]. Conservative estimates suggest that approximately 950 metric tons of quinolones were used in salmon aquaculture in Chile between 2000 and 2008, and approximately 1500 metric tons of tetracycline and 478 metric tons of florfenicol were used for this purpose between 2000 and 2007 [23–25].

Antimicrobial agents are usually administered to salmon mixed with food [19,22]. Uningested food and fish feces containing unabsorbed antimicrobials and secreted antimicrobial metabolites in the water and sediment in the environment of salmon farming sites often retain their antimicrobial activity and can remain in the aquatic environment for variable periods of time depending on their initial concentrations, biodegradability, and physical and chemical characteristics [19,26–28]. Such materials can select for antimicrobial resistant bacteria in the sediment and water column and can often influence microbial diversity not only by eliminating susceptible bacteria but also by acting on other susceptible microorganisms such as microalgae [22,26,29,30].

Selection of antimicrobial-resistant bacteria in the marine environment could have detrimental impacts on piscine and human health by facilitating transfer of antimicrobial resistance genetic determinants from environmental marine microbes to fish pathogens and terrestrial bacteria including human and animal pathogens [19,22,23,31]. It is clearly evident that bacteria from marine and terrestrial ecosystems can share antimicrobial resistance genes and that some emerging antimicrobial resistance genes in human pathogens may have an aquatic bacterial origin [32–35]. For example, the fish pathogen *Yersinia ruckerii*, the cause of enteric redmouth disease, shares an antimicrobial resistance plasmid and antimicrobial resistance genes with the plague bacillus, *Yersinia pestis* [36]. This sharing of movable genetic elements and antimicrobial resistance genes between bacteria of different ecological niches potentially endangers treatment of human patients [22,32–34,36]. Such genetic and epidemiological findings strongly suggest that the aquatic and terrestrial ecosystems are not isolated regarding the dissemination of antimicrobial resistance genes among their bacterial populations, probably as the result of horizontal gene transfer [22,37].

The high level of antimicrobial use in salmon aquaculture in Chile could have negative impacts on environmental biodiversity and terrestrial animal and human health by selecting for bacteria in the marine environment containing antimicrobial resistance genes. We therefore compared numbers of culturable bacteria and antimicrobial resistant bacteria for three antimicrobials used extensively in Chilean salmon aquaculture (oxytetracycline, oxolinic acid, and florfenicol) in the marine sediment adjacent to salmon aquaculture pens and at a control site 8 km distant with no observed aquaculture or other human activities.

Results

Antimicrobials in sediment samples from aquaculture and control sites

Traces of flumequine, a fluoroquinolone antimicrobial, were present in four sediment samples at the aquaculture site (Fig. 1): two in December, 2008, and two in January, 2009. Flumequine was also present in four sediment samples from the control site, 8 km from the aquaculture site (Fig. 1): one in December, 2008, two in January, 2009 and one in April, 2009. Oxytetracycline, oxolinic acid, and florfenicol were not detected in any of the 36 total samples examined (data not shown).

Culturable bacteria in sediment samples from aquaculture and control sites

The total numbers of culturable bacteria in sediments from aquaculture and control sites varied significantly over the course of a year ($P<0.001$, two-way ANOVA, rank transformed data), fluctuating between approximately 1×10^3 and 1×10^5 colony forming units (cfu) g^{-1} (Fig. 2A). There were highly significant differences in culturable bacterial numbers between aquaculture and control sites over the entire study period ($P<0.001$, rank transformation test), with bacterial numbers significantly higher in late spring (November, 2008) and high summer through winter (January through July, 2009) than at other times ($P<0.05$, Student-Newman-Keuls post-test). Sensitivity analyses using only data consistent with a dilution series or using all 258 data points produced similar results. Results obtained using a standard two-way ANOVA of log-transformed data were also consistent with this non-parametric analysis.

Sampling of sediments at intermediate distances between aquaculture (0.0 km) and control (8.0 km) sites in November, 2008 (Fig. 2B), revealed significant differences in numbers of culturable bacteria ($P<0.001$, two-way ANOVA, rank transformed data) which did not decrease until 1 km from the aquaculture site ($P<0.05$, Student-Newman-Keuls post-test). Sensitivity analyses using only data consistent with a dilution series or using all 258 data points produced similar results. Results obtained using ANOVA of log-transformed data were also consistent with non-parametric analysis.

Antimicrobial-resistant bacteria in sediment samples from aquaculture and control sites

Measurements of the antimicrobial resistant fraction (ARF) of bacteria cultured from aquatic sediments are useful for comparing changes in antimicrobial resistance in this environment [38]. ARF to oxytetracycline (Fig. 3A) and oxolinic acid (Fig. 3B) were significantly different between aquaculture and control sites over the entire period of study ($P<0.001$, two-way ANOVA, rank transformed data). For both these antimicrobials, there was no significant interaction between time of year and nature of the study site (aquaculture or control). ARF to oxytetracycline was significantly lower during spring, 2008 (September-November, 2008) than during the following summer (December, 2008–January, 2009), before rising significantly the following spring (September, 2009) ($P<0.05$, Student-Newman-Keuls post-test). ARF to oxolinic acid (Fig. 3B) also was significantly higher in early spring both years (September, 2008; September, 2009) than in late spring, fall and winter (November, 2008; April, 2009: July, 2009), before rising to intermediate levels in high summer (December, 2008; January, 2009) ($P<0.05$, Student-Newman-Keuls post-test). Although ARF to florfenicol also varied significantly throughout the year (Fig. 3C), the interaction of time and study site was significant ($P<0.002$, two-way ANOVA, rank transformed data), and significant differences between aquaculture and control sites were only seen in late spring and high summer (November, 2008, through January, 2009) ($P<0.05$, Student-Newman-Keuls post-test). Sensitivity analyses using only data consistent with a dilution series or using all 258 data points produced similar results. Results obtained with parametric analysis of log-transformed data were again consistent with this non-parametric analysis.

Sampling in November, 2008, at intermediate distances between aquaculture and control sites revealed that ARF to oxytetracycline (Fig. 4A), oxolinic acid (Fig. 4B), and florfenicol (Fig. 4C) showed significant decreases from aquaculture to control sites for all three antimicrobials (oxytetracycline, $P=0.008$; oxolinic acid, $P=0.002$; florfenicol, $P=0.015$, two-way ANOVA, rank transformed data). In the case of oxytetracycline and oxolinic acid (Figs. 4A, 4B), ARF was only significantly lower 8 km from the aquaculture site ($P<0.05$, Student-Newman-Keuls post-test). Significant elevations of ARF to florfenicol were maintained 0.5 km from the aquaculture site, but were significantly lower by 1 km (Fig. 4C). Sensitivity analyses using only data consistent with a dilution series or using all 258 data points yielded similar results. Results from parametric analysis of log-transformed data were again consistent with this non-parametric analysis.

Detection of antimicrobial resistance genes in bacteria from aquaculture and control sites

The presence of culturable antimicrobial-resistant bacteria in sediments from aquaculture and control sites suggested the presence of antimicrobial resistance determinants in these bacteria. PCR confirmed the presence of genes mediating

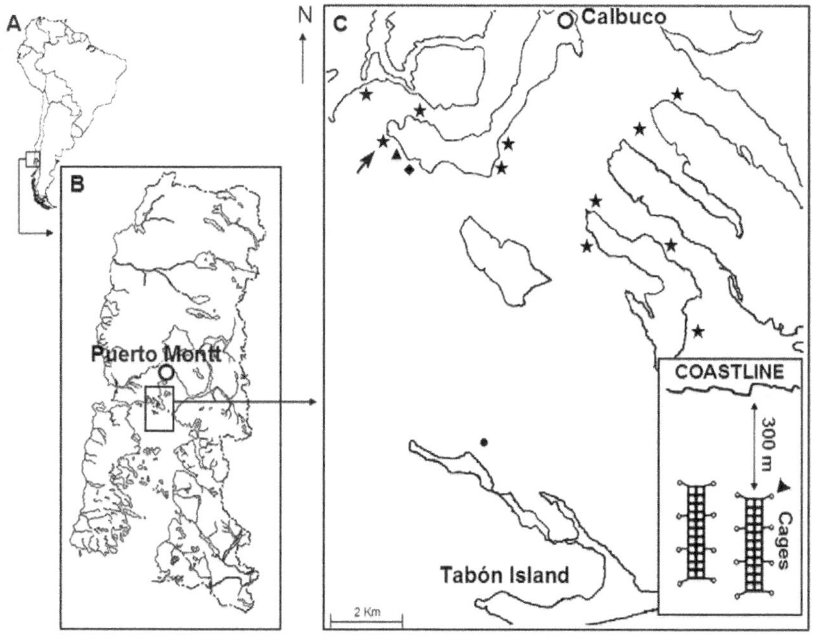

Figure 1. Locations of aquaculture and sites of sampling in the Calbuco archipelago, Chile. Salmon farming sites are indicated by stars. The "aquaculture site" sampled in the present study (arrowhead, inset) was 20 m from the salmon farm indicated by arrow. Other sites sampled in the present study were located 0.5 km (solid triangle), 1 km (solid diamond) and 8 km (solid circle) from the aquaculture site. The latter site was off the coast of Tabón Island, an island with no aquacultural activities or other human activity, and is referred to as the "control site" in the text.

resistance to oxytetracycline, oxolinic acid, and florfenicol in bacteria from sediment samples that had not been selected for antimicrobials. Plasmid-mediated quinolone resistance (PMQR) genes were studied because these plasmid-mediated resistances are potentially transmissible and some of them appear to originate from aquatic bacteria [32,33,39,40]. Moreover, they have recently begun to disseminate among terrestrial animal and human pathogens and are readily detected by PCR [40]. Unselected isolates of bacteria from aquaculture and control sites (24 from each site) contained plasmid-encoded genes for resistance to quinolones, including *qnrA*, *qnrB*, *qnrS*, *oqxA* and *aac(6')-Ib-cr* (Fig. 5, Table 1). Several of these bacteria also harbored *tetA*, *tetB*, *tetK*, *tetM*, and *floR* genes (Fig. 5, Table 1). Some bacteria harbored multiple antimicrobial resistant determinants (Table 2). The combination of PMQR and tetracycline resistance genes was the most frequent (8), followed by PMQR and florfenicol resistance genes (3) and finally PMQR, tetracycline and florfenicol resistance genes (3). Several isolates also generated a positive signal (confirmed by DNA sequencing) for the *int1* gene encoding integrase 1 (Table 1), suggesting the presence of type 1 integrons in these bacteria [41–45].

Species identification of bacteria harboring antimicrobial resistance genes from aquaculture and control sites

PCR amplification of 16S rRNA genes in eight bacterial isolates from the aquaculture site identified two isolates of *Sporosarcina* sp., two isolates of *Arthrobacter* sp. and one isolate *Vibrio* sp. Bacterial isolates from the control site included one *Pseudoalteromonas* sp. isolate and two isolates of *Vibrio* sp. (Table 2). The 16S rDNA sequences of these amplicons were >99% identical to those in GenBank (E value of 0.0) (Table 2). These observations are not consistent with the possibility that the bacteria in which antimicrobial resistance genes were present were human and terrestrial animal pathogens contaminating Chilean coastal waters [46,47].

Discussion

We studied marine sediments from two sites in the Calbuco Archipelago in southern Chile to determine numbers of culturable and antimicrobial-resistant bacteria they contained. One was a site 20 m from a salmon aquaculture facility (Fig. 1), the other, off the coast of an island 8 km distant from the aquaculture site, the only local island with no aquaculture activities, few human dwellings, and no water sources or discharges of solids into the sea (Fig. 1). This latter site was thus expected to be pristine and a suitable control site. Surprisingly, residues of flumequine, a quinolone with potential cross resistance with oxolinic acid, were present in sediments at both sites, most likely carried there by marine currents from the many other aquaculture sites in the area that use this antimicrobial and that have been in operation over the past 10 years (Fig. 1) [23–25]. Flumequine has been used as widely as oxolinic acid in aquaculture in Chile; approximately 548 metric tons were used between 2000 and 2007 [23–25]. Although the control site was not as pristine as it was originally thought to be, sediments from the aquaculture site still contained significantly larger numbers of culturable bacteria than sediments from the control site (Fig. 2A), with increased bacterial numbers being present up to 1 km from the aquaculture site (Fig. 2B). These findings essentially confirm previous reports on the ability of aquaculture activities to increase culturable bacterial numbers [48,49]. Sediments from the aquaculture site also contained increased numbers of culturable bacteria resistant to tetracycline, oxolinic acid and florfenicol; this effect persisted for distances up to 1 km from the aquaculture site (Fig. 4). The presence of flumequine residues in the sediment from the apparently pristine control site (Fig. 1) and significant antimicrobial resistance at distances up to 1 km from the aquaculture site (Fig. 4) suggest that

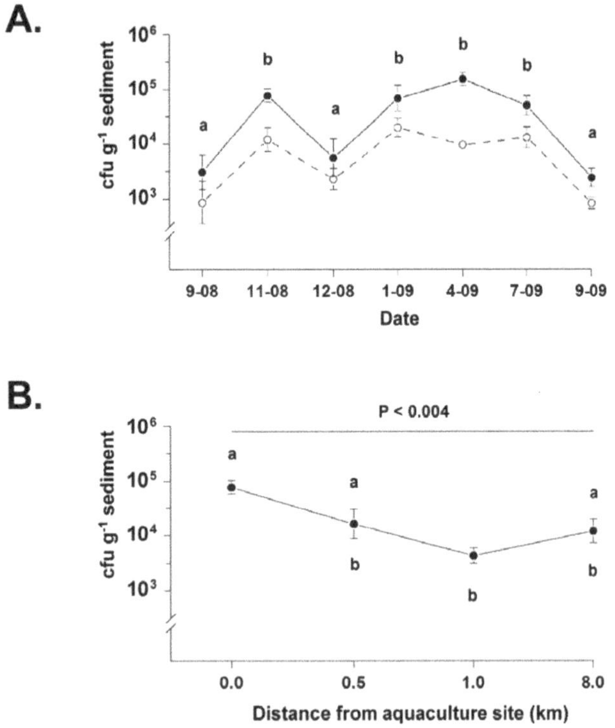

Figure 2. Culturable bacteria in sediment samples at aquaculture and control sites taken at various time points. A. Colony forming units (cfu) g^{-1} sediment (mean ± SE) in samples taken from September, 2008, to September, 2009, were significantly higher at the aquaculture site (closed circles) than at the control site (open circles) at all time points (P<0.001); different lower case letters indicate significant differences (P<0.05). A total of 66 samples were taken, 33 from the aquaculture site and 33 from the control site. B. Cfu g^{-1} sediment (mean ± SE) taken in November, 2008, at the aquaculture site (0.0 km) and at sites 0.5, 1.0 and 8.0 km (control site) distant from it. Aquaculture and control sites correspond to sites shown in Fig. 2A for this date; five samples were taken from each additional site studied. Different lower case letters indicate significant differences (P<0.05, see text for details of statistical analysis).

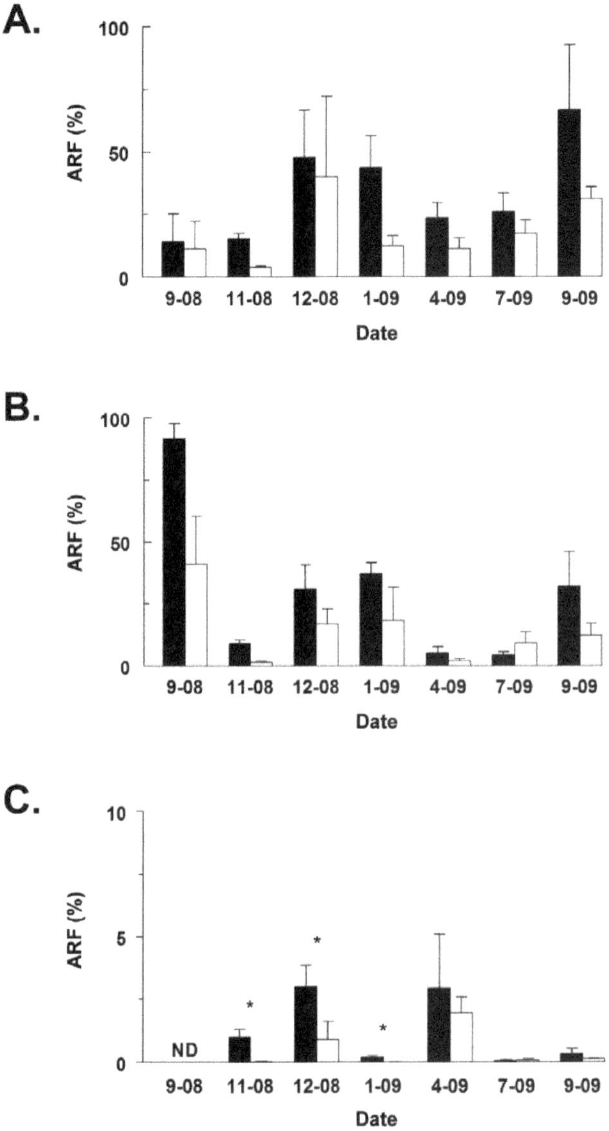

Figure 3. Antimicrobial resistant bacteria in sediment samples from aquaculture and control sites. Antimicrobial resistance fraction (ARF) (mean ± SE) of culturable bacteria to (A) oxytetracycline and (B) oxolinic acid in sediments from aquaculture (solid bars) and control (open bars) sites from September, 2008, to September, 2009, were significantly different between aquaculture and control sites over the entire study period (P<0.001). ARF to (C) florfenicol in sediments from aquaculture and control sites were significantly different only in November, 2008, December, 2008, and January, 2009. *, P<0.05, see text for details of statistical analysis. A total of 66 samples were taken, 33 from the aquaculture site and 33 from the control site.

excessive use of antimicrobials in salmon aquaculture sites may not only have an effect on marine sediments directly under and close to aquaculture pens but also at some distance from where these activities take place as a result of transport by water currents of both unchanged antimicrobials and their antimicrobially active metabolites. Previous work has not detected antimicrobials beyond 30 to 50 m from the aquaculture site but the amounts of antimicrobials used in those situations were a fraction of those used in Chile [26,29,50–52]. This suggests that the size of the area impacted by aquacultural activities with regard to antimicrobial resistance is related to the amounts of antimicrobials used in these activities.

Obtaining accurate counts of culturable bacteria in marine sediments is complicated by incomplete dispersal of particulates and their attached bacteria. This incomplete dispersal leads to erratic values in dilution series. Although clinical studies frequently employ various criteria to ensure the quality of data to be analyzed [53,54], our study is one of the first if not the first in this area to use explicit criteria to assure the quality of the data to be analyzed. The fact that similar conclusions were obtained in multiple sensitivity analyses confirms the validity of this approach.

It has been suggested that the significantly larger numbers of culturable antimicrobial-resistant bacteria demonstrable in sediment of aquaculture sites relative to control sites may be the result of changes produced by excess organic matter passing into the environment from uningested fish food and feces rather than from antimicrobial use per se [55–57]. Unfortunately, there is no experimental evidence to support this hypothesis. It is difficult to develop a scenario based on current concepts of microbial genetics and physiology that could explain preferential stimulation of growth of antimicrobial-resistant bacteria by organic matter alone unless this matter also contained other chemical entities such as

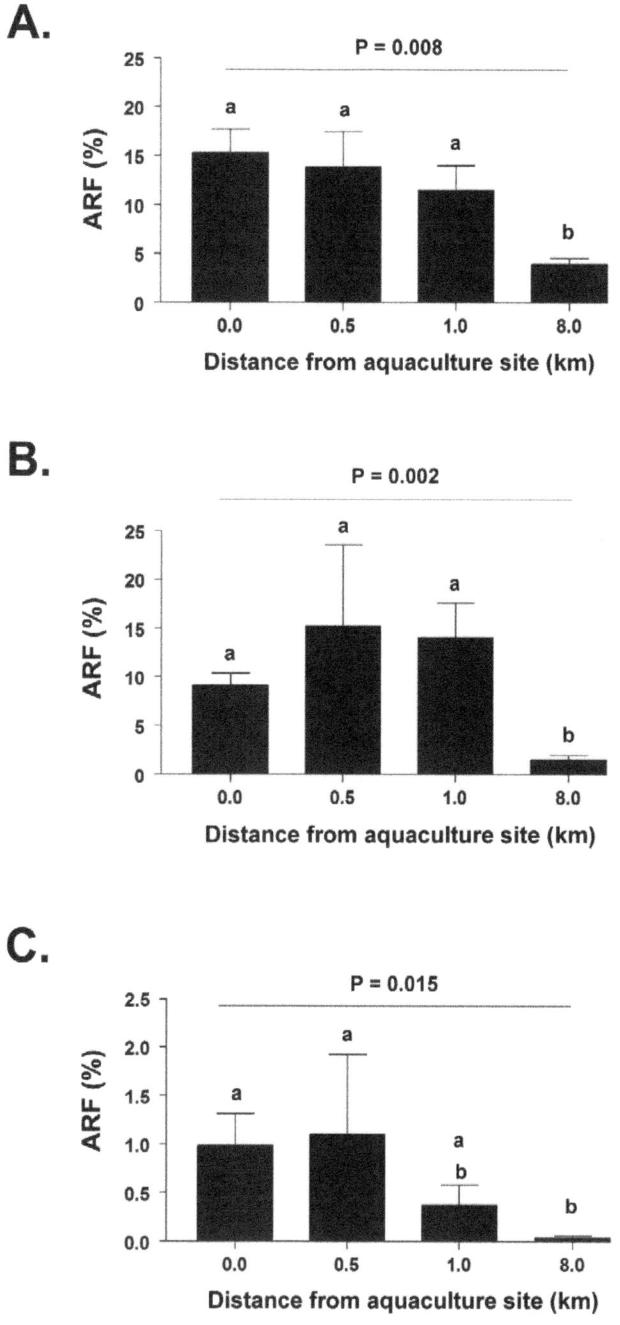

Figure 4. Variation in ARF to selected antimicrobials with distance from aquaculture site. ARF (mean ± SE) in November, 2008, to (A) oxytetracycline, (B) oxolinic acid, and (C) florfenicol in sediments at aquaculture site (0.0 km) and at sites 0.5, 1.0 and 8.0 km (control site) distant from it. ARF for aquaculture and control sites correspond to ARF shown in Fig. 3 for this date. Five samples were taken from each additional site studied. ARF to each antimicrobial were significantly greater at the aquaculture than at the control site (probabilities indicated for each antimicrobial). Different lower case letters within each panel indicate significant differences between ARF ($P<0.05$, see text for details of statistical analysis).

metal ions, disinfectants or metabolites that could co-select for metabolite utilization and ubiquitous antimicrobial resistance genes linearly integrated in mobile genetic units such as plasmids, transposons and integrons throughout Bacteria and Archea [58,59].

An important limitation of this research is the lack of information concerning the use and frequency of application of antimicrobials during aquaculture activities before and during this study. This stems from the proprietary nature of this information and from the general lack of publicly available and well-organized data on antimicrobial use in the Chilean aquaculture industry. This dual lack critically limits our ability to relate antimicrobial use to the observed relative increase in antimicrobial-resistant bacteria and might also be responsible for our failure to detect antimicrobials other than flumequine in the sediments. The marked increase in antimicrobial-resistant bacteria in the spring and summer when activity at aquaculture sites customarily increases, the presence of flumequine residues and the increased ARF to antimicrobials known to be heavily used in this industry [23] does however suggest a possible relationship between them. It has been postulated that antimicrobials administered to fish by food do not remain in the sediment, thus decreasing their ability to exert selective pressure upon antimicrobial-resistant bacteria [51], However, the present and previous work indicate that antimicrobials remain in the sediment at concentrations able to exert selective pressure there [52,60,61].

Approximately half of unselected culturable marine bacterial species from both aquaculture and control sites harbored antimicrobial resistance genes (Table 1); the antimicrobial resistances detected in these bacteria are probably mediated by these genes. Because *tetM* tetracycline resistance gene and other antimicrobial resistance genes have been demonstrated in ancient (30,000 years before the present) bacterial DNA extracted from terrestrial permafrost in Alaska [62], the effect of antimicrobial use in salmon aquaculture on marine sediments is most likely restricted to selecting those bacteria able to survive in their presence. However, the numerically similar frequencies of antimicrobial resistance genes at both sites is certainly consistent with the presence of antimicrobial residues at both sites, and again suggests that the control site was not as pristine as it was originally thought to be.

There are several caveats regarding the observed bacterial resistance phenotypes and genotypes. Because we only sequenced three amplicons for the *aac(6')-Ib-cr* gene, we cannot be certain that the five amplicons detected have the mutation which mediates quinolone acetylation [63]. Furthermore, oxytetracycline, oxolinic acid, and florfenicol resistance phenotypes can be encoded by a multiplicity of chromosomal and plasmid genes and not only by the ones studied in the present work [40,64–66]. Because we did not perform an exhaustive investigation of alternative antimicrobial resistance genes for quinolone, tetracyclines and chloramphenicol, did not search for the presence of genes mediating resistance to other antimicrobials, and studied only culturable bacteria, we are probably underestimating the resistome present in marine bacteria at the aquaculture and control sites. This underestimation could lower the chance of detecting any differences regarding these genes between these sites. Interestingly, a few of the strains studied also harbored an integron type 1, a genetic element usually associated with multiple antimicrobial resistance cassettes and known to be present in bacteria from aquatic sediments impacted by human activity [41–44].

Bacteria from the marine environment where salmon aquaculture takes place contain antimicrobial resistance genes towards antimicrobials used extensively in this activity. This confirms previous work indicating that plasmid-mediated quinolone resistance genes are present in aquatic bacteria [32,33,39] and that these aquatic bacteria could well be the original source for

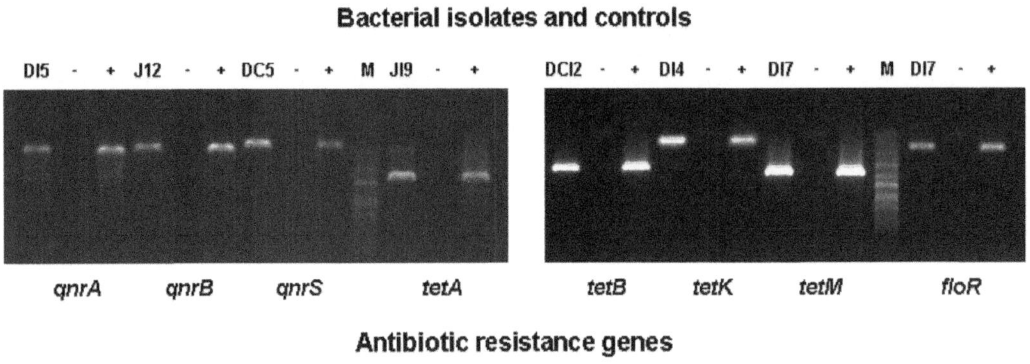

Figure 5. Antimicrobial resistance genes in unselected marine bacterial isolates and controls. Detection of antimicrobial resistance genes in bacteria cultured from marine sediments obtained from December, 2008, to January, 2009. *qnr*, *tet* and *floR* genes were detected by PCR as described in Material and Methods with primers in Table 3. D15, J12, DC5, J19, DC12, D14 and D17 are bacterial isolates from sediment. −, negative control (*E. coli* DH5α). +, positive controls (Table 3). M, molecular weight markers.

dissemination of these determinants in human pathogens [22,32,33,39]. Such bacteria will have a selective advantage after introduction of antimicrobials into their environment [22,31,37,41]. The relevance of increased antimicrobial-resistant bacteria in sediments of salmon aquaculture sites for the emergence of antimicrobial resistance in fish and human pathogens is unknown [22,23,56,57]. The significant increase in antimicrobial-resistant bacterial populations to oxytetracycline, oxolinic acid, and florfenicol in aquaculture sediments suggests they could be a potential source for antimicrobial resistance genes in fish and human pathogens as a result of horizontal gene transfer [18,22,23,36,67,68]. This problem could be exacerbated in Chile because of major contamination of seawater with antimicrobial-resistant animal and human pathogens [46,47]. Horizontal gene transfer of antimicrobial resistance genetic elements and mutagenesis may also be stimulated by microbial stress triggered by the presence of sub-inhibitory concentrations of antimicrobials such as flumequine in the sediment [68–70].

The present preliminary study in a single salmon aquaculture and a single control site suggests, as has been previously demonstrated, that salmon aquaculture activities in Chile have the potential to alter concentrations of culturable bacteria in marine sediments and increase the proportion of antimicrobial-resistant bacteria to three major classes of antimicrobials used in clinical medicine [29,49,61]. The spatial limitation of the present study hampers an immediate generalization of its conclusions to other aquacultural and control sites in Chilean coastal waters. Additional studies are thus necessary to confirm these results and to identify the dynamics of these processes more carefully. The presence of resistance genes to these antimicrobials in marine bacteria and residual antimicrobials in the sediment suggests that the increase in antimicrobial resistance results from selection of bacteria in this environment. The cautionary principle also suggests that use of antimicrobials in salmon aquaculture in Chile needs to be controlled and reduced because this increase has the potential to generate other problems of food safety and industrial health [20,22,23,71,72]. We expect that the new sanitary scenario instituted in response to the epidemic of infectious salmon anemia will result in a significant reduction in antimicrobial use and particularly in avoidance of quinolone antimicrobials because of their relevance to human health [18,23,73].

Materials and Methods

Location of aquaculture and control sampling sites

The two sites studied were located in the Calbuco archipelago in southern Chile, Region X, near the town of Calbuco (41°48′S, 73°11′W) (Fig. 1). At least 11 salmon farming sites have been in operation in this area over the past 10 years with an estimated annual production of approximately 15,000 metric tons. One of these farms consists of two salmon culture units with 22 and 24 pens located approximately 300 m from the coastline at a water depth of 45 m (arrow, Fig. 1). It can produce approximately 1,200 metric tons of salmon annually. Because of a confidentiality agreement, neither the exact identity of the farm, the biomass of fish cultured, nor the identity and amounts of antimicrobials used before and during the period of study can be revealed. Areas directly under the salmon pens are protected by anti-predator nets and are inaccessible to divers. For this reason, sediment samples from this site ("aquaculture site") were obtained from an area approximately 20 m from the outer southeastern corner of the pens next to a buoy (arrowhead, lower inset of Fig. 1). This

Table 1. Antimicrobial resistance genes present in marine sediment bacteria from aquaculture and control sites in Chile.

		Antimicrobial resistance genes to													
		Tetracycline				Quinolones								Florfenicol	
Site	No. of Strains	*tetA*	*tetB*	*tetK*	*tetM*	*qnrA*	*qnrB*	*qnrC*	*qnrD*	*qnrS*	*qepA*	*oqxA*	*aac(6′)-Ib-cr*	*floR*	*int1*
Aquaculture	24	4	5	5	2	2	1	0	0	3	0	4	4	4	3
Control	24	3	7	4	0	2	1	0	0	5	0	3	1	2	1

Table 2. Bacterial isolates identified by 16S rRNA gene sequence analysis.

Species	Genes	Compared to GenBank No.	% of identity	Site of isolation
Sporosarcina sp.	*tetK, floR, qnrA, qnrS*	FJ425906.1	>99	Aquaculture
Arthrobacter sp.	*qnrA, tetB*	EF550164.1	>99	Aquaculture
Sporosarcina sp.	*oqxA, aac(6')-1b-cr*	EU204977.1	>99	Aquaculture
Arthrobacter sp.	*qnrS, tetA*	JF799958.1	>99	Aquaculture
Vibrio sp.	*qnrB, tetK*	DQ146994.1	>99	Aquaculture
Pseudoalteromonas sp.	*qnrA, tetB*	FJ497709.1	>99	Control
Vibrio sp.	*qnrS*	EU195936.1	>99	Control
Vibrio sp.	-	JN128258.1	>99	Control

distance (<50 m) and the amount of organic matter at this site (>3.5%, unpublished data), indicates that its sediments were influenced by aquacultural activity [74,75]. The other site ("control site") was located 250 m off the northern coast of Tabón Island at a water depth of 30 m (Fig. 1). Tabón Island is 6 km long and close to 200 m wide and is situated 8 km south of the aquaculture site. It is the only local island without aquacultural activities, has few human dwellings, no sources of water and lacks solid discharges into the sea. Water currents at both sites vary between 15 and 18 cm s^{-1} during tidal flood and ebb, respectively, suggesting a high dispersal of materials in the water column and from sediments at both sites. Water surface temperature and salinity in this area vary from 9°C and 28‰ (parts per thousand) in the winter to 18°C and 32‰ in the summer. Both sampling sites are located in open access areas for which entrance and sampling permits are not required by Chilean regulations.

Sediment samples were taken from a circumscribed area of superficial sediment at the aquaculture and control sites by scuba divers using 15 cm diameter PVC plastic core samplers. After the sediment had been obtained, the sampler was closed by the diver with a plastic cap to avoid contamination. No endangered or protected organisms were captured in course of obtaining these sediment samples. The dates for sampling were arbitrarily chosen to encompass a full year with an emphasis on sampling during the austral spring/summer when aquacultural activities are concentrated. Sampling was performed seven times over a 12 month period: September, November and December, 2008; and January, April, July and September, 2009. Three samples were taken in September, 2008; five samples were taken at each of the other times resulting in 33 samples from the aquaculture site and 33 samples from the control site for a total of 66 samples from both sites. In November 2008, five samples were also taken at sites 0.5 km and 1 km from the aquaculture site for a total of 10 samples from these additional sites (Fig. 1).

Measurement of antimicrobials in sediment samples

The presence in sediment samples of oxytetracycline, oxolinic acid, flumequine, and florfenicol was determined at the Instituto de Farmacia, Universidad Austral, Valdivia, Chile, by HPLC using standard protocols at fixed UV/Vis wave lengths [60,76,77]. These assays were done on four sediment samples taken from the aquaculture site and four sediment samples taken from the control site on each of four dates: December, 2008, January, 2009, April, 2009, and July, 2009. Two sediment samples each taken 0.5 and 1 km from the aquaculture site in November, 2008, were also tested for these antimicrobials. A total of 36 samples were tested for antimicrobials.

Bacterial cultures

Culturable antimicrobial-susceptible and -resistant bacteria were determined by suspending 0.1 g of the top 2-cm of each sediment sample in 0.9 ml phosphate buffered saline, pH 7.4, within 2–3 hours after collection; this suspension was then sonicated (Elma Transsonic 310, Singen, Germany) for 5 minutes to ensure detachment of bacteria from sediment particles. A 10-fold dilution series (undiluted to 10^{-5}) of this suspension was plated on Marine agar plates (Difco) containing no antimicrobials or oxytetracycline, 150 µg ml^{-1} (AppliChem GmbH, Darmstadt, Germany), or oxolinic acid, 10 µg ml^{-1} (Sigma-Aldrich GmbH, Steinheim, Germany), or florfenicol, 30 µg ml^{-1} (Sigma-Aldrich GmbH). Stock solutions of antimicrobials were kept frozen at −20°C and thawed immediately before use. Plates were incubated for 7 days at 20°C and the number of visible colonies were counted. These data were used to calculate cfu g^{-1} sediment = (-total plate counts×dilution factor), and ARF (in percent) for each antimicrobial = (total plate counts with antimicrobial×dilution factor/total plate counts without antimicrobial×dilution factor)×100 [38]. Isolated colonies of bacteria growing on Marine agar plates with and without antimicrobials were selected and stored frozen at −80°C in 96-well microtiter plates in 36% glycerol for later study of antimicrobial resistance genes.

Detection of plasmid-mediated antimicrobial resistance genes

Cultures of marine sediment samples obtained from December, 2008, and January, 2009, were transported in Marine soft agar in 1.5 ml Eppendorf microtubes and restreaked on Marine agar (9.0 cm diameter Petri dishes) containing antimicrobials to ascertain clonality. One isolated colony of each was grown for further studies and stored in 30% glycerol at −80°C. Bacterial cultures were kept at 4°C on marine agar plates with antimicrobials for daily manipulations while experiments were in progress. The following antimicrobial resistance genes were studied by PCR using primers shown in Table 3. Quinolone resistance genes: topoisomerase protection *qnrA*, *qnrB*, *qnrC*, *qnrD* and *qnrS* genes [40,78–80]; the putative enzymatic inactivation gene *aac(6')-Ib-cr* [40,63]; and efflux pump genes *qepA* and *oqxA* [81,82]. Tetracycline resistance genes: efflux pump genes *tetA*, *tetB* and *tetK* [64,65,83]; and the ribosome protection gene, *tetM* [64,65]. Florfenicol resistance gene, *floR*, encoding an efflux pump [66,84]. The *intI* gene encoding integrase 1 [41,45] was also examined. Table 3 also contains information regarding strains used as positive and negative controls in these PCR reactions. Single colonies of oxytetracycline, oxolinic acid, and florfenicol resistant strains were picked from plates kept at 4°C and inoculated into

Table 3. Primers used in this study.

Gene	Primer	Sequence (5′→3′)	Amplicon (bp)	Positive control	References
qnrA	qnrA1ROB	ATTTCTCACGCCAGGATTTG	516	pMG252	[79]
	qnrA2ROB	GATCGGCAAAGGTTAGGTCA			
qnrB	qnrB1ROB	GATCGTGAAAGCCAGAAAGG	469	pMG298	[79]
	qnrB2ROB	ACGATGCCTGGTAGTTGTCC			
qnrC	qnrC-F	GGGTTGTACATTTATTGA	447	pDNA qnrC	[80]
	qnrC-R	TCCACTTTACGAGGTTCT			
qnrD	qnrD fw	CGAGATCAATTTACGGGGAATA	592	pDNA qnrD	[78]
	qnrD rv	AACAAGCTGAAGCGCCTG			
qnrS	qnrS1ROB	ACGACATTCGTCAACTGCAA	417	pMG306	[79]
	qnrS2ROB	TAAATTGGCACCCTGTAGGC			
qepA	qepA-F	CGTGTTGCTGGAGTTCTTC	403	pAT851	[82]
	qepA-R	CTGCAGGTACTGCGTCATG			
oqxA	oqxAF	CTCGGCGCGATGATGCT	392	DNA	[81]
	oqxAR	CCACTCTTCACGGGAGACGA			
aac(6′)-Ib-cr	Aac(6′)-1bXbal	CAGCTCTAGAATTTTTAAGCGTGCAT	620	pMG298	[63]
	Aac(6′)-1b2R	ATATGCGAATTCTTAGGCATCACTGC			
tetA	tetAf_A3	GCCTCCTGCGCGATCTGG	848	pEDtetA2	[65]
	tetAr_A2	CGAAGCAAGCAGGACCATG			
tetB	tetB_BF	CAGTGCTGTTGTTGTCATTAA	571	pEDtetB1	[65]
	tetB_BR	GCTTGGAATACTGAGTGTAA			
tetK	tetKf	TCGATAGGAACAGCAGTA	169	pT181	[83]
	tetKr	CAGCAGATCCTACTCCTT			
tetM	tetM_M6	GTTTATCACGGAAGYGC	687	pJFP76	[65]
	tetM_M4	GAAGCCCAGAAAGGATTYGGT			
floR	flo_f	AATCACGGGCCACGCTGTATC	215	pAB5S9	[66]
	flo_r	CGCCGTCATTCTTCACCTTC			
int1	intI1F	GTTCGGTCAAGGTTCTGG	890		[45]
	intI1R	CGTAGAGACGTCGGAATG			
16S rDNA	16sRNAf	AGAGTTTGATCCTGGCTCAG	variable		[88]
	16sRNAr1	ACGGCTACCTTGTTACGACTT			

10 ml Marine Broth (Difco, BD, Franklin Lakes, NJ, USA) with ciprofloxacin HCl (ICN, Aurora, Ohio, USA), 0.05 μg ml^{-1}, oxytetracycline hydrochloride (Sigma-Aldrich, St. Louis, MO, USA), 150 μg ml^{-1}, or florfenicol (Sigma-Aldrich, St. Louis, MO, USA), 30 μg ml^{-1}, and cultured at 20°C until late log phase. Cultures were centrifuged at 200× g for 15 min to remove suspended matter. The supernatant was centrifuged at 7,000× g for 10 min to pellet the bacteria, pellets were washed once with phosphate buffered saline, pH 7.4, and resuspended in 400 μl 10 mM TrisCl-10 mM EDTA for storage at −80°C until DNA was extracted [85]. Total DNA was extracted from 200 μl of bacteria by adding 10% SDS in sodium phosphate buffer, pH 8.0, to 1.25% SDS (wt/vol), incubating at 70°C for 30 min, and then disrupting bacteria with three freeze-thaw cycles: two minutes liquid N_2-10 min 70°C (once), two minutes liquid nitrogen-10 min 100°C (twice). After centrifugation of the viscous solutions at 4,000× g for 10 min, the supernatant was transferred to a clean tube used for PCR screening of antimicrobial resistance genes [86]. All PCR assays were done using MasterCycler Gradient, Eppendorf, Germany. Multiplex PCR for *qnrA/B/S* was conducted in 25 μl reaction volumes with GoTaq Flexi DNA polymerase, 0.8 units (Promega, Madison, WI, USA); 1× Green GoTaq Flexi Buffer; $MgCl_2$, 2 mM; dNTP, 0.15 mM each; primers, 0.5 μM each. Initial denaturation at 95°C for 3 min was followed by 35 cycles of 95°C, 20 sec-54°C, 30 sec-72°C, 40 sec; final extension of 72°C-7 min. For detection of the other PMQR, *tet* and *floR* genes, reaction volumes were 12.5 μl, and PCR was performed with Choice TaqBlue DNA polymerase 0.5 units (Denville Scientific Inc., Metuchen, NJ); 1× reaction buffer with $MgSO_4$, 15 mM; dNTP, 0.25 mM each; primers 0.5 μM each. Initial denaturation at 3 min at 95°C was followed by 35 cycles of denaturation at 95°C-20 sec; annealing for 30 sec at various temperatures for each group of primers; extension time at 72°C was dependent on fragment length, being 30 sec for a 500 bp fragment (see Table 3); final extension, 7 min-72°C. Amplicons were detected in 1% agarose gel with ethidium bromide, viewed and recorded in an Alpha Imager AIC, Alpha Innotech, Japan. Identity of amplicons was ascertained by comparison with positive controls and by DNA sequencing (GENEWIZ, Inc, South Plainfield, NJ, USA) of at least one amplicon of each gene (data not shown). DNA sequences were identified by BLAST analysis

against the non-redundant nucleotide sequence database at GenBank.

Species identification of marine bacteria containing antimicrobial resistance genes

Identification of marine bacteria containing antimicrobial resistance genes was done by PCR amplification of 16S ribosomal genes [87,88] using primers 16S rRNAf and 16S rRNAr1 (Table 3). The amplicons obtained were approximately 1500 bp and spanned 99% of 16S rRNA genes. Amplicons were sequenced and were identified by BLAST analysis against the non-redundant nucleotide sequence database at GenBank.

Statistical analysis

To insure data quality and to exclude erratic values in quantitation of colonies the following three criteria were used. 1. If colony counts were consistent with a dilution series (roughly monotonic and decreasing with increasing dilutions), cfu g^{-1} were calculated using the plate with the lowest dilution with <160 colonies. 2. If colony counts were consistent with a dilution series but no plate had <160 colonies, cfu g^{-1} were calculated using the plate with the lowest dilution with >160 colonies. 3. If colony counts were not consistent with a dilution series (suggesting incomplete dispersal in the undiluted material), cfu g^{-1} were calculated using the plate with the lowest dilution consistent with a dilution series with <160 colonies. Data not meeting these criteria (5 of 258 points) were excluded from analysis. For computational purposes, plates with no colonies were imputed a value of 1, a value at the non-detect level. Data were analyzed by two-way ANOVA using rank transformed data [89] and a Student-Newman-Keuls post-hoc test as appropriate. Values of $P \leq 0.05$ were considered significant. A sensitivity analysis was done.

Acknowledgments

We thank Mónica Maldonado for laboratory work and Cristian Valenzuela and Carlos Aranda for methodological suggestions during the experimental work; Drs. R. Schroeder and H. Dölz, Instituto de Farmacia, Universidad Austral, Valdivia, Chile, for the HPLC determinations of antimicrobials in marine sediments; Dr. G. A. Jacoby, Lahey Clinic Inc, Burlington, MA, USA, for *Escherichia coli* J53 pMG252, J53 pMG298 and J53 pMG306 and for *K. pneumoniae* with *oqxA*; Dr. D.C. Hooper, Massachusetts General Hospital, Boston, MA, USA, for *P. mirabilis* 06-489 and *E. coli* 5-59; Dr. J. L. Martinez-Martinez, H.U. Marques de Valdecilla, Santander, Spain, for *E. coli* strains TG1 p2007057 and pAT851; Drs. S. Levy and L. McMurry, Tufts Medical School, Boston, MA, USA, for *E. coli* DH5α pET*tetA2*, *E. coli* DH5α pET*tetB1* and *S. aureus* pT181; Dr. T. Kitten, The Philips Institute of Oral and Craniofacial Molecular Biology, VCU School of Dentistry, Richmond, VA, USA, for *E. coli* DH10B pJFP76; Dr. E. Giraud, INRA, Tours Research Center, Nouzilly, France, for *E. coli* TG1 pAB5S9; and Dr. Henning Sørum, Norwegian School of Veterinary Science, Oslo, Norway, and Dr. Les Burridge, Fisheries and Oceans Canada, St. Andrews Biological Station, New Brunswick, Canada, for their careful reading of the manuscript and many helpful suggestions. We also thank Mariam Hernández and Harriett V. Harrison for their invaluable help during preparation of the manuscript.

Author Contributions

Conceived and designed the experiments: AHB FCC. Performed the experiments: AT MAM LAH AL LI AHB. Analyzed the data: AHB FCC AT LI FM HPG. Wrote the paper: FCC AHB HPG AT.

References

1. Costa-Pierce BA (2010) Sustainable ecological aquaculture systems: the need for a new social contract for aquaculture development. Mar Technol Soc J 44: 88–112.
2. Tett P (2008) Fish farm wastes in the ecosystem. In: Holmer M, Black K, Duarte CM, Marbà N, Karakassis I, editors. Aquaculture in the Ecosystem. Berlin Heidelberg, German: Springer-Verlag. pp. 1–46.
3. Moore PR, Evenson A, Luckey TD, McCoy E, Elvehjemande CA, et al. (1946) Use of sulfasuxidine, streptothricin, and streptomycin in nutritional studies with the chick. J Biol Chem 165: 437–441.
4. Grumbles LC, Delaplane JP, Higgins TC (1948) Sulfaquinoxaline in the control of *Eimeria tenella* and *Eimeria necatrix* in chickens on a commercial broiler farm. Science 107: 196.
5. Jukes TH, Stokstad ELR, Taylor RR, Cunha TJ, Edwards HM, et al. (1950) Growth promoting effect of aureomycin on pigs. Arch Biochem 26: 324–330.
6. Marshall BM, Levy SB (2011) Food animals and antimicrobials: impacts on human health. Clin Microbiol Rev 24: 718–733.
7. Barnes EM (1958) The effect of antibiotic supplements on the faecal streptococci (Lancefield group D) of poultry. Br Vet J 114: 333–344.
8. Dibner JJ, Richards JD (2005) Antibiotic growth promoters in agriculture: history and mode of action. Poult Sci 84: 634–643.
9. Castanon JI (2007) History of the use of antibiotic as growth promoters in European poultry feeds. Poult Sci 86: 2466–2471.
10. Marshall BM, Levy SB Food animals and antimicrobials: impacts on human health. Clin Microbiol Rev 24: 718–733.
11. Wegener HC (2003) Antibiotics in animal feed and their role in resistance development. Curr Opin Microbiol 6: 439–445.
12. Aarestrup FM, Duran CO, Burch DGS (2008) Antimicrobial resistance in swine production. Anim Health Res Rev 9: 135–148.
13. Aarestrup FM, Jensen VF, Emborg HD, Jacobsen E, Wegener HC (2010) Changes in the use of antimicrobials and the effects on productivity of swine farms in Denmark. Am J Vet Res 71: 726–733.
14. Asche F, Roll KH, Tveterås S (2008) Future trends in aquaculture: productivity growth and increased production. In: Holmer M, Black K, Duarte CM, Marbà N, Karakassis I, editors. Aquaculture in the Ecosystem: Springer. pp. 271–292.
15. Barton JR, Fløysand A (2010) The political ecology of Chilean salmon aquaculture, 1982–2010. A trajectory from economic development to global sustainability Global Environ Change. pp. 739–752.
16. Bravo S, Midtlyng PJ (2007) The use of fish vaccines in the Chilean salmon industry 1999–2003. Aquaculture 270: 36–42.
17. Bustos PA, Young ND, Rozas MA, Bohle HM, Ildefonso RS, et al. (2011) Amoebic gill disease (AGD) in Atlantic salmon *(Salmo salar)* farmed in Chile. Aquaculture 110: 281–288.
18. Heuer OE, Kruse H, Grave K, Collignon P, Karunasagar I, et al. (2009) Human health consequences of use of antimicrobial agents in aquaculture. Clin Infect Dis 49: 1248–1253.
19. Burridge L, Weis JS, Cabello F, Pizarro J, Bostick K (2010) Chemical use in salmon aquaculture: A review of current practices and possible environmental effects. Aquaculture 306: 7–23.
20. Buschmann AH, Cabello F, Young K, Carvajal J, Varela DA, et al. (2009) Salmon aquaculture and coastal ecosystem health in Chile: analysis of regulations, environmental impacts and bioremediation systems. Ocean Coast Manage 52: 243–249.
21. Buschmann AH, Riquelme VA, Hernández-González MC, Varela DA, Jiménez JE, et al. (2006) A review of the impacts of salmon farming on marine coastal ecosystems in the southeast Pacific. ICES J Mar Sci 63: 1338–1345.
22. Sørum H (2006) Antimicrobial drug resistance in fish pathogens. In: Aarestrup FM, editor. Antimicrobial Resistance in Bacteria of Animal Origin. Washington, DC: ASM Press. pp. 213–238.
23. Millanao BA, Barrientos HM, Gomez CC, Tomova A, Buschmann A, et al. (2011) Uso inadecuado y excesivo de antibióticos: salud pública y salmonicultura en Chile.[Injudicious and excessive use of antibiotics: public health and salmon aquaculture in Chile]. Rev Med Chil 139: 107–118.
24. Barrientos M (2006) Estudio cualitativo y cuantitativo de las quinolonas y fluoroquinolonas importadas y autorizadas para uso y disposicion en medicina y en veterinaria en Chile, en el periodo 2002–2005. Consideraciones sobre su impacto para la salud pública y el medio ambiente. [Qualitative and quantitative study of quinolines and fluorquinolines imported and authorized for veterinary and medical use in Chile, 2002–2005. Considerations of its impact on public health and the environment]. Universidad Austral de Chile, Valdivia, Chile. Tesis Electrónicas UACh website. Available: http://cybertesis.uach.cl/tesis/uach/2006/fcb2751e/doc/fcb2751e.pdf. Accessed 2012 Jul 13. 122 p.
25. Millanao A (2002) Estudio cualitativo y cuantitativo de las quinolonas y fluoroquinolonas importadas y autorizadas para uso y disposicion en medicina y en veterinaria en Chile, en el periodo 1998–2001. Consideraciones sobre su impacto para la salud pública y el medio ambiente. [Qualitative and quantitative study of quinolines and fluorquinolines imported and authorized for medical and veterinary use and disposition in Chile, 1998–2001. Considerations of its impact on public health and the envrionment]. Universidad Austral de Chile, Valdivia, Chile. Tesis Electrónicas UACh website. Available: http://cybertesis.uach.cl/tesis/uach/2002/fcm645e/doc/fcm645e.pdf. Accessed 2012 Jul 13. 121 p.

26. Capone DG, Weston DP, Miller V, Shoemaker C (1996) Antibacterial residues in marine sediments and invertebrates following chemotherapy in aquaculture. Aquaculture 145: 55–75.
27. Hektoen H, Berge JA, Hormazabal V, Yndestad M (1995) Persistence of antibacterial agents in marine sediments. Aquaculture 133: 175–184.
28. Herwig RP, Gray JP, Weston DP (1997) Antibacterial resistant bacteria in superficial sediments near salmon net-cage farms in Puget Sound, Washington. Aquaculture 149: 163–283.
29. Coyne R, Smith P, Moriarty C (2001) The fate of oxytetracycline in the marine environment of a salmon cage farm. Mar Environ Health 3: 1–24.
30. Holten Lützhøft H-C, Halling-Sørensen B, Jørgensen SE (1999) Algal toxicity of antibacterial agents applied in Danish fish farming. Arch Environ Contam Toxicol 436: 1171–1175.
31. DePaola A, Peeler JT, Rodrick GE (1995) Effect of oxytetracycline-medicated feed on antibiotic resistance of gram-negative bacteria in catfish ponds. Appl Environ Microbiol 61: 2335–2340.
32. Cattoir V, Poirel L, Aubert C, Soussy CJ, Nordmann P (2008) Unexpected occurrence of plasmid-mediated quinolone resistance determinants in environmental *Aeromonas* spp. Emerg Infect Dis 14: 231–237.
33. Cattoir V, Poirel L, Mazel D, Soussy CJ, Nordmann P (2007) *Vibrio splendidus* as the source of plasmid-mediated QnrS-like quinolone resistance determinants. Antimicrob Agents Chemother 51: 2650–2651.
34. Pan JC, Ye R, Wang HQ, Xiang HQ, Zhang W, et al. (2008) *Vibrio cholerae* O139 multiple-drug resistance mediated by *Yersinia pestis* pIP1202-like conjugative plasmids. Antimicrob Agents Chemother 52: 3829–3836.
35. Rhodes G, Huys G, Swings J, McGann P, Hiney M, et al. (2000) Distribution of oxytetracycline resistance plasmids between aeromonads in hospital and aquaculture environments: implication of Tn*1721* in dissemination of the tetracycline resistance determinant Tet A. Appl Environ Microbiol 66: 3883–3890.
36. Welch TJ, Fricke WF, McDermott PF, White DG, Rosso ML, et al. (2007) Multiple antimicrobial resistance in plague: an emerging public health risk. PLoS One 2: e309.
37. Baquero F, Martinez JL, Canton R (2008) Antibiotics and antibiotic resistance in water environments. Curr Opin Biotechnol 19: 260–265.
38. Schmidt AS, Bruun MS, Dalsgaard I, Pedersen K, Larsen JL (2000) Occurrence of antimicrobial resistance in fish-pathogenic and environmental bacteria associated with four Danish rainbow trout farms. Appl Environ Microbiol 66: 4908–4915.
39. Saga T, Kaku M, Onodera Y, Yamachika S, Sato K, et al. (2005) *Vibrio parahaemolyticus* chromosomal *qnr* homologue *VPA0095*: demonstration by transformation with a mutated gene of its potential to reduce quinolone susceptibility in *Escherichia coli*. Antimicrob Agents Chemother 49: 2144–2145.
40. Strahilevitz J, Jacoby GA, Hooper DC, Robicsek A (2009) Plasmid-mediated quinolone resistance: a multifaceted threat. Clin Microbiol Rev 22: 664–689.
41. Jacobs L, Chenia HY (2007) Characterization of integrons and tetracycline resistance determinants in *Aeromonas* spp. isolated from South African aquaculture systems. Int J Food Microbiol 114: 295–306.
42. Rosewarne CP, Pettigrove V, Stokes HW, Parsons YM (2010) Class 1 integrons in benthic bacterial communities: abundance, association with Tn*402*-like transposition modules and evidence for coselection with heavy-metal resistance. FEMS Microbiol Ecol 72: 35–46.
43. Schmidt AS, Bruun MS, Dalsgaard I, Larsen JL (2001) Incidence, distribution, and spread of tetracycline resistance determinants and integron-associated antibiotic resistance genes among motile aeromonads from a fish farming environment. Appl Environ Microbiol 67: 5675–5682.
44. Sørum H, L'Abee-Lund TM, Solberg A, Wold A (2003) Integron-containing IncU R plasmids pRAS1 and pAr-32 from the fish pathogen *Aeromonas salmonicida*. Antimicrob Agents Chemother 47: 1285–1290.
45. Xu H, Davies J, Miao V (2007) Molecular characterization of class 3 integrons from *Delftia* spp. J Bacteriol 189: 6276–6283.
46. Miranda CD, Zemelman R (2001) Antibiotic resistant bacteria in fish from the Concepcion Bay, Chile. Mar Pollut Bull 42: 1096–1102.
47. Silva J, Zemelman R, Mandoca MA, Henriquez M, Merino C, et al. (1987) Antibiotic-resistant gram negative bacilli isolated from sea water and shellfish. Possible epidemiological implications. Rev Latinoam Microbiol 29: 165–169.
48. Bissett A, Bowman J, Burke C (2006) Bacterial diversity in organically-enriched fish farm sediments. FEMS Microbiol Ecol 55: 48–56.
49. Navarro N, Leakey RJG, Black KD (2008) Effect of salmon cage aquaculture on the pelagic environment of temperate coastal waters: seasonal changes in nutrients and microbial community. Mar Ecol Prog Ser 361: 47–58.
50. Kerry J, Coyne R, Gilroy D, Hiney M, Smith P (1996) Spatial distribution of oxytetracycline and elevated frequencies of oxytetracycline resistance in sediments beneath a marine salmon farm following oxytetracycline therapy. Aquaculture 145: 31–39.
51. Smith P (1996) Is sediment deposition the dominant fate of oxytetracycline used in marine salmonid farms: a review of available evidence. Aquaculture 146: 157–169.
52. Weston DP (2000) Ecological effects of the use of chemicals in aquaculture. In: Arthur JR, Lavilla-Pitogo CR, Subasinghe RP, editors. Use of Chemicals in Aquaculture in Asia: Proceedings of the Meeting on the Use of Chemicals in Aquaculture in Asia 20–22 May 1996, Tigbauan, Iloilo, Philippines. Tigbauan, Iloilo, , Philippines: Aquaculture Department, Southeast Asian Fisheries Development Center. pp. 20–30.
53. Council NR (2010) The Prevention and Treatment of Missing Data in Clinical Trials. Washington, D.C.: National Academies Press. 162 p.
54. Pepe MS (2003) The Statistical Evaluation of Medical Tests for Classificaiton and Prediction. Oxford Statistical Science Series 28 Oxford University Press. p. 318.
55. Kapetanaki M, Kerry J, Hiney M, O'Brien C, Coyne R, et al. (1995) Emergence, in oxytetracycline-free marine mesocosms, of microorganisms capable of colony formation on oxytetracycline-containing media. Aquaculture 134: 227–236.
56. Smith P, Hiney MP, Samuelsen OB (1994) Bacterial resistance to antimicrobial agents used in fish farming: A critical evaluation of method and meaning. Annu Rev Fish Dis 4: 273–313.
57. Smith P (2008) Antimicrobial resistance in aquaculture. Rev Sci Tech 27: 243–264.
58. Akinbowale OL, Peng H, Grant P, Barton MD (2007) Antibiotic and heavy metal resistance in motile aeromonads and pseudomonads from rainbow trout *(Oncorhynchus mykiss)* farms in Australia. Int J Antimicrob Agents 30: 177–182.
59. McIntosh D, Cunningham M, Ji B, Fekete FA, Parry EM, et al. (2008) Transferable, multiple antibiotic and mercury resistance in Atlantic Canadian isolates of *Aeromonas salmonicida* subsp. *salmonicida* is associated with carriage of an IncA/C plasmid similar to the *Salmonella enterica* plasmid pSN254. J Antimicrob Chemother 61: 1221–1228.
60. Björklund H, Bondestam J, Bylund G (1990) Residues of oxytetracycline in wild fish and sediments from fish farms. Aquaculture 86: 359–367.
61. Samuelsen OB, Torsvik V, Ervik A (1992) Long-range changes in oxytetracycline concentration and bacterial resistance toward oxytetracycline in a fish farm sediment after medication. Sci Total Environ 114: 25–36.
62. D'Costa VM, King CE, Kalan L, Morar M, Sung WW, et al. (2011) Antibiotic resistance is ancient. Nature 477: 457–461.
63. Robicsek A, Strahilevitz J, Jacoby GA, Macielag M, Abbanat D, et al. (2006) Fluoroquinolone-modifying enzyme: a new adaptation of a common aminoglycoside acetyltransferase. Nat Med 12: 83–88.
64. Chopra I, Roberts M (2001) Tetracycline antibiotics: mode of action, applications, molecular biology, and epidemiology of bacterial resistance. Microbiol Mol Biol Rev 65: 232–260.
65. Miranda CD, Kehrenberg C, Ulep C, Schwarz S, Roberts MC (2003) Diversity of tetracycline resistance genes in bacteria from Chilean salmon farms. Antimicrob Agents Chemother 47: 883–888.
66. Singer RS, Patterson SK, Meier AE, Gibson JK, Lee HL, et al. (2004) Relationship between phenotypic and genotypic florfenicol resistance in *Escherichia coli*. Antimicrob Agents Chemother 48: 4047–4049.
67. Canton R (2009) Antibiotic resistance genes from the environment: a perspective through newly identified antibiotic resistance mechanisms in the clinical setting. Clin Microbiol Infect 15 Suppl 1: 20–25.
68. Hastings PJ, Rosenberg SM, Slack A (2004) Antibiotic-induced lateral transfer of antibiotic resistance. Trends Microbiol 12: 401–404.
69. Baquero F (2009) Environmental stress and evolvability in microbial systems. Clin Microbiol Infect 15 Suppl 1: 5–10.
70. Dorr T, Lewis K, Vulic M (2009) SOS response induces persistence to fluoroquinolones in *Escherichia coli*. PLoS Genet 5: e1000760.
71. Fortt ZA, Cabello CF, Buschmann RA (2007) Residues of tetracycline and quinolones in wild fish living around a salmon aquaculture center in Chile. Rev Chilena Infectol 24: 14–18.
72. Samuelsen OB, Lunestad BT, Husevåg B, Hølleland T, Ervik A (1992) Residues of oxolinic acid in wild fauna following medication in fish farms. Dis Aq Org 12: 111–119.
73. Cabello FC (2006) Heavy use of prophylactic antibiotics in aquaculture: a growing problem for human and animal health and for the environment. Environ Microbiol 8: 1137–1144.
74. Carroll ML, Cochrane S, Fieler R, Velvin R, White P (2003) Organic enrichment of sediments from salmon farming in Norway: environmental factors, management practices, and monitoring techniques. Aquaculture 226: 165–180.
75. Soto D, Norambuena F (2004) Evaluation of salmon farming effects on marine systems in the inner seas of southern Chile: a large-scale mensurative experiment. J Appl Ichthyol 20: 493–501.
76. Pouliquen H, Gouelo D, Larhantec M, Pilet N, Pinault L (1997) Rapid and simple determination of oxolinic acid and oxytetracycline in the shell of the blue mussel (*Mytilus edulis*) by high-performance liquid chromatography. J Chromatogr B Biomed Sci Appl 702: 157–162.
77. Sørensen LK, Hansen H (2001) Determination of oxolinic acid in marine sediment by HPLC with fluorescence detection. J Liq Chromat Rel Tech 24: 2469–2476.
78. Cavaco LM, Hasman H, Xia S, Aarestrup FM (2009) *qnrD*, a novel gene conferring transferable quinolone resistance in *Salmonella enterica* serovar Kentucky and *Bovismorbificans* strains of human origin. Antimicrob Agents Chemother 53: 603–608.
79. Robicsek A, Strahilevitz J, Sahm DF, Jacoby GA, Hooper DC (2006) *qnr* prevalence in ceftazidime-resistant *Enterobacteriaceae* isolates from the United States. Antimicrob Agents Chemother 50: 2872–2874.
80. Wang M, Guo Q, Xu X, Wang X, Ye X, et al. (2009) New plasmid-mediated quinolone resistance gene, *qnrC*, found in a clinical isolate of *Proteus mirabilis*. Antimicrob Agents Chemother 53: 1892–1897.

81. Kim HB, Wang M, Park CH, Kim EC, Jacoby GA, et al. (2009) *oqxAB* encoding a multidrug efflux pump in human clinical isolates of *Enterobacteriaceae*. Antimicrob Agents Chemother 53: 3582–3584.
82. Minarini LA, Poirel L, Cattoir V, Darini AL, Nordmann P (2008) Plasmid-mediated quinolone resistance determinants among enterobacterial isolates from outpatients in Brazil. J Antimicrob Chemother 62: 474–478.
83. Ng LK, Martin I, Alfa M, Mulvey M (2001) Multiplex PCR for the detection of tetracycline resistant genes. Mol Cell Probes 15: 209–215.
84. Bolton LF, Kelley LC, Lee MD, Fedorka-Cray PJ, Maurer JJ (1999) Detection of multidrug-resistant *Salmonella enterica* serotype Typhimurium DT104 based on a gene which confers cross-resistance to florfenicol and chloramphenicol. J Clin Microbiol 37: 1348–1351.
85. Bey BS, Fichot EB, Dayama G, Decho AW, Norman RS (2010) Extraction of high molecular weight DNA from microbial mats. Biotechniques 49: 631–640.
86. Miller DN, Bryant JE, Madsen EL, Ghiorse WC (1999) Evaluation and optimization of DNA extraction and purification procedures for soil and sediment samples. Appl Environ Microbiol 65: 4715–4724.
87. Griffen AL, Beall CJ, Firestone ND, Gross EL, Difranco JM, et al. (2011) CORE: a phylogenetically-curated 16S rDNA database of the core oral microbiome. PLoS One 6: e19051.
88. Weisburg WG, Barns SM, Pelletier DA, Lane DJ (1991) 16S ribosomal DNA amplification for phylogenetic study. J Bacteriol 173: 697–703.
89. Conover WJ, Iman RL (1981) Rank transformations as a bridge between parametric and nonparametric statistics. Am Stat 35: 124–133.

17

Ultradeep 16S rRNA Sequencing Analysis of Geographically Similar but Diverse Unexplored Marine Samples Reveal Varied Bacterial Community Composition

Chairmandurai Aravindraja[9], **Dharmaprakash Viszwapriya**[9], **Shunmugiah Karutha Pandian***

Department of Biotechnology, Alagappa University, Karaikudi, Tamil Nadu, India

Abstract

Background: Bacterial community composition in the marine environment differs from one geographical location to another. Reports that delineate the bacterial diversity of different marine samples from geographically similar location are limited. The present study aims to understand whether the bacterial community compositions from different marine samples harbour similar bacterial diversity since these are geographically related to each other.

Methods and Principal Findings: In the present study, 16S rRNA deep sequencing analysis targeting V3 region was performed using Illumina bar coded sequencing. A total of 22.44 million paired end reads were obtained from the metagenomic DNA of Marine sediment, Rhizosphere sediment, Seawater and the epibacterial DNA of Seaweed and Seagrass. Diversity index analysis revealed that Marine sediment has the highest bacterial diversity and the least bacterial diversity was observed in Rhizosphere sediment. Proteobacteria, Actinobacteria and Bacteroidetes were the dominant taxa present in all the marine samples. Nearly 62–71% of rare species were identified in all the samples and most of these rare species were unique to a particular sample. Further taxonomic assignment at the phylum and genus level revealed that the bacterial community compositions differ among the samples.

Conclusion: This is the first report that supports the fact that, bacterial community composition is specific for specific samples irrespective of its similar geographical location. Existence of specific bacterial community for each sample may drive overall difference in bacterial structural composition of each sample. Further studies like whole metagenomic sequencing will throw more insights to the key stone players and its interconnecting metabolic pathways. In addition, this is one of the very few reports that depicts the unexplored bacterial diversity of marine samples (Marine sediment, Rhizosphere sediment, Seawater) and the host associated marine samples (Seaweed and Seagrass) at higher depths from uncharacterised coastal region of Palk Bay, India using next generation sequencing technology.

Editor: Jonathan H. Badger, J. Craig Venter Institute, United States of America

Funding: This study was supported by a Research grant from Department of Biotechnology (DBT), the Government of India (GOI) (Grant number BT/PR11994/NDB/52/134/2009). Financial support provided to CA and DV in the form of Junior Research Fellowship from Department of Biotechnology (DBT), the Government of India (GOI) and Council of Scientific and Industrial Research (CSIR) (09/688(0017)/2011-EMR-I, India, respectively is gratefully acknowledged. The authors sincerely acknowledge the computational and bioinformatics facility provided by the Alagappa University Bioinformatics Infrastructure Facility (funded by DBT, GOI; Grant No, BT/BI/25/001/2006). The funders had no role in study design, data collection and analysis, decision to publish, or preparation of the manuscript.

Competing Interests: The authors have declared that no competing interests exist.

* E-mail: sk_pandian@rediffmail.com

[9] These authors contributed equally to this work.

Introduction

Marine environment, which constitutes the largest portion of earth, comprises highly diverse and complex microbial communities. Marine bacteria are important for maintenance of carbon dynamics in marine ecosystem. The presence of variety of heterotrophic bacteria and their importance is very well recognized for sustained ecological and biogeochemical cycle in marine environment [1–3].

Both culture dependent and culture independent methods have been widely deployed to characterize heterotrophic bacterial composition and its distribution in marine ecosystem. Since culture dependent methods fail to delineate the true bacterial diversity in the given environment, culture independent methods including T-RFLP, SSCP and DGGE have been widely used to study the bacterial diversity [4–6]. Though these techniques provide a bird's eye view of bacterial community composition, ultra deep sequencing using next generation technology revolutionized our understanding of complex microbial community in a broader way.

Availability of unprecedented 16S rRNA gene sequences in public databases provides a testament to the deployment of Next generation sequencing technology in studying bacterial diversity. Though 454 (Roche) had been widely utilized to study the

Table 1. Overview of sequence information.

Sample	Barcode	No. of raw reads	No. of high quality reads[a]	No. of reads that passed CD-HIT DUP[b]	No. of reads with predicted rRNA like sequences	No. of reads with identified rRNA sequences
		(in millions)				
R-SED	CGTACT	4.24	3.79	1.10	0.54	94151
M-SED	CTGATC	4.78	4.37	0.54	0.29	71112
SR	CACTGT	3.20	2.84	0.78	0.38	46245
SW	GACTGA	5.12	3.60	0.69	0.34	68503
SG	CGTACT	5.10	4.63	0.82	0.42	80316

[a]More than 70% of bases in a read with >20 phred score and reads which are of low quality were trimmed and used.
[b]Number of reads after chimera and duplicate removal.

bacterial diversity, low sequencing cost, high depth coverage of Illumina (Solexa) sequencing technology has surpassed the utility of 454 in microbial diversity analysis [7–9]. In addition, enhanced sample throughput using bar coding, advancement in Illumina sequencing platform (MiSeq, HiSeq) provide a pre-eminent view of microbial composition than analyzed with other sequencing technologies [10]. Unprecedented data using Illumina with high taxonomic resolution gain insights to understand even the rare microbes in different environmental habitats [11,12].

Plethora of information is available about the bacterial community composition of different marine habitats along the coastal regions of western countries [13–16]. Few reports are available about the bacterial diversity along the coastal region in India. Lack of in-depth analysis of bacterial diversity were observed in these studies as most of the works were restricted based on culture dependent and other low throughput techniques [6,17,18]. Hence a considerable effort has been taken in the present study to explore variability in bacterial community composition in hitherto uncharacterized geographically related marine environmental samples through high throughput sequencing. Comparative assessment of heterotrophic bacterial community of these marine samples were carried out using bar-coded Illumina sequencing of hypervariable V3 region of 16S rRNA gene. This study will ameliorate our understanding of the variation in distribution of microbial assemblages in these geographically related marine samples in a better way and also throw more insight to the biotechnological potential of previously unexplored marine habitats.

Results and Discussion

Generation of multimillion Illumina reads and production of high quality reads

To delineate the bacterial community composition among the geographically linked marine samples, we carried out high throughput Illumina paired end sequencing. Rhizosphere sediment (*Rhizophora apiculata*), marine sediment, seawater, seaweed (*Gracilaria* sp.) and seagrass (*Cymodacea* sp.) of Karankadu coastal region, Palk Bay, India are the five different geographically linked marine samples chosen for the present study. The bacterial diversity in Palk Bay is hitherto unearthed although few reports are there for its wide range of biotechnological applications [19,20]. Targeting the hypervariable V3 region, a total of 22.44 million paired end raw reads were obtained for five marine samples using Illumina GAIIx platform. Averages of 4.48 million reads were obtained for each sample. Stringent measures were taken to rule out the erroneous reads. Raw reads with non ATGC characters, primer containing sequences and possible adapter contaminated sequences were trimmed and taken for subsequent analysis. In order to eliminate the low quality reads, reads with more than 70% of bases with Phred value of greater than 20 alone were taken into account. Since chimeric reads tend to inflate bacterial diversity, they were removed using CD-HIT DUP and nearly 3.93 million chimeric pruned reads were obtained from all the five marine samples [21]. These stringent measures removed nearly 87% of the initial reads which is in accordance with other reports that filter 50–80% of initial reads [8,9,22] (Table 1).

Diversity Index and Taxonomic assignment at Phylum level

Taxonomy assignment was done using MG-RAST pipeline [23]. For all five samples the rarefaction curves have reached near plateau, indicating that the sampling depth and sequencing coverage were good (Figure 1). Both the Shannon weiner index and Simpson index showed that marine sediment had the highest diversity while rhizosphere sediment had the lowest (Table 2). The relative distribution of each phylum varied among the marine samples. The predominant phyla observed in all the five marine samples were Proteobacteria, Actinobacteria and Bacteroidetes. This result supports the fact that the representative members from these phyla are cosmopolitan and are not specific to the specific habitat. This is the first report that documents the presence of different bacterial phyla on the surface of seaweed and seagrass. The relative abundance of each phylum varied among rhizosphere sediment, marine sediment, seawater, seagrass and seaweed and remains distinct to each other (Figure 2). This result was also complemented by the heat map analysis in which these marine samples do not cluster together (Figure 3). The relative abundance of firmicutes (*Bacillus* and *Halobacillus*) was higher in sediment sample and rather it was lower in other marine samples. This report is in total agreement with earlier reports wherein the higher abundance of firmicutes in sediment is attributed to human fecal contamination [24,25]. In addition, other possible reasons for the abundance could be the ability of *Bacillus* sp to form endospore [26]; Firmicutes are better competitors in the natural environment since they are widely recognized as producers of enzyme inhibitors, antibiotics and its metabolic and physiologic versatility [27]. Furthermore, seawater contains high proportion of cyanobacteria which was almost rare in other marine samples. One may hypothesize that since most of the cyanobacteria are photosyn-

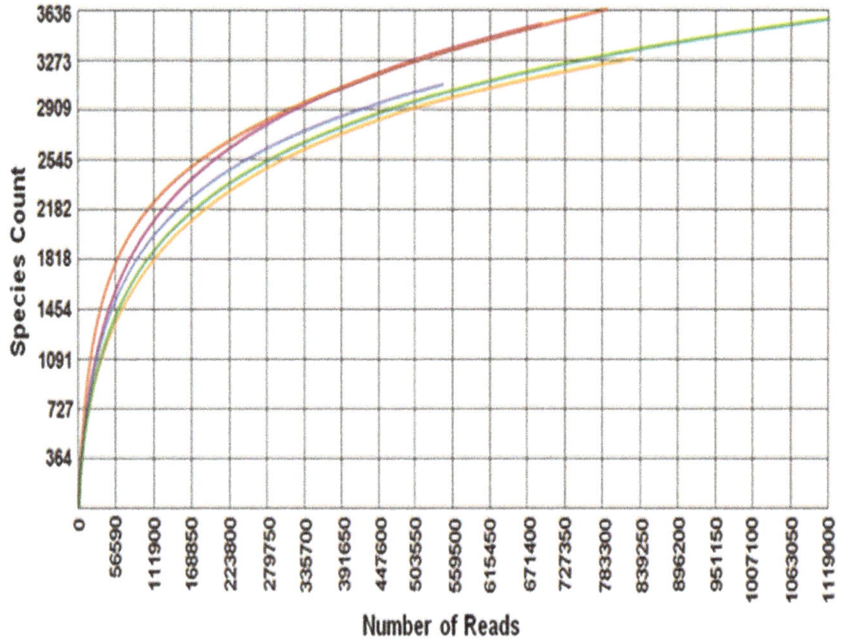

Figure 1. Rarefaction analysis for the observed species. The rarefaction curves for all the marine samples reached the near plateau phase representing good sampling depth. The samples are R-SED (Green), M-SED (Blue), SR (Red), SW (Violet) and SG (Orange).

thetic, it is freely floating in seawater and it may not carry out photosynthesis effectively if it is associated with other marine samples. In general, the relative abundance of each phylum among the marine samples was highly variable.

Distribution of bacterial communities at the higher taxonomic level

We assessed the relative distribution of bacterial taxa at the level of class and genus for each sample. Except seawater, Alphaproteobacteria was abundant in all other marine samples. Though the high abundance of Alphaproteobacteria has been well documented in seawater, the relative low abundance of this class in seawater in the present study may be explained by difference in various ecophysiological factors, which is not the focus of the present study. The representative members of genera *Agrobacterium, Ochrobactrum, Bartonella, and Ruegeria* were most often present in all the marine samples. Nearly equal distribution of Gammaproteobacteria was observed in rhizosphere sediment, seawater, seaweed and seagrass samples whereas increased abundance was observed in sediment sample. The dominant genera of Gammaproteobacteria like *Halorhodospira, Dyella, Vibrio, Methylococcus, Halothiobacillus, Klebsiella, E.coli, Enterobacter and Acinetobacter* were present in all the marine samples in various levels. This observation is in parallel with the other global studies in which Gammaproteobacteria was found to be widely distributed in different marine niches supported by its huge phylogenetic and phenotypic diversity [28]. Several intermediate abundant taxa like Actinobacteria, Flavobacteria and Optitutae were present in all the marine samples. *Desulfovibrio*, the most dominant genus from Delta proteobacteria was found to be abundant in seaweed. The other members of the class like *Desulfovibrio, Desulfobulbus, Desulfosarcina, Sulfobacterium and Bacteriovorax* were scarce in all the samples. Several reports documented the presence of members of Deltaproteobacteria in marine environment which mainly acts as sulfate reducers [29] (Figure 4).

Table 2. Diversity indices of the marine samples.

Sample	Diversity Index	
	Shannon Index	Simpson Index
R-SED	3.49	0.20
M-SED	4.30	0.04
SR	3.63	0.15
SW	3.75	0.12
SG	4.10	0.10

At phylum level, most of the bacterial taxon was recurrently found in all the marine samples, but at higher taxonomic level (genus and species level) the relative contribution of the bacterial taxa of the community varied from one sample to other. This suggests that though all the samples are geographically linked to each other, the composition of bacterial taxon at the genus and species level gets varied among each other. The reasons for this characterization include the fact that each marine sample provides different nutrient composition and varied ecological conditions for specific bacterial taxon to reside and proliferate. Further analysis was carried out to understand how these marine samples are distinct to each other in terms of both abundant and rare species diversity. The details of taxonomic affiliation for all the samples have been provided in Table S1 to S5. These tables provide the detailed view of taxonomic level distribution of sequences from phylum to strain level and their corresponding number of hits in descending order of their relative abundance (%). The hits are based on clustering the sequences at 97% identity against RNA databases implemented in MG- RAST (namely RDP, Greengenes, SILVA LSU and SSU). Graphical representation of the relative abundance of bacterial diversity from phylum to species can be visualized using Krona chart in Figure S1 to S5.

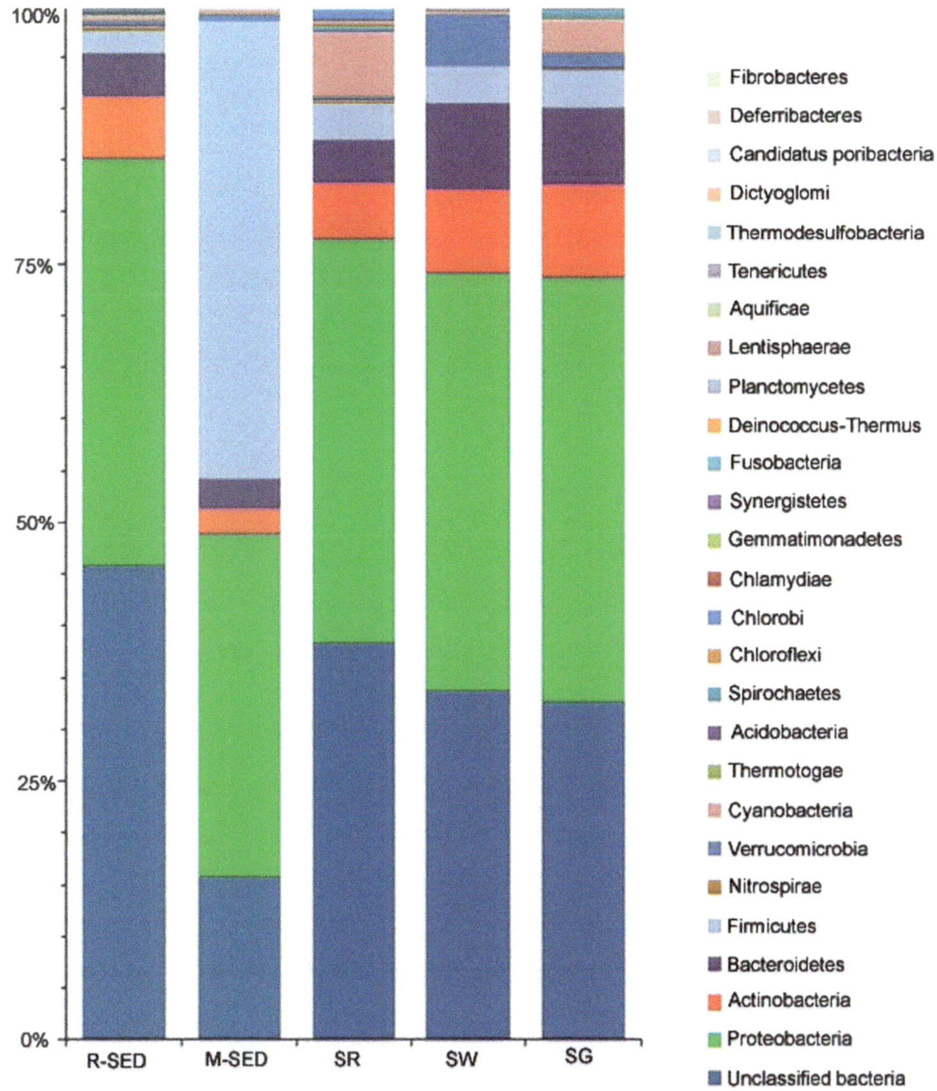

Figure 2. Relative abundance of bacterial phyla across the samples. The percentage of sequences is plotted on Y-axis. Unclassified bacteria were reads without recognizable match in the searched database. Except for M-SED, Proteobacteria is the predominant phylum in all the samples. In M-SED, Firmicutes is the predominant phylum.

Analysis of abundant and rare species diversity in marine samples

Both abundant and rare species constitute the bacterial population in all the marine samples. Based on the previous reports [11,30] the rare species may be defined as the frequency of species with <0.01% of the total population and the rest of the population are considered to be abundant. According to this the proportion of abundant and rare species ranges from 28.12% to 37.7% and 62.3% to 71.18% respectively in all the marine samples (Figure 5A). Venn diagram was constructed to predict the number of unique and shared species among the abundant and rare species in five marine samples. It was observed that few of the rare species were shared commonly among the samples (Figure 5B) and most of the species which were unique to a particular sample are rare (Figure S6). But in case of abundant species, most of the abundant species were shared among the samples and only few are unique (Figure 5C). Details of the unique and rare species corresponding to each marine sample are given in Table S6. It has been a long belief that abundant species that is represented by few species in the bacterial population plays an important role in marine habitat. But the present report provides a testament that rare species also play an important role in biogeochemical cycle and other functional role in the marine environment. In addition, the rare species which are unique to a particular sample may play important roles than previously predicted. Further research regarding the importance of rare species in the marine environment is needed to fully understand its role in the marine ecosystem.

In summary, we report that, though these marine samples are geographically linked to each other, the bacterial community compositions are distinct to each other at higher taxonomic level. This suggested that, distribution of bacterial community composition was very specific to specific habitat even though these habitats are present in similar geographical location. In addition, though high throughput sequencing using illumina depicts the known bacterial community of marine samples in the present study, the unclassified bacteria constitute a wide range of 15% to

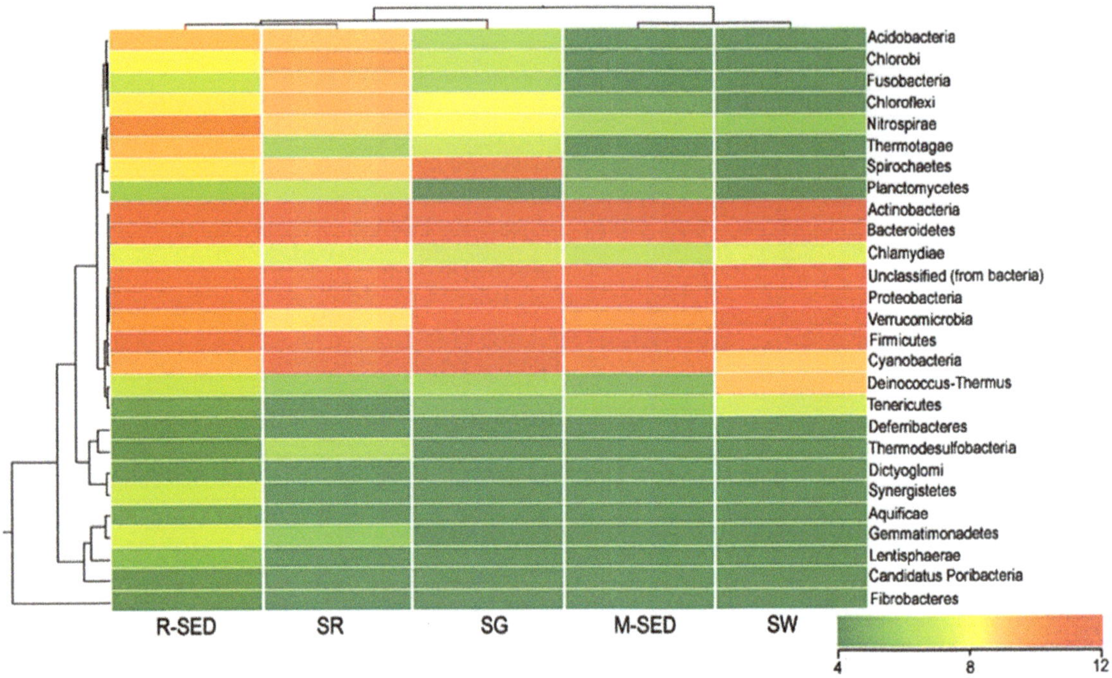

Figure 3. Heat map analysis of the marine samples. The marine samples do not cluster together indicating that the bacterial diversity among the samples is varied and distinct.

45% for all the marine samples. Further studies like whole metagenome sequencing holds promise to delineate the uncultured bacterial diversity of these samples. Moreover, this is the first study that sheds light on hitherto unexplored bacterial diversity of geographically similar marine samples in Palk Bay region which hold greater promise for broad range of biotechnological applications.

Materials and Methods

Ethics Statement

No specific permissions were required for the present study described here. The location is neither privately-owned nor protected in any way and this study did not involve any endangered or protected species.

Sample collection

Marine sediment (M-SED), rhizosphere sediment (R-SED) (*Rhizophora apiculata*), sea water (SR), seaweed (SW) (*Gracilaria* sp.) and seagrass (SG) (*Cymodacea* sp.) were collected from Karankadu Coastal area (8° 28′N lat. and 77° 41′E long.), Palk Bay, Tamil Nadu, India. The sampled seaweed and sea grass are the most predominant ones in the sampling site. All samples except seawater were collected in sterile plastic bags. Seawater was collected in sterile can and initially filtered to remove the debris followed by subsequent filtration with 0.22 micron filter (Millipore, Massachusetts, US). All the other samples were transported to laboratory within two hours in ice and later stored at −80°C until processing.

DNA extraction

DNA from sediment and rhizosphere sediment were extracted according to the lyzozyme method as described [31]. DNA extraction from filters was done using standard procedure [13]. DNA extraction from the surface of seaweed and sea grass were done as previously described [32]. Extracted DNA was visualized by agarose gel electrophoresis and quantified using nanodrop spectrophotometer (Shimadzu, Singapore).

Amplicon library construction

Library construction involved two PCR reactions (Kapa HiFi Hot start, Kapa Biosystems, Massachusetts, US). The first reaction targeted the V3 region using primers 341F, 5′-CCTACGGGAGGCAGCAG-3′ and 518R, 5′-ATTACCGCGGCTGCTGG-3′ with the initial amount of 20 ng of DNA. Cycle conditions were an initial denaturation at 98°C for 3min, followed by 20 cycles at 98°C for 30 sec, 60°C for 30 sec, 72°C for 30 sec, and ended with a final extension step at 72°C for 3 min. The amplified products of 170–200 bp were purified by gel elution using minelute columns (Qiagen, Hilden, Germany). For the subsequent cycle of PCR, 150 ng of the amplified sample was used as template and modified primers (which included V3 specific sequences and adapter sequences) were used as suggested [33]. These modified primers had four degenerate bases in the forward primer to enhance base calling accuracy, cluster density and aids in identifying unique clusters. The reverse primer contains a 6-bp barcode sequence for multiplexing. Similar PCR conditions were used as the initial protocol except the number of cycles were reduced to six. The amplified products were cleaned up using Agencourt Ampure XP SPRI beads (Beckman Coulter, California, US). The prepared library was validated for its quality by running an aliquot through High Sensitivity Bioanalyzer Chip (Agilent technologies, California, US) and quantified using nanodrop spectrophotometer.

Illumina sequencing

PCR products with unique indices from each library were taken in equal nanogram quantities and subjected to 100-

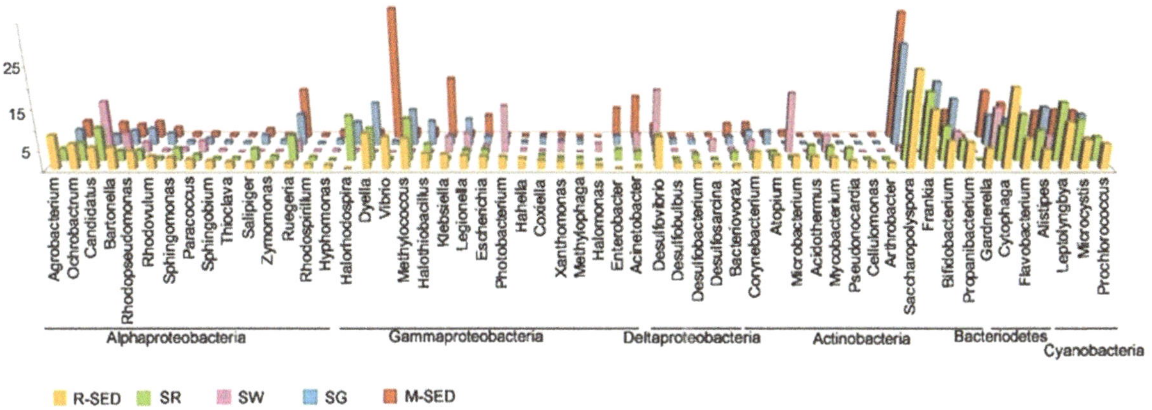

Figure 4. Comparison of bacterial communities among the samples at genus level. The members of the phyla Alphaproteobacteria, Gammaproteobacteria, Deltaproteobacteria, Actinobacteria, Bacteriodetes and Cyanobacteria which are above 1% are compared among the samples. As evident all the samples had varying levels of different genera.

nucleotide paired-end multiplex sequencing using Illumina GAIIx sequencer at Genotypic Technology Pvt. Ltd (Bangalore, India). All the samples except R-SED were pooled in a single lane. R-SED amplicons were sequenced in another lane combined with other samples not related to this work. Image analysis and base calling were done using Illumina Analysis pipeline (Version 2.2).

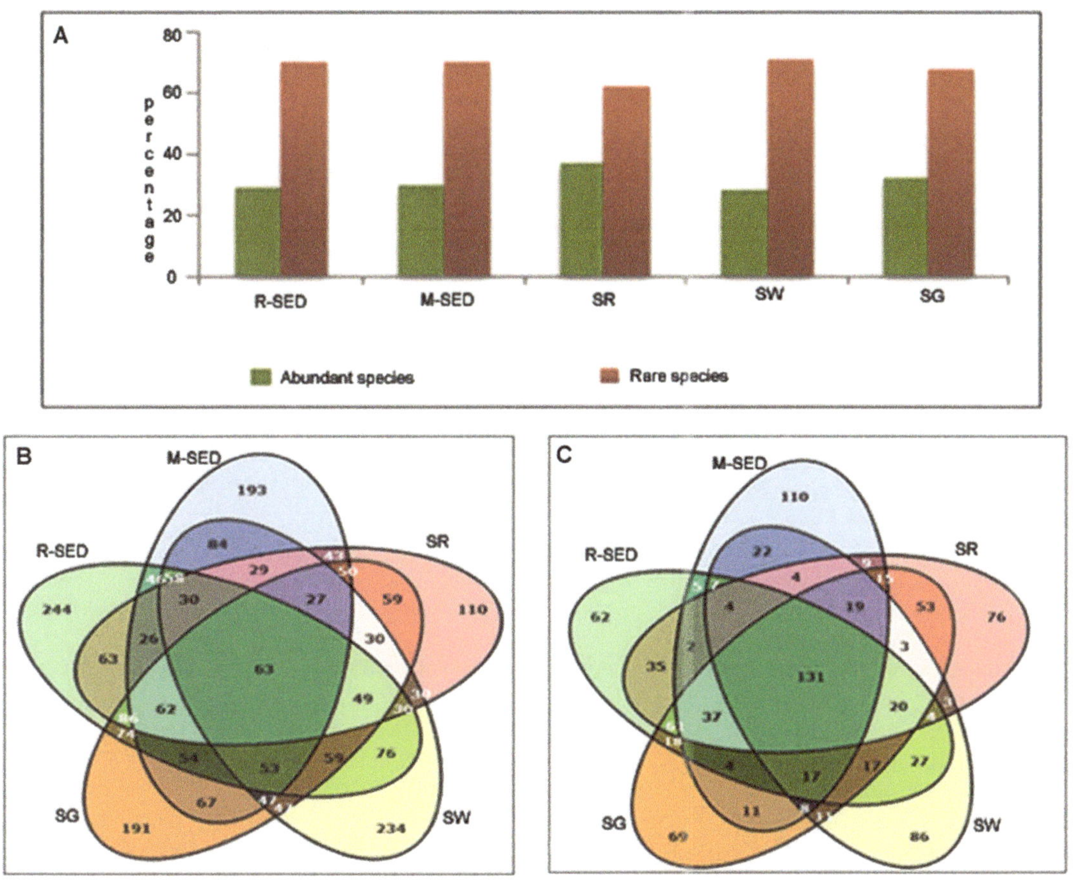

Figure 5. Analysis of abundant and rare species among the marine samples. A. **Percentage of abundant and rare species across the samples.** In all the five samples when compared to abundant species the percentage of rare species is significantly high indicating the importance of rare species in a community. B. **Shared and specific rare species among the marine samples.** Most of the rare species are unique to particular sample and only a few are shared by all the samples. C. **Shared and specific abundant species among the marine samples.** Most of the abundant species are common to all the samples and only a few are unique to particular sample.

Initial Processing of sequence reads

Demultiplexing was done using Consensus Assessment of Sequence and Variation (CASAVA). Only high quality reads with more than 70% of bases with Phred Score greater than 20 were considered significant and taken for subsequent analysis. Reads with adapter sequences were removed. The primer sequences, the barcode and the degenerate bases were removed using an automated Perl code generating processed reads.

Data analysis

Duplicates and chimeras from processed reads were removed using CD-HIT DUP with the minimum length of common sequence shared between a chimeric read and each of its parents set as 20 bases [34]. Abundance ratio between a parent read and a chimeric read was set to 1. The resulting dataset was pre-screened using uclust for a minimum of 70% identity to ribosomal sequences and then clustered at 97% identity against RNA databases implemented in MG- RAST (namely RDP, Greengenes, SILVA LSU and SSU) [35]. Taxonomic assignment from phylum level to strain level was assigned based on the hits. Abundance graphs were plotted based on the number of hits. Heatmap and rarefaction curve were plotted using MG-RAST. The Venn diagram was made with Venn diagram plotter jquery.venny, a tool developed by genotoul bioinfo (http://bioinfor.genotoul.fr). Diversity index was calculated using SPADE software [36]. Krona graphs were plotted using Krona [37].

Nucleotide sequence accession number

Paired end Illumina sequence data from this study were submitted to the NCBI Sequence Read Archive (SRA) under accession number SRP018711.

Supporting Information

Figure S1 Graphical representation of the relative abundance of bacterial diversity from phylum to species level of M-SED can be visualized in this file using Krona visualization tool. This file will help the readers to understand the relative distribution and abundance of the complex bacterial community composition of the sample through simple graphical representation.

Figure S2 Graphical representation of the relative abundance of bacterial diversity from phylum to species level of R-SED can be visualized in this file using Krona visualization tool.

Figure S3 Graphical representation of the relative abundance of bacterial diversity from phylum to species level of SR can be visualized in this file using Krona visualization tool.

Figure S4 Graphical representation of the relative abundance of bacterial diversity from phylum to species level of SW can be visualized in this file using Krona visualization tool.

Figure S5 Graphical representation of the relative abundance of bacterial diversity from phylum to species level of SG can be visualized in this file using Krona visualization tool.

Figure S6 This file is a bar graph describing the percentage of abundant and rare species among the unique species of the five marine samples.

Table S1 This file contains the details of taxonomic level distribution for M-SED sample. Details of phylum to strain level distribution and their corresponding number of hits are given by clustering the sequences at 97% identity against RNA databases implemented in MG- RAST (namely RDP, Greengenes, SILVA LSU and SSU).

Table S2 This file contains the details of taxonomic level distribution for R-SED sample.

Table S3 This file contains the details of taxonomic level distribution for SR sample.

Table S4 This file contains the details of taxonomic level distribution for SW sample.

Table S5 This file contains the details of taxonomic level distribution for SG sample.

Table S6 This file contains the details of shared and specific abundant species among the five samples in the first sheet and the details of shared and specific rare species among the samples in the second sheet.

Acknowledgments

The authors are grateful to Dr. Raja C Mugasimangalam, Dr. Vidya Niranjan, Mr. Suyash Agarwal, Mrs. Rashmi Praveen, Mr. Girishwaran Pragash and Mr. ArunPrasath Sekar of Genotypic Technology Pvt. Ltd., Bangalore for their guidance in Illumina sequencing and preliminary data analysis. The authors also thank Prof. Kalai Mathee of Florida International University, Miami for critically reading the manuscript.

Author Contributions

Conceived and designed the experiments: SKP CA DV. Performed the experiments: CA DV. Analyzed the data: CA DV. Contributed reagents/materials/analysis tools: SKP CA DV. Wrote the paper: CA DV SKP.

References

1. Whitman WB, Coleman DC, Wiebe WJ (1998) Prokaryotes: the unseen majority. Proc Natl Acad Sci USA 95: 6578–6583.
2. Austen MC, Lambshead PJD, Hutchings PA, Boucher G, Snelgrove PVR, et al. (2002) Biodiversity links above and below the marine sediment water interface that may influence community stability. Biodivers Conserv 11: 113–136.
3. Zinger L, Amaral-Zettler LA, Fuhrman JA, Horner-Devine MC, Huse SM, et al. (2011) Global patterns of bacterial beta-diversity in seafloor and seawater ecosystems. PLOS ONE 6: e24570.
4. Muyzer G, De Waal EC, Uitterlinden AG (1993) Profiling of complex microbial populations by denaturing gradient gel electrophoresis analysis of polymerase chain reaction-amplified genes coding for 16S rRNA. Appl Environ Microbiol 59: 695–700.
5. Smalla K, Oros-Sichler M, Milling A, Heuer H, Baumgarte S, et al. (2007) Bacterial diversity of soils assessed by DGGE, T-RFLP and SSCP fingerprints of PCR-amplified 16S rRNA gene fragments: Do the different methods provide similar results? J Microbiol Meth 69: 470–479.

6. Nithya C, Pandian SK (2012) Evaluation of Bacterial Diversity in Palk Bay Sediments Using Terminal-Restriction Fragment Length Polymorphisms (T-RFLP). Appl Biochem Biotechnol 167: 1763–1777.
7. Claesson MJ, Wang Q, O'Sullivan O, Greene-Diniz R, Cole JR, et al.(2010) Comparison of two next-generation sequencing technologies for resolving highly complex microbiota composition using tandem variable 16S rRNA gene regions. Nucleic Acids Res 38: e200–e200.
8. Gloor GB, Hummelen R, Macklaim JM, Dickson RJ, Fernandes AD, et al. (2010) Microbiome profiling by Illumina sequencing of combinatorial sequence-tagged PCR products. PLOS ONE 5: e15406.
9. Caporaso JG, Lauber CL, Walters WA, Berg-Lyons D, Lozupone CA, et al. (2011) Global patterns of 16S rRNA diversity at a depth of millions of sequences per sample. Proc Natl Acad Sci USA 108: 4516–4522.
10. Caporaso JG, Lauber CL, Walters WA, Berg-Lyons D, Huntley J, et al. (2012) Ultra-high-throughput microbial community analysis on the Illumina HiSeq and MiSeq platforms. ISME J 6: 1621–1624.
11. Galand PE, Casamayor EO, Kirchman DL, Lovejoy C (2009) Ecology of the rare microbial biosphere of the Arctic Ocean. Proc Natl Acad Sci USA 106: 22427–22432.
12. Gobet A, Baer SI, Huse SM, van Beusekom JEE, Quince C, et al. (2011) Diversity and dynamics of rare and of resident bacterial populations in coastal sands. ISME J 6: 542–553.
13. Venter JC, Remington K, Heidelberg JF, Halpern AL, Rusch D, et al. (2004) Environmental genome shotgun sequencing of the Sargasso Sea. Science 304: 66–74.
14. Sogin ML, Morrison HG, Huber JA, Welch DM, Huse SM, et al. (2006) Microbial diversity in the deep sea and the underexplored "rare biosphere". Proc Natl Acad Sci USA 103: 12115–12120.
15. Huber JA, Welch DBM, Morrison HG, Huse SM, Neal PR, et al. (2007) Microbial population structures in the deep marine biosphere. Science 318: 97–100.
16. Huse SM, Dethlefsen L, Huber JA, Welch DM, Relman DA, et al. (2008) Exploring microbial diversity and taxonomy using SSU rRNA hypervariable tag sequencing. PLOS Genet 4: e1000255.
17. Nithyanand P, Manju S, Pandian SK (2011) Phylogenetic characterization of culturable actinomycetes associated with the mucus of the coral *Acropora digitifera* from Gulf of Mannar. FEMS Microbiol Lett 314: 112–118.
18. Nithyanand P, Pandian SK (2009) Phylogenetic characterization of culturable bacterial diversity associated with the mucus and tissue of the coral *Acropora digitifera* from the Gulf of Mannar. FEMS Microbiol Ecol 69: 384–394.
19. Nithya C, Aravindraja C, Pandian SK (2010) *Bacillus pumilus* of Palk Bay origin inhibits quorum-sensing-mediated virulence factors in Gram-negative bacteria. Res Microbiol 161: 293–304.
20. Nithya C, Devi MG, Pandian SK (2011) A novel compound from the marine bacterium *Bacillus pumilus* S6-15 inhibits biofilm formation in Gram-positive and Gram-negative species. Biofouling 27: 519–528.
21. Haas BJ, Gevers D, Earl AM, Feldgarden M, Ward DV, et al. (2011) Chimeric 16S rRNA sequence formation and detection in Sanger and 454-pyrosequenced PCR amplicons. Genome Res 21: 494–504.
22. Degnan PH, Ochman H (2011) Illumina-based analysis of microbial community diversity. ISME J 6: 183–194.
23. Meyer F, Paarmann D, D'Souza M, Olson R, Glass EM, et al. (2008) The metagenomics RAST server-a public resource for the automatic phylogenetic and functional analysis of metagenomes. BMC Bioinformatics 9: 386.
24. Janssen PH (2006) Identifying the dominant soil bacterial taxa in libraries of 16S rRNA and 16S rRNA genes. Appl Environ Microbiol 72: 1719–1728.
25. Wu CH, Sercu B, Van De Werfhorst LC, Wong J, DeSantis TZ, et al. (2010) Characterization of coastal urban watershed bacterial communities leads to alternative community-based indicators. PLOS ONE 5: e11285.
26. Vreeland RH, Rosenzweig WD, Powers DW (2000) Isolation of a 250 million-year-old halotolerant bacterium from a primary salt crystal. Nature 407: 897–900.
27. Pandey S, Sree A, Dash SS, Sethi DP, Chowdhury L (2013) Diversity of marine bacteria producing beta-glucosidase inhibitors. Microb Cell Fact 12: 35.
28. Williams KP, Gillespie JJ, Sobral BW, Nordberg EK, Snyder EE, et al. (2010) Phylogeny of gammaproteobacteria. J Bacteriol 192: 2305–2314.
29. Madrid VM, Taylor GT, Scranton MI, Chistoserdov AY (2001) Phylogenetic diversity of bacterial and archaeal communities in the anoxic zone of the Cariaco Basin. Appl Environ Microbiol 67: 1663–1674.
30. Pedros-Alio C (2006) Marine microbial diversity: can it be determined? Trends Microbiol 14: 257–263.
31. Jiang YX, Wu JG, Yu KQ, Ai CX, Zou F, et al. (2011). Integrated lysis procedures reduce extraction biases of microbial DNA from mangrove sediment. J Biosci Bioeng 111: 153–157.
32. Burke C, Kjelleberg S, Thomas T (2009) Selective extraction of bacterial DNA from the surfaces of macroalgae. Appl Environ Microbiol 75: 252–256.
33. Bartram AK, Lynch MDJ, Stearns JC, Moreno-Hagelsieb G, Neufeld JD (2011) Generation of multimillion-sequence 16S rRNA gene libraries from complex microbial communities by assembling paired-end Illumina reads. Appl Environ Microbiol 77: 3846–3852.
34. Li W, Fu L, Niu B, Wu S, Wooley J (2012) Ultrafast clustering algorithms for metagenomic sequence analysis. Brief. Bioinform 13: 656–668.
35. Glass EM, Meyer F (2011) The Metagenomics RAST Server: A Public Resource for the Automatic Phylogenetic and Functional Analysis of Metagenomes. Handbook of Molecular Microbial Ecology I: Metagenomics and Complementary Approaches. 325–331.
36. Hong SH, Bunge J, Jeon SO, Epstein SS (2006) Predicting microbial species richness. Proc Natl Acad Sci USA 103: 117–122.
37. Ondov B, Bergman N, Phillippy A (2011) Interactive metagenomic visualization in a Web browser. BMC Bioinformatics 12: 385.

18

Rebuilding Biodiversity of Patagonian Marine Molluscs after the End-Cretaceous Mass Extinction

Martin Aberhan[1]*, **Wolfgang Kiessling**[1,2]

1 Museum für Naturkunde, Leibniz Institute for Evolution and Biodiversity Science, Berlin, Germany, **2** GeoZentrum Nordbayern, Paläoumwelt, Universität Erlangen–Nürnberg, Erlangen, Germany

Abstract

We analysed field-collected quantitative data of benthic marine molluscs across the Cretaceous–Palaeogene boundary in Patagonia to identify patterns and processes of biodiversity reconstruction after the end-Cretaceous mass extinction. We contrast diversity dynamics from nearshore environments with those from offshore environments. In both settings, Early Palaeogene (Danian) assemblages are strongly dominated by surviving lineages, many of which changed their relative abundance from being rare before the extinction event to becoming the new dominant forms. Only a few of the species in the Danian assemblages were newly evolved. In offshore environments, however, two newly evolved Danian bivalve species attained ecological dominance by replacing two ecologically equivalent species that disappeared at the end of the Cretaceous. In both settings, the total number of Danian genera at a locality remained below the total number of late Cretaceous (Maastrichtian) genera at that locality. We suggest that biotic interactions, in particular incumbency effects, suppressed post-extinction diversity and prevented the compensation of diversity loss by originating and invading taxa. Contrary to the total number of genera at localities, diversity at the level of individual fossiliferous horizons before and after the boundary is indistinguishable in offshore environments. This indicates an evolutionary rapid rebound to pre-extinction values within less than *ca* 0.5 million years. In nearshore environments, by contrast, diversity of fossiliferous horizons was reduced in the Danian, and this lowered diversity lasted for the entire studied post-extinction interval. In this heterogeneous environment, low connectivity among populations may have retarded the recolonisation of nearshore habitats by survivors.

Editor: Donald J. Colgan, Australian Museum, Australia

Funding: The study was supported by the Deutsche Forschungsgemeinschaft (grant number: Ab 109/8-1; www.dfg.de). The funders had no role in study design, data collection and analysis, decision to publish, or preparation of the manuscript.

Competing Interests: The authors have declared that no competing interests exist.

* Email: martin.aberhan@mfn-berlin.de

Introduction

The study of mass extinction events can reveal patterns of selective survival and the dynamics of ecosystem recovery over evolutionary time scales [1–3]. Apart from developing models of the processes involved in recovery dynamics [2], [4], such information can be gleaned from empirical studies of the geological past.

The mass extinction at the Cretaceous-Palaeogene (K-Pg) boundary is among one of the best studied Phanerozoic bioevents [5]. Geographically, the scope of studies encompasses local, regional, and global scales [6–8]. Yet, our understanding is still limited about how global signals of diversity dynamics arose as a result of the accumulation of events at local and regional scales. A general challenge is the reciprocal relationship between the geographical extent of a study and its temporal resolution. The 11-myr binning resolution of global occurrence compilations of marine macrofauna [8] is too coarse to uncover the dynamics of post-extinction diversity rebounds, whereas local-scale analyses may lack general validity.

Here we construct high-resolution diversity trajectories from benthic marine molluscs at two fossiliferous K-Pg boundary sites in Patagonia and incorporate information on the evolutionary and biogeographical characteristics of the fauna and on species abundances. The two studied sites represent marginal to shallow marine (nearshore) and middle shelf environments (offshore). We focus on evolutionary patterns and longer-term diversity trends across the K-Pg interval. In particular, we test two null hypotheses: Within the first *ca* 4 million years (myr) after the extinction event (1) the diversity of mollusc assemblages is indistinguishable from that of pre-extinction faunas; (2) the transient loss of diversity through global extinctions and local extirpations was compensated for by gains via immigrating and newly originating taxa. We demonstrate that hypothesis (1) is only partly supported, and hypothesis (2) can be rejected by empirical evidence.

Materials and Methods

No permits were required for the described study, which complied with all relevant regulations.

Sampling localities

We analysed two stratigraphic sections in the Argentine part of Patagonia. The first one is a composite section from the adjacent localities Bajada del Jagüel (BJG, 38°6′S, 68°23′W) and Opaso (38°8′S, 68°24′W), located in the Neuquén Basin north of the city of Neuquén. The second section, close to Estancia San Ramón in the middle Chubut River valley (42°41′S, 69°51′W), is located in the Cañadón Asfalto Basin. During latest Cretaceous–earliest

Palaeocene times the sea in both areas formed narrow embayments, which opened to the Atlantic Ocean to the east [9].

The sections at BJG and Opaso consist of monotonous, middle shelf siliciclastic mudstones of the Jagüel Formation, representing open marine sedimentation below, or close to, the storm wave base [10]. In the following they are referred to as 'offshore'. Because the section at Opaso virtually lacks a Maastrichtian record to which the Danian samples could be compared, the faunal data from BJG and Opaso were arranged chronologically according to their distance from the K-Pg boundary and integrated into a single succession (i.e. our offshore section). We thereby assumed equal sedimentation rates at these close-by localities. Faunal samples were derived from 11.4 metres below the K-Pg boundary to 27.5 metres above the boundary.

In San Ramón, faunal sampling extends over 188 m of varied sedimentary rocks of the Lefipán Formation. The Maastrichtian to early Palaeocene succession comprises siliciclastic mudstones, siltstones, sandstones, and conglomerates. Litho- and biofacies indicate deposition in a marginal to shallow marine, tide-dominated deltaic setting with tidal bars, prodelta, and delta plain environments, which are represented in both the Maastrichtian and the Danian [9]. In this paper, the aforementioned deltaic setting is referred to as 'nearshore'. Salinity was an important factor affecting the composition and ecological structure of the fauna in these environments. The uppermost Maastrichtian is characterized by virtually monospecific assemblages of the bivalve genus *Corbicula* owing to oligohaline to mesohaline salinity, whereas more fully marine conditions existed below and above this interval [9]. These very low-diversity brackish-water assemblages were excluded from our analyses because they represent a different environment.

Biostratigraphy and interval lengths

In BJG, the K-Pg boundary is at the erosive base of a coarse-grained, tuffaceous sandstone bed [10], [11]. Calcareous nannofossils prove the presence of the first Cenozoic nannofossil zone (NP1 or NNTp1) in the upper part of the sandstone layer, and younger subzones in the mudstones overlying the sandstone [10]. The mudstones below the sandstone are late Maastrichtian in age [10], [11]. According to Keller et al. [11], the 11.4 metres of section below the K-Pg boundary, which yielded the late Maastrichtian quantitative samples in our study, are characteristic of the nannofossil zone CC26 and the *A. mayaroensis* and *P. hariaensis* foraminiferal zones (zonal names according to [12]). In Opaso, the K-Pg boundary sandstone is exposed about half a metre above the base of the sampled section. Micro- and nannofossils indicate the presence of early (NP1-NP3?/P1a-b) to late Danian (NP4/P1c) biozones [13]. Applying the ages given in Gradstein et al. [12], the samples of BJG and Opaso extend from ~2 myr before the K-Pg boundary to ~4 myr after the boundary. Because the earliest Danian planktonic foraminiferal zones P0 and Pα, the latest Maastrichtian *P. hantkeninoides* foraminiferal zone, and impact tracers are missing, the sections at BJG and Opaso are stratigraphically incomplete [10], with possibly up to 500 kyr of the earliest Danian and 300 kyr of the latest Maastrichtian missing [10], [11]. In summary, the amount of time represented by the Danian strata is about twice that of the Maastrichtian strata.

In the nearshore San Ramón section, a distinct boundary layer is not preserved. However, the presence of latest Maastrichtian and earliest Danian dinoflagellate markers and age-diagnostic palynomorphs indicate a very minor hiatus [14]. A more detailed biostratigraphical subdivision is not available in this very shallow-water setting. Because depositional environments are the same below and above the boundary, and because the sampled Maastrichtian succession (100 metres) is similar in thickness to that of the Danian (88 metres), we tentatively assume roughly equal time spans for the studied pre- and post-extinction intervals.

Analytical methods

We quantitatively sampled 157 single beds of uniform lithology for macroinvertebrates, comprising a total of 5,547 individuals. Bivalves and gastropods dominate the macrofauna. Taxa were identified as precisely as possible, mostly at species level. Our analyses focus on genus diversity of bivalves and gastropods. We elevated subgenera to genus rank and omitted all taxa not resolvable at least to the genus level. Because most genera are represented by a single species at the studied localities, the number of taxa is almost identical at genus and species level. Following standard procedures [15] the number of individuals of bivalves was determined as the sum of articulated specimens plus the larger number of right or left valves of disarticulated specimens plus one-half the number of fragments which could not be assigned to either right or left valves but make up more than 50% of a valve. In one special case – *Entolium membranaceum,* an equivalved and equilateral pectinoid with symmetrical auricles – distinction of the right and left valves was impossible, thus numbers of specimens were halved to yield numbers of individuals. Gastropods were counted when more than 50% of the shell was preserved.

To avoid large variations in sample size, we pooled samples of successive fossiliferous beds until at least 100 individuals were reached or until the lithology changed markedly. In the offshore section, this resulted in 24 pooled samples with a mean size of 129.1±7.1 (standard error) individuals. The 28 pooled nearshore samples had a mean of 63.1±5.5 individuals.

We focus on two measures of diversity. The first one determines the standardised number of genera for each of the pooled samples. We refer to this measure as sample-level diversity [16]. We standardised sample-level diversity using the 'shareholder quorum' subsampling method developed by Alroy [8], [17]. This method involves random sampling of individuals to equal frequency coverage, the shareholder quorum. In our analyses the quorum was set to 0.8 and our values are averages of 100 iterations. The shareholder quorum is superior to traditional subsampling methods (e.g., rarefaction) in that variations in evenness are explicitly taken into account. Therefore, it is less biased by different abundance distributions and represents a nearly unbiased estimate of richness.

Our second measure of diversity aims at comparing the total number of Danian genera in a section with that of Maastrichtian genera in the same section. This measure is termed total (Maastrichtian or Danian) diversity. To this end, the genera of the pooled and standardised samples were summed up over consecutively younger samples for the Maastrichtian and for the Danian part of the section, respectively. Genera are added to the count of total diversity at the sampling level at which they first appeared in the succession and are considered to persist up to the K-Pg boundary (for Maastrichtian total diversity) or to the youngest Danian sample (for Danian total diversity). This procedure constructs sample-based accumulation curves [18], whereby the sequence of samples is ordered in time via stratigraphy. This approach assumes that diversity accumulates during both the Maastrichtian and the Danian. While confidence intervals on stratigraphic ranges of genera support this approach, we emphasize that this procedure is thought to provide theoretical 'genus accumulation curves', which allow for a comparison of Maastrichtian and Danian diversity at various sampling levels.

With reference to the K-Pg boundary, we assigned each genus to one of five categories: (1) going extinct globally; (2) disappearing in Patagonia but continuing elsewhere; (3) surviving; (4) immigrating; and (5) originating. The reference area for categories 2–5 is Patagonia because geographical restriction to localities may produce strong sampling biases. We categorised genera as immigrants when they occurred in the Danian sections of our study area, had no Cretaceous record from Patagonia, and had been reported outside of Patagonia either before the Danian or within a part of the Danian which is stratigraphically older than our own field records. We categorised genera as originating in Patagonia when they are endemic to this region and first appeared in the Danian, or when a Danian occurrence in our sections is stratigraphically older than records from anywhere else in the world. For each pooled and standardised sample we calculated proportional diversity of the various categories. For the purpose of assigning genera to the five evolutionary/biogeographical categories, geographical and stratigraphical range data were retrieved from the Paleobiology Database (http://paleobiodb.org), Sepkoski's compendium [19], and numerous monographs on Cretaceous and Palaeogene molluscs. The main sources of information and specific references that were used for the categorisation of taxa are provided in Tables 1 and 2. A full documentation of primary data is presented in Tables S1 and S2.

Extinction intensities are expressed as percent extinctions of genera using the data presented in Tables 1 and 2. For comparison with our Patagonian sites, percent global extinctions of molluscan genera were also calculated for three other well-studied K-Pg boundary localities: Brazos River, Texas [20]; Braggs, Alabama [21]; and Stevns Klint, Denmark [22]. These calculations used the primary data provided in the respective publications.

Finally, we compared the relative abundances of the most common Maastrichtian species with their abundances in the Danian and vice versa at both sections. We summed up the total abundances of each species in the Danian and Maastrichtian parts of our respective sections and calculated their proportional abundance in each pooled sample. The proportional abundance of the most common species (those in the upper 90th percentile of all proportional abundances) was then ordered by decreasing proportional abundance and compared with the proportions in the stratigraphically older (for Danian abundances) or younger (for Maastrichtian abundances) parts of the sections. The bivalve *Corbicula*, which is very abundant in the low-diversity brackish-water assemblages, was excluded from this analysis.

All analyses were performed in the R programming environment (www.r-project.org). The field-collected Patagonian specimens are housed at the Departamento de Ciencias Geológicas, Universidad de Buenos Aires, Argentina.

Results

Genus-level extinction rates

Intensities of end-Cretaceous extinctions may vary markedly across local, regional, and global scales. In our nearshore section the percentage of global extinctions of benthic molluscan genera is 21.7%. In the offshore section 32.1% of genera became extinct. These values are comparable to those of well-studied Northern Hemisphere boundary sections such as Braggs, Alabama (30.5% extinction of bivalves and gastropods [21]) or Stevns Klint, Denmark (22.5% extinction of bivalves [22]), but markedly higher than at Brazos River, Texas (10.7% extinction of bivalves and gastropods [20]). Local extirpations (60.9% in our nearshore section, 53.6% in the offshore section) are much greater than global extinctions. These higher values are the cumulative result of local extirpations, global extinctions, and sampling effects.

Maastrichtian and Danian diversity compared

Sample-level diversity in offshore environments fluctuated without a consistent temporal trend (Figure 1). The diversity of even the first quantitative Danian samples is mostly within the range of Maastrichtian assemblages. In contrast, Danian sample-level diversity in nearshore habitats is, on average, distinctly lower than that of pre-extinction assemblages (Figure 1). This pattern is evident in both fine-grained and coarse-grained lithologies and persists until the end of the studied time interval, suggesting contrasting diversity patterns between nearshore and offshore environments. Despite comparable extinction intensities, recovery of sample-level diversity was only achieved offshore.

Total diversity in the Danian is significantly and consistently lower than total Maastrichtian diversity at the same sampling level (Figure 2). This pattern holds for both nearshore and offshore environments, and points to a distinct difference in diversity structure between Maastrichtian and Danian faunas.

The contributions of surviving, immigrating, and originating genera

In both environments, the most diverse components of Maastrichtian assemblages are genera that survived the K-Pg boundary event (Figure 3). Immigrating and newly evolving genera play only subordinate roles in the composition of Danian samples. Their occurrence is often sporadic and without obvious trends (Figure 3). The reduced total diversity in the Danian (Figure 2) indicates that immigrants and newly evolved taxa could not compensate for the loss of Maastrichtian diversity at the K-Pg boundary.

Abundance shifts of dominant species across the K-Pg boundary

In both environmental settings, the dominant Maastrichtian species became extinct at the K-Pg boundary (Figure 4). In the nearshore environment, this affected the five most abundant Maastrichtian species. Offshore, three of the four most abundant species did not survive the boundary, only the oyster *Pycnodonte vesicularis*, the second most common Maastrichtian species in the offshore environment, survived. However, with only a single specimen found in the Danian, it can be considered as ecologically extinct. It is best classified as a 'Dead Clade Walking', a term coined to characterise lineages that survived mass extinctions only to remain marginal or decline in the aftermath [23].

A second intriguing abundance pattern is seen in species that dominate the Danian assemblages. These were usually rare in the Maastrichtian part of the respective stratigraphic section (Figure 5). The three most abundant nearshore species, collectively making up almost 60% of Danian individuals, are survivors with low abundances during the Maastrichtian. In the offshore environment, three of the five most abundant Danian species are survivors that also changed their relative abundances from rare to common. The two dominant Danian offshore species – the oyster *Pycnodonte burckhardti* and the pectinid *Delectopecten neuquenensis* – are Patagonian endemics that originated in the early Danian.

Discussion

Incumbency effects and recovery with minor origination

The persistent dominance of survivors (Figure 3) points to the effectiveness of survival mechanisms [24] and contrasts with recovery models in which a short-term survival phase is followed by diversification owing to origination of new taxa [25]. At any

Table 1. Categorisation of molluscs at San Ramòn.

Class	Genus	Range in section	Category	Additional notes and references on records in the	
				Cretaceous	**Palaeogene**
Bivalvia	*Acesta*	Maa	2		
Gastropoda	*Acteon*	Maa	3		[50]
Bivalvia	*Anomia*	Maa	3		See Table 2
Bivalvia	*Aphrodina*	Danian	3	Valanginian of Argentina [51], [52]	
Bivalvia	*Arcomytilus*	Danian	4		
Bivalvia	*Astarte*	Danian	3	Hauterivian of Argentina [53]	Palaeocene of Argentina [54]
Bivalvia	*Australoneilo*	Maa – Danian	3		
Gastropoda	*Austroaporrhais*	Maa	3		Danian of Patagonia [54], [55]
Bivalvia	*Austrotrigonia*	Maa	1		
Bivalvia	*Bicorbula*	Maa – Danian	3		
Bivalvia	*Brachidontes*	Maa	3		Eocene of Patagonia [56], [57]
Bivalvia	*Camptochlamys*	Maa	1	Casadío et al. [58]	
Bivalvia	*Corbicula*	Maa – Danian	3		
Gastropoda	*Cryptorhytis*	Maa	3		See Table 2
Bivalvia	*Cucullaea*	Maa	3		Danian of Patagonia (unpublished data of authors)
Bivalvia	*Cucullastis*	Danian	4	Late Cretaceous of New Zealand [59]	
Bivalvia	*Cuspidaria*	Danian	4		
Bivalvia	*Eriphyla*	Maa – Danian	3		
Bivalvia	*Etea*	Maa	2		
Gastropoda	*Euspira*	Maa – Danian	3		
Bivalvia	*Glycymeris*	Danian	3	Maastrichtian of Argentina (unpublished data of authors from Sierra Huantraico)	
Gastropoda	*Heteroterma*	Maa – Danian	3		
Bivalvia	*Inoperna*	Maa	1		
Bivalvia	*Lahillia*	Maa	3		Eocene of Patagonia [60]
Bivalvia	*Ledina*	Maa – Danian	3		
Bivalvia	*Leionucula*	Maa – Danian	3		
Bivalvia	*Linearia*	Maa	2		Eocene of USA [61]
Bivalvia	*Lyriochlamys*	Maa	1		
Bivalvia	*Meretrix*	Maa – Danian	3		
Bivalvia	*Mesocallista*	Maa	1		
Bivalvia	*Modiolus*	Maa – Danian	3		
Gastropoda	*Nacella*	Danian	3	See Table 2	
Bivalvia	*Neilo*	Maa	3		Danian [62], Eocene [60] and Oligocene [50] of Patagonia
Bivalvia	*Nucula*	Maa	3		Danian of Patagonia [26]
Bivalvia	*Ostrea*	Maa – Danian	3		
Bivalvia	*Pachymya*	Maa	1		
Bivalvia	*Pacificor*	Maa – Danian	3		
Bivalvia	*Pacitrigonia*	Maa	1		
Bivalvia	*Panopea*	Maa – Danian	3		
Gastropoda	*Parasyrinx*	Maa	2		
Bivalvia	*Pholadomya*	Maa	2		
Bivalvia	*Phygraea*	Maa	3		See Table 2
Bivalvia	*Pinna*	Maa – Danian	3		
Gastropoda	*Priscaphander*	Maa	3		See Table 2

Table 1. Cont.

Class	Genus	Range in section	Category	Additional notes and references on records in the	
				Cretaceous	Palaeogene
Bivalvia	*Protagelus*	Maa	1		
Gastropoda	*Pseudamaura*	Danian	4		[54]
Bivalvia	*Pseudolimea*	Maa	3		See Table 2
Gastropoda	*Pseudotylostoma*	Danian	5		Genus endemic to Patagonia
Bivalvia	*Pteromyrtea*	Maa - Danian	3		
Gastropoda	*Pyropsis*	Maa	2		
Bivalvia	*Rinetrigonia*	Maa	1		
Bivalvia	*Saccella*	Danian	3	Maastrichtian of Argentina (unpublished data of authors from Sierra Huantraico)	
Bivalvia	*Solyma*	Maa	1		
Bivalvia	*Spineilo*	Danian	5		Danian records from Patagonia are from the lower part of the Danian (this study; foraminiferal zone P1b of [63]), whereas oldest records from New Zealand [64] are of late Danian age [26]. Palaeocene records from Russia appear to be late Danian or younger [65]; see also [66]
Gastropoda	*Struthioptera*	Maa - Danian	3		
Bivalvia	*Tikia*	Maa - Danian	3		
Gastropoda	*Turritella*	Maa - Danian	3		

Alphabetical list of genera (including subgenera) of bivalves and gastropods from the late Maastrichtian and Danian at San Ramón, western Argentina, and their assignment to evolutionary/biogeographical categories.
Evolutionary/biogeographical categories: 1 = going extinct globally in the late Maastrichtian; 2 = disappearing regionally from Patagonia in the late Maastrichtian; 3 = surviving the K-Pg boundary in Patagonia; 4 = immigrating to Patagonia in the Danian; 5 = originating in Patagonia during the Danian. Maa = Maastrichtian.

given sampling level, originating and invading genera could not compensate for the loss of genera caused by extinctions (Figure 2). A similar observation was reported for Antarctic molluscs at the K-Pg boundary, where few new genera were observed, and only later in the Danian did the number increase [26].

Delayed recovery is frequently explained by unfavourable environments, either as protracted impact-related effects, or as specific regional effects [27–29]. However, local environmental stress fails to explain the Danian *versus* Maastrichtian difference in total diversity in the offshore section where there is no evidence for a shift to more unfavourable environments. Similarly, in the nearshore section the range of sedimentary environments and overall environmental heterogeneity in the Danian and in the Maastrichtian match well [9]. Furthermore, previous work from our working group that used the same raw data from the same nearshore section showed that even Danian assemblages from stable, subtidal environments, characterised by diverse trophic and life habit groups, did not reach Maastrichtian diversity levels [9]. This is exemplified by comparing the Danian *Struthioptera-Venericardia* faunal association with the Maastrichtian *Struthioptera-Panopea* association of Scasso et al. [9]. Both associations are dominated by infaunal guilds but also have relatively high proportions of epifaunal molluscs. In both associations, trophic resources have been subdivided among suspension-feeders, deposit-feeders, and carnivores, suggesting biologically controlled communities [9]. Yet, the diversity of samples forming the Danian association remains distinctly below the diversity of the Maastrichtian samples. Thus, unfavourable habitats cannot be held responsible for the differences among pre- and post-extinction diversity at either of the two studied sections.

An alternative explanation for the delayed recovery would be low immigration rates caused either by geographic isolation of the basins or by a general shortage in the supply of taxa from outside Patagonia. However, there is no evidence of isolation because the Patagonian marine basins opened to the South Atlantic Ocean to the east, which was a large open ocean with pelagic sedimentation during K-Pg boundary times [9], [10]. Based on general physical oceanographic principles, wind driven surface currents, tidally induced amphidromic circulation, and thermohaline circulation are assumed to have been well developed in the studied areas, all of which would have contributed to the exchange of waters with surrounding regions. The presence in the Neuquén Basin of diverse planktonic foraminifera [11], a plankton group only found in high diversity in marine waters of normal salinity, provides direct evidence for such exchange with open marine waters. Palaeogeographic reconstructions of the marine basins in Patagonia for the Maastrichtian–Palaeocene interval show direct shallow water connections between the basins [9]. Also, with the closely located South American shelf to the north, the Antarctic shelf to the south, and refugia along the margin of the southern-Pacific [26], several potential areas were available as a source for new immigrants. Therefore, geographic isolation and a shortage in the supply of larvae are unlikely scenarios for the delayed recovery in diversity.

We suggest that the numbers of new taxa and of taxa invading from outside Patagonia were low due to the presence of survivors

Table 2. Categorisation of molluscs at Bajada del Jagüel and Opaso.

Class	Genus	Range in sections	Category	Additional notes and references on records in the	
				Cretaceous	**Palaeogene**
Gastropoda	*Acteon*	Maa	3		[50]
Bivalvia	*Ambigostrea*	Maa	1		
Bivalvia	*Amphidonte*	Maa	1		
Bivalvia	*Anomia*	Maa – Danian	3		
Bivalvia	*Aphrodina*	Danian	3	Valanginian of Argentina [51]	
Bivalvia	*Arcomytilus*	Danian	4		
Bivalvia	*Australoneilo*	Maa – Danian	3		
Bivalvia	*Bicorbula*	Danian	3	See Table 1	
Bivalvia	*Brachidontes*	Maa	3		Eocene of Patagonia [56], [57]
Bivalvia	*Camptochlamys*	Maa	1	[58]	
Bivalvia	*Cosmetodon*	Maa – Danian	3		
Gastropoda	*Cryptorhytis*	Maa – Danian	3		
Bivalvia	*Cucullastis*	Danian	4	Late Cretaceous of New Zealand [59]	
Bivalvia	*Delectopecten*	Danian	4	Early Cretaceous of Greenland [67]	Danian of Patagonia [62], [68]
Bivalvia	*Disparilia*	Maa – Danian	3		
Bivalvia	*Entolium*	Maa	1		
Bivalvia	*Glycymeris*	Danian	3	Maastrichtian of Argentina (unpublished data of authors from Sierra Huantraico)	
Bivalvia	*Gryphaeostrea*	Danian	3	Aguirre-Urreta [69]	del Río et al. [62], [66]
Gastropoda	*Heteroterma*	Danian	3	See Table 1	
Bivalvia	*Jagolucina*	Danian	5		Thanetian of Australia [70]; Eocene of France [71]
Bivalvia	*Leionucula*	Danian	3	See Table 1	[66]
Bivalvia	*Lima*	Maa	2		
Bivalvia	*Linearia*	Maa	2		Eocene of USA [61]
Bivalvia	*Lyriochlamys*	Maa	1		
Bivalvia	*Mesocallista*	Maa	1		
Bivalvia	*Modiolus*	Maa – Danian	3		
Gastropoda	*Nacella*	Maa – Danian	3		
Bivalvia	*Neilo*	Maa	3		Eocene [60] and Oligocene [50] of Patagonia
Bivalvia	*Nicaniella*	Danian	4		
Bivalvia	*Nuculana*	Maa – Danian	3		
Bivalvia	*Ostrea*	Maa – Danian	3		
Bivalvia	*Pacificor*	Danian	3	See Table 1	
Bivalvia	*Panopea*	Danian	3	See Table 1	
Gastropoda	*Perissodonta*	Danian	5		Danian records from Patagonia are from the lower part of the Danian (this study; foraminiferal zone P1b of [63]), whereas oldest Palaeocene records from New Zealand [64] are late Danian [26]
Bivalvia	*Phelopteria*	Maa	1	[72]	
Bivalvia	*Phygraea*	Maa – Danian	3		
Bivalvia	*Pinna*	Danian	3	See Table 1	
Gastropoda	*Priscaphander*	Danian	3	See Table 1	
Bivalvia	*Protagelus*	Maa	1		
Bivalvia	*Pseudolimea*	Maa – Danian	3		

Table 2. Cont.

Class	Genus	Range in sections	Category	Additional notes and references on records in the	
				Cretaceous	Palaeogene
Bivalvia	*Pteromyrtea*	Maa - Danian	3		
Gastropoda	*'Pyropsis'*	Danian	4		Species belonging to a probably new genus referred to as *'Pyropsis'* have been discussed in [26], [73], [74]; oldest records are from Antarctica
Bivalvia	*Roudairia*	Maa	1		
Bivalvia	*Saccella*	Danian	3	Maastrichtian of Argentina (unpublished data of authors from Sierra Huantraico)	
Bivalvia	*Spineilo*	Danian	5		See Table 1
Bivalvia	*Spondylus*	Maa	2		
Gastropoda	*Struthioptera*	Danian	3	See Table 1	
Gastropoda	*Taioma*	Danian	4	Late Cretaceous of Antarctica [74]	[74], [75]
Bivalvia	*Tikia*	Maa - Danian	3		
Gastropoda	*Tornatellaea*	Danian	4		Palaeocene of Argentina [54], [75]

Alphabetical list of genera (including subgenera) of bivalves and gastropods from the late Maastrichtian and Danian at Bajada del Jagüel and Opaso, western Argentina, and their assignment to evolutionary/biogeographical categories.
Evolutionary/biogeographical categories: 1 = going extinct globally in the late Maastrichtian; 2 = disappearing regionally from Patagonia in the late Maastrichtian; 3 = surviving the K-Pg boundary in Patagonia; 4 = immigrating to Patagonia in the Danian; 5 = originating in Patagonia during the Danian. Maa = Maastrichtian.

within Patagonia impeding their progress. This phenomenon is known as incumbency or priority effects, in which existing organisms prevent the successful establishment of newly evolving or immigrating species [30], [31]. Resistance to invasions or to the establishment of newly evolved taxa occurs because these 'would-be usurpers' are initially less well adapted than established species, and the latter keep incipient populations of new species from taking hold [32], [33]. A closely related concept is that of pre-emptive competition [34], emphasizing the importance of being 'first in the field'. Incumbency can be expected to play a role when resources are limited and communities are saturated with species [15]. Mass extinctions are generally thought to reduce or remove the incumbency effect [35]. Our analyses, however, suggest that despite significant extinctions, the removal of incumbents did not entail an increase in immigration or origination rates to counterbalance diversity reduction. Rather, the number of

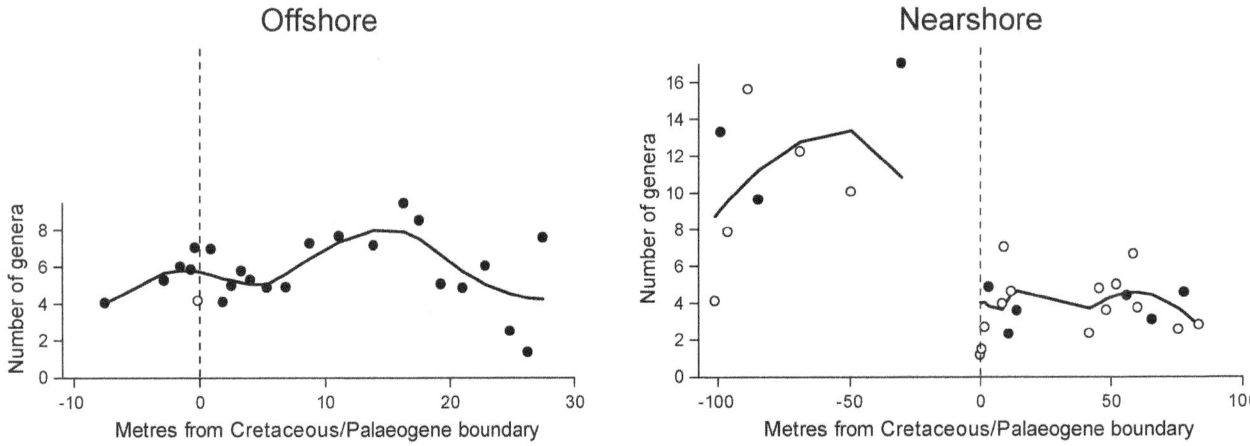

Figure 1. Sample-level diversity of benthic marine molluscs in two Cretaceous-Palaeogene boundary sections in Patagonia. Standardised genus diversity is based on the shareholder quorum subsampling method [17]. Dashed vertical lines indicate position of the Cretaceous-Palaeogene boundary; values on x-axis refer to metres before (Cretaceous, negative values) or after the boundary. Curved solid line is LOESS smoothing with a smoothing parameter of 0.5. Black dots indicate samples from fine-grained sedimentary rocks (siliciclastic mudstones and siltstones), open circles indicate samples from coarse-grained sedimentary rocks (sandstones and conglomerates). Nearshore, the uppermost *ca* 26 metres of the Cretaceous are characterised by low-diversity brackish-water assemblages, which are not included in the figure.

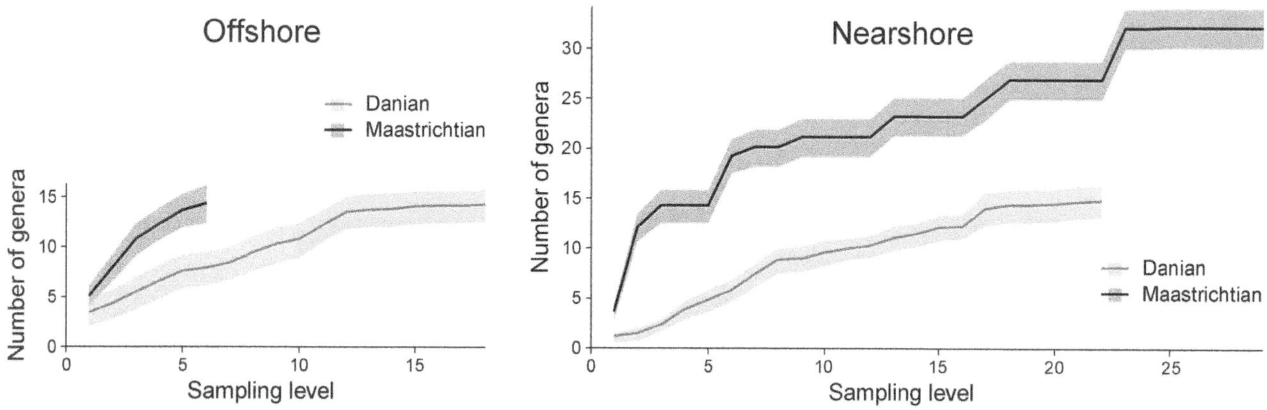

Figure 2. Comparing Cretaceous (Maastrichtian) and Palaeogene (Danian) accumulation curves for the estimated number of molluscan genera. Sampling level refers to consecutively numbered fossiliferous samples, arranged from stratigraphically oldest to youngest, and shows the cumulative number of genera at the respective level (see text for details on pooling and standardising of samples). Grey areas indicate the standard deviations around the means. In both offshore and nearshore habitats total diversity in the Danian is consistently lower than total Maastrichtian diversity at any given sampling level.

survivors seems to have remained sufficiently high to impose constraints on the establishment of immigrants and originating taxa, consistent with incumbency. We consider incumbency as an

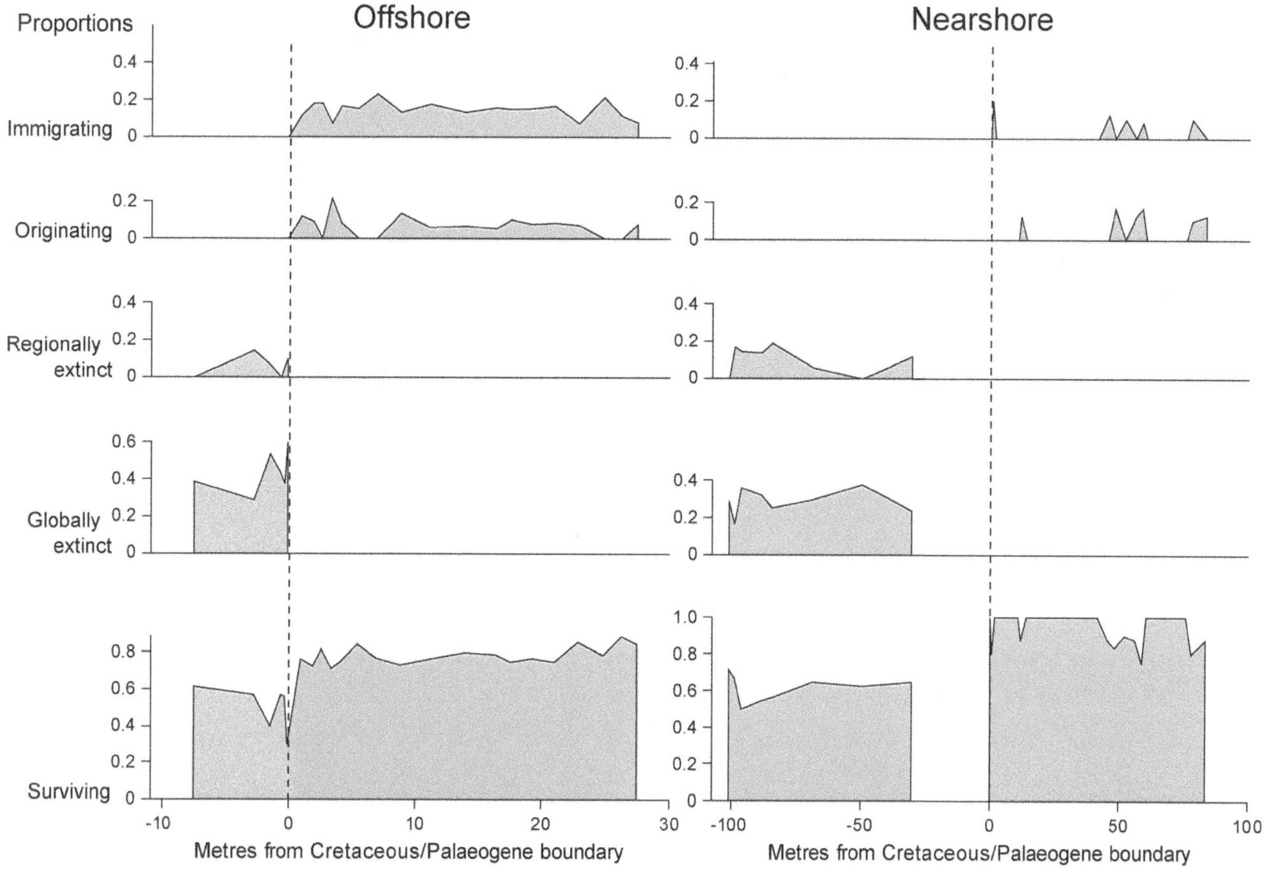

Figure 3. Proportional diversity of molluscan genera in five evolutionary/biogeographical categories across the Cretaceous-Palaeogene boundary. Proportions are shown for an offshore (panels on the left) and a nearshore boundary section (panels on the right) in Patagonia. Dashed vertical lines indicate position of the Cretaceous-Palaeogene boundary; values on x-axis refer to metres before (Cretaceous, negative values) or after the boundary. Nearshore, the uppermost *ca* 26 metres of the Cretaceous are characterised by low-diversity brackish-water assemblages, which are not included in the figure. Post-extinction faunas are dominated by survivors. Immigrating and originating genera play subordinate roles in the composition of Danian samples.

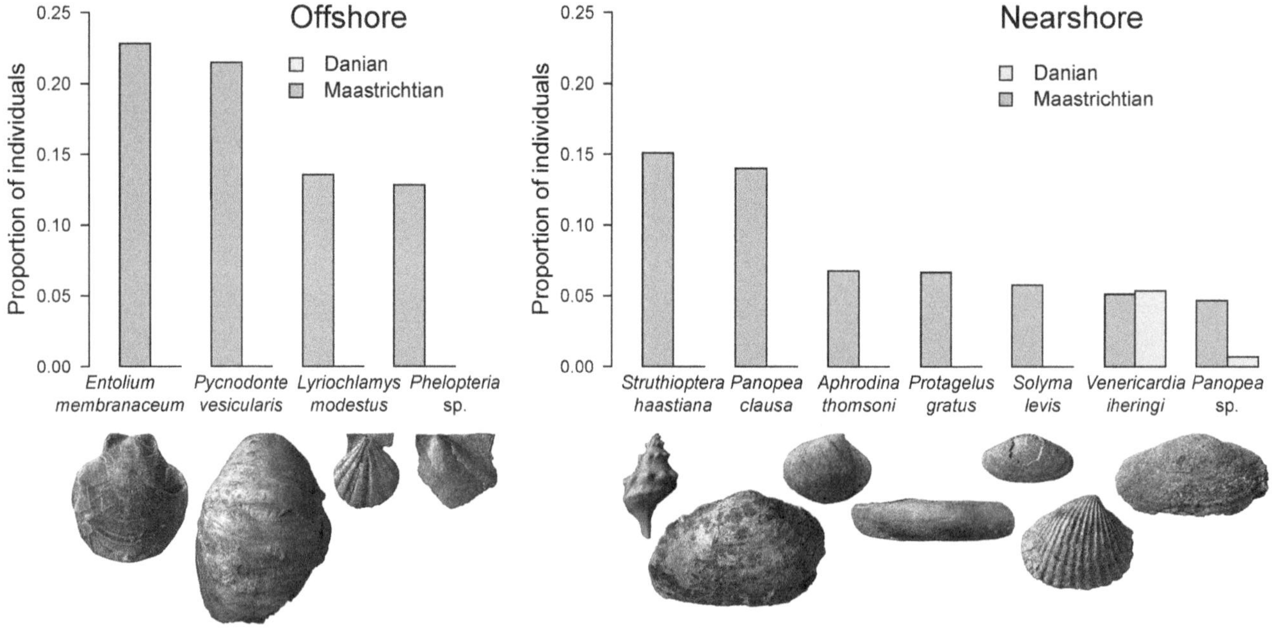

Figure 4. Proportional abundance of most common Maastrichtian molluscs contrasted with their abundances in the Danian. The displayed Maastrichtian species are those in the upper 90th percentile of the pooled samples of that time. Their Danian proportions are plotted alongside. The majority of the dominant Maastrichtian species did not survive into the Danian.

important mechanism for the delayed recovery in both onshore and offshore environments of Patagonia.

The extinction of dominants and the rise of rare species

In both studied environments the most common Maastrichtian species became extinct at the K-Pg boundary (Figure 4). This observation appears to be in contrast with ecological studies which suggest that high abundance promotes survival [36]. It can, however, be reconciled with the finding that post-Palaeozoic marine bivalves with high abundances exhibit extinction rates elevated over those of moderately common genera [37]. Also, results of a regional study of North American bivalves showed that

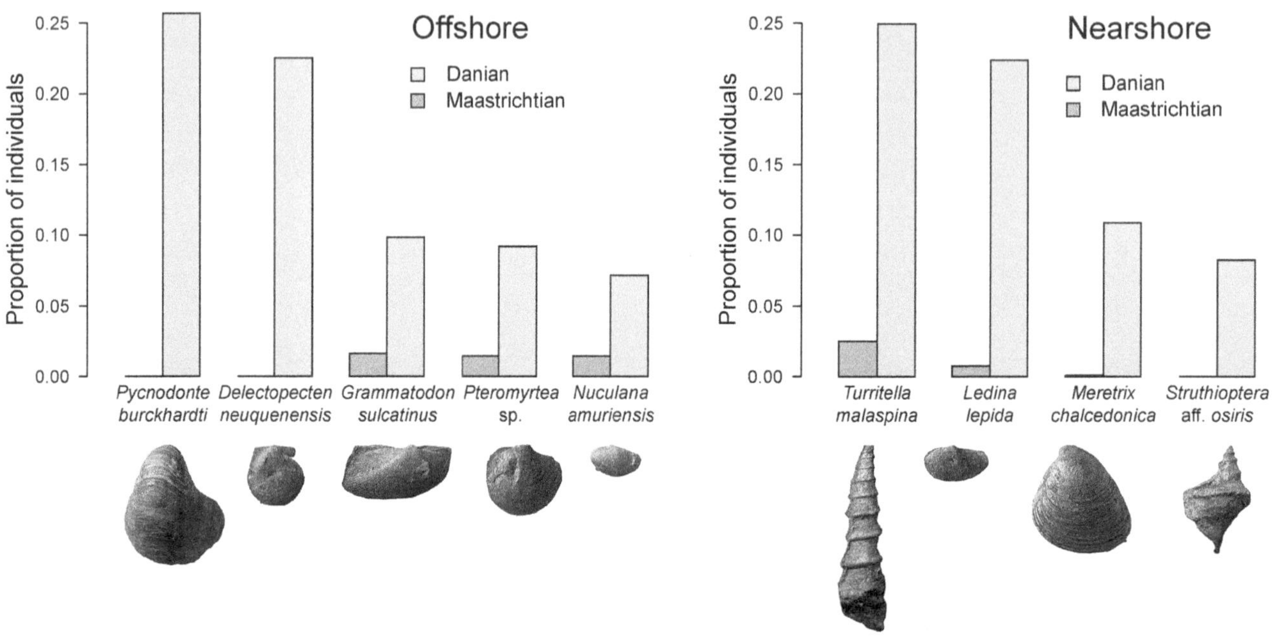

Figure 5. Proportional abundance of most common Danian molluscs contrasted with their abundances in the Maastrichtian. The displayed Danian species are those in the upper 90th percentile of the pooled samples of that time. Their Maastrichtian proportions are plotted alongside. Most of the dominant Danian species are survivors that were rare during the Maastrichtian. The two most abundant Danian offshore species are Patagonian endemics that newly originated in the early Danian.

abundant taxa were no more likely to survive the K-Pg mass extinction than rare ones [38].

Interestingly, survivors that dominate the Danian assemblages were rare during the Maastrichtian (Figure 5). We suggest that although being resistant to environmental stress, the new dominants were poor competitors before the boundary event. Once they were released from competition by the removal of dominants, the relative competitive ability of survivors increased. This suggests a particular type of incumbency, where numerically subordinate taxa arose to ecological dominance in the aftermath of a major environmental perturbation. Modelling of competitive interactions provides a potential mechanism for such reorganisations of ecosystems [39]: The loss of a competitively superior species releases otherwise unavailable resources, allowing the competitively inferior species to expand its realised niche. Thus, less abundant species may not only provide invasion resistance by their aggregate effects on resource utilization [40], but can be important by gaining dominance after environmental turmoil. Because distinct inversions in abundance across mass extinctions are known from other ecosystems [41] this may indicate a more general macroecological pattern.

The diversity-based conclusions stress the importance of survivors in post-extinction assemblages (Figure 3). This finding is generally confirmed by our abundance-based analyses with one important modification. In the offshore environment, the two top-ranking Danian species are newly evolved bivalves that originated in the early Danian of Patagonia (Figure 5). Their dominance indicates that newly originating post-extinction taxa can have high numbers of individuals even though overall numbers of originating taxa are low. There is a striking taxonomical, morphological, and ecological analogy between these two Danian species and the two most abundant Maastrichtian species in the offshore environment (Figures 4, 5). All four species are epifaunal suspension feeders and it seems that, after its extinction, the Maastrichtian pectinoid *Entolium membranaceum* was replaced by the Danian pectinoid *Delectopecten neuquenensis*, whereas the oyster *Pycnodonte vesicularis* was replaced by the congeneric *Pycnodonte burckhardti*. Although end-Cretaceous extinctions did not remove entire ecological guilds it is plausible that they created enough amount of open ecospace into which taxa with equivalent ecological traits could evolve.

The pace of recovery in sample-level diversity

The mean sample-level diversity remained unchanged across the K-Pg boundary in the studied Patagonian offshore environments at BJG and Opaso (Figure 1). The early Danian hiatus of *ca* 500 kyr at BJG [11] provides an upper limit for the duration necessary to restore sample-level diversity. The stratigraphically more complete section at Brazos River, east Texas, represents an environment similar to our offshore setting. Sample-level diversity at Brazos started to rebound *ca* 1 myr after the boundary [6] and full local diversity recovery to Cretaceous levels was attained after about 2 myr [42]. Thus, complete recovery of sample-level diversity to pre-extinction values took longer there than in Patagonia. Similarly, another study concluded that the local richness of offshore mollusc assemblages of the U.S. Gulf Coastal Plain recovered to Cretaceous values *ca* 2.7 myr after the event [43]. In conclusion, the time necessary to recover to pre-extinction diversity in mollusc-dominated offshore ecosystems varies and may take less than *ca* 0.5 myr according to our results from Patagonia and up to *ca* 2.7 myr according to data from the U.S. Gulf Coastal Plain. In contrast to offshore settings, Danian sample-level diversity did not recover to Maastrichtian levels in the Patagonian nearshore environments. We have shown from the data that there is a prolonged period of incomplete recovery in nearshore environments, although a lack of more detailed chronostratigraphic information prevents us from quantifying its duration more precisely.

Divergence of Maastrichtian and Danian diversity structure

Accumulation of total Danian diversity slowed down over time (Figure 2), seemingly compatible with a model of diversity-dependent niche-filling and a gradual approach to carrying capacity. These curves, however, should not be taken as accurate reflections of the rate at which post-extinction diversity builds up because they build on the assumption that (1) all genera found in any sample in the Danian are also present in all younger Danian samples, and (2) any genera not found in older samples represent true absences.

Geological time is critical for our interpretation that Figure 2 indicates systematic differences between Maastrichtian and Danian biodiversity. For example, the underlying (re-)population dynamics may be the same for each interval, but an apparent difference could be generated if Maastrichtian strata represented a much longer time span than Danian strata. Because the opposite is true in the offshore section (Danian samples encompass about twice as much time as Maastrichtian samples) any potential bias would operate only in the reverse direction. Interval lengths are less clear in the nearshore section but we assume roughly equal amounts of time (see above).

The reduced total diversity in the Danian and the longer-term stability of sample-level diversity across the boundary in offshore habitats are no contradiction. Jointly these patterns imply that, on average, Danian offshore assemblages retained, or rapidly regained, their Maastrichtian sample-level diversity despite being recruited from an overall smaller pool of Danian taxa. This indicates a shift in the partitioning of sample-level diversity and total diversity.

Erwin [44] pointed out that positive feedbacks may increase carrying capacity during biotic recovery. An example of a positive feedback in mollusc-dominated ecosystems may be the Cenozoic radiation of predatory gastropods and the resulting escalatory trends ([43] and references therein). If such effects were present in our area, they must have occurred subsequent to the studied time interval.

Nearshore–offshore differentials in recovery

In contrast to the offshore environment, the reduction in nearshore Danian sample-level diversity was pronounced and lasted over the entire studied interval. Differential evolutionary dynamics along onshore–offshore gradients – well-known in post-Palaeozoic benthic marine invertebrates, with preferential origination occurring onshore [45], [46] – is inadequate to account for the observed difference in our study. Not only would this mechanism favour the recovery of onshore diversity – which is just the opposite of the pattern we observe – but originations in general played only a minor role during recovery in our study areas. The nearshore–offshore difference might have arisen from unpredictable, stochastic processes, consistent with the contingent recovery model [47] which emphasises the importance of chance in the recovery of marine plankton from the end-Cretaceous extinction. A more mechanistic explanation focuses on the role of survivors as the most diverse group. We suggest that the recolonisation by survivors from nearby localities may have been facilitated in the spatially and temporally more homogeneous and continuous offshore environment. Under such conditions we expect that populations are well connected with each other. In ecology, high levels of connectivity within metapopulations are often considered to be beneficial for recovery after local depletion or extirpation [48]. In contrast, species inhabiting the more variable nearshore environments tend to have more patchy distributions. Their supposedly lower connectivity among populations may have hampered the recolonisation by survivors in nearshore settings.

Conclusions

We analysed so far underrepresented K-Pg boundary sections from the Southern Hemisphere, which contribute to a more complete understanding of recovery processes. Overall, our results corroborate and extend earlier work finding interregional and environmental variation in recovery dynamics in benthic [3], [7] and planktonic marine systems [28], [29]. Diversity of the post-extinction interval was controlled by taxa that survived the extinction event in Patagonia. We emphasise the importance of biotic interactions and conclude that survivors had negative effects on the potential origination of new taxa and the invasion of exotic genera through incumbency. Furthermore, survivors constituted the pool of taxa from which the majority of new dominants were recruited after the most abundant Maastrichtian species went extinct at the boundary. Such distinct shifts in the abundances of survivors are a feature these benthic marine ecosystems share with terrestrial plant communities across mass extinction boundaries [49]. Restoration of sample-level diversity of mollusc assemblages can be rapid (< *ca* 500 kyr in offshore habitats). In contrast, total Danian diversity remained below that of the Maastrichtian, despite a marked Palaeogene increase in the global diversity of gastropods and bivalves [8].

Acknowledgments

We thank Dave Jablonski, Dave Lazarus, Jocelyn Sessa, Sonny Walton, and an anonymous reviewer for valuable discussions and constructive comments on the manuscript. We thank Francisco Medina, Roberto Scasso, Henning Scholz, and Sven Weidemeyer for assistance in the field, and Elke Siebert for support in figure preparation. This is Paleobiology Database publication number 200.

Author Contributions

Analyzed the data: WK MA. Wrote the paper: MA. Designed the study: MA. Interpreted results: MA WK. Prepared figures: WK MA. Improved final manuscript: WK.

References

1. McKinney ML (1997) Extinction vulnerability and selectivity: Combining ecological and paleontological views. Annu Rev Ecol Syst 28: 495–516.
2. Erwin DH (2001) Lessons from the past: Biotic recoveries from mass extinctions. Proc Natl Acad Sci USA 98: 5399–5403. (doi:10.1073/pnas.091092698).
3. Aberhan M, Weidemeyer S, Kiessling W, Scasso RA, Medina FA (2007) Faunal evidence for reduced productivity and uncoordinated recovery in Southern Hemisphere Cretaceous/Paleogene boundary sections. Geology 35: 227–230. (doi:10.1130/G23197A.1).
4. Solé RV, Saldaña J, Montoya JM, Erwin DH (2010) Simple model of recovery dynamics after mass extinction. J Theor Biol 267: 193–200. (doi:10.1016/j.jtbi.2010.08.015).
5. Schulte P, Alegret L, Arenillas I, Arz JA, Barton PJ, et al. (2010) The Chicxulub asteroid impact and mass extinction at the Cretaceous-Paleogene boundary. Science 327: 1214–1218. (doi:10.1126/science.1177265).
6. Hansen TA, Farrell BR, Upshaw B (1993) The first 2 million years after the Cretaceous-Tertiary boundary in east Texas: Rate and paleoecology of the molluscan recovery. Paleobiology 19: 251–265.
7. Jablonski D (1998) Geographic variation in the molluscan recovery from the end-Cretaceous extinction. Science 279: 1327–1330.
8. Alroy J (2010) The shifting balance of diversity among major marine animal groups. Science 329: 1191–1194. (doi:10.1126/science.1189910).
9. Scasso RA, Aberhan M, Ruiz L, Weidemeyer S, Medina FA, et al. (2012) Integrated bio- and lithofacies analysis of coarse-grained, tide-dominated deltaic environments across the Cretaceous/Paleogene boundary in Patagonia, Argentina. Cretaceous Research 36: 37–57. (doi:10.1016/j.cretres.2012.02.002).
10. Scasso RA, Concheyro A, Kiessling W, Aberhan M, Hecht L, et al. (2005) A tsunami deposit at the Cretaceous-Paleogene boundary in the Neuquén Basin of Argentina. Cretaceous Research 26: 283–297. (doi:10.1016/j.cretres.2004.12.003).
11. Keller G, Adatte T, Tantawy AA, Berner Z, Stinnesbeck W, et al. (2007) High stress late Maastrichtian – early Danian palaeoenvironment in the Neuquén Basin, Argentina. Cretaceous Research 28: 939–960. (doi:10.1016/j.cretres.2007.01.006).
12. Gradstein FM, Ogg JG, Schmitz MD, Ogg GM (2012) The geologic time scale 2012. Oxford: Elsevier. 1144 p.
13. Papú OH, Prámparo MB, Náñez C, Concheyro A (1999) Palinología y micropaleontología de la formación Jagüel (Maastrichtiano-Daniano), perfil Opaso, cuenca Neuquina, Argentina. Servicio Geológico Minero Argentino, Anales 33: 17–31.
14. Barreda VD, Cúneo NR, Wilf P, Currano ED, Scasso RA, et al. (2012) Cretaceous/Paleogene floral turnover in Patagonia: drop in diversity, low extinction, and a *Classopollis* spike. PLoS One 7: 1–8. (doi:10.1371/journal.pone.0052455).
15. Patzkowsky ME, Holland SM (2007) Diversity partitioning of a Late Ordovician marine biotic invasion: controls on diversity in regional ecosystems. Paleobiology 33: 295–309.
16. Kowalewski M, Kiessling W, Aberhan M, Fürsich FT, Scarponi D, et al. (2006) Ecological, taxonomic, and taphonomic components of the post-Paleozoic increase in sample-level species diversity of marine benthos. Paleobiology 32: 533–561.
17. Alroy J (2010) Fair sampling of taxonomic richness and unbiased estimation of origination and extinction rates. In: Alroy J, Hunt G, editors. Quantitative Methods in Paleobiology. The Paleontological Society Papers 16. 55–80.
18. Gotelli NJ, Colwell RK (2001) Quantifying biodiversity: procedures and pitfalls in the measurement and comparison of species richness. Ecol Lett 4: 379–391.
19. Sepkoski JJ (2002) A compendium of fossil marine animal genera. Bulletins of American Paleontology 363: 1–563.
20. Hansen T, Farrand RB, Montgomery HA, Billman HG, Blechschmidt G (1987) Sedimentology and extinction patterns across the Cretaceous-Tertiary boundary interval in east Texas. Cretaceous Research 8: 229–252. (doi:10.1016/0195-6671(87)90023-1).
21. Bryan JR, Jones DS (1989) Fabric of the Cretaceous–Tertiary marine macrofaunal transition at Braggs, Alabama. Palaeogeogr Palaeoclimatol Palaeoecol 69: 279–301. (doi:10.1016/0031-0182(89)90170-3).
22. Heinberg C (1999) Lower Danian bivalves, Stevns Klint, Denmark: Continuity across the K/T boundary. Palaeogeogr Palaeoclimatol Palaeoecol 154: 87–106. (doi:10.1016/S0031-0182(99)00088-7).
23. Jablonski D (2002) Survival without recovery after mass extinctions. Proc Natl Acad Sci USA 99: 8139–8144. (doi:10.1073/pnas.102163299).
24. Harries PJ, Kauffman EG, Hansen TA (1996) Models for biotic survival following mass extinction. In: Hart MB, editor. Biotic recovery from mass extinction events. London: The Geological Society. 41–60.
25. Erwin DH (1998) The end and the beginning: recoveries from mass extinctions. Trends Ecol Evol 13: 344–349.
26. Stilwell JD (2003) Patterns of biodiversity and faunal rebound following the K–T boundary extinction event in Austral Palaeocene molluscan faunas. Palaeogeogr Palaeoclimatol Palaeoecol 195: 319–356. (doi:10.1016/S0031-0182(03)00364-X).
27. Coxall HK, D'Hondt S, Zachos JC (2006) Pelagic evolution and environmental recovery after the Cretaceous-Paleogene mass extinction. Geology 34: 297–300. (doi:10.1130/G21702.1).
28. Jiang S, Bralower TJ, Patzkowsky ME, Kump LR, Schueth JD (2010) Geographic controls on nannoplankton extinction across the Cretaceous/Palaeogene boundary. Nat Geosci 3: 280–285. (doi:10.1038/ngeo775).
29. Hull PM, Norris RD (2011) Diverse patterns of ocean export productivity change across the Cretaceous-Paleogene boundary: New insights from biogenic barium. Paleoceanography 26: 1–10. (doi:10.1029/2010PA002082).
30. Valentine JW, Jablonski D, Krug AZ, Roy K (2008) Incumbency, diversity, and latitudinal gradients. Paleobiology 35: 169–178.
31. Algar AC, Mahler DL, Glor RE, Losos JB (2013) Niche incumbency, dispersal limitation and climate shape geographical distributions in a species-rich island adaptive radiation. Glob Ecol Biogeogr 22: 391–402. (doi:10.1111/geb.12003).

32. Sheehan PM (2008) Did incumbency play a role in maintaining boundaries between Late Ordovician brachiopod realms? Lethaia 41: 147–153.
33. Vermeij GJ, Roopnarine PD (2013) Reining in the Red Queen: the dynamics of adaptation and extinction reexamined. Paleobiology 39: 560–575. (doi:10.1666/13009).
34. Hallam A (1990) Biotic and abiotic factors in the evolution of early Mesozoic marine molluscs. In: Ross RM, Allmon WD, editors. Causes of evolution – a paleontological perspective. Chicago: The University of Chicago Press. 249–269.
35. Bambach RK, Knoll AH, Sepkoski JJ (2002) Anatomical and ecological constraints on Phanerozoic animal diversity in the marine realm. Proc Natl Acad Sci USA 99: 6854–6859. (doi:10.1073/pnas.092150999).
36. Gaston KJ, Blackburn TM, Greenwood JJD, Gregory RD, Quinn RM, et al. (2000) Abundance–occupancy relationships. J Appl Ecol 37 (Suppl. 1): 39–59.
37. Simpson C, Harnik PG (2009) Assessing the role of abundance in marine bivalve extinction over the post-Paleozoic. Paleobiology 35: 631–647.
38. Lockwood R (2003) Abundance not linked to survival across the end-Cretaceous mass extinction: Patterns in North American bivalves. Proc Natl Acad Sci USA 100: 2478–2482. (doi:10.1073/pnas.0535132100).
39. Roopnarine PD (2006) Extinction cascades and catastrophe in ancient food webs. Paleobiology 32: 1–19.
40. Lyons KG, Brigham CA, Traut BH, Schwartz MW (2005) Rare species and ecosystem functioning. Conserv Biol 19: 1019–1024. (doi:10.1111/j.1523-1739.2005.00106.x).
41. McElwain JC, Popa ME, Hesselbo SP, Haworth M, Surlyk F (2007) Macroecological responses of terrestrial vegetation to climatic and atmospheric change across the Triassic/Jurassic boundary in East Greenland. Paleobiology 33: 547–573.
42. Hansen TA, Kelley PH, Haasl DM (2004) Paleoecological patterns in molluscan extinctions and recoveries: comparison of the Cretaceous–Paleogene and Eocene–Oligocene extinctions in North America. Palaeogeogr Palaeoclimatol Palaeoecol 214: 233–242. (doi:10.1016/j.palaeo.2004.06.017).
43. Sessa JA, Bralower TJ, Patzkowsky ME, Handley JC, Ivany LC (2012) Environmental and biological controls on the diversity and ecology of Late Cretaceous through early Paleogene marine ecosystems in the U.S. Gulf Coastal Plain. Paleobiology 38: 218–239.
44. Erwin DH (2007) Increasing returns, ecological feedback and the Early Triassic recovery. Palaeoworld 16: 9–15. (doi:10.1016/j.palwor.2007.05.013).
45. Jablonski D (2005) Evolutionary innovations in the fossil record: the intersection of ecology, development, and macroevolution. J Exp Zool 304B: 504–519. (doi:10.1002/jez.b.21075).
46. Kiessling W, Aberhan M (2007) Environmental determinants of marine benthic biodiversity dynamics through Triassic-Jurassic times. Paleobiology 33: 414–434.
47. Hull PM, Norris RD, Bralower TJ, Schueth JD (2011) A role for chance in marine recovery from the end-Cretaceous extinction. Nat Geosci 4: 856–860. (doi:10.1038/ngeo1302).
48. Hughes TP, Carpenter S, Rockström J, Scheffer M, Walker B (2013) Multiscale regime shifts and planetary boundaries. Trends Ecol Evol 28: 389–395. (doi:10.1016/j.tree.2013.05.019).
49. McElwain JC, Punyasena SW (2007) Extinction events and the plant fossil record. Trends Ecol Evol 22: 548–557. (doi:10.1016/j.tree.2007.09.003).
50. del Río CJ (2004) Tertiary marine molluscan assemblages of eastern Patagonia (Argentina): a biostratigraphic analysis. J Paleontol 78: 1097–1122.
51. Lazo DG (2003) Taxonomy, facies relationships and palaeobiology of bakevelliid bivalves from the Lower Cretaceous of west-central Argentina. Cretaceous Research 24: 765–788. (doi:10.1016/j.cretres.2003.09.001).
52. Lazo DG (2007) Early Cretaceous bivalves of the Neuquén Basin, west-central Argentina: notes on taxonomy, palaeobiogeography and palaeoecology. Geol J 42: 127–142. (doi:10.1002/gj.1080).
53. Kauffman EG, Leanza HA (2004) A remarkable new genus of Mytilidae (Bivalvia) from the Lower Cretaceous of Southwestern Gondwanaland. J Paleontol 78: 1187–1191. (doi:10.1666/0022-3360(2004)078).
54. del Río CJ (2002) Moluscos del Terciario marino. Geología y Recursos Naturales de Santa Cruz 2: 1–22.
55. Zinsmeister WJ, Griffin M (1995) Late Cretaceous and Tertiary aporrhaid gastropods from the southern rim of the Pacific Ocean. J Paleontol 69: 692–702.
56. Chiesa JO, Camacho HH (2001) Invertebrados marinos Eocenos de la parte inferior de la Formacíon Río Foyel, Provincia de Río Negro, Argentina. Rev Esp Paleontol 16: 299–316.
57. Chiesa JO, Parma SG, Camacho HH (1995) Invertebrados marinos de la Formación El Chacay (Eoceno), provincia de Santa Cruz, Argentina. Sistemática y bioestratigrafía. Academia Nacional de Ciencias Exactas, Físicas y Naturales, Monografías 11: 17–68.
58. Casadío S, Griffin M, Parras A (2005) *Camptonectes* and *Plicatula* (Bivalvia, Pteriomorphia) from the Upper Maastrichtian of northern Patagonia: palaeobiogeographic implications. Cretaceous Research 26: 507–524. (doi:10.1016/j.cretres.2005.01.009).
59. Woods H (1917) The Cretaceous faunas of the north-eastern part of the South Island of New Zealand. New Zealand Geological Survey, Palaeontological Bulletin 4: 1–41.
60. Casadío S, Griffin M, Marenssi S, Net L, Parras A, et al. (2009) Paleontology and sedimentology of Middle Eocene rocks in Lago Argentino area, Santa Cruz Province, Argentina. Ameghiniana 46: 27–47.
61. Squires RL, Goedert JL (1994) New species of Eocene small to minute mollusks from the Crescent Formation, Black Hills, southwestern Washington. Veliger 37: 253–266.
62. del Río CJ, Concheyro A, Martínez SA (2011) The Maastrichtian – Danian at General Roca (Patagonia, Argentina): a reappraisal of the chronostratigraphy and biostratigraphy of a type locality. Neues Jahrb Geol Paläontol Abh 259: 129–156. (doi:10.1127/0077-7749/2011/0103).
63. Berggren WA, Kent DV, Swisher CC, Aubry M-P (1995) A revised Cenozoic geochronology and chronostratigraphy. In: Berggren WA, et al., editors. Geochronology, time scales and global stratigraphic correlation. SEPM Special Publications 54: 129–212.
64. Harrington HJ (1958) Geology of Kaitangata Coalfield. New Zealand Geological Survey Bulletin 59: 1–131.
65. Kafanov AI, Ogasawara K (2004) Neogene and Paleogene molluscan (Bivalvia) cenozones of the Russian Northeast. Science Reports of the Institute of Geoscience, University of Tsukuba, Section B 25: 9–45.
66. del Río CJ, Martínez SA, Stilwell JD, Concheyro A (2007) Paleontology of the Cerros Bayos section, Roca Formation (Danian), La Pampa Province, Argentina. Alcheringa 31: 241–269. (doi:10.1080/03115510701484713).
67. Waller TR, Marincovich L (1992) New species of *Camptochlamys* and *Chlamys* (Mollusca: Bivalvia: Pectinidae) from near the Cretaceous/Tertiary boundary at Ocean Point, North Slope, Alaska. J Paleontol 66: 215–227.
68. del Río CJ, Beu AG, Martínez SA (2008) The pectinoidean genera *Delectopecten* Stewart, 1930 and *Parvamussium* Sacco, 1897 in the Danian of Northern Patagonia, Argentina. Neues Jahrb Geol Paläontol Abh 249: 281–295.
69. Aguirre-Urreta MB (1989) The Cretaceous decapod Crustacea of Argentina and the Antarctic Pensinsula. Palaeontology 32: 499–552.
70. Darragh TA (1997) Gastropoda, Scaphopoda, Cephalopoda, and new Bivalvia of the Paleocene Pebble Point Formation, Victoria, Australia. Proceedings of the Royal Society of Victoria 109: 57–108.
71. Chavan A (1946) Nouvelles notes sur les *Jagonia* (Lamellibranches). Bulletin du Muséum National d'Histoire Naturelle Paris, sér. 2, 18: 345–347.
72. Lazo DG, Damborenea SE (2011) Barremian bivalves from the Huitrín Formation, west-central Argentina: taxonomy and paleoecology of a restricted marine association. J Paleontol 85: 721–749.
73. Stilwell JD (2003) Macropalaeontology of the *Trochocyathus-Trematotrochus* band (Paleocene/Eocene boundary), Dilwyn Formation, Otway Basin, Victoria. Alcheringa 27: 245–275. (doi:10.1080/03115510308619107).
74. Stilwell JD, Zinsmeister WJ, Oleinik AE (2004) Early Paleocene mollusks of Antarctica: systematics, paleoecology and paleobiogeographic significance. Bulletins of American Paleontology 367: 1–89.
75. Griffin M, Hünicken MA (1994) Late Cretaceous-Early Tertiary gastropods from southwestern Patagonia, Argentina. J Paleontol 68: 257–274.

Cumulative Human Impacts on Mediterranean and Black Sea Marine Ecosystems: Assessing Current Pressures and Opportunities

Fiorenza Micheli[1]*, Benjamin S. Halpern[2], Shaun Walbridge[3], Saul Ciriaco[4], Francesco Ferretti[1], Simonetta Fraschetti[5], Rebecca Lewison[6], Leo Nykjaer[7], Andrew A. Rosenberg[8]

1 Hopkins Marine Station, Stanford University, Pacific Grove, California, United States of America, **2** National Center for Ecological Analysis and Synthesis, Santa Barbara, California, United States of America, and the Center for Marine Assessment and Planning, University of California Santa Barbara, Santa Barbara, California, United States of America, **3** ESRI, Redlands, California, United States of America, **4** Miramare Marine Protected Area, Trieste, Italy, **5** Laboratory of Marine Biology, Università del Salento, Lecce, Italy, **6** Coastal and Marine Institute Laboratory, San Diego State University, San Diego, California, United States of America, **7** European Commission, Joint Research Centre, Institute for Environment and Sustainability, Ispra, Italy, **8** Union of Concerned Scientists, Cambridge, Massachusetts, United States of America

Abstract

Management of marine ecosystems requires spatial information on current impacts. In several marine regions, including the Mediterranean and Black Sea, legal mandates and agreements to implement ecosystem-based management and spatial plans provide new opportunities to balance uses and protection of marine ecosystems. Analyses of the intensity and distribution of cumulative impacts of human activities directly connected to the ecological goals of these policy efforts are critically needed. Quantification and mapping of the cumulative impact of 22 drivers to 17 marine ecosystems reveals that 20% of the entire basin and 60–99% of the territorial waters of EU member states are heavily impacted, with high human impact occurring in all ecoregions and territorial waters. Less than 1% of these regions are relatively unaffected. This high impact results from multiple drivers, rather than one individual use or stressor, with climatic drivers (increasing temperature and UV, and acidification), demersal fishing, ship traffic, and, in coastal areas, pollution from land accounting for a majority of cumulative impacts. These results show that coordinated management of key areas and activities could significantly improve the condition of these marine ecosystems.

Editor: James P. Meador, Northwest Fisheries Science Center, NOAA Fisheries, United States of America

Funding: This work was supported by the Oak Foundation and the Pew Charitable Trust. SF was also supported by the European Community's 7th Framework Programme (FP7/2007–2013) under Grant Agreement No. 287844 for the project 'Towards COast to Coast NETworks of marine protected areas (from the shore to the high and deep sea), coupled with sea-based wind energy potential (COCONET)'. The funders had no role in study design, data collection and analysis, decision to publish, or preparation of the manuscript.

Competing Interests: One of the co-authors' (SW) affiliation with a commercial company (ESRI) does not alter the authors' adherence to all the PLOS ONE policies on sharing data and materials.

* E-mail: micheli@stanford.edu

Introduction

Ecosystem-based management (EBM) and Marine Spatial Planning (MSP) are widely being pursued as strategies to achieve the sustainable flow of marine ecosystem services [1,2]. These comprehensive marine management frameworks are now mandated in some nations around the world [3], including Canada (Canada's Ocean Act of 1996), the USA (The National Ocean Policy of 2010), and Australia (http://www.gbrmpa.gov.au/zoning-permits-and-plans/zoning).

Member states of the European Union (EU) are also committed to adopting an ecosystem approach to marine management, including marine spatial planning. As mandated by the EU Marine Strategy Framework Directive (MSFD 2008), all European states should assess the environmental status of their territorial waters by July 2014, and develop strategies to achieve "good environmental status" by 2020 (GES, http://ec.europa.eu/environment/water/marine/es.htm). Marine spatial planning is also part of the EU Integrated Maritime Policy (IMP) of 2011 (http://ec.europa.eu/maritimeaffairs/policy/). At the regional scale, all the 21 Mediterranean nations have ratified the UNEP's Mediterranean Action Plan (MAP) to move Mediterranean marine management towards an ecosystem approach (ECAP), and to expand into non-EU Mediterranean waters the same conservation and management measures implemented in EU waters.

The Mediterranean and Black Sea ecosystems have been threatened by historical and current pressures e.g [4,5,6] which have led to major shifts in marine ecosystems and widespread conflict among marine users [7,8,9,10]. Because of such intense pressure from multiple uses and stressors, the Mediterranean is characterized as a sea "under siege" [6,11], and here, as in other intensely used ocean areas, an EBM approach has been recommended as a better management alternative to current sectoral management [1].

MAP ECAP, the EU MSFD and IMP provide an unprecedented opportunity to implement comprehensive and coordinated management of multiple uses and activities affecting the Mediterranean and Black Seas. These initiatives have no formal links, but a common timeline has recently been agreed upon [12]. It is expected that the implementation of the MAP ECAP will provide a platform for harmonization of national marine strategies of all Mediterranean countries (EU and non-EU) on a regional scale.

The development of basin-wide plans requires information on current impacts that can inform effective marine policy over the next years. Some relevant research has been conducted in this region. Coll et al. [6] have mapped cumulative impacts to key taxa and this assessment highlighted several areas of concern. However, a comprehensive analysis focused on whole ecosystems is lacking and quantification of the intensity and distribution of cumulative impacts in the whole Mediterranean and Black Sea that directly connects to ecosystem goals and priorities is critically needed.

We apply an approach developed to assess and map cumulative human impacts [13] that was previously applied to other marine regions, including the US EEZ e.g [14,15], western Canada [16], and the North (http://harmony.dmu.dk/) and Baltic Seas [17]. Our goals are to: (1) quantify and map cumulative impacts to the entire Mediterranean and Black Sea to provide the data needed to guide and inform the development of effective marine policy; and (2) propose and apply a tool for assessing the environmental status of the territorial waters of EU member states. In particular, we identify the most and least impacted ecoregions and ecosystems within the Mediterranean and Black Sea, the top threats affecting EU territorial waters and the entire basin, and the areas that represent top priorities for EBM and conservation efforts. These analyses are aimed at supporting coordinated and comprehensive actions across the basins, ensuring GES consider all impacts and are relevant at both the national and the regional scales.

Methods

We used a cumulative impact model that follows a 4-step process [13]. We first assembled spatial datasets for $n=22$ anthropogenic drivers (D_i) and $m=17$ ecosystems (E_j) (Table S1 and Figs. S2–S11 in File SI). All but the climatic drivers have direct correspondence with the MSFD's good environmental status (GES) descriptors (Table S2 in File SI). Second, we log[X+1]-transformed and rescaled between 0–1 each driver layer to put them on a single, unitless scale that allows direct comparison, and converted ecosystem data into 1 km^2 presence/absence layers. Third, we calculated cumulative impact scores (I_C) for each 1 km^2 pixel as $I_C=\sum_{i=1}^{n}\frac{1}{m}\sum_{j=1}^{m}D_i*Ej*\mu_{i,j}$, where $\mu_{i,j}$ is the impact weight for anthropogenic driver i and ecosystem j, and $1/m$ produces an average impact score across ecosystems [13]. Impact weights were estimated using expert judgment to quantify vulnerability of ecosystems to human drivers of ecological change [18]. Although these weights are not specific to the Mediterranean, regional experts were included in those analyses, and weights have proven fairly consistent for other regional assessments [19,20]. The use of expert judgment instead of direct empirical assessments to calculate impact weights greatly increases uncertainty of our impact scores. Empirical quantification of the ecological impacts of a suite of drivers is currently unavailable and filling this gap is a critical need within the Mediterranean and other regions [10,18]. Despite these limitations, there is a long history in the decision sciences of assessing how to set priorities (e.g., rank threats) by using the best available scientific judgments when data are scarce and uncertainty exists. Teck et al. [19] critically examined uncertainties associated with expert judgment and showed the robustness of our approach to eliciting expert opinions for informing cumulative impact assessments for the California Current. Based on these results, we used the same approach and vulnerability weights here. Regardless, it is critical that uncertainties associated with a lack of empirical data on ecosystem vulnerability are communicated clearly, especially when integrating cumulative impact mapping into decision making [21].

Cumulative impact to individual ecosystems (I_E) was calculated as, $I_E=\sum_{i=1}^{n}D_i*E_j*\mu_{i,j}$ and impact of individual drivers across all ecosystem types (I_D) was calculated as $I_D=\sum_{i=1}^{m}D_i*E_j*\mu_{i,j}$. Finally, we used the same thresholds used in Halpern et al. [13] to designate ecologically meaningful categories of the cumulative impact scores, i.e. ecosystems that are subject to: very high (Ic>15.52); high (12–15.52); medium high (8.47–12); medium (4.95–8.47); low (1.4–4.95); and very low impact (<1.4). These thresholds were based on empirical data on the condition of ecosystems containing coral reefs, seagrass beds, mangroves and surrounding matrix of soft bottom habitats [13]. In Halpern et al. [13], cumulative impact scores were translated into estimates of ocean condition by using linear regression to compare estimates of the current condition of 16 regions containing coral reefs from around the world [23] to the average cumulative impact score for all cells containing coral reefs in those regions. The statistically significant linear regression equation was then used to translate impact scores into categories of ocean condition. We used these bins of cumulative impact scores for describing the condition of Mediterranean marine ecosystems and for color-coding all figures. Categorization of cumulative impact scores in very low to very high impact classes does not imply a value judgment, but is aimed instead at enabling analyses and visual display of results.

We used these thresholds for categorizing level of threat, even though they were derived in different regions and ecosystem types, because a similar in-situ verification of the Mediterranean ecosystem condition currently does not exist. The sensitivity of our results to key steps in this process and further details on the validation method were analyzed in previous articles and results were shown to be robust to variation in weights and thresholds [13,19]. However, the assumption that thresholds derived from tropical coastal ecosystems apply to Mediterranean habitats and to offshore and deep ecosystems was not directly validated. Empirical ground-truthing of the relationship between cumulative impact scores and ecosystem condition remains a top priority within the Mediterranean and worldwide [13,14,21].

We first calculated and mapped Ic over the entire Mediterranean and Black Sea basins. We also calculated impact scores separately for each ecoregion, sensu [22] (see Table 1), and for each of four categories of drivers: climatic (temperature and UV increase, and acidification), land-based (nutrient input, organic pollution, urban runoff, risk of hypoxia and coastal population density), sea-based (commercial shipping, invasive species, oil spills and oil rigs), and fishing (all fishing gears and types) (Table S1 in File SI).

Secondly, we calculated and mapped Ic only for the territorial waters of the EU member states, (up to 12 nm from the coastline) where more data are available. Analyses were conducted at two different spatial scales – only for EU member states and for the entire Mediterranean and Black Sea – to account for the spatial scale of available data layers. All 22 drivers were used in the EU analysis but only 18 in the Mediterranean and Black Sea-wide analysis because data on coastal erosion, renourishment, engineering, and urbanization trends were available only for EU member states (Table S1 in File SI). Per-pixel Ic scores ranged between 0–24.5 in the Mediterranean-wide analysis, and 0.2–97.3 for the EU analysis. Sources and methods used to develop each data layer are detailed in the SOM (Text S1 in File SI).

Table 1. Average, maximum, SD and CV of the cumulative impact scores within the seven Mediterranean ecoregions and the Black Sea.

Ecoregion	avg. *Ic*	max *Ic*	SD	CV
Alboran Sea	9.1	20.2	2.0	22.4
Levantine Sea	8.9	19.2	2.5	27.8
Aegean Sea	8.6	18.5	2.6	29.9
Adriatic Sea	8.4	19.0	3.4	40.9
Tunisian Plateau/Gulf of Sidra	8.3	21.0	2.6	31.1
Ionian Sea	8.3	24.5	1.7	20.4
Western Mediterranean	7.7	22.4	1.9	24.7
Black Sea	6.1	16.4	2.2	35.9

Results

Mediterranean and Black Sea

The map of cumulative human impacts highlights the widespread distribution of drivers, and resulting impacts, throughout the Mediterranean and Black Sea (Fig. 1). Regions of medium-high to very high impact (20.5% of the total area) are found within the Alboran Sea, the Gulf of Lyons, the Sicily Channel and Tunisian Plateau, the Adriatic Sea, off the coasts of Egypt and Israel, along the coasts of Turkey, and within the Marmara and Black Sea (Fig. 1). Areas of very-low to low impact account for a total 13.6% of the total surface area, and are present within the central Tyrrhenian Sea, parts of the northern and central Adriatic Sea, the southern Levantine Sea, and the eastern and western sides of the Black Sea (Fig. 1). A majority (65.9%) of the Mediterranean and Black Sea are subject to medium cumulative impact.

When analyzed by ecoregion, the Alboran and Levantine ecoregions have the highest average cumulative impact, the western Mediterranean and Black Sea the lowest (Table 1), although areas of high impact exist even within these ecoregions (Fig. 1). In fact, the greatest per-pixel *Ic* scores were seen within the western Mediterranean and in the Ionian Sea, and the Adriatic and Black Seas exhibited the greatest variability in impact scores (Table 1), comprising areas of both very high and low impact (Fig. 1).

Pelagic and benthic offshore ecosystems have the greatest average cumulative impact (I_E; Table S3 in File SI). However, this is partly driven by their large extent: the maximum pixel-level values of the I_E scores, indicative of locally high impacts, are observed in intertidal habitats (particularly salt marshes, suspension-feeding reefs, rocky shores and mud flats), and in nearshore sublittoral and continental shelf hard bottoms that are affected by both sea-based and land-based drivers (Table S3 in File SI).

EU member states

Quantification of the percent of national waters of EU member states in different impact categories (Table 2) reveals that a majority (60–99%) of waters within 12 nm of the coastline are subject to medium-high to very high impact. 0–20% of national waters are subject to low or very low impact, and this percent is less than 10% for a majority of nations (Table 2). Average *Ic* (averaged by nation) range from 6.2–15.3, corresponding to medium to high impact (Table S4 in File SI), and show no correlation with the area of national waters ($R^2 = 0.0005$, NS).

Cumulative impact maps show high spatial variation in impact scores, with heavily impacted systems (in red) found within the waters of all nations, and small areas with relatively low impact (in blue or green) along the coasts of most nations (Fig. 2). The addition of four coastal drivers to this analysis (coastal erosion, renourishment, engineering, and urbanization trends) results in greater *Ic* scores in most locations compared to the Mediterranean-wide map (Figs. 1–2).

Relative contribution and spatial distribution of drivers

Drivers associated with climate change (SST and UV increase, and acidification), demersal fishing, and shipping result in the greatest average impact on Mediterranean and Black Sea ecosystems (Fig. 3a). These drivers, along with coastal hypoxia, were found to exert the greatest impact within territorial waters of the EU nations (Fig. 3b), followed by coastal population density, invasive species, land-based pollution (inorganic pollution, pesticide and fertilizer runoff) and modification of the coastline (through coastal erosion and engineering). The lowest estimated impacts, both within EU waters and for the whole basin, are associated with oil spills and rigs (Fig. 3).

Driver categories differ in their distribution across the Mediterranean and Black Sea (Fig. 4; Table S1 in File SI). Climatic drivers are broadly distributed but have the greatest impact scores in the eastern Mediterranean. Fishing affects all coastal areas, as well as most of the Sicily Channel and the Alboran, Adriatic, and Aegean Sea. Sea-based activities result in the highest scores in the western and southern regions, and land-based activities broadly affect coastal areas, as well as large portions of the Adriatic and Black Sea.

All ecoregions are affected by multiple drivers, but the relative importance of different drivers varies among ecoregions (Fig. 5). Climatic drivers have the greatest contribution to the average cumulative impact score of all ecoregions, though this contribution is lower for the Alboran and Adriatic Seas (Fig. 5a). When climatic drivers are not included (Fig. 5b), demersal fishing, hypoxia and pollution from land-based activities are major contributors to high cumulative impacts to the Adriatic and Black Sea, and demersal fishing and shipping are major contributors in the other ecoregions.

Discussion

Our analysis highlights that 20% of the entire Mediterranean and Black Sea and 60–99% of the territorial waters of EU member states are subject to high impact, while less than 20% has low impact and very few areas, less than 1%, remain relatively unaffected by human activities. Cumulative impact varies greatly across and within ecoregions and countries. Highly impacted areas are found in the Alboran Sea, the Gulf of Lyons, the Sicily Channel and Tunisian Plateau, the Adriatic Sea, off the coasts of Egypt and Israel, along the coasts of Turkey, and within the Marmara and Black Sea (Figs. 1 and 2). Contemporarily, we highlight areas characterized by low cumulative human impacts off Croatia, Albania, Italy, Tunisia and Egypt, in offshore areas of the central Tyrrhenian and Black Sea, and in several small areas along the coasts of most countries (Figs. 1 and 2). These areas represent important opportunities for conservation aimed at preventing future degradation.

Some of the highly impacted areas we identified in this analysis coincide with the areas of conservation concern identified by Coll et al. [6], including portions of the Northern Adriatic Sea, the Sicily Channel, the inner Ionian Sea, the Aegean Sea, and the Gulf of Lyons. These areas emerge as clear priorities for future

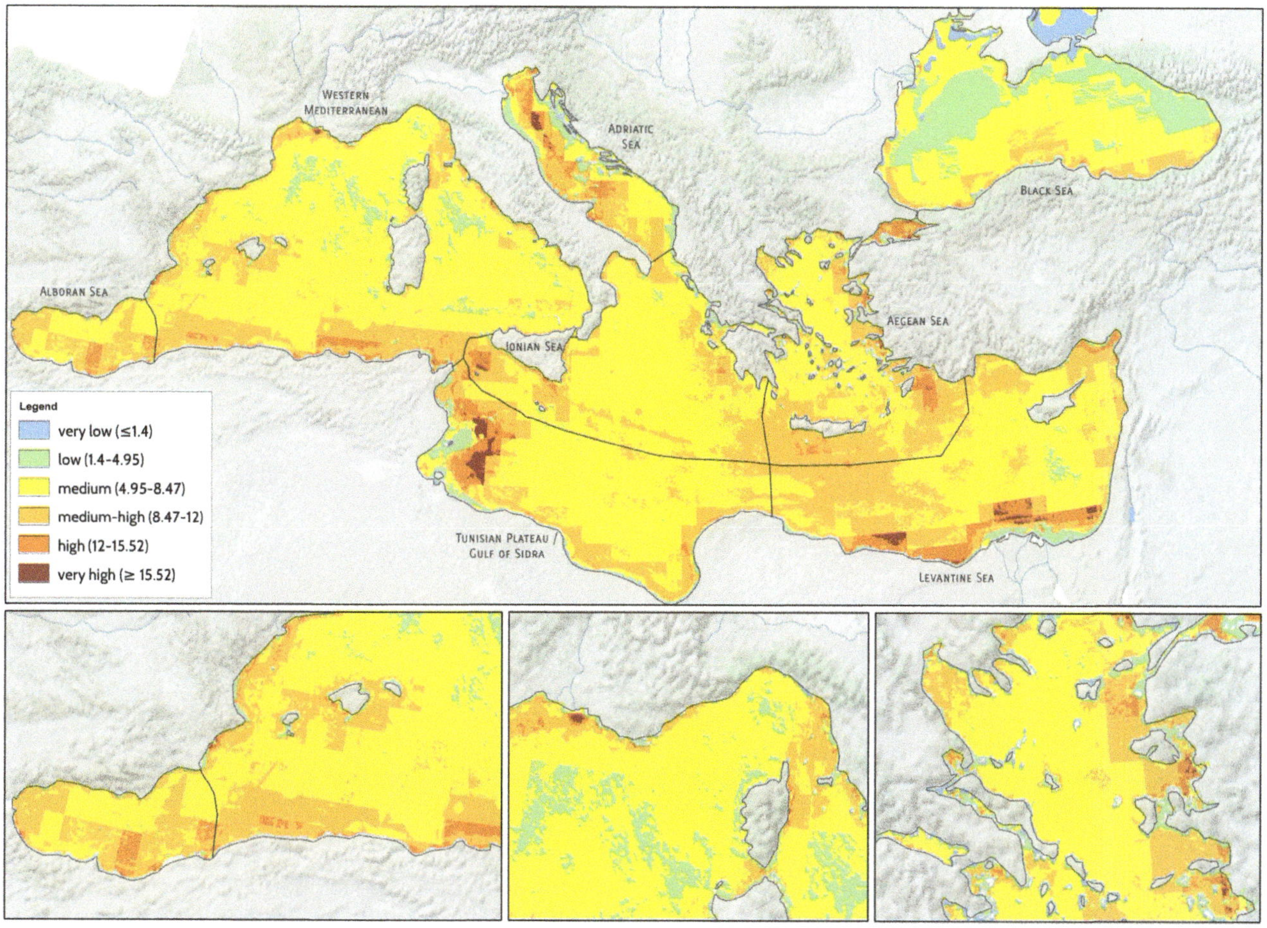

Figure 1. Spatial distribution of cumulative impacts to marine ecosystems of the Mediterranean and Black Sea. Inserts at the bottom show larger views of the Alboran (left), Northern Tyrrhenian (center), and Aegean Sea (right). Colors correspond to the different impact categories listed in the legend.

management action and protection. However, our analyses highlight additional highly impacted areas where multiple drivers overlap with vulnerable habitats, e.g. in the central Adriatic Sea, the Alboran Sea, the Tunisian Plateau, and in the southern Levantine Sea. These areas may also represent high priorities for management and conservation action. By using habitats as proxies

Table 2. Percent of national territorial waters of Mediterranean and Black Sea EU member states within different impact categories: very high impact (I_c>15.52); high impact (12–15.52); medium-high impact (8.47–12); medium impact (4.95–8.47); low impact (1.4–4.95); and very low impact (<1.4).

Country	area (km^2)	very low	low	med	med-high	high	very high
Slovenia	266.2	0.0	10.2	22.7	0.0	0.8	66.3
Cyprus	95,833.6	1.8	4.4	0.7	12.2	63.3	17.6
France	480,103.7	0.4	4.1	26.2	26.7	27.6	14.9
Italy	700,184.6	0.0	6.3	14.0	44.6	21.4	13.7
Spain	744,352.6	0.0	6.6	7.2	40.0	33.9	12.2
Bulgaria	48.050.1	5.3	14.9	15.7	33.6	22.5	8.0
Greece	615,025.4	0.8	9.0	9.7	51.3	21.3	7.9
Monaco	390.6	0.0	0.0	0.8	60.2	32.8	6.3
Malta	68,240.6	0.9	2.9	37.5	34.8	19.0	4.8
Romania	41,509.7	1.4	16.5	20.5	47.5	12.8	1.2

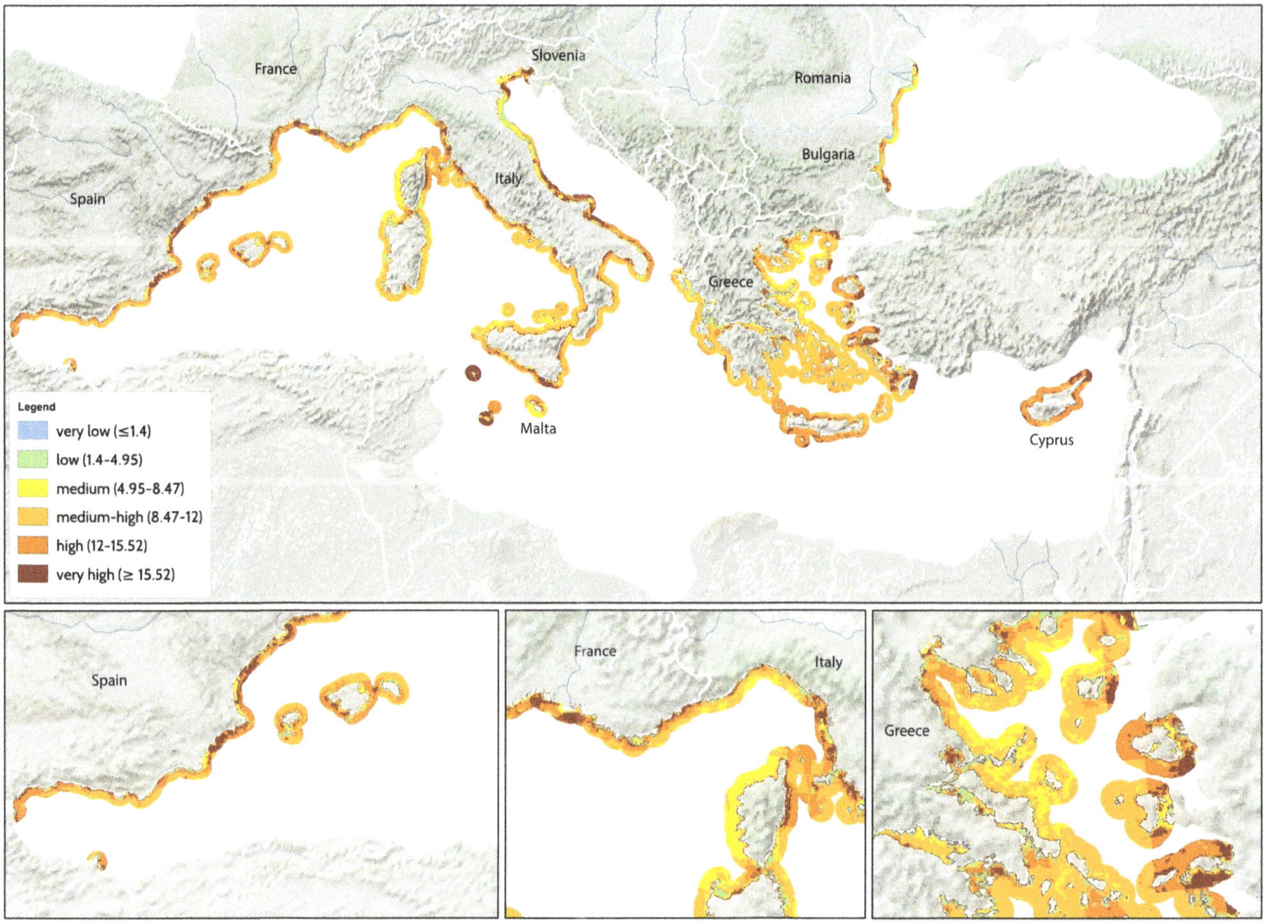

Figure 2. Spatial distribution of cumulative impacts to the territorial waters of EU member states. Inserts at the bottom show larger views of the Alboran (left), Northern Tyrrhenian (center), and Aegean Seas (right). Colors correspond to the different impact categories listed in the legend. Territorial waters extend 12 nm from the coastline.

for biodiversity instead of species data, our approach allows for the identification of areas of conservation concern in relatively data poor regions, such as the southern and eastern Mediterranean Sea, where species data are still scarce [6]. Taken together, results from Coll et al.'s [6] and our study show how multiple, complementary approaches are needed to assess cumulative impact and to direct research and management efforts to the areas that most urgently need them.

High cumulative impact scores were always associated with multiple drivers, supporting the need for coordinated, comprehensive plans addressing all drivers of ecosystem change across the Mediterranean and Black Sea. However, our results also highlight opportunities for a major reduction of cumulative impact by prioritizing a subset of the drivers for policy action. Demersal fisheries, ship traffic, and, in some coastal areas, fertilizer run-off and resulting hypoxia are major contributors to the cumulative impacts we have analyzed. While it is well known that these activities are important pressures on ecosystems worldwide, this analysis shows that reducing the effects of trawling, ship traffic and nutrient loading from some land-based activities could lead to large reduction in cumulative impact, relative to addressing other drivers. Furthermore, different management policies will be most effective in different regions. For example, in the Sicily Channel and Alboran Sea, a better environmental status could be achieved by focusing on spatial plans for fisheries and commercial shipping, and in the northern Adriatic Sea and the Black Sea on fisheries and pollution from land. In all regions, high spatial variability in the distribution of human pressures and vulnerable ecosystems provide opportunities to reduce cumulative impact through spatial planning of current and emerging activities.

In accord with previous global and regional impacts research [13–15] climatic drivers (temperature and UV increase, and ocean acidification) were found to cause the largest potential impacts. While climatic drivers cannot be removed through local management action, their spatial distribution highlights the areas where climate mitigation and adaptation is most critical, e.g., in the eastern Mediterranean Sea.

The estimated high impacts of major climatic and human drivers are influenced by both their widespread distribution (particularly climatic stressors), and the high vulnerability of multiple ecosystem types to these pressures e.g., [18]. In contrast, oil spills and rigs had the lowest estimated impacts, both within EU waters and for the whole basin. These highly harmful marine stressors showed in the Mediterranean a smaller effect because of their limited current spatial extent overlapping with habitats relatively resistant to oil-associated threats. Episodic and unpredictable, oil spills do not represent a chronic stressor to these ocean regions although a large-scale spill would clearly be catastrophic and damaging for these ecosystems.

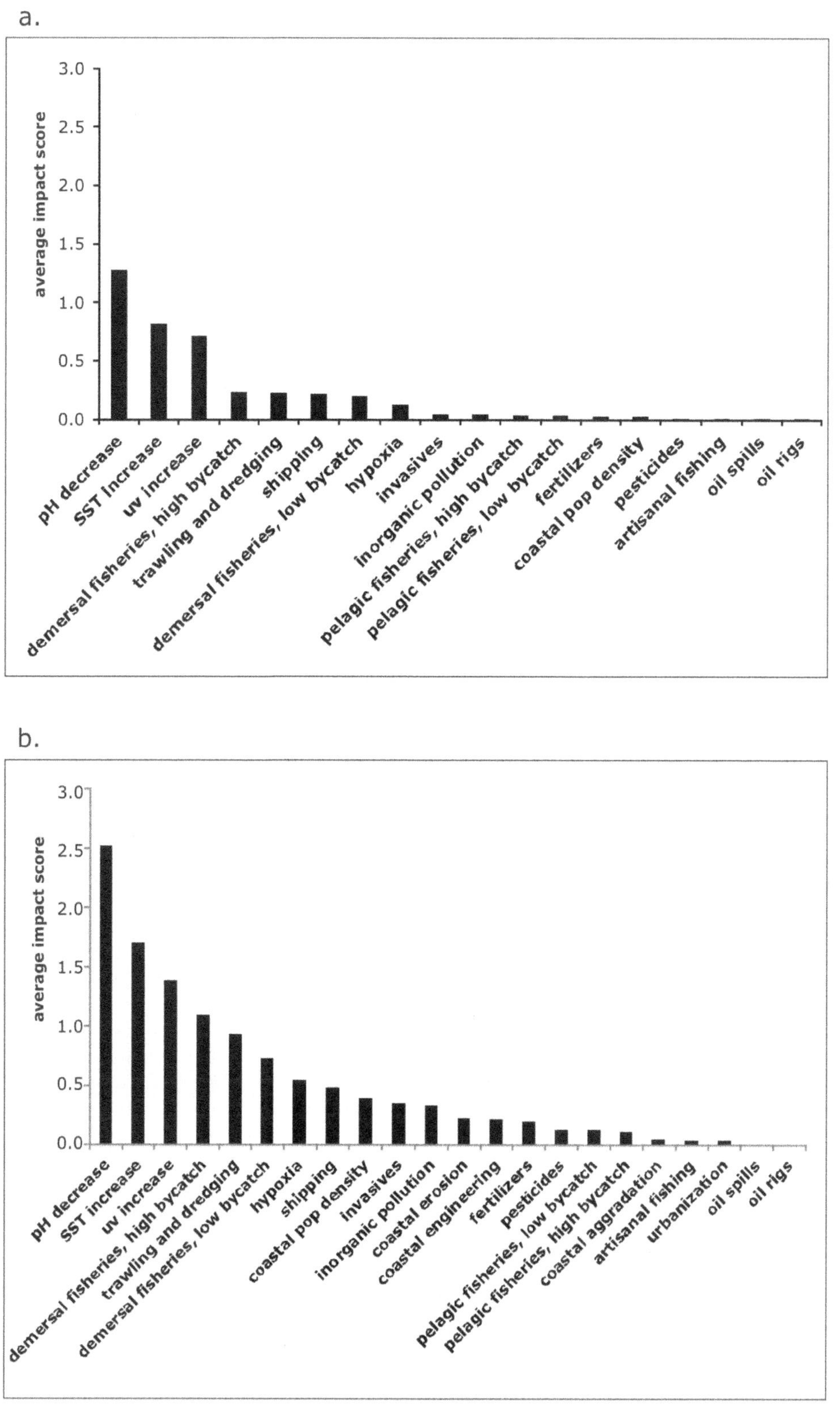

Figure 3. Average Impact scores of drivers. Average impact scores for each driver are reported within (a) the entire Mediterranean and Black Sea, and (b) the territorial waters of EU member states.

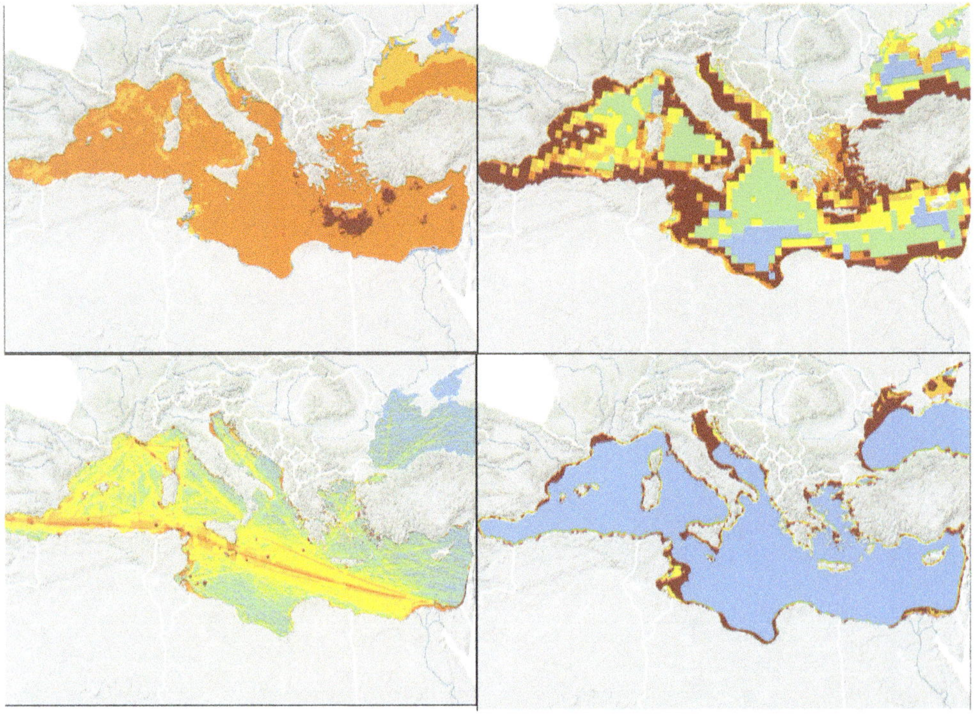

Figure 4. Spatial distribution of cumulative impact of driver categories. Driver categories are: climate (i.e. the combined cumulative impact of temperature and UV increase, and acidification; top left), fishing (all fishing types combined; top right), sea-based drivers (commercial shipping, invasive species, oil spills and oil rigs; bottom left) and land-based drivers (nutrient input, organic pollution, urban runoff, risk of hypoxia and coastal population density; bottom right). Color scales correspond to highest (red) to lowest (blue) cumulative impact, within each panel. Different scales were used in each panel to better highlight spatial patterns for each driver category.

Cumulative impact analyses and mapping can help inform policy reform and management plans by highlighting top priorities and possible opportunities for initiating EBM and MSP, and by providing imagery to facilitate communication of issues and opportunities to policy makers, environmental managers, conservationists, businesses and the public. However, several issues require attention when interpreting results from our analyses. First, major sources of uncertainty remain to be addressed in future empirical studies and modeling efforts [13,16,21]. Empirical information on how ecosystems change in response to different combinations and intensities of drivers is still scarce [10]. A better understanding of if or where ecosystems experience non-linear responses to cumulative impact and thresholds of resistance would be particularly valuable for setting management targets or limits. At present, non-linearities are difficult to anticipate and interpret and adaptive management responses that are robust to unexpected outcomes are needed. Data on the consequences of non-linear behaviors have never been included in Mediterranean analyses and rarely in extra Mediterranean areas [24–26]. Direct empirical assessments of the vulnerability of different ecosystems, in addition to expert judgments, and of the relationship between cumulative impact scores and ecosystem conditions are critically needed. Second, the quality of available data is widely variable, with a great need for improved spatial information on the distribution of different ecosystems, direct measures of fishing effort, and the distribution and effects of important drivers such as marine litter, bycatch, and aquaculture. The effects of data gaps can be seen by comparing the Mediterranean wide and EU analyses (Figs. 1–2): addition of only four data layers in the latter analysis results in greater cumulative impact to coastal areas, indicating that impact is underestimated in the broader analysis. As additional spatial data e.g., [27] become available, they could be incorporated in new iterations of these analyses. It will be especially important to perform even more detailed analyses using all locally available information in the top priority areas highlighted here. Third, our use of fisheries catch data [13] (Fig. S1 in File SI) instead of effort may lead to overestimation of impact in highly productive regions (e.g., the Alboran Sea) and underestimation in regions, such as parts of the Adriatic Sea, subject to intense historical fishing pressure and thus currently depleted [4,28]. Improved access to effort data is a key priority: despite the Mediterranean Sea is a public trust resource and fishery-related data are collected with public funding, much of these data remain confidential. When spatially-explicit effort data become available they can and should be incorporated into future assessments. Fourth, our approach results in a static view of impact. Analyses of trajectories of change [9,28] and tools such as InVEST (http://www.naturalcapitalproject.org) that simulate consequences of different management actions into the future would improve management plans and MSP at local and regional scales. Our analysis serves as a baseline against which future actions can be measured. Finally, and most critically, institutions and social processes play a key role in advancing marine management that no amount of data and technical sophistication can replace. Securing social and political acceptance of conservation and management initiatives and establishing effective processes for their implementation is critical [29]. Direct engagement of scientists and conservation practicioners in the planning process, analysis of social and economic costs and benefits of different management options,

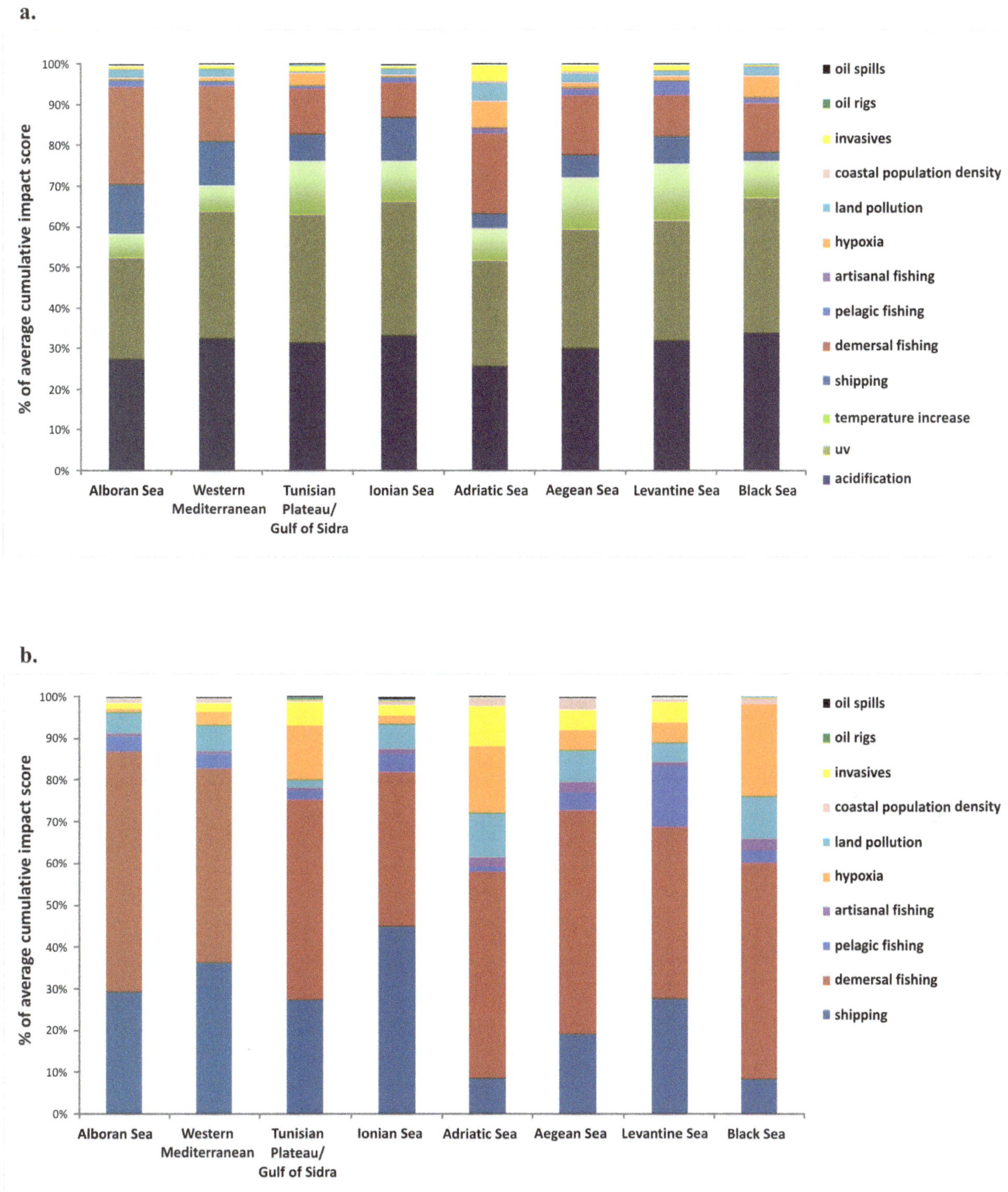

Figure 5. Percent contribution of different drivers to the average cumulative impact score of each ecoregion. Percent contributions are reported with (a) and without (b) climatic drivers. Some drivers (different demersal and pelagic fisheries, and different types of pollution from land) are aggregated to show relative contributions more clearly.

and involvement of diverse stakeholders will be essential to the successful implementation of marine spatial plans. Advancing marine conservation and management will require these fundamental participatory processes [29].

Achieving the environmental goals that nations have committed to, within the Mediterranean region and worldwide, will require ongoing and comprehensive monitoring of impacts in conjunction with new policies that balance biodiversity protection with human uses. Within the EU the European Commission has stimulated the development of MSP among the EU member states, and the debate is ongoing on how to implement the key principles of MSP [30,31]. In parallel to these efforts, MAP is striving to expand the development of MSP into non-EU Mediterranean waters.

The impact assessment approach described here allows for a transparent, repeatable and updatable synthesis and integration of disparate information, facilitating communication and discussion of policy options and alternatives at different spatial scales. Cumulative impact assessment highlights that coordinated management of key areas and activities could significantly enhance the environmental condition of intensely used marine regions. Therefore cumulative impact assessment could be considered as one of the valuable tools for achieving the objectives of the EU maritime policy and MAP.

Supporting Information

File SI Detailed description of methods and spatial data layers, additional summary statistics of cumulative impact scores, and comparison of individual impact data layers from the global cumulative impact analysis [13] with the new layers included in this analysis. Text S1. Methods; **Table S1.** List and characteristics of spatial datalayers; **Table S2.** Correspondence between descriptors of Good Environmental Status and the data layers used in this analysis; **Table S3.** Summary statistics of cumulative impact score for individual ecosystems; **Table S4.** Summary statistics of cumulative impact scores within territorial waters of Mediterranean and Black Sea EU member states; **Figure S1.** Correlation between fishing effort and catch for pelagic longlines; **Figure S2.** Modeled distribution of marine invasive species from Halpern et al. (2008); **Figure S3.** New data layer on the distribution of harmful invasive species in Mediterranean coastal waters; **Figure S4.** Modeled pollution at sea layer (Halpern et al. 2008); **Figure S5.** Occurrence and magnitude of accidents resulting in oil release; **Figure S6.** Frequency of positive temperature anomalies (Halpern et al. 2008); **Figure S7.** Annual average rate of increase in SST; **Figure S8.** Risk of hypoxia; **Figure S9.** Coastal erosion and aggradation along the coastlines of Mediterranean EU countries; **Figure S10.** Coastal armoring along the coastlines of Mediterranean EU countries; **Figure S11.** Urbanization trends within Mediterranean EU countries.

Acknowledgments

We thank P. Christy, A. Stips, F. Briand, M. Orbach, N. Hoepffner, A. White, G. Notarbartolo di Sciara, T. Agardy, D. Hyrenbach, B. Galil, C. Osio, P. Guidetti, A. Zenetos, A. Vella, J. O'Leary, S. Sokolow, K. Nickols, G. De Leo, and five anonymous reviewers for useful comments and suggestions, and D. Maldini for helping with data mining.

Author Contributions

Conceived and designed the experiments: FM BH. Analyzed the data: SW FM. Contributed reagents/materials/analysis tools: SW BH FM SC LN RL. Wrote the paper: FM BH SW FF SF RL LN AR.

References

1. Crowder LB, Norse EA (2008) Essential ecological insights for marine ecosystem-based management and marine spatial planning. Marine Policy 32(5): 772–778.
2. Ruckelshaus M, Klinger T, Knowlton N, DeMaster DP (2008) Marine Ecosystem-based Management in Practice: Scientific and Governance Challenges. BioScience 58: 53–63.
3. Rosenberg AA, Mooney-Seus M, Kiessling I, Mogensen CB, O'Boyle R, et al. (2009) Lessons from national level implementation in North America and beyond. In: McLeod KO, Leslie H, editors. Ecosystem-based management of the ocean. Washington DC: Island Press. 392p.
4. Lotze HK, Lenihan HS, Borque BJ, Bradbury RH, Cooke RG, et al. (2006) Depletion, Degradation, and Recovery Potential of Estuaries and Coastal Seas. Science 312: 1806–1809.
5. Lejeusne C, Chevaldonne' P, Pergent-Martini C, Boudouresque CF, Perez T (2010) Climate change effects on a miniature ocean: the highly diverse, highly impacted Mediterranean Sea. Trends in Ecology & Evolution 25(4): 250–60.
6. Coll M, Piroddi C, Albouy C, Ben Rais Lasram F, Cheung WWL, et al. (2011) The Mediterranean Sea under siege: spatial overlap between marine biodiversity, cumulative threats and marine reserves. Global Ecology and Biogeography 21(4): 465–480.
7. Airoldi L, Beck MW (2007) Loss, status and trends for coastal marine habitats of Europe. Oceanography and Marine Biology: An Annual Review 45: 345–405.
8. Abdulla A, Gomei M, Hyrenbach D, Notarbartolo di Sciara G, Agardy T (2008) Challenges facing a network of representative marine protected areas in the Mediterranean: prioritizing the protection of underrepresented habitats. ICES Journal of Marine Science 66(1): 22–28.
9. Ferretti F, Myers RA, Serena F, Lotze H (2008) Loss of large predatory sharks from the Mediterranean Sea. Conservation Biology 22: 952–964.
10. Claudet J, Fraschetti S (2010) Human-driven impacts on marine habitats: a regional meta-analysis in the Mediterranean Sea. Biological Conservation 143(9): 2195–2206.
11. Galil BS (2008) Alien species in the Mediterranean Sea - which, when, where, why? Hydrobiologia 597(1): 105–116.
12. UNEP (2012) Support to the Barcelona Convention for the implementation of the ecosystem approach, including the establishment of MPAs in open seas areas, including deep sea. Contribution Agreement N°21.0401/2008/519114/SUB/D2. FINAL REPORT – April 2012. 50pp.
13. Halpern BS, Walbridge S, Selkoe KA, Kappel CV, Micheli F, et al. (2008) A global map of human impact on marine ecosystems. Science 319: 948–952.
14. Halpern BS, Kappel CV, Selkoe KA, Micheli F, Ebert CM, et al. (2009) Mapping cumulative human impacts to California Current marine ecosystems. Conservation Letters 2(3): 138–148.
15. Selkoe KA, Halpern BS, Ebert CM, Franklin EC, Selig ER, et al. (2009) A map of human impacts to a "pristine" coral reef ecosystem, the Papahanaumokuakea Marine National Monument. Coral Reefs 28: 635–650.
16. Ban NC, Alidina HM, Ardron JA (2010) Cumulative impact mapping: Advance, relevance and limitations to marine management and conservation, using Canada's Pacific waters as a case study. Marine Policy 34: 876–886.
17. Korpinen S, Meski L, Andersen JH, Laamanen M (2011) Human pressures and their potential impact on the Baltic Sea ecosystem. Ecological Indicators 15: 105–114.
18. Halpern BS, Selkoe KA, Micheli F, Kappel CV (2007) Evaluating and ranking the vulnerability of global marine ecosystems to anthropogenic threats. Conservation Biology 21: 1301–1315.
19. Teck SJ, Halpern BS, Kappel CV, Micheli F, Selkoe KA, et al. (2010) Using expert judgment to estimate marine ecosystem vulnerability in the California Current. Ecological Applications 20(5): 1402–1416.
20. Kappel CV, Halpern BS, Selkoe KA, Cooke RM (2011) Eliciting expert knowledge of ecosystem vulnerability to human stressors to support comprehensive ocean management. In: Perera AH, Drew CA, Johnson CJ, editors. Expert knowledge and its application in landscape ecology. New York: Springer. 307pp.
21. Halpern HS, Fujita R (2013) Assumptions, challenges, and future directions in cumulative impact analysis. Ecosphere, in press.
22. Spalding MD, Fox HE, Allen GR, Davidson N, Ferdana ZA, et al. (2007) Marine ecoregions of the world: a bioregionalization of coastal and shelf areas. BioScience 57: 573–583.
23. Pandolfi JM, Bradbury RH, Sala E, Hughes TP, Bjorndal KA, et al. (2005) Global Trajectories of the Long-Term Decline of Coral Reef Ecosystems. Science 301: 955–958.
24. Brown CJ, Saunders MI, Possingham HP, Richardson AJ (2013) Managing for interactions between local and global stressors of ecosystems. PLOS ONE 8: e65765
25. Large SI, Fay G, Friedland KD, Link JS (2013) Defining trends and thresholds in responses of ecological indicators to fishing and environmental pressures. Ices Journal of Marine Science 70: 755–767
26. Plaganyi EE, Bell JD, Bustamante RH, Dambacher JM, Dennis DM, et al. (2011) Modelling climate-change effects on Australian and Pacific aquatic ecosystems: a review of analytical tools and management implications. Marine and Freshwater Research 62: 1132–1147
27. Trujillo P, Piroddi C, Jacquet J (2012) Fish Farms at Sea: The Ground Truth from Google Earth. PLoS ONE 7: 30546.
28. Ferretti F, Osio GC, Jenkins CJ, Rosenberg A, Lotze HK (2013) Long-term change in a meso-predator community in response to prolonged

and heterogeneous human impact. Sci Rep. 3: 1057. doi:10.1038/srep01057
29. Portman ME, Notarbartolo di Sciara G, Agardy T, Katsanevakis S, Possingham HP, et al. (2013) He who hesitates is lost: Why conservation in the Mediterranean Sea is necessary and possible now. Marine Policy 4: 270–279.
30. Commission of the European Communities (2008) Roadmap for maritime spatial planning: achieving common principles in the EU. COM 2008 791.
31. Barale V, Schaefer NJ (2011) Maritime spatial planning: opportunities & challenges in the framework of the EU integrated maritime policy. Journal of Coastal Conservation 15: 237–245.

20

Proteomic Analysis of Salt-Responsive Proteins in the Leaves of Mangrove *Kandelia candel* during Short-Term Stress

Lingxia Wang[1,2]⁹, Xiao Liu[2]⁹, Meng Liang[2], Fanglin Tan[3], Wenyu Liang[4], Yiyong Chen[2], Yongxiang Lin[2], Li Huang[3], Jianhong Xing[2], Wei Chen[1,2]*

1 Key Laboratory of Ministry of Education for Genetics, Breeding and Multiple Utilization of Crops, Fujian Agriculture and Forestry University, Fuzhou, Fujian, China, **2** School of Life Sciences, Fujian Agriculture and Forestry University, Fuzhou, Fujian, China, **3** Fujian Academy of Forestry, Fuzhou, Fujian, China, **4** School of Life Sciences, Ningxia University, Yinchuan, Ningxia, China

Abstract

Salt stress is a major abiotic stress that limits crop productivity in many regions of the world. A comparative proteomic approach to identify salt stress-responsive proteins and to understand the molecular mechanisms was carried out in the woody halophyte *Kandelia candel*. Four-leaf-old *K. candel* seedlings were exposed to 150 (control), 300, 450, and 600 mM NaCl for 3 days. Proteins extracted from the leaves of *K. candel* seedlings were separated by two-dimensional gel electrophoresis (2-DE). More than 900 protein spots were detected on each gel, and 53 differentially expressed protein spots were located with at least two-fold differences in abundance on 2-DE maps, of which 48 were identified by matrix-assisted laser desorption ionization time-of-flight/time-of-flight mass spectrometry (MALDI-TOF-TOF/MS). The results showed that *K. candel* could withstand up to 450 mM NaCl stress by up-regulating proteins that are mainly involved in photosynthesis, respiration and energy metabolism, Na^+ compartmentalization, protein folding and assembly, and signal transduction. Physiological data, including superoxide dismutase (SOD) and dehydroascorbate reductase (DHAR) activities, hydrogen peroxide (H_2O_2) and superoxide anion radicals (O_2^-) contents, as well as Na^+ content and K^+/Na^+ ratios all correlated well with our proteomic results. This study provides new global insights into woody halophyte salt stress responses. Identification of differentially expressed proteins promotes better understanding of the molecular basis for salt stress reduction in *K. candel*.

Editor: Keith R. Davis, University of Louisville, United States of America

Funding: This work was supported by the National Natural Science Grant of China (Award no. 31070542) and the Provincial Natural Science Grant of Fujian, China (Award no. 2010J01067). The funders had no role in study design, data collection and analysis, decision to publish, or preparation of the manuscript.

Competing Interests: The authors have declared that no competing interests exist.

* E-mail: weichen909@163.com

⁹ These authors contributed equally to this work.

Introduction

Salinity is one of the major abiotic stresses that pose a severe threat to agricultural productivity in many global regions [1]. Salinity can cause a variety of changes in the metabolism of higher plants, such as photosynthetic and respiratory repression, osmotic stress, ion toxicity, oxidative stress, and nutrient deficiencies, leading to a decrease in the growth and productivity of plants [2–9].

Mangroves, which are intertidal plants that form unique communities along tropical and subtropical coastal areas, constitute a model for salt-tolerant xylophytes [10,11]. Several mangrove tree species display optimal growth in the presence of moderate salt stress conditions, whereas severe salinity restricts their growth [12]. The most striking feature of mangroves is their ability to tolerate salt up to seawater concentrations (~500 mM NaCl); this has attracted the attention of many scientists interested in elucidating this phenomenon [13,14]. Previous studies indicated that mangroves have evolved diverse strategies to respond to high salinity in their environment. Some species actively excrete excess salt by means of specialized salt glands in their leaves, while other species excrete salt by ultrafiltration at the root cell membranes of cortical cells [11]. Hotta et al. [15] reported that the response mechanism of mangroves to salt may be linked to changes in vacuolar size. Moreover, accumulation of compatible solutes and induction of antioxidative enzymes constitute other biochemical mechanisms of mangroves against high salinity [13]. Some reports studied the effect on photosynthetic parameters such as the rate of photosynthesis, stomatal conductance [12,16], and chloroplast structure and function [17]. In addition, some studies investigated the changes in gene expressions under salt stress [18–21]. Tanaka et al. [22] reported that the gene of vacuolar Na^+/H^+ antiporter from *Bruguiera sexangula* played an important role in cellular salinity adjustments. Miyama et al. [23] identified 14,842 expressed sequence tags from leaves and roots of *B. gymnorrhiza* under high salinity or hormone treatments. The mRNA expressions of Cu–Zn SOD, catalase and ferritin in response to salt stress in *A. marina* were studied and their role in oxidative stress response was confirmed [24]. The transcriptional response to high salinity and hyperosmotic stress in *B. gymnorrhiza* was investigated using the microarray approach. Eight hundred sixty-five genes showed significant differential expression under salt and osmotic stress [25]. Proteomic responses to salinity have been studied in various herbaceous plants including rice [1,26–28], canola [4], soybean [29], wheat [30], barley [31], the model plant *Arabidopsis thaliana* [32,33], cucumber [34], and the facultative model halophyte *Salicornia europaea* [35]. Understanding how different levels of protein abundance form the basis of salt tolerance mechanisms in mangroves will shed new light and give a new dimension to salt stress research. To date, few studies have been dedicated to observing the effects of salt stress on salt-tolerant xylophytes. However, two studies reported effects of long term salt stress (12 d and 45 d, respectively) on the abundance of proteins in mangrove, *B. gymnorrhiza* (L.) Lam [36,37]. Tada and Kashimura [36] reported that fructose-1,6-bisphosphate (FBP) aldolase and osmotin were found related to salt tolerance in *B. gymnorrhiza* roots. Zhu et al. [37] reported that photosynthesis-related proteins and antioxidant enzymes in the leaves were up-regulated by 200 mM NaCl and down-regulated by 500 mM NaCl suggesting that the salt-responsive mechanisms in *B. gymnorrhiza* at these NaCl concentrations were different.

K. candel is one of the most important species of mangrove around the coasts of South and Southeast Asia from western India to Borneo. It has developed various ways to adapt and respond to changes in concentrations of salinity in the environment. However, to our knowledge, no information is available describing how *K. candel* responds to short-term (3 days) exposure to NaCl at the level of protein expression. In the present study, proteome analysis was used to identify the mechanisms responsible for the response of *K. candel* to salt stress. The objective of the current study is to characterize specific metabolic and regulatory mechanisms of salt tolerance in *K. candel*. These proteomic data contribute significantly to our understanding of protein alteration in this salt-tolerant xylophyta. The results give insights into the molecular mechanism in responses to salinity stress in *K. candel* seedlings.

Results

Changes in SOD and DHAR activities and H_2O_2 and O_2^- contents

Induction of antioxidant enzymes is an important protective mechanism for minimizing oxidative damage from salt stress. Analysis of SOD indicated significant changes in activity upon salt treatment. The SOD activity peaked in response to 450 mM NaCl (356% of the control) (Figure 1A), whereas the DHAR activity peaked at 300 mM (155% of the control) (Figure 1B). Furthermore, at 600 mM NaCl, the SOD levels were still significantly higher than the control (150 mM) value, whereas, with DHAR (Figure 1B), the 600 mM value is not significantly different from the 150 mM activity. High salinity concentration leads to the production of large amounts of H_2O_2 and O_2^-. An increase in the content of H_2O_2 was detected with increasing salinity levels (Figure 1C). H_2O_2 content reached up to 198% at 600 mM NaCl when compared with the control (Figure 1C). However, O_2^- contents obviously increased only by 25% under 600 mM salt treatment, but no significantly different between 450 mM and 600 mM salt-treated seedling leaves (Figure 1D).

Na^+ and K^+ ion content analysis

Na^+ concentrations in *K. candel* leaves were affected by salinity. Compared with the control plants, the Na^+ concentration in leaves increased slightly after salt treatment for 3 d (Figure 2A). Salinity did not significantly affect the K^+ content in *K. candel* leaves under salt treatments in this study (Figure 2A). Additionally, K^+/Na^+ ratios only decreased slightly with increasing salt content (Figure 2B).

Proteomic and hierarchical clustering analysis

Proteins extracted from *K. candel* leaves following NaCl treatment were separated by 2-DE. More than 900 protein spots were detected in each gel by ImageMaster software (Figure 3). Fifty-three differentially expressed protein spots showed two-fold increases in abundance. MS analysis of the protein spots resulted in the identification of 48 proteins. The identified proteins are listed in Table 1 and Figure 4. In this study, all identified proteins were categorized into eight functional groups using the MapMan ontology as shown in Figure 5.

A total of 48 salt-responsive proteins were submitted to a hierarchical clustering to group proteins showing similar expression patterns under salt treatment. Two main clusters were formed (Figure 6): one cluster included 13 proteins whose abundance decreased under 600 mM NaCl treatment compared with the non-saline; the other cluster contained 35 up-regulated proteins. Examples for proteins that were decreased under severe salinity conditions are a chlorophyll a-b binding protein (CAB) 2 (spot no. 7), chlorophyll A/B binding protein (CAB) (spot no. 37), oxygen evolving enhancer protein 2 (OEE2) (spot no. 43), whose abundance increased under 300–450 mM NaCl treatment but decreased under 600 mM NaCl treatment, respectively. These changes in protein abundance may imply that photosynthesis is down-regulated under high salt conditions. From the increased cluster tree, most of the proteins were up-regulated with the increase of salinity. These proteins are mainly involved in carbohydrate metabolism and energy, protein metabolism, and signal transduction. For instance, the proteins involved in carbohydrate metabolism and energy (spot nos. 4, 6, 17, 18, 19, 20, 22, 25, 27, 34 and 39) were increased significantly by salt adaptation treatment compared with control. In addition, this cluster also comprised other salt inducible proteins, such as photosynthesis related proteins (spot nos. 13, 14, 16, 31 and 35) and a detoxifying and antioxidant related enzyme (spot no. 45).

Immunoblot analysis for DHAR and HSC70

In the current study, the accuracy of 2-DE analysis was further validated by immunoblot analysis. Proteins of *K. candel* leaves were separated by one-dimensional SDS-PAGE, and immunoblot analysis was performed for DHAR and heat shock cognate protein 70 (HSC70) (Figure 7). In agreement with the changes in protein abundance observed by 2-DE, DHAR showed an increased amount in response to 300 and 450 mM NaCl treatment (Figure 7A). HSC70 immunoblot analysis revealed an increase in the amount of cross-reacting polypeptide bands in response to high salinity (Figure 7B).

Discussion

Although there have been two previous reports about proteins related to salt tolerance in mangrove (*B. gymnorrhiza*), only 3 and 10 proteins were identified, respectively. Tada and Kashimura [36] identified FBP-aldolase, osmotin, and a novel protein in the root. Zhu et al. [37] reported that salt tolerance to 200 mM NaCl was due to increased expression of RuBisCO activase, glutathione transferase, and heat shock proteins (HSPs), which improved the salt tolerance of *B. gymnorrhiza*. A reduction in the abundance of the above proteins was observed after treatment with 500 mM NaCl. Our study has revealed that additional important proteins are related to salt tolerance in the mangrove plant *K. candel*. Forty-

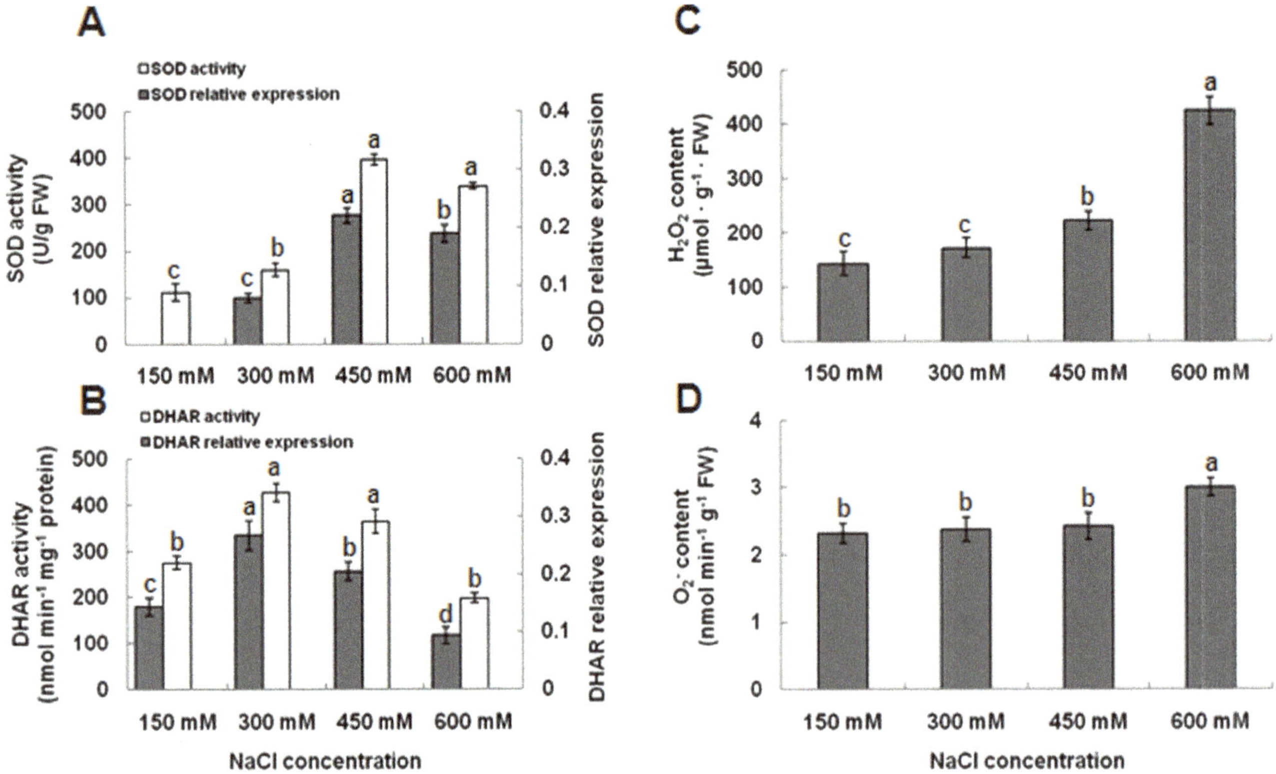

Figure 1. Effect of salinity on the activites of SOD and DHAR and contents of H_2O_2 and O_2^- in *K. candel* leaves. Values (means ± SD) were determined from three independent experiments (n = 3) after plants had been treated with 150, 300, 450, and 600 mM NaCl for 3 d.

eight differentially expressed proteins were identified that are known to be involved in several cellular processes such as photosynthesis, respiratory and energy metabolism, detoxification and antioxidation, as well as signal transduction.

Proteins involved in photosynthesis

Photosynthesis is sensitive to be affected by salinity among the primary metabolism in high plant [38]. Proteomic analysis revealed that a several proteins associated with photosynthesis was differentially expressed upon salt stress (Figure 8). As shown in Table 1, 11 enzymes involved in photosynthesis were positively identified. Of these, 3 enzymes (spot nos. 7, 37, 43) have important roles for light-dependent reactions; the other 8 enzymes (spot nos. 4, 5, 6, 14, 16, 13, 31, 35) take part in the Calvin cycle. All 11 proteins were up-regulated under 300 mM and 450 mM NaCl stress. For light-dependent reaction related proteins, the expression level of OEE2 (spot no. 43), CAB 2, chloroplastic (spot no. 7) and CAB (spot no. 37) were slightly up-regulated by exposure to 300 mM and 450 mM NaCl but decreased dramatically in response to severe stress (600 mM NaCl). Previous studies reported that OEE, consisting of OEE1, OEE2, and OEE3, play an important role in the light-induced oxidation of water in

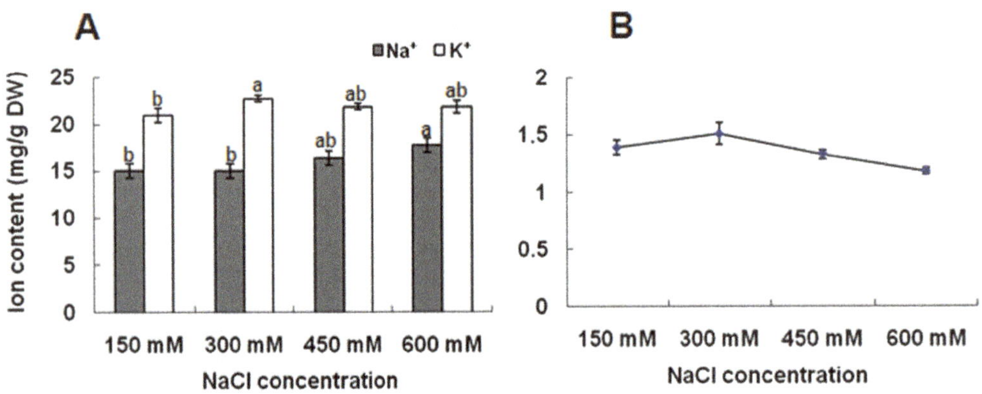

Figure 2. Effect of salinity on ionic contents in *K. candel* leaves. (A) Na^+ and K^+ concentration in leaves; (B) K^+/Na^+ ratio in leaves. Values (means ± SD) were determined from five independent experiments (n = 5) after plants had been treated with 150, 300, 450, and 600 mM NaCl for 3 d.

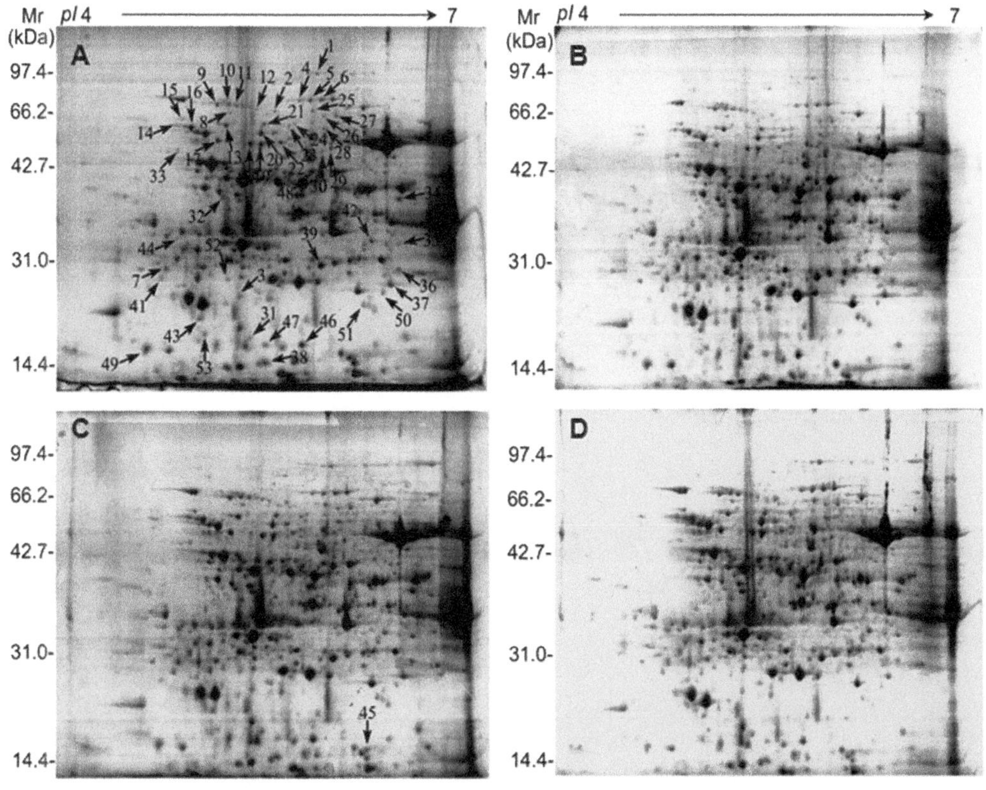

Figure 3. Representative 2-DE gel images of *K. candel* leaves treated with NaCl. An equal amount (1.5 mg) of total proteins was loaded on each gel strip (pH 4–7). After IEF, 12.5% SDS-PAGE gels were used for second dimension separation. Protein spots were visualized using CBB staining. The 2-DE maps of proteins from leaves of the control samples (150 mM NaCl) and the NaCl-treated samples (300, 450, and 600 mM) are presented. Proteins differentially regulated in response to salinity are numbered in pairs of control and NaCl-treated maps. Arrows indicate the 48 spots that showed significant changes in the control and NaCl-treated samples. Identification of these protein spots by MALDI-TOF-TOF-MS/MS is marked with arrows in (A) and given in Table 1.

photosystem II (PSII) of plants [39]. However, these subunits in PSII complex can be easily dissociated under salt stress [39]. Up-regulation of OEE2 has been reported in *Bruguiera gymnorrhiza* in response to salt stress, which suggests that it might be needed to repair the injury of the PSII complex and to maintain the oxygen evolution reaction [36]. CAB 2 (spot no. 7), a component of the light-harvesting complex of photosystem I (PSI), facilitates light absorption and transfer of the excitation energy to reaction centers for the reduction of $NADP^+$ to NADPH [40]. As to Calvin cycle related proteins, RuBisCO large subunit (spot nos. 31 and 35), RuBisCO large subunit-binding protein subunit beta (spot no. 13) and RuBisCO subunit binding-protein alpha subunit (spot nos. 14 and 16) were all significantly increased in abundance under all salt treatments. RuBisCO subunit binding-protein alpha subunit is important in RuBisCO complex assembly [41]. RuBisCO large subunit-binding protein subunit beta (ruba) (spot no. 13) is related to RuBisCO activation [42]. The chloroplast-localized protein transketolase (TK) (TargetP [43]; spot nos. 4, 5, and 6), which located in chloroplast by subcellular prediction tool, which is involved in the regeneration phase of the Calvin cycle, was also up-regulated under salt stress. The above results suggested that *K. candel* could withstand up to 450 mM NaCl stress possibly through the up-regulation proteins related to the light-dependent reactions, which in turn provides adequate amounts of energy equivalents necessary for the Calvin cycle and other processes important during salt stress. In addition, the up-regulation of the Calvin cycle produces more photosynthetic products, such as sucrose and starch, to improve plant salt tolerance [44].

Proteins related to carbohydrate and energy metabolism

Large amounts of energy are needed for the growth and development of *K. candel* under salt stress. This energy is mainly produced through carbohydrate metabolism, such as glycolysis (EMP) and tricarboxylic acid cycle (TCA) [27,45,46]. Fifteen differential expressed proteins in *K. candel* leaves under salt stress catalyze several steps in these pathways (Figure 8). 2,3-bisphosphoglycerate-independent phosphoglycerate mutase (PGAM) (spot no. 27) and fructokinase-1 (FRK) (spot no. 32) are enzymes of EMP. PGAM was up-regulated up to 600 mM NaCl stress while FRK reached maximum abundance under 450 mM salt concentration. FRK catalyzes the transfer of a phosphate group from ATP to fructose in glycolysis and is the most important gateway in the control of sugar influx into EMP [47]. Furthermore, three protein spots representing aconitate hydratase (AH) (spot no. 1), succinate dehydrogenase (SD) (spot no. 25), and malate dehydrogenase (MD) (spot no. 34) were found to be up-regulated. The first two enzymes are key enzymes in the TCA cycle. Thus, the increase of FRK under NaCl along with the other enzymes of EMP-TCA in this study would contribute to glucose breakdown for energy generation to cope with salt stress. In addition, mitochondrial F1–ATPase beta subunit (spot nos. 18 and 19), ATP synthase CF1 alpha subunit (spot no. 20), and F1 ATPase (spot no. 22), involving in ATP synthesis [42], were up-

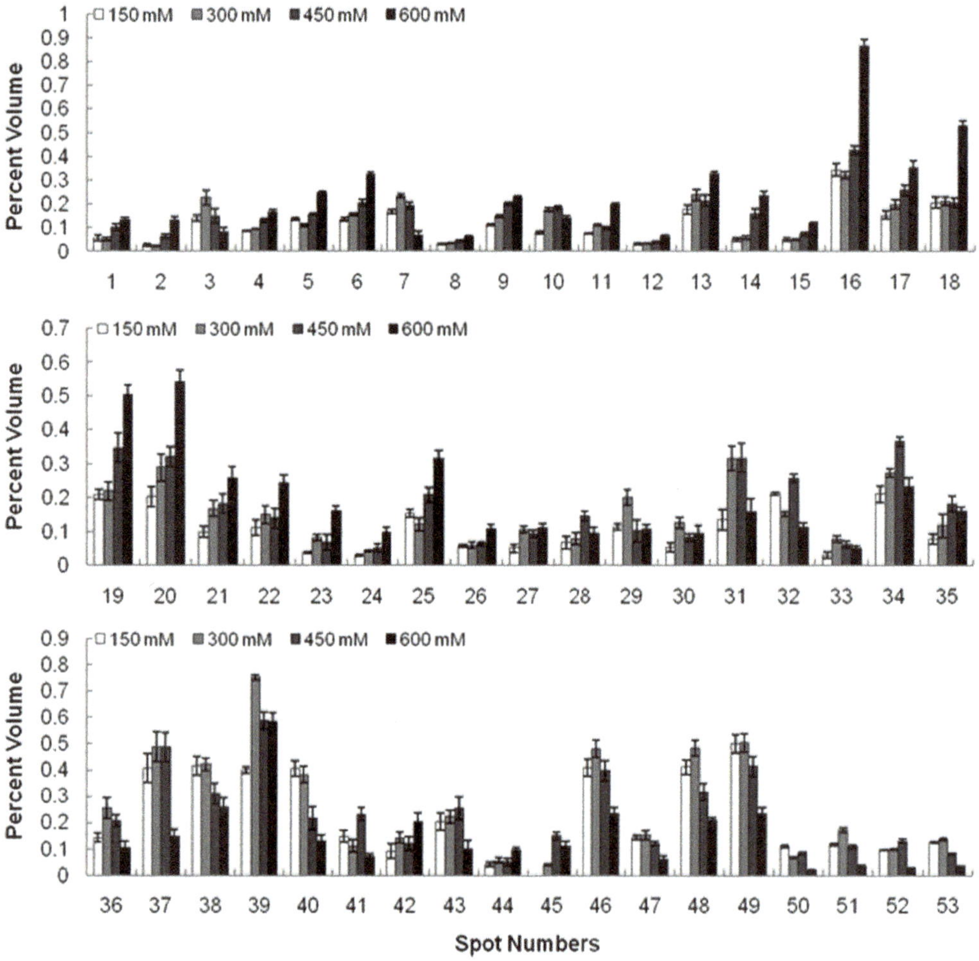

Figure 4. Relative expression levels of 53 differentially expressed proteins of *K. candel* leaves under salt stress. Changes in protein expression under salt stress were calculated by Image Master software 5.0. Mean of relative protein abundance and standard error. Three treatments including control (150, 300, 450, 600 mM NaCl for 72 h) were performed.

regulated under salt stress in this study. ATP, the main source of energy, is indispensable for many metabolism pathways in higher plants. The energy requirements in response to external stress may considerably increase [48]. The up-regulated ATP synthesis

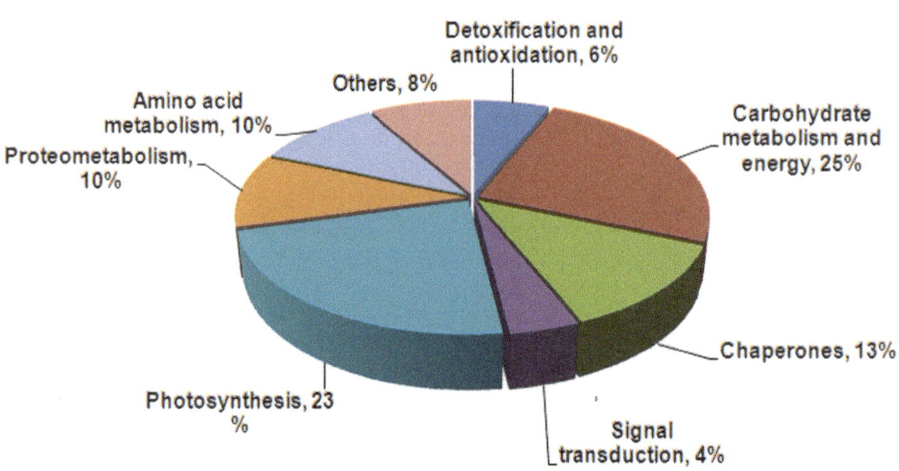

Figure 5. Functional classification of differentially expressed proteins identified in the seedling leaves of *K. candel* under salt stress. The classification is based on KEGG (http://www.kegg.jp/kegg/pathway.html) and previous literature.

Table 1. Protein identities and their relative changes under salt stress in *Kandelia candel* leaves from 2D-Gel analysis.

Spot No[a]	Protein Name	Species	Accession No[b]	Matched peptide sequences (m/z)[c]	Theor.[d] MW(kDa)/p*I*	Exp.[d] MW(kDa)/p*I*	PS/IS[e]	MP/TP[f]	Cov%[g]
Photosynthesis									
13	RuBisCO large subunit-binding protein subunit beta, chloroplast precursor (60 kDa chaperonin subunit beta) (CPN-60 beta)	*Pisum sativum*	P08927	DLINILEDAIR(1284.7329) SQYLDDIAILTGGTVIR(1835.0303)	62.95/5.85	59.15/5.34	214/135	2/20	34%
7	Chlorophyll a–b binding protein 2,chloroplastic	*Populus euphratica*	P84990	NRELEVIHSR(1252.7225) SAPQSIWYGPDRPK(1601.8837)	37.99/8.2	27.02/4.89	158/116	2/4	100%
14	rubisco subunit binding-protein alpha subunit, ruba, putative	*Ricinus communis*	XP_002534347	LADAVGLTLGPR(1182.682)	53.17/5.25	58.4/4.95	99/79	1/8	13%
16	Rubisco subunit binding-protein alpha subunit, ruba, putative	*Ricinus communis*	XP_002534347	LADAVGLTLGPR(1182.6986)	53.17/5.25	57.58/5.08	139/95	1/11	23%
31	ribulose-1,5-bisphosphate carboxylase/oxygenase large subunit	*Kandelia candel*	AAB66915	ALRLEDLR(985.5867) DTDILSAFR(1037.5345) LEDLRIPPAYSK(1401.792) TFQGPPHGIQVER (1465.7908) LTYYTPDYETKDTDILSAFR(2412.2302)	48.07/6.34	18.38/5.51	419/327	5/8	20%
35	ribulose-1,5-bisphosphate carboxylase/oxygenase large subunit	*Kandelia candel*	AAB66915	DNGLLLHIHR(1187.6913) EITAGFVDLLR(1233.7106) TFQGPPHGIQVER(1465.7932)	48.07/6.34	31.00/6.66	212/184	3/12	13%
37	chlorophyll A/B binding protein, putative	*Ricinus communis*	XP_002524566	QYFLGLEK(997.5484) WLAYGEIINGR(1291.7159)	29.36/6.85	25.69/6.61	149/130	2/5	18%
43	Oxygen evolving enhancer protein 2	*Bruguiera Gymnorhiza*	BAA96364	EVEFPGQVLR(1173.6053) QYYFLSVLTR(1289.6636)	17.53/4.91	20.82/5.14	107/94	2/3	18%
4	plastid transketolase	*Nicotiana tabacum*	ACF60500	TPSILALSR(957.5609) NPYWFNR(996.4607)	80.05/6.16	74.56/5.90	120/84	2/12	17%
5	plastid transketolase	*Nicotiana tabacum*	ACF60500	TPSILALSR(957.5534) NPYWFNR(996.4529) ALPTYTPESPADATR(1589.7661)	80.05/6.16	74.56/5.98	152/115	3/12	15%
6	transketolase 1	*Capsicum annuum*	CAA75777	NPYWFNR(996.4579) ALPTYTPESPADATR(1589.7644)	80.05/6.16	74.56/6.06	154/128	2/12	13%
Carbohydrate metabolism and energy									
1	aconitate hydratase 3	*Citrus clementina*	CBE71057	QVEIPFKPAR(1184.6525) INPLVPVDLVVDHSVQVDVAR (2284.2341)	98.04/5.89	98.1/6.04	181/165	2/11	10%
17	enolase	*Zea mays*	ACG31732	AAVPSGASTGVYEALELRDGGSDYLGK (2683.3953)	47.9/5.67	52.88/5.28	111/85	1/6	17%
18	mitochondrial F1-ATPase beta subunit	*Dimocarpus longan*	ACJ06633	VVDLLAPYQR(1173.6497) AHGGFSVFAGVGER(1390.6788) VGLTGLTVAEHFR(1399.7584) FTQANSEVSALLGR(1492.7637) IPSAVGYQPTLSTDLGGLQER (2202.1514)	59.81/6.18	52.5/5.51	571/520	5/12	27%
19	mitochondrial F1-ATPase beta subunit	*Dimocarpus longan*	ACJ06633	VVDLLAPYQR(1173.6666) AHGGFSVFAGVGER(1390.7012) VGLTGLTVAEHFR(1399.7795) FTQANSEVSALLGR(1492.7859) IPSAVGYQPTLSTDLGGLQER (2202.1875)	59.81/6.18	51.31/5.58	554/503	5/12	24%
20	ATP synthase CF1 alpha subunit	*Citrus sinensis*	YP_740460	ERIEQYNR(1107.5598) LIESPAPGIISR(1252.7411) IAQIPVSEAYLGR(1416.808) EAYPGDVFYLHSR(1553.7662) IVNIGTVLQVGDGIAR(1624.9561) VINALAKPIDGRGEISASESR (2183.219)	55.45/5.09	56.8/5.58	606/429	6/25	35%

Table 1. Cont.

Spot No[a]	Protein Name	Species	Accession No[b]	Matched peptide sequences (m/z)[c]	Theor.[d] MW(kDa)/p/	Exp.[d] MW(kDa)/p/	PS/IS[e]	MP/TP[f]	Cov%[g]
22	F1 ATPase	*Pisum sativum*	BAA20135	VGLTGLTVAEHFR(1399.7831)	60.11/6.63	52.5/5.75	106/54	1/13	23%
25	succinate dehydrogenase, putative	*Ricinus communis*	XP_002530482	AASTILATGGYGR(1237.6605) LGANSLLDIVVFGR(1473.856) AVIELENYGLPFSR(1607.8613)	68.46/6.18	66.15/6.00	262/241	3/8	15%
26	Os01g0817700 (Putative phosphoglycerate mutase)	*Oryza sativa* Japonica Group	NP_001044625	VHILTDGR(910.5146) FGHVTFFWNGNR(1481.7295) LPSHYLVSPPEIER(1636.8873)	60.75/5.42	63.07/6.11	237/219	3/9	15%
27	2,3-bisphosphoglycerate independent phosphoglycerate mutase	*Ricinus communis*	XP_002519975	FGHVTFFWNGNR(1481.7295) LPSHYLVSPPEIER(1636.8873)	61.15/5.52	59.96/5.81	217/163	2/15	18%
32	Fructokinase-1	*Oryza sativa Indica Group*	A2WXV8	TALAFVTLR(991.6175) APGGAPANVAIAVAR (1334.7853)	34.71/5.14	36.98/5.31	186/142	2/9	32%
34	Malate dehydrogenase, cytoplasmic	*Mesembryantheum crystallinum*	O24047	NVIIWGNHSSTQYPDVNHATVK (2480.1733)	35.48/6.00	37.02/6.41	77/38	1/3	15%
39	triosephosphate isomerase	*Gossypium hirsutum*	ACJ11723	FFVGGNWK(954.5039)	27.19/6.00	25.33/6.04	85/48	1/7	35%
45	chloroplast superoxide dismutase	*Bruguiera gymnorhiza*	CAM98444	AFVVHELEDDLGKGGHELSLTTGNAGGR (2879.5146)	23.21/6.29	15.51/6.29	183/163	1/3	22%
Detoxification and antioxidation									
36	Dehydroascorbate reductase 1	*Actinidia deliciosa*	ADB85570	EYVIAGWEPK(1191.5714)	23.80/5.17	27.46/6.65	76/61	1/4	18%
2	vacuolar H^+-ATPase	*Malus×domestica*	ABO33173	LHEDLTSGFR(1174.604) LAADTPLLTGQR(1255.723) VGHDNLIGEIIR(1335.7715) EASIYTGITIAEYFR(1733.9414) LTTFEDSEKESEYGYVR(2053.0273)	68.88/5.37	66.2/5.70	586/407	5/28	47%
Chaperones									
9	non-cell-autonomous heat shock cognate protein 70	*Cucurbita maxima*	AAN86274	NALENYAYNMR(1358.6045) STVHDVVLVGGSTR(1426.7528) TTPSYVAFTDSER(1473.6743) ARFEELNMDLFR(1540.7422) NAVVTVPAYFNDSQR(1680.8284) EQVFSTYSDNQPGVLIQVYEGER (2658.294)	70.77/5.13	70.88/5.28	672/545	6/23	42%
10	High molecular weight heat shock protein	*Malus×domestica*	AAF34134	NALENYAYNMR(1358.632) TTPSYVAFTDTER (1487.7235) NAVVTVPAYFNDSQR(1680.8619) EQVFSTYSDNQPGVLIQVYEGER (2658.268)	71.17/5.17	72.44/5.35	440/332	4/21	40%
11	Hsc70	*Solanum lycopersicum*	AAB42159	NALENYAYNMR(1358.632) TTPSYVAFTDTER (1487.7235)	71.47/5.18	70.88/5.42	429/332	4/19	29%
				NAVVTVPAYFNDSQR(1680.8619) EQVFSTYSDNQPGVLIQV YEGER(2658.3374)					
12	heat shock protein, putative	*Ricinus communis*	XP_002518324	FESLVNHLIER(1356.7566) AVITVPAYFNDAQR(1564.8694) GVNPDEAVAMGAAIQGGILR(1939.085) GVNPDEAVAMGAAIQGGILR(1955.0647)	71.08/6.10	66.2/5.56	276/230	4/14	23%
21	heat shock protein 60	*Prunus dulcis*	AAN63805	APGFGENRK(975.5055) MISTSEEIAQVGTISANGER(2093.0415)	57.73/5.26	58.4/5.61	214/123	2/19	37%

Table 1. Cont.

Spot No[a]	Protein Name	Species	Accession No[b]	Matched peptide sequences (m/z)[c]	Theor.[d] MW(kDa)/pI	Exp.[d] MW(kDa)/pI	PS/IS[e]	MP/TP[f]	Cov%[g]
46	hypothetical protein OsI_01440 (Peptidyl-prolyl cis-trans isomerase)	*Oryza sativa* Indica Group	EEC70431	TPWLDNR(901.4586)	25.36/8.05	18.16/5.86	86/48	1/7	32%
Signal transduction									
30	calreticulin, putative	*Ricinus communis*	XP_002514817	TLVLQYSIR(1092.6418) AREEEEAQR(1117.5186)	49.78/5.81	47.40/6.06	89/74	2/9	14%
44	14-3-3-like protein	*Zea mays*	NP_001105677	DSTLIMQLLR(1189.6669) DSTLIMQLLR(1205.6462) IVSSIEQKEESR(1404.7404) TVDVEELTVEER (1418.7095)	28.87/4.82	31.26/4.96	258/188	4/13	41%
Proteometabolism									
3	proteasome subunit beta type 6,9, putative	*Ricinus communis*	XP_002527995	TVIINSEGVTR(1188.6373) SGSAADSQIVSDYVR(1554.7103) YFLHQHTIQLGQPATVK(1981.0228)	24.85/5.17	23.03/5.45	365/316	3/8	50%
15	disulfide isomerase	*Gloeospermum blakeanum*	ACZ95473	SASGNLVQYDGDR(1381.61)	10.92/4.44	62.08/5.01	77/66	1/2	15%
8	Cell division protein ftsH, putative	*Ricinus communis*	XP_002520373	FLEYLDKDR(1198.5884) TPGFSGADLANLLN EAAILAGR (2171.1414) KVDLFENGTIAIVEA VSPELGNR (2471.3274)	75.33/6.43	66.2/5.38	568/446	3/23	34%
38	eukaryotic translation initiation factor 5A	*Gossypium hirsutum*	ADG27839	TYPQQAGTIR(1134.6205) TYPQQAGTIRK(1262.715)	17.40/5.61	14.84/5.63	116/91	2/5	24%
42	Protein kinase domain containing protein, expressed	*Oryza sativa* (japonica cultivar-group)	ABA94760	AAMQQRGR(917.4733) VASNCLATMDNKR (1422.6827) NVIGPQSRTKPAPHLQFR(2046.1355)	80.25/9.98	31.00/6.44	99/0	3/21	28%
Amino acid metabolism									
24	phosphoglycerate dehydrogenase	*Gossypium hirsutum*	ACJ11736	EVFESSGGR(967.4564) GGVIDEEALVR (1157.6221) GLGMHVIAHDPYAPADR (1835.8821)	63.78/7.14	60.74/5.85	197/171	3/10	16%
28	Ketol-acid reductoisomerase, chloroplastic	*Pisum sativum*	O82043	DLFHLLPDAFK(1315.6791) GILLGAVHGIVESLFR(1680.9719) GILLGAVHGIVESLFRR(1837.0664) QIGVIGWGSQGPAQAQNLR(1980.0358) EINGAGINSSFGVHQDVDGR(2071.9685)	62.81/6.62	58.4/6.09	366/306	5/14	23%
40	3-isopropylmalate dehydratase, putative	*Ricinus communis*	XP_002530662	NSVATGEIYPLETEVR(1777.9103)	26.91/6.43	25.91/5.14	118/111	1/2	13%
29	AlaT1	*Vitis vinifera*	AAZ43369	EVAEFIGRR(1076.583) GVMQILNTIIR (1257.7366) GVMQILNTIIR(1273.717)	53.66/6.54	47.40/6.14	467/376	5/17	31%
				HYLSLTSGGLGAYSDSR(1783.8708) IIFTNVGNPHALGQKPLTFPR(2320.3147)					
48	glutamine synthetase	*Cucumis melo*	AAX35343	HKDHISAYGEGNER(1612.736)	47.63/8.06	40.49/5.97	133/101	1/9	22%
Others									
23	predicted protein	*Populus trichocarpa*	XP_002322147	QLFIDGEWR(1163.5789) QLFIDGEWREPVLK (1729.9249)	54.88/5.25	58.37/5.56	114/108	2/3	5%
33	unknown	*Glycine max*	ACU22776	AAYNNPER(934.4539) NSQQFQALR(1091.5814)	41.17/4.8	48.18/5.00	121/99	2/6	21%

Table 1. Cont.

Spot No[a]	Protein Name	Species	Accession No[b]	Matched peptide sequences (m/z)[c]	Theor.[d] MW(kDa)/pI	Exp.[d] MW(kDa)/pI	PS/IS[e]	MP/TP[f]	Cov%[g]
41	Unknown	*Populus trichocarpa×Populus deltoides*	ABK96657	GNELWYGPDR(1206.589) VDFKEPVWFK (1294.7012) GNELWYGPDRVK(1433.7603)	24.36/4.77	26.13/4.85	134/106	3/6	17%
47	hypothetical protein	*Vitis vinifera*	XP_002271	VAGAAADILGAASHYGK (1571.8383)	14.97/6.69	18.16/5.61	129/119	1/2	24%

[a]Numbering corresponds to the 2-DE gel in Figure 3.
[b]The number of the predicted protein in NCBInr.
[c]The matched peptide sequences.
[d]Molecular mass and *pI* theoretical and experimental.
[e]Protein score and ion score.
[f]Number of matched peptides and total searched peptides.
[g]Percentage of predicated protein sequence with matched sequence.

indicates that ATP formation is one of the strategies of plants to cope with salt stress. The results above indicated that EMP-TCA activity coupling with ATPase in *K. candel* leaves was enhanced suggesting that the respiratory metabolism in *K. candel* was increased under salt stress. Increased EMP-TCA activity and ATP synthesis imply that salt stress forces the plant to remobilize energy to cope with salt stress [49]. Together with the amplification of the light reactions of photosynthesis, this may insure the continuous generation of ATP and NAD(P)H necessary to mediate the enhanced salt resistance in *K. candel* [35,50]. Thus, the species seem to derive its ability to improve stress tolerance through the adjustment of its-energy metabolism.

Proteins associated with detoxification and antioxidation

Vacuolar H^+-ATPase transports protons across the tonoplast, resulting in the formation of a proton gradient. The resulting gradient provides the driving force for active Na^+ transport into the vacuole [51], and prevents the cytoplasm from reaching toxic Na^+ levels. This sequestration of Na^+ into the vacuole is a crucial and effective strategy for reducing Na^+ concentration in cytoplasm and regulating cell osmosis in plant [34]. In the present study, both the Na^+ concentration (Figure 2A) and the expression of vacuolar H^+-ATPase (spot 2) were increased (Figure 4) and K^+/Na^+ ratios (Figure 2B) only decreased slightly with increasing salt content as the NaCl treatment progressed in leaves under salt stress. These results were consistent with the reports of a study by Wang et al. [35]. In the present study, *K. candel* is likely to have an ability to sequester Na^+ into vacuole and maintain K^+ homeostasis under NaCl stress. This active Na^+ efflux requires a H^+ gradient across the vacuolar membrane generated by stimulating protein expression of the vacuolar H^+-ATPase [52], suggesting that up regulation of vacuolar H^+-ATPase might play a vital role in salinity tolerance of *K. candel*.

Salt stress induces reactive oxygen species (ROS) accumulation which may lead to oxidative stress, and high NaCl concentration is harmful to plant cell components [6,53]. In plants, however, ROS may be considered as signaling molecules to increase antioxidant enzymes for adapting to high salt levels [4]. With further increases in salt levels, the detoxification roles of oxidant-tolerant enzymes may dominate ROS signaling effects, indicating that the plant can use detoxifying and antioxidant enzymes to respond to high salinity [35]. Spot no. 45 showed homology with chloroplast SOD. The function of SOD is to catalyze the conversion of O_2^- to H_2O_2 and O_2 during various stresses, which is deemed to be one of the first lines of defense against free radical damage in plant cells [54]. Accumulation of SOD in response to salt stress has been reported to play a protective role in canola [4] and *S. europaea* [35]. In this study, SOD levels increased in response to salinity, with a more pronounced increase at 450 mM NaCl and 600 mM NaCl (Figure 4). As a result, O_2^- content was not significantly increased at 300 and 450 mM NaCl, and only slightly increased under high salt stress (600 mM NaCl), suggesting that salt-tolerant *K. candel* an increased capacity for scavenging O_2^- (Figure 1D).

Furthermore, we detected an increase in the levels of DHAR1 (spot no. 36) in response to 300 and 450 mM NaCl in our proteomics (Figure 4). This increase was confirmed using Western blot analysis (Figure 7A) esearch with privious ions. The previous report indicated that this protein was up-regulated under salt stress [31]. Increases in DHAR abundance may be a result of H_2O_2 removal through the production of ascorbic acid during stress [55]. DHAR is frequently designated as an enzyme to protect against oxidative stress in plants [56]. Previous studies indicated that the antioxidative defense system as a whole was induced during salt stress for scavenging ROS [6,57]. Mandhania et al.

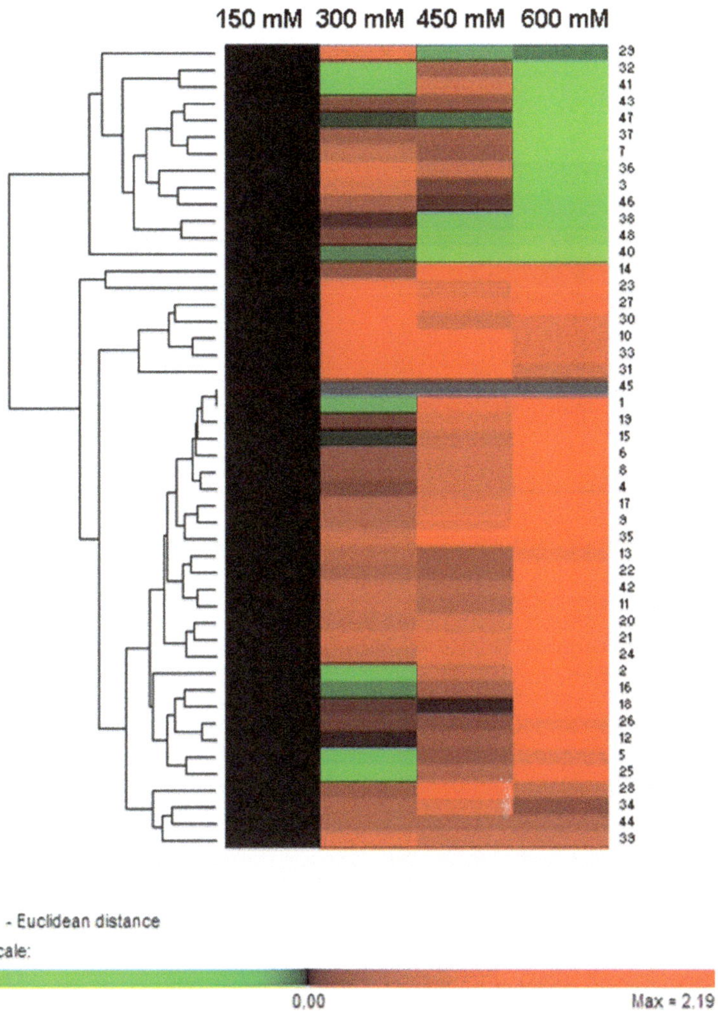

Figure 6. Hierarchical clustering analysis of the 48 differentially expressed proteins from *K. candel* leaves under different salinity levels. The rows represent individual proteins. The proteins that increased and decreased in abundance are indicated in red and green, respectively. Proteins not detected on control gels are indicated in gray. The intensity of the colors increases as the expression differences increase, as shown in the bar at the bottom.

[58] and Yan et al. [59] reported that antioxidant enzymes are up-regulated at the protein level by salt stress in both, wheat and rice. In the present study, H_2O_2 content remained low during treatments with up 300 mM to 450 mM NaCl, but increased obviously at 600 mM NaCl (Figure 1C). This is consistent with our findings that SOD and DHAR activities and protein levels increase during exposure of up to 450 mM NaCl (Figure 1A and B). These results support the idea that functionally active antioxidant enzymes, SOD and DHAR, could be used to enhance resistance up to 450 mM salt stress in *K. candel.*

Chaperones

Salt stress results in protein misfolding or unfolding, which injures plant cells. To avoid these, cells produce molecular chaperones, such as the members of HSPs, which assist protein folding or assembly and prevent irreversible protein aggregation by maintaining native conformations during salt stress [32,34,35,45]. Under adverse conditions, HSPs can protect plants against stress by refolding proteins to reestablish normal protein conformation and maintain cellular homeostasis [60]. Previous studies reported that HSC 70 was implicated in a variety of cellular processes, including the folding of nascent chain polypeptides or the import/translocation of mitochondrial or chloroplast precursor proteins [60,61]. Another report indicated that HSP 60, a mitochondrial chaperone, plays a vital role in the transport of proteins from the cytoplasm into the mitochondrial matrix and in the refolding of proteins, thus preventing protein aggregation when the mitochondria are subjected to stress [62]. In this study, Non–cell-autonomous heat shock cognate protein 70 (spot no. 9), high molecular weight HSP (spot no. 10), HSC 70 (spot no. 11), HSP, putative (spot no. 12), and HSP 60 (spot no. 21) were identified. These proteins were all up-regulated by exposure to high salinity, suggesting that the proteins play a crucial role in aiding the folding and assembly proteins under salt tolerance in *K. candel* seedlings (Figure 8). These results are similar with the report by Zhu et al. [37] in mangroves plant, *Bruguiera gymnorhiza*. This was further confirmed by an increase in protein abundance in response to salt-treatment as determined by Western blot (Figure 7B).

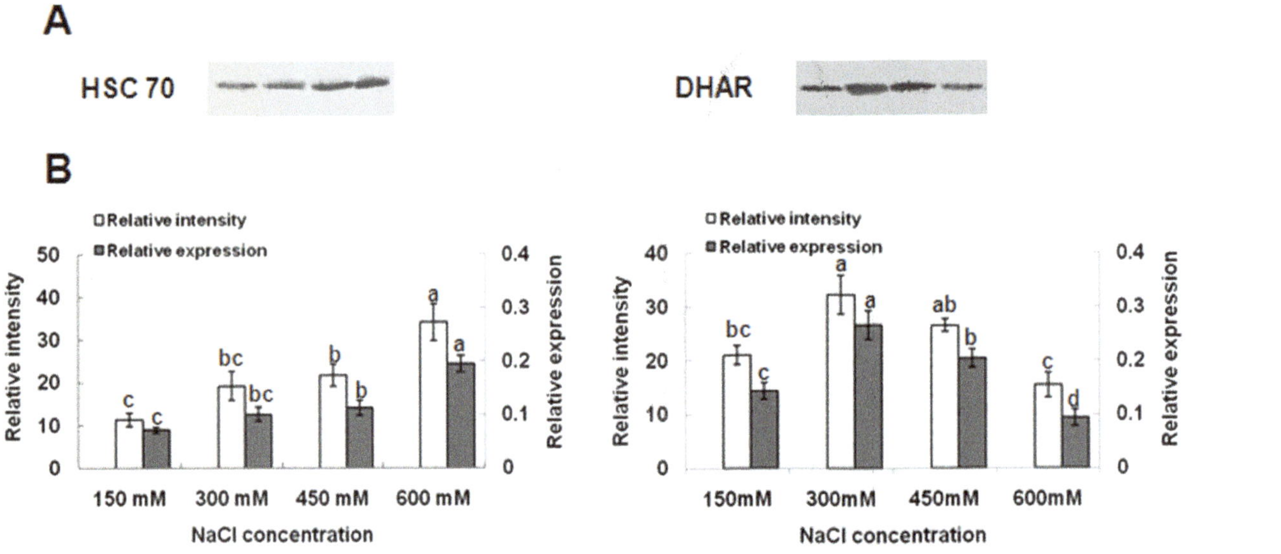

Figure 7. Western blot analysis of HSC 70 and DHAR expression patterns and the relative expressions on protein levels in *K. candel* leaves. (A) Antibodies against HSC 70 and DHAR were used to detect the change of protein levels in leaves in response to salt stress treatment of the plants. 50 µg protein samples were loaded in each lane, separated on 12% SDS-PAGE gel, followed by Western blotting and visualized with DAB as described in Materials and Methods section; (B) Gray analysis of the results by Quantity One software and the relative expression of HSC 70 and DHAR (spots 11 and 36) were shown. Bars represent the means of the relative intensity of the protein stains of three biological replicates of the control and salt-stressed treatments (150, 300, 450 and 600 mM NaCl).

Proteins involved in signal transduction

One of the proteins that increased abundantly during salt stress is a 14-3-3-like protein (spot no. 44) (Figure 4). In plants, the 14-3-3 proteins, a highly conserved family, are known to be involved in responses to diverse stresses including salinity [49,63]. The biological roles of 14-3-3 complexes are in the regulation of primary metabolism, signal transduction, and subcellular and defense reactions [64]. They are also recognized as positive regulators of plasma membrane H^+-ATPase in the regulation of ion transport and cytoplasmic pH [65,66]. Moreover, 14-3-3 proteins have been implicated in various signal transduction pathways through controlling the activities of kinases and phosphatases [67,68], which suggests that 14-3-3 proteins regulate multiple pathways involved in salt stress responses in higher plants [49]. As shown in Figure 4, calreticulin (CRT) (spot no. 30) was up-regulated in response to salt treatment. CRT, one of the most important calcium-binding proteins, is involved in calcium signaling in the endoplasmic reticulum during the stress response in plants [60,69–72]. Therefore, the results indicated that up-regulation of 14-3-3-like proteins and CRT might play roles in signal transduction in *K. candel* under salt stress.

Proteometabolism

Disulfide isomerase (DI) (spot no. 15) is an enzyme that participates in disulfide bonds formation and breakage between cysteine residues when proteins folding, which was up-regulated in this study (Figure 8) [73,74]. These reactions will lead to the rearrangement of disulfide bonds in a single protein that exist intra-molecularly. Proteasome, which involves in regulating the particular protein concentration, can degrade unneeded or damaged proteins in plant cells. In the present study, proteasome subunit beta type 6, 9, putative (spot no. 3) was up-regulated by 300 and 450 mM NaCl but down-regulated by 600 mM NaCl, suggesting that degradation of unneeded, damaged, and misfolded proteins by the proteasome pathway was active in plant resistance to salt toxicity under at least 450 mM concentration.

Conclusions

In the present study, we provided a comprehensive proteome dynamics of the leaves in the woody halophyte *K. candel* under salt stress. Forty eight differentially expressed proteins, showing more than a 2-fold change in abundance, were identified. As a result, we gained new information about proteins in *K. candel* and their role in the stress response. First, proteins involved in light-dependent reactions, Calvin cycle and respiration were up-regulated to improve salt resistance under moderate NaCl treatment (300–450 mM), but some of them were down-regulated under high salt stress (600 mM NaCl). Their function could be to maintain photosynthetic electron flow and to provide energy equivalents necessary for repairs. The second strategy involved the up-regulation of proteins leading to an energy remobilization in *K. candel* under all NaCl treatments. Third, vacuolar H^+-ATPase played a vital role in Na^+ detoxification in the plant cell, while up-regulation of SOD and DHAR prevented the accumulation of ROS. Our proteomic data revealed that a series of exquisite biochemical mechanisms for salt tolerance of *K. candel* which enable this species to withstand up to 450 mM NaCl stress. The proteome result was corroborated by immunoblots of the representative proteins and measurements of SOD and DHAR activities, as well as the content determinations of H_2O_2, O_2^-, Na^+, and K^+/Na^+. It is noteworthy that several novel salt-responsive proteins are identified in this study compared with previous reports in mangrove plants. This study has allowed us to expand our knowledge of the mechanisms by which woody halophytes respond to salinity and may assist in designing and developing more salt-tolerant plants in the future.

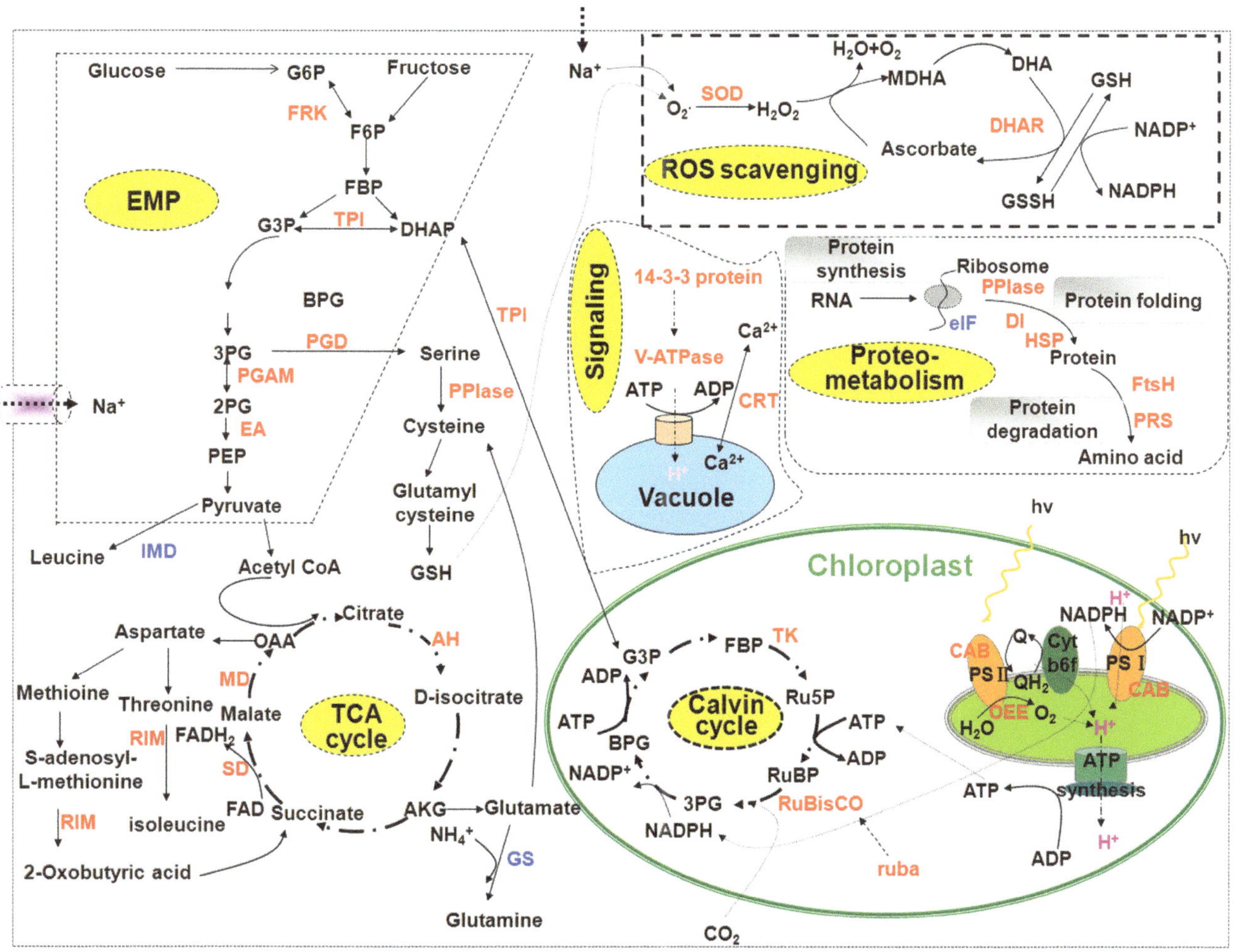

Figure 8. Schematic presentation of a mechanism for salt tolerance in *K. candel*. Most differentially expressed proteins were integrated, and were indicated in red (up-regulated at least under 450 mM NaCl treatment) or blue (down-regulated), respectively. Abbreviations: ADP, adenosine diphosphate; AKG, oxoglutarate; BPG, 1,3-bisphosphoglycerate; cytb6f, cytochrome b6f; DHA, dehydroascrobate; DHAP, dihydroxyacetone phosphate; EA, enolase; eIF, eukaryotic translation initiation factor; F6P, fructose-6-phosphate; $FADH_2$, reduced flavin adenine dinucleotide; FtsH, Cell division protein ftsH; G3P, glyceraldehydes-3-phosphate; G6P, glucose-6-phosphate; GS, glutamine synthetase; GSH, reduced glutathione; GSSH, oxidized glutathione; IMD, isopropylmalate dehydratase; MDHA, monodehydroascorbate; MDHAR, MDHA reductase; $NADP^+$/NADPH, nicotinamide adenine dinucleotide phosphate; OAA, oxaloacetic acid; PEP, phosphoenolpyruvate; PG, phosphoglycolate; PGD, phosphoglycerate dehydrogenase; PPlase, peptidyl-prolyl cis-trans isomerase; PRS, proteasome; Q, quinone; R5P, ribose-5-phosphate; Ru5P, ribulose-5-phosphate; RuBisCO, ribulose-1,5-bisphosphate carboxylase/oxygenase; RuBP, ribulose-1,5-bisphosphate; RIM, reductoisomerase; TPI, triosephosphate isomerase.

Materials and Methods

Plant growth conditions

Mature hypocotyls of *Kandelia candel* were obtained from the estuary of Luoyang river (24°58′N, 118°39′E), Quanzhou city, Fujian province, China, which features salinity levels ranging from 8‰ to 20‰. We conducted the study in the experimental zone of the mangrove reserve that obeys the Article 18 of Regulations of the People's Republic of China on Nature Reserves. Similar-sized hypocotyls were chosen for the experiment, which was conducted in a sand culture system under the following conditions: a thermoperiod of 30/25°C (day/night), a photoperiod of 16 h/8 h (day/night), light intensity ranging from 1,200 µmol m^{-2} s^{-1} to 1,250 µmol m^{-2} s^{-1}, and a relative humidity of 70%. Since *K. candel* is a salt-tolerant plant and naturally grows best in sea water with a salt concentration of approximately 150 mM, the hypocotyls were cultivated in Hoagland solution (Macronutrients: 5 mM $Ca(NO_3)_2$, 5 mM KNO_3, 2 mM $MgSO_4{\cdot}7H_2O$, 1 mM KH_2PO_4 and 0.1 mM EDTA-Na_2; Micronutrients: 46 µM H_3BO_3, 9 µM $MnCl_2{\cdot}4H_2O$, 0.3 µM $CaSO_4{\cdot}5H_2O$, 0.7 µM $ZnSO_4{\cdot}7H_2O$ and 0.08 µM $(NH4)_6Mo_7O_{24}$ •$4H_2O$) containing 150 mM NaCl in rectangular plastic trays (30 cm×42 cm×14 cm) until four leaves had fully expanded. Subsequently, the seedlings were subjected for 3 d to 150 (control), 300, 450, and 600 mM NaCl in a split plot design with three replicates. The leaves were harvested and kept at −80°C. Three independent biological replicates were prepared for physiological and proteomic analyses.

Na^+ and K^+ ion determination

Leaves of seedlings were weighed, washed in distilled water and dried in a 80°C oven until a constant weight. Dry powder (0.3 g) of each sample was digested in concentrated H_2SO_4. Na^+ and K^+ content in the leaves were determined by the flame emission method as described by Chuang et al. [75] using a flame photometer (FP 640, Shanghai, China). Five biological replicates were extracted for each treatment.

Enzyme extraction and activity assay

Leaf samples (1 g) were ground into powder in liquid N_2. Four volumes of 50 mM potassium phosphate buffer (pH 7.0) containing 1 mM EDTA, 2% (w/v) polyvinylpyrrolidone (PVP), and 0.05% Triton X-100 were added and homogenized following the procedures described by Gossett et al. [76]. Centrifugation at 10,000×g for 15 min at 4°C, the supernatant was applied to measure enzyme activity as follows. SOD activity was estimated using the inhibition of photochemical reduction of nitroblue tetrazolium (NBT) at 560 nm [77]. One unit of SOD activity was defined as 50% inhibition of NBT photoreduction DHAR activity was analyzed by the method of the previous report [78]. The activities of DHAR were expressed by measuring the ascorbic acid in absorbance at 265 nm. Protein content was measured as previously described by Bradford [79].

Analysis of H_2O_2 and O_2^- content

For H_2O_2 content, leaf tissues (1 g) were ground in liquid N_2 and then homogenized in 5 ml cold acetone. After centrifugation at 3,000 g, 4°C for 10 min, the supernatants were used for H_2O_2 content assays. According to the method of Patterson et al. [80], H_2O_2 content was assayed by analyzing the production of titanium–hydroperoxide complex at 410 nm. To determine the O_2^- content, 1 g of leaf tissue was ground to a fine powder in liquid nitrogen and then homogenized in 5 mL of of extraction buffer (pH 7.8) containing 50 mM sodium phosphate, 5 mM EDTA, and 1% (w/v) PVP. After centrifugation at 10,000 g, 4°C for 20 min, the supernatants were collected and the O_2^- content was detected O_2^- contents was assayed using the oxidation of hydroxylamine at 530 nm as described by Wang and Luo [81].

Protein extraction

Total leaf proteins were extracted using a phenol extraction procedure [82,83]. Leaf tissues (3 g) of each sample were ground with liquid nitrogen with mortar and pestle, suspended in 12 mL of cooled extraction buffer containing 50 mM L-ascorbic acid, 100 mM KCl, 50 mM disodium tetraborate decahydrate, 0.5% (v/v) Triton X-100, 2% (v/v) β-mercaptoethanol, 2% PVP, 1 mM phenylmethylsulfonyl fluoride and 100 mM Tris-HCl, pH 8.0. An equal volume of ice-cold Tris-HCl-saturated phenol (pH 8.0) was added to the solution, vortexed and the suspension re-homogenized on ice prior to centrifugation at 5,500×g for 10 min at 4°C. Subsequently, the phenolic phase was transferred to second tube, six-fold 100 mM ammonium acetate/methanol added for an overnight incubation at −20°C, followed by centrifugation at 4°C, 20,000×g for 20 min. The pellet was rinsed with cold methanol and washed with acetone containing 0.07% (v/v) β-mercaptoethanol twice. The pellet was collected and lyophilized at −20°C. Protein concentration was determined by the Bradford assay with bovine serum albumin as the standard [79].

Two-dimensional gel electrophoresis of proteins

Two-dimensional gel electrophoresis was undertaken according to the method of Liang et al. [84] and Liu et al. [85] with the Ettan IPGphor system (GE Healthcare, Little Chalfont, UK). The leaf protein (1,500 μg) was loaded onto IPG strips (24 cm, linear gradient pH 4–7; GE Healthcare) in a rehydration tray for 12 h. Isoelectric focusing (IEF) was performed under the following conditions: 200 V for 1 h, 500 V for 1 h, 1 kV for 1 h, gradient to 8 kV for 0.5 h, and, finally, 8 kV for a total of 48,000 Vhs at 20°C. After IEF, the IPG strips were equilibrated in equilibration buffer [50 mM Tris–HCl, pH 8.8, 6 M urea, 30% (v/v) glycerol, and 2% (w/v) SDS] containing 1% (w/v) dithiothreitol for 15 min and then in equilibration buffer containing 2.5% (w/v) iodoacetamide for 15 min. The strips were transferred onto vertical 12% SDS-PAGE gels, and proteins separated at 10 mA/gel for 30 min, followed by 30 mA/gel over night. After electrophoresis, gels were visualized using Coomassie brilliant blue (CBB) R-250 staining. The CBB-stained 2-D gels were scanned using a UMAX Power Look 2100×L scanner (Maxium Technologies, Taipei, China) and analyzed using Imagemaster^{TM} 2D Platinum software version 5.0 (GE Healthcare). Relative comparison of the intensity abundance between control and NaCl-treated (three replicate samples for each group) was performed using the Student's test. The protein spots with distinct differences were regarded to have at least a 2.0-fold amount of change.

In-gel tryptic digestion of proteins and MALDI-TOF-TOF/MS analysis

Protein spots were manually excised from preparative CBB-stained gels and digested with sequencing-grade modified trypsin (Promega, Madison, WI, USA) following the manufacturer's protocol, then incubated at 37°C for 12 h. Tryptic peptides were redissolved in 0.8 μl of matrix solution [α-cyano-4-hydroxycinnamic acid (Sigma, St. Louis, MO, USA) in 0.1% trifluoroacetic acid, 50% acetonitrile] before application to the MALDI plate.

After air dying, samples were analyzed using a 4700 MALDI-TOF-TOF Proteomics Analyzer (Applied Bio-systems, Foster City, CA, USA). Combined MS and MS/MS spectra were submitted to MASCOT (V2.1, Matrix Science, London, UK) by GPS Explorer software (V3.6, Applied Biosystems). Database searches were used the following parameters: NCBInr database (release date: 2010.07.01); taxonomy of green plant; trypsin digest with one missing cleavage; no fixed modifications; MS tolerance of 100 ppm; MS/MS tolerance of 0.6 Da; and possible oxidation of methionine. MASCOT protein scores (based on combined MS and MS/MS spectra) greater than 75 were considered statistically significant ($P \leq 0.05$). The individual MS/MS spectrum with the most statistically significant (confidence interval of 95%) ion score (based on MS/MS spectra) was accepted.

Hierarchical cluster analyses

The expression profiles of differential proteins were analyzed through two-way hierarchical clustering according to the PermutMatrix software (Figure 6). Rows were mean centered, and Euclidean distance and Average Linkage were used for data aggregation.

Immunoblot analysis

Protein samples (50 μg/lane) were separated using 12% one-dimensional SDS-PAGE gel electrophoresis, transferred onto nitrocellulose membranes, and incubated at room temperature for 2 h with rabbit polyclonal antibodies raised against either DHAR or HSC70 proteins (Agrisera, Sweden) at 1:5,000 dilution. After washing three times with TBST buffer (0.01 M TBS, 0.1% Tween-20, pH 7.6), the membranes were exposed for 2 h at room temperature to horseradish peroxidase-conjugated goat anti-rabbit IgG at 1:300 dilution. Positive signals were visualized with 3, 3′-diaminobenzidine.

Statistical analysis

Data from repeated measurements are shown as mean ± SD. Comparison of differences among the groups was carried out using Student's test. Significance was defined as $P \leq 0.05$.

Acknowledgments

We would like to thank Dr. Susanne Hoffmann-Benning of the Department of Biochemistry and Molecular Biology at Michigan State University (USA) for critical review of the manuscript.

Author Contributions

Conceived and designed the experiments: WC. Performed the experiments: XL LW YC FT LH. Analyzed the data: LW XL ML YL JX. Wrote the paper: WC LW XL WL.

References

1. Song Y, Zhang CJ, Ge WN, Zhang YF, Burlingame AL, et al. (2011) Identification of NaCl stress-responsive apoplastic proteins in rice shoot stems by 2D-DIGE. J Proteomics 74: 1045–1067.
2. Allakhverdiev SI, Nishiyama Y, Suzuki I, Tasaka Y, Murata N (1999) Genetic engineering of the unsaturation of fatty acids in membrane lipids alters the tolerance of *Synechocystis* to salt stress. Proc Natl Acad Sci U S A 96: 5862–5867.
3. Askari H, Edqvist J, Hajheidari M, Kafi M, Salekdeh GH (2006) Effects of salinity levels on proteome of *Suaeda aegyptiaca* leaves. Proteomics 6: 2542–2554.
4. Bandehagh A, Salekdeh GH, Toorchi M, Mohammadi A, Komatsu S (2011) Comparative proteomic analysis of canola leaves under salinity stress. Proteomics 11: 1965–1975.
5. Kosová K, Vítámvás P, Prášil IT, Renaut J (2011) Plant proteome changes under abiotic stress–Contribution of proteomics studies to understanding plant stress response. J Proteomics 74: 1301–1322.
6. Meloni DA, Oliva MA, Martinez CA, Cambraia J (2003) Photosynthesis and activity of superoxide dismutase, peroxidase and glutathione reductase in cotton under salt stress. Environ Exp Bot 49: 69–76.
7. Parida AK, Das AB (2005) Salt tolerance and salinity effects on plants: a review. Ecotoxicol Environ Saf 60: 324–349.
8. Tuteja N (2007) Mechanisms of high salinity tolerance in plants. Methods Enzymol 428: 419.
9. Zhu JK (2001) Plant salt tolerance. Trends Plant Sci 6: 66–71.
10. Huang GY, Wang YS, Ying GG, Dang AC (2012) Analysis of type 2 metallothionein gene from mangrove species (*Kandelia candel*). Trees 26: 1537–1544.
11. Parida AK, Jha B (2010) Salt tolerance mechanisms in mangroves: a review. Trees 24: 199–217.
12. Parida AK, Das AB, Mittra B (2004) Effects of salt on growth, ion accumulation, photosynthesis and leaf anatomy of the mangrove, *Bruguiera parviflora*. Trees 18: 167–174.
13. Takemura T, Hanagata N, Sugihara K, Baba S, Karube I, et al. (2000) Physiological and biochemical responses to salt stress in the mangrove, *Bruguiera gymnorrhiza*. Aquat Bot 68: 15–28.
14. Dasgupta N, Nandy P, Sengupta C, Das S (2011) Protein and enzymes regulations towards salt tolerance of some Indian mangroves in relation to adaptation. Trees 26: 377–391.
15. Hotta M, Nemoto S, Mimura T (2000) Re-evaluation of role of vacuole during salt adaptation in higher plant cells. Plant Cell Physiol 41: 79.
16. Santiago LS, Lau TS, Melcher PJ, Steele OC, Goldstein G (2000) Morphological and physiological responses of Hawaiian *Hibiscus tiliaceus* populations to light and salinity. Int J Plant Sci 161: 99–106.
17. Parida A, Das AB, Mittra B (2003) Effects of NaCl stress on the structure, pigment complex composition, and photosynthetic activity of mangrove *Bruguiera parviflora* chloroplasts. Photosynthetica 41: 191–200.
18. Liang S, Fang L, Zhou RC, Tang T, Deng SL, et al. (2012) Transcriptional homeostasis of a mangrove species, *Ceriops tagal*, in saline environments, as revealed by microarray analysis. PLoS One 7: e36499.
19. Wong YY, Ho CL, Nguyen PD, Teo SS, Harikrishna JA, et al. (2007) Isolation of salinity tolerant genes from the mangrove plant, *Bruguiera cylindrica* by using suppression subtractive hybridization (SSH) and bacterial functional screening. Aquat Bot 86: 117–122.
20. Huang W, Fang XD, Li GY, Lin QF, Zhao WM (2003) Cloning and expression analysis of salt responsive gene from *Kandelia candel*. Biol Plant 47: 501–507.
21. Mehta PA, Sivaprakash K, Parani M, Venkataraman G, Parida AK (2005) Generation and analysis of expressed sequence tags from the salt-tolerant mangrove species *Avicennia marina* (Forsk) Vierh. Theor Appl Genet 110: 416–424.
22. Tanaka Y, Fukuda A, Nakamura A, Yamada A, Saito T, et al. (2000) Molecular cloning and characterization of mangrove Na^+/H^+ antiporter cDNA. Plant Cell Physiol 41: s27.
23. Miyama M, Shimizu H, Sugiyama M, Hanagata N (2006) Sequencing and analysis of 14,842 expressed sequence tags of burma mangrove, *Bruguiera gymnorrhiza*. Plant Sci 171: 234–241.
24. Jithesh MN, Prashanth SR, Sivaprakash KR, Parida A (2006) Monitoring expression profiles of antioxidant genes to salinity, iron, oxidative, light and hyperosmotic stresses in the highly salt tolerant grey mangrove, *Avicennia marina* (Forsk.) Vierh. by mRNA analysis. Plant Cell Rep 25: 865–876.
25. Miyama M, Tada Y (2008) Transcriptional and physiological study of the response of Burma mangrove (*Bruguiera gymnorhiza*) to salt and osmotic stress. Plant Mol Biol 68: 119–129.
26. Sarhadi E, Bazargani MM, Sajise AG, Abdolahi S, Vispo NA, et al. (2012) Proteomic analysis of rice anthers under salt stress. Plant Physiol Biochem 58: 280–287.
27. Abbasi FM, Komatsu S (2004) A proteomic approach to analyze salt-responsive proteins in rice leaf sheath. Proteomics 4: 2072–2081.
28. Kim DW, Rakwal R, Agrawal GK, Jung YH, Shibato J, et al. (2005) A hydroponic rice seedling culture model system for investigating proteome of salt stress in rice leaf. Electrophoresis 26: 4521–4539.
29. Ma HY, Song LR, Shu YJ, Wang S, Niu J, et al. (2012) Comparative proteomic analysis of seedling leaves of different salt tolerant soybean genotypes. J Proteomics 75: 1529–1546.
30. Guo GF, Ge P, Ma CY, Li XH, Lv DW, et al. (2012) Comparative proteomic analysis of salt response proteins in seedling roots of two wheat varieties. J Proteomics 75: 1867–1885.
31. Fatehi F, Hosseinzadeh A, Alizadeh H, Brimavandi T, Struik PC (2012) The proteome response of salt-resistant and salt-sensitive barley genotypes to long-term salinity stress. Mol Biol Rep 39: 6387–6397.
32. Pang QY, Chen SX, Dai SJ, Chen YZ, Wang Y, et al. (2010) Comparative proteomics of salt tolerance in *Arabidopsis thaliana* and *Thellungiella halophila*. J Proteome Res 9: 2584–2599.
33. Jiang YQ, Yang B, Harris NS, Deyholos MK (2007) Comparative proteomic analysis of NaCl stress-responsive proteins in *Arabidopsis* roots. J Exp Bot 58: 3591–3607.
34. Du CX, Fan HF, Guo SR, Tezuka T, Li J (2010) Proteomic analysis of cucumber seedling roots subjected to salt stress. Phytochemistry 71: 1450–1459.
35. Wang XC, Fan PX, Song HM, Chen XY, Li XF, et al. (2009) Comparative proteomic analysis of differentially expressed proteins in shoots of *Salicornia europaea* under different salinity. J Proteome Res 8: 3331–3345.
36. Tada Y, Kashimura T (2009) Proteomic analysis of salt-responsive proteins in the mangrove plant, *Bruguiera gymnorhiza*. Plant Cell Physiol 50: 439–446.
37. Zhu Z, Chen J, Zheng HL (2012) Physiological and proteomic characterization of salt tolerance in a mangrove plant, *Bruguiera gymnorrhiza* (L.) Lam. Tree Physiol 32: 1378–1388.
38. Munns R, James RA, Läuchli A (2006) Approaches to increasing the salt tolerance of wheat and other cereals. J Exp Bot 57: 1025–1043.
39. Sugihara K, Hanagata N, Dubinsky Z, Baba S, Karube I (2000) Molecular characterization of cDNA encoding oxygen evolving enhancer protein 1 increased by salt treatment in the mangrove *Bruguiera gymnorrhiza*. Plant Cell Physiol 41: 1279–1285.
40. Wan XY, Liu JY (2008) Comparative proteomics analysis reveals an intimate protein network provoked by hydrogen peroxide stress in rice seedling leaves. Mol Cell Proteomics 7: 1469–1488.
41. Komatsu S, Yamada E, Furukawa K (2009) Cold stress changes the concanavalin A-positive glycosylation pattern of proteins expressed in the basal parts of rice leaf sheaths. Amino Acids 36: 115–123.
42. Chen FG, Zhang S, Jiang H, Ma WJ, Korpelainen H, et al. (2011) Comparative proteomics analysis of salt response reveals sex-related photosynthetic inhibition by salinity in *Populus cathayana* cuttings. J Proteome Res 10: 3944–3958.
43. Emanuelsson O, Nielsen H, Brunak S, von Heijne G (2000) Predicting subcellular localization of proteins based on their N-terminal amino acid sequence. J Mol Biol 300: 1005–1016.
44. Wang XC, Chang LL, Wang BC, Wang D, Li PH, et al. (2013) Comparative proteomics of *Thellungiella halophila* leaves from plants subjected to salinity reveals the importance of chloroplastic starch and soluble sugars in halophyte salt tolerance. Mol Cell Proteomics 12: 2174–2195.
45. Ndimba BK, Chivasa S, Simon WJ, Slabas AR (2005) Identification of *Arabidopsis* salt and osmotic stress responsive proteins using two-dimensional difference gel electrophoresis and mass spectrometry. Proteomics 5: 4185–4196.
46. Höper D, Bernhardt J, Hecker M (2006) Salt stress adaptation of *Bacillus subtilis*: A physiological proteomics approach. Proteomics 6: 1550–1562.
47. Yu JJ, Chen SX, Wang T, Sun GR, Dai SJ (2013) Comparative Proteomic Analysis of *Puccinellia tenuiflora* Leaves under Na_2CO_3 Stress. Int J Mol Sci 14: 1740–1762.
48. Rasoulnia A, Bihamta MR, Peyghambari SA, Alizadeh H, Rahnama A (2011) Proteomic response of barley leaves to salinity. Mol Biol Rep 38: 5055–5063.
49. Zhang H, Han B, Wang T, Chen SX, Li HY, et al. (2011) Mechanisms of plant salt response: insights from proteomics. J Proteome Res 11: 49–67.
50. Zhao Q, Zhang H, Wang T, Chen SX, Dai SJ (2013) Proteomics-based investigation of salt-responsive mechanisms in plant roots. J Proteomics 82: 230–253.
51. Batelli G, Verslues PE, Agius F, Qiu QS, Fujii H, et al. (2007) SOS2 promotes salt tolerance in part by interacting with the vacuolar H^+-ATPase and upregulating its transport activity. Mol Cell Biol 27: 7781–7790.
52. Zhang JL, Shi HZ (2013) Physiological and molecular mechanisms of plant salt tolerance. Photosynth Res 115: 1–22.
53. Kumar Swami A, Alam SI, Sengupta N, Sarin R (2011) Differential proteomic analysis of salt stress response in *Sorghum bicolor* leaves. Environ Exp Bot 71: 321–328.
54. Maserti BE, Del Carratore R, Croce CM, Podda A, Migheli Q, et al. (2011) Comparative analysis of proteome changes induced by the two spotted spider

mite *Tetranychus urticae* and methyl jasmonate in citrus leaves. J Plant Physiol 168: 392–402.
55. Eltayeb AE, Kawano N, Badawi GH, Kaminaka H, Sanekata T, et al. (2006) Enhanced tolerance to ozone and drought stresses in transgenic tobacco overexpressing dehydroascorbate reductase in cytosol. Physiol Plant 127: 57–65.
56. Huang GJ, Chen HJ, Chang YS, Lu TL, Lin YH (2008) Sweet potato storage root thioredoxin h2 with both dehydroascorbate reductase and monodehydroascorbate reductase activities. Bot Stud 49: 1–7.
57. Abogadallah GM (2010) Antioxidative defense under salt stress. Plant signaling & behavior 5: 369–374.
58. Mandhania S, Madan S, Sawhney V (2006) Antioxidant defense mechanism under salt stress in wheat seedlings. Biol Plant 50: 227–231.
59. Yan SP, Tang ZC, Su WA, Sun WN (2005) Proteomic analysis of salt stress-responsive proteins in rice root. Proteomics 5: 235–244.
60. Wang WX, Vinocur B, Shoseyov O, Altman A (2004) Role of plant heat-shock proteins and molecular chaperones in the abiotic stress response. Trends Plant Sci 9: 244–252.
61. Hartl FU (1996) Molecular chaperones in cellular protein folding. Nature 381: 571–580.
62. Rospert S, Looser R, Dubaquié Y, Matouschek A, Glick BS, et al. (1996) Hsp60-independent protein folding in the matrix of yeast mitochondria. EMBO J 15: 764.
63. Wang WF, Shakes DC (1996) Molecular evolution of the 14-3-3 protein family. J Mol Evol 43: 384–398.
64. You XR, Wang LX, Liang WY, Gai YH, Wang XY, et al. (2012) Floral reversion mechanism in longan (*Dimocarpus longan* Lour.) revealed by proteomic and anatomic analyses. J Proteomics 75: 1099–1118.
65. Palmgren MG (1998) Proton gradients and plant growth: role of the plasma membrane H^+-ATPase. Adv Bot Res 28: 1–70.
66. Finnie C, Borch J, Collinge DB (1999) 14-3-3 proteins: eukaryotic regulatory proteins with many functions. Plant Mol Biol 40: 545–554.
67. Aitken A (1995) 14-3-3 proteins on the MAP. Trends Biochem Sci 20: 95.
68. Camoni L, Harper JF, Palmgren MG (1998) 14-3-3 proteins activate a plant calcium-dependent protein kinase (CDPK). FEBS Lett 430: 381–384.
69. Johnson S, Michalak M, Opas M, Eggleton P (2001) The ins and outs of calreticulin: from the ER lumen to the extracellular space. Trends Cell Biol 11: 122–129.
70. Katz A, Waridel P, Shevchenko A, Pick U (2007) Salt-induced changes in the plasma membrane proteome of the halotolerant alga *Dunaliella salina* as revealed by blue native gel electrophoresis and nano-LC-MS/MS analysis. Mol Cell Proteomics 6: 1459–1472.
71. Agrawal GK, Rakwal R, Yonekura M, Kubo A, Saji H (2002) Proteome analysis of differentially displayed proteins as a tool for investigating ozone stress in rice (*Oryza sativa* L.) seedlings. Proteomics 2: 947–959.
72. Persson S, Rosenquist M, Svensson K, Galvão R, Boss WF, et al. (2003) Phylogenetic analyses and expression studies reveal two distinct groups of calreticulin isoforms in higher plants. Plant Physiol 133: 1385–1396.
73. Wilkinson B, Gilbert HF (2004) Protein disulfide isomerase. Biochim Biophys Acta 1699: 35–44.
74. Gruber CW, Čemažar M, Heras B, Martin JL, Craik DJ (2006) Protein disulfide isomerase: the structure of oxidative folding. Trends Biochem Sci 31: 455–464.
75. Chuang FS, Sarbeck JR, John PS, Winefordner JD (1973) Flame spectrometric determination of sodium, potassium and calcium in blood serum by measurement of microsamples. Mikrochim Acta 61: 523–531.
76. Gossett DR, Millhollon EP, Lucas M (1994) Antioxidant response to NaCl stress in salt-tolerant and salt-sensitive cultivars of cotton. Crop Sci 34: 706–714.
77. Hyland K, Voisin E, Banoun H, Auclair C (1983) Superoxide dismutase assay using alkaline dimethylsulfoxide as superoxide anion-generating system. Anal Biochem 135: 280–287.
78. Cakmak I, Marschner H (1992) Magnesium deficiency and high light intensity enhance activities of superoxide dismutase, ascorbate peroxidase, and glutathione reductase in bean leaves. Plant Physiol 98: 1222–1227.
79. Bradford MM (1976) A rapid and sensitive method for the quantitation of microgram quantities of protein utilizing the principle of protein-dye binding. Anal Biochem 72: 248–254.
80. Patterson BD, MacRae EA, Ferguson IB (1984) Estimation of hydrogen peroxide in plant extracts using titanium (IV). Anal Biochem 139: 487–492.
81. Wang AG, Luo GH (1990) Quantitative relation between the reaction of hydroxylamine and superoxide anion radicals in plants. Plant Physiol Commun 6: 55–57.
82. Saravanan RS, Rose JK (2004) A critical evaluation of sample extraction techniques for enhanced proteomic analysis of recalcitrant plant tissues. Proteomics 4: 2522–2532.
83. Weng ZX, Wang LX, Tan FL, Huang L, Xing JH, et al. (2013) Proteomic and physiological analyses reveal detoxification and antioxidation induced by Cd stress in *Kandelia candel* roots. Trees: 1–13.
84. Liang WY, Zhou YW, Wang LX, You XR, Zhang YP, et al. (2012) Ultrastructural, physiological and proteomic analysis of *Nostoc flagelliforme* in response to dehydration and rehydration. J Proteomics 75: 5604–5627.
85. Liu H, Liu YZ, Zheng SQ, Jiang JM, Wang P, et al. (2010) Comparative proteomic analysis of longan (*Dimocarpus longan* Lour.) seed abortion. Planta 231: 847–860.

21

The Future of the Oceans Past: Towards a Global Marine Historical Research Initiative

Kathleen Schwerdtner Máñez[1,2]*, Poul Holm[3], Louise Blight[4], Marta Coll[5,6], Alison MacDiarmid[7], Henn Ojaveer[8], Bo Poulsen[9], Malcolm Tull[10]

1 Department of Social Sciences, Leibniz Center for Tropical Marine Ecology, Bremen, Germany, **2** Asia Research Center, Murdoch University, Murdoch, Western Australia, Australia, **3** Trinity Long Room Hub Arts and Humanities Research Institute, Trinity College Dublin, Dublin, Ireland, **4** World Wildlife Fund (WWF) -Canada, Vancouver, British Columbia, Canada, **5** Renewable Marine Resources Department, Institute of Marine Science, Barcelona, Spain, **6** Laboratoire Écosystèmes Marins Exploités, Sète Cedex, France, **7** Marine Ecology, National Institute of Water and Atmospheric Research, Kilbirnie, Wellington, New Zealand, **8** Estonian Marine Institute, University Tartu, Tartu, Estonia, **9** Faculty of Social Sciences, Aalborg University, Aalborg Ø, Denmark, **10** Murdoch Business School, Murdoch University, Murdoch, Western Australia, Australia

Abstract

Historical research is playing an increasingly important role in marine sciences. Historical data are also used in policy making and marine resource management, and have helped to address the issue of shifting baselines for numerous species and ecosystems. Although many important research questions still remain unanswered, tremendous developments in conceptual and methodological approaches are expected to contribute to a comprehensive understanding of the global history of human interactions with life in the seas. Based on our experiences and knowledge from the "History of Marine Animal Populations" project, this paper identifies the emerging research topics for future historical marine research. It elaborates on concepts and tools which are expected to play a major role in answering these questions, and identifies geographical regions which deserve future attention from marine environmental historians and historical ecologists.

Editor: Richard K. F. Unsworth, Seagrass Ecosystem Research Group, Swansea University, United Kingdom

Funding: Fellowships to the BIOWEB project and by the Spanish National Program 'Ramon y Cajal'. HO was funded by the Estonian Ministry of Education and Research (grant SF0180005s10). We thank HMAP for paying the publication costs of this paper. The funders had no role in study design, data collection and analysis, decision to publish, or preparation of the manuscript.

Competing Interests: The authors have declared that no competing interests exist.

* Email: kathleen.schwerdtner@zmt-bremen.de

Introduction

In the last fifteen years marine science has taken a historical turn. Looking much further back in time than most previous studies, global initiatives such as the History of Marine Animal Populations (HMAP) and the Sea Around Us programmes, as well as individual studies have investigated diverse socio-ecological systems, from coastal European marshes to Pacific islands. In a parallel move, environmental historians have undertaken the study of human engagement with the underwater realm in a sea change for history. Historians, archaeologists, economists, sociologists and geographers, have engaged with marine scientists in an interdisciplinary effort to bring together the study of human and underwater worlds. Thanks to this collaborative effort of marine and human sciences, researchers have not only identified but, for many regions and species, resolved the problem of shifting baselines (a term popularised by Daniel Pauly in his seminal 1995 paper [1]) by pushing back the chronological limits of our knowledge [2].

While there have been tremendous advancements in marine historical research, these are yet to be assimilated into an integrated seamless view of the global history of human interactions with life in the oceans (Figure 1). Furthermore, many important questions still remain unanswered. The following are what we consider to be the five key areas of research inquiry for a global marine historical research agenda:

i. What did the sea look like before human exploitations? So far, almost all the accessible information relates to after first human contact, and by implication very little is known about pristine life in marine waters, with the possible exception of the present-day ecosystems of the most southerly Antarctic waters, and the abyssal deeps. A better understanding of unexploited seas would serve as an important ecological baseline against which the impacts of human activities could be assessed.

ii. What is the relative importance of key drivers of environmental change over historical time frames? There is good evidence for several abiotic, climatic and human factors affecting life in the sea, but knowledge about their relative importance through time and their interactions remains poor.

iii. What has been the significance of marine resources for human societies over time? The physiological benefits of a marine diet are well documented but the economic and more intangible implications for historical societies have not been teased out.

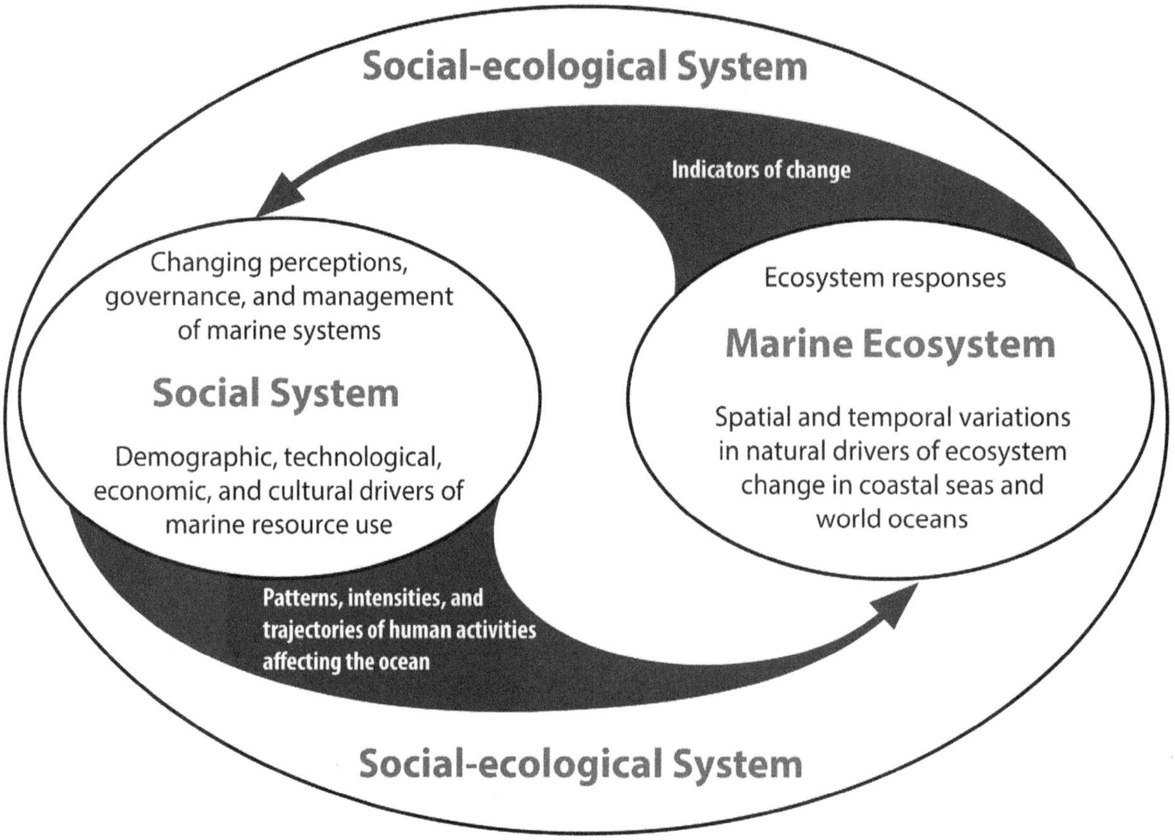

Figure 1. Integrated view of the global history of human interactions with life in the oceans.

iv. In cultural terms, societies have valued the goods and services provided by the sea very differently across various cultures and through time. What are the circumstances that have encouraged societies to exploit or give up the oceans?

v. How may historical information be used for better ocean governance and management in the future? While there are examples of the usefulness of historical baselines for fisheries management, the wider potential of historical research for ecosystem and seabed management or the development of human coastal societies remains largely unexplored.

Beginning with an elaboration of these emerging research topics in the sections below, this paper aims to develop a new research agenda for marine historical research. We also discuss potential applications and further development of existing conceptual and methodological approaches, namely the ecosystem service concept, the use of indicators and modelling approaches, new molecular methods, advanced oral history, and the need for gendered historical research. We then discuss the current coverage of historical research, and suggest further expansion in terms of regions and ecosystems to ensure a true global perspective. Finally, we summarize our thoughts on the research agenda, and suggest the foundation of a global research network: the Oceans Past Initiative (OPI) to assist in coordinating research efforts worldwide.

Impacts on Marine Systems Over Time

2.1 Untangling drivers of change

Little is known about what the oceans looked like before humans began to affect marine environments. The study of sediment cores as a library of past DNA provides one possible way of understanding what used to be swimming in the water columns above sea floors in a pre-exploitation era [3]. This could enable researchers to obtain a better understanding of pristine (pre-human) ocean life. It may also shed some light on when human activities started to impact marine systems, and how these effects can be differentiated from natural drivers of change. Similarly, a small handful of modern systems may yet be unaffected (or relatively so) by human impacts. These areas include the abyssal deeps, and the most southerly Antarctic waters [4]; the degree to which these provide clues about the functioning of other intact ecosystems prior to exploitation is currently unknown.

However, the vast majority of extant marine systems have been shaped by both natural and anthropogenic influences. Understanding the changes that these marine systems have undergone requires untangling the underlying drivers of these changes. Drivers are natural or human-induced factors that directly or indirectly force changes in ecosystems [5]. While direct drivers influence ecosystem components and processes themselves, indirect drivers alter one or more direct drivers. Physical and biological drivers such as changes in temperature regime or nutrient input, resource extraction, or the introduction of alien invasive species act directly upon marine systems and their component organisms [6–10]. Indirect drivers include human demographics, economic, sociocultural, and political processes, and scientific and technological innovation [11]. The majority of changes in ecosystems are caused by multiple interacting anthropogenic and/- or natural drivers. These interactions can be cumulative, antagonistic, synergistic, or otherwise. Decoupling the effects of single drivers to distinguish human influences from natural variability is a key interest of both scientists and ecosystem managers, and should be an integrative part of a research agenda that addresses historical change.

Determining impacts of natural drivers on pristine marine systems would improve our understanding of ecosystem functions and the role played by anthropogenic influences. Analysing the hard parts of organisms is one way to obtain information to infer the interaction of environmental drivers and ecological responses. For example, the shells of molluscs and otoliths (fish ear bones) are primarily constructed from calcium carbonate ($CaCO_3$). Stable isotopes of oxygen, carbon and nitrogen have a long history of use as geological and biological tracers [12–14]. If fish otoliths, shellfish remains or other biological samples can be obtained from a known time in the past (e.g., from dated midden horizons, museum collections), then the stable isotope and band increment evidence for environmental temperature and/or organismal growth and maturation for that period can compared to the same measures from modern individuals to examine long-term change in parameters of interest, and potential effects of human activities. Stable isotope analysis can also help researchers to examine different aspects of climate, food web structures, and other characteristics of past marine ecosystems [15]. For example, sequences of temperature estimates based on δ18O in fish otoliths can be applied to infer season at capture, which contributes to a better understanding of past fishing patterns [16]. The technique has also been used for the study of marine species and human diet. Recently, Orton and colleagues have shown the growing role of cod imports for medieval London. Using cod remains from excavations, they proof the use of this technique for the indication of changes in human diet [17]. Naturally occurring climate variations have an impact on the productivity of marine ecosystems [18]. A number of investigations have documented substantial fluctuations in fish populations related to climate forcing [19,20]. There is also well-established evidence for major climatic influences on phyto- and zooplankton communities [21,22]. Recent studies have identified ecosystem-level shifts in various ecosystems around the globe [21,23–25]. These shifts are a consequence of the interplay between climate forcing and various human impacts, primarily resource exploitation and pollution/eutrophication. Detailed investigations of reef growth and species composition in an Australian coral reef have shown that environmental degradation caused by human impacts has added additional pressure on the natural historical instability of these reefs [26]. Results from another Australian reef indicate that sediment and nutrient input following European settlement prohibited the recovery of *Acropora* assemblages after a series of acute disturbance events (SST anomalies, cyclones and flood events) [27]. One approach to detangling the impacts of such combined natural and anthropogenic drivers is the examination of paleontological, climatological, archaeological and historical evidence from localities affected by similar climate events but with differing levels of human impacts [28]. Such an approach could also provide a fresh perspective on natural variability.

Required research. Untangling natural and anthropogenic causes of environmental change to estimate the extent of human impacts on the oceans should be one of the major tasks for historical research in the near future. Coordinated multidisciplinary research activities around the globe that focus on the same time period in areas of contrasting human exploitation of marine resources will assist in distinguishing the effects of natural and human drivers of marine environmental change.

2.2. The role of humans

Examination of the earliest human interactions with the marine environment in most parts of the world is beset by a major problem. The rise in sea level after the most recent ice-age has obscured or destroyed much of the earliest evidence of human exploitation along the margins of ancient shores [29]. Although in some areas such as the North Sea some evidence of early exploitation has been recovered from submarine deposits it has often been reworked by waves and currents making interpretation difficult [30]. However, there is evidence for an intensive use of marine resources by early modern humans in Pinnacle Point (South Africa) some 164,000 years ago. Gathering shellfish seems to have been part of a response to sea level fluctuations, which forced these people to expand their home ranges and to follow the shifting position of the coast [31]. Almost contemporarily, Neanderthals gathered shellfish in the coastal waters of Torremolinos in Southern Spain [32]. Archeological studies have revealed a general shift in marine resource exploitation from inshore towards offshore areas. Marine fish became increasingly important - as early as 1,500 years ago - on the Californian coast [33]. Similar patterns have been discovered in the Wadden Sea [34].

But for most of human history, the extent of anthropogenic impacts on natural environments was largely restricted to local and regional levels. Within the last 200 years, human actions reached the global level by affecting processes such as global nutrient and water cycles. The traces of human activities are not restricted to populated areas anymore, but can be found everywhere in the ocean [35]. For example, plastic debris is accumulating at the shorelines of even the most remote islands and in the deep sea [36]. Johan Rockström and colleagues [37] claim that human actions have accelerated to such an extent that they might push the Earth beyond its "planetary boundaries". The term planetary boundary refers to thresholds between alternate states of the global system. Once crossed, the system enters a new state, and irreversible environmental change might be unavoidable. By now, the existence of thresholds has been posited for numerous ecological and social-ecological systems, including the marine realm (see the thresholds database of the Resilience Alliance and the Santa Fee Institute, http://www.resalliance.org/index.php/thresholds_database). The world has entered a geological new era: the Anthropocene.

In this era, an important line of research will be the study of the interaction of different human activities and their impacts on specific marine areas in the past. Cross-regional comparisons of magnitude and direction of these changes to marine life in response to multiple human pressures could also be extremely helpful in teasing apart multiple long-term drivers. For example, it was only by the late 1990s that more or less the full range of harvestable fish and invertebrates in all ecological strata of Southeast Asian waters was caught [38]. Near-pristine marine systems could still be found in some parts of the Indonesian Archipelago until the 1960s, a time at which Atlantic waters had already been affected by industrial fishing activities for more than two centuries. In particular, data from large, isolated, late settled islands or regions such as Madagascar, New Zealand, and Antarctica, where significant human impacts became relevant much later than in the seas bordering the major continental landmasses, can play an important role in distinguishing among natural climatic variability and human drivers (including anthropogenic climate change). Recent work in Australian coral reefs has shown that it is indeed possible to differentiate natural variability from human-induced environmental change [27].

Despite the importance of similar drivers acting at the global level, most ecosystem changes are caused by a set of interactions that are more or less unique to a particular place [11]. To understand how resource and socio-economic endowments, in combination with other factors, have created specific exploitation patterns, bioregional histories of marine environments need to be developed. This will also provide a more comprehensive view of "what once was", e.g., historical baselines of marine animal populations, enhancing our ability to articulate desirable ecological states and management recommendations [39].

In this line of research, specific attention must be paid to the role of globalisation and changing consumption patterns. Processes of globalisation have had a significant influence on marine resource exploitation for hundreds of years. The quest for marine resources was an important aspect of European expansion into Asia, Africa and the Americas. Arctic marine mammals, sea turtles in the Caribbean, and cod from Iceland to Newfoundland are just some examples of valuable sought-after species [40]. Similarly, Asian, particularly Chinese food and medicine markets, have driven a search for marine species on a regional to global scale over the past few centuries. For example, the development of sea cucumber fishing and trading by people from Makassar, Indonesia, can be traced from its beginning more than 300 years ago to the industrialization of the fishery in the 20th century and the depletion of sea cucumbers in the 1980s [41]. Economic gain was only one side of the story. Changing preferences for certain products, the exploitation of new areas, and technical innovations all played a role in a widening search for marine resources.

Required research. How has the pattern of marine resource exploitation proceeded in each area settled by humans, and which species were affected? Was the unfolding of events in Africa fundamentally different from histories in Asia, Australia, the Americas, and Europe and recently settled remote islands? To address this will require integration of the earliest archaeological records with historical and modern information sources to determine patterns of marine resource use over the entire period of human settlement of coastal margins. Studying the role of globalisation and changing consumption patterns may improve our understanding of resource use by linking serial depletion on a global scale to local demands. We also need research into changing consumers preferences, especially towards more sustainably harvested marine products. Do such changes occur on a short or long-term scale, are they caused by public advice, by economic forces, or by other cultural factors that are to date poorly understood?

Altered Marine Ecosystems

3.1. Extinctions, species declines, and habitat changes

Extinctions and resource depletions in the marine realm have been evaluated by a number of studies (eg., [41,42,43]). We now have a relatively precise overview of globally extinct marine mammals, birds, fish, and molluscs [42]. Many species have become extirpated, or nearly so, across part of their range, such as is the case with several shark and ray species in the Mediterranean Sea [43]. Others have been depleted to such an extent that they cannot fulfil their former role in the ecosystem, as in the case of reduced abundances of some filter feeders [44]. To truly understand the extent of these changes, long-term biodiversity loss should be studied in different regions of the oceans over various periods of time. This includes the reconstruction of decadal, centennial, and millennial dynamics of marine animal populations [45,46]. A number of studies have provided important knowledge in this respect, for example on changes to shark populations in the West Atlantic [47], or on numbers of large predatory fish in different seas [48]. Most of the historical studies to date have concentrated on exploited higher trophic levels of marine animals, while intermediate and lower trophic levels have largely been ignored [49]. In large part this is because the evidence about large exploited species is more readily available from archaeological and historical sources. But there is also evidence for a sequential exploitation of invertebrate species in the past [41]. How this has shaped the status of ecosystems should be an important research topic in future.

Changes in coastal and marine habitats are a related research topic that deserves more attention. Although this line of work has been less prominent that changes in species, important contributions have already been made. For example, Lotze and colleagues reconstructed time lines, causes, and consequences of change in several estuaries and coastal seas worldwide and found similar patterns of habitat destruction, water quality degradation, and the influence of invasive species [44]. Other work has also focused on sub-tidal algae forests in the Adriatic Sea [50], or provided a two-century perspective on changes in a Scottish coastal area [51].

Required research. In the future, a major research avenue may be to link geospatial data and historical maps showing coastal and marine habitats to information about past distribution and abundance of species derived from archaeological, historical, and genetic sources. Such research could help us to understand indirect causes and effects of habitat change and be of value to marine and coastal management. Focussed research on species across levels, including intermediate and lower levels, is required to describe the patterns and magnitudes of their exploitation, preferably from the period of first human settlements to the present day.

3.2. Recoveries of populations, habitats and ecosystems

There are many examples around the world where previously affected species, habitats and ecosystems have or are responding to new management interventions. For example, over the last 40 years of protection populations of New Zealand fur seal (*Arctocephalus fosteri*) have started to rebound after 600 years of exploitation first by Polynesian settlers and later by European sealers [52]. Interestingly, the same protection regime has made little impact on the recovery of New Zealand sea lions (*Phocarctos hookeri*). In this species the return of maturing females to their beach of birth has limited the expansion of breeding populations into their historic range [53].

Research on ecological recoveries has a very clear link to providing input to decision-makers and managers, because it brings a historical perspective into the present management of marine resources and tells them what has worked so far and what not. Given that the approach to managing marine systems is evolving towards ecosystem-based adaptive management, a long-term, historical perspective to ecosystem function and change is very much needed [54,55]. An interesting example of such work is a recent reconstruction of social-ecological interactions over the last 700 years in Hawaiian coral reefs. It reports previously undetected recovery periods related to a complex set of changes in underlying social systems resulting in the release of reefs from direct anthropogenic stressors [56].

Required research. Future investigations should continue to assess the adaptive capacity and resilience of degraded ecosystems and depleted stocks to recover from human impacts. There is already work done on how marine resources have recovered from previous exploitation-related declines [45,57], but we need a stronger focus on which species, ecosystem traits, and management activities might facilitate recoveries.

Perceptions, Governance, and Management of Marine Systems

How people perceive and value marine environments and the resources they provide determines individual and collective preferences, actions and strategies in the marine realm. Historical descriptions of coastal and marine environments are prime examples of documented perceptions of the past. In the absence of other data, they have been successfully used to establish abundance changes of marine species in different time periods [58,59].

Different perceptions and valuation systems also underlie the institutional structures that govern and manage marine systems. Research on governance structures must be linked with research on perceptions and values to understand what drives and has driven approaches to marine resource exploitation and its management in different periods of time. For analytical clarity, it is important to distinguish between governance and management. Governance describes a social function centered on efforts to steer the actions of humans toward achieving desirable outcomes and avoiding undesirable ones. It covers the fundamental goals, institutional processes and structures which are the basis for planning and decision-making, and sets the stage within which management occurs. Management refers to the process by which human and material resources are used to achieve a defined goal within a known institutional structure [60].

The notion of the sea as a seemingly endless source of resources has long dominated marine governance, or rather the relative absence of institutions governing the sea. However, there are also examples of long-enduring traditional management regimes which have regulated the exploitation of valuable or scarce resources such as pearl oysters, sea cucumbers, or *Trochus* spp. shells in many coastal areas in the Pacific and parts of Southeast Asia, long before modern fisheries management was invented. For example, the Indonesian regime of "sasi laut" regulated access to marine resources by placing temporal and spatial harvesting restrictions on them. Similarly regulations in European inshore waters such as the Limford sustained fisheries for hundreds of years [61]. While such community-based management is certainly not a panacea, studies around the world have found that traditional community-based governance regimes contain elements that may support sustainable resource use [62,63].

Required research. How have different local perceptions of marine systems affected their governance and management, and how have these evolved over time? Can we identify elements of these historical, community-based approaches to ocean governance and management that are also relevant to modern marine management?

Emerging methodological approaches

A number of emerging methodological approaches are expected to have a considerable influence on future marine historical research. They include conceptual developments such as the application of the ecosystem service concept in historical analyses or the consideration of social issues like gender and equity issues, but also new biophysical tools, for example different kinds of molecular and stable isotope analyses. These approaches are not necessarily new to science, but it is their application to historical research that is expected to improve our ability to analyse and evaluate changes in marine systems and their organisms over time.

5.1 Applying the ecosystem service concept in historical analyses

Ecosystem services are an array of potential benefits derived from specific ecological components and processes, ranging from the provision of fish to the capacity of the ocean to buffer climate change. Environmental change has clearly altered the level of ecosystem services provided by marine systems, but to what extent is largely unknown. There is a need to improve our understanding of how past changes in marine systems have affected marine ecosystem productivity and marine ecosystem functioning, and how this has influenced the ability of marine systems to provide ecosystem services.

The ecosystem service concept values nature in relation to human uses. By acknowledging the role of ecosystems as providers of essential goods and services, it links ecosystem functions with the economy and social spheres, including livelihoods and well-being [64]. The ecosystem service concept thus provides an approach to assess the value of marine systems, although this does not have to be a monetary valuation. More importantly, it offers a standardized method to compare these values over time, by showing how changes in ecological functions have caused changes in benefits to society [65].

Although much historical research is clearly related to ecosystem services and benefits, the concept of ecosystem services itself has hardly been used in the discipline. One exception is the decline in marine mammals in New Zealand that followed the well-documented onset of Māori sealing soon after initial settlement, and European whaling in the early 19th century [66]. Both cultures viewed many species of marine mammals as valuable commodities to be harvested, so the numbers of these mammals declined precipitously as a result [67,68]. Now, these species are protected and their value to New Zealand as a provisioning service has declined to zero. Instead, marine mammals are currently prized for their spiritual and existence value; people enjoy directly viewing them from land, sea, and air; and they are appreciated as subjects for research and educational activities. Moreover, whales are now recognised as having important roles in ecosystem regulation [69]. Studies like these illustrate the usefulness of the ecosystem service concept as a common framework for analysing the impacts of environmental change on societies.

5.2 Developing indicators

Indicators are characteristics of ecological and social systems or their components, such as species, populations, networks, and social groups, which ideally indicate a certain state or dynamic of a system that is otherwise difficult to measure and evaluate. Because ecological and social systems are inherently complex, the use of indicators helps to describe them and their changes in simpler terms. If chosen well, indicators can track changes over time and across species or regions, and can inform managers and the general public.

One of the most fundamental ecological indicators of historical change is a change in population abundance. This can be measured as a decrease or increase in the number of individuals, their biomass, average size or age, as well as an expansion or contraction of their distribution over time [70]. These indicators have been applied to a variety of records from the past. For example, archaeological records from shell middens revealed declines in the relative abundance, size and age of white sturgeon from 2600 to 700 years ago based on their bone frequency and dentary width [71,72]. Historical whaling maps and log books have been used to reconstruct the rapid depletion of right whales (*Eubalaena japonica*) in the North Pacific by 19th-century whalers

[73]. Fisheries catch and effort data throughout the Mediterranean have been used to analyse declines in the catch-per-unit-effort (CPUE) of sharks since the 19th century [74]. In rarer cases, it has been possible to extend such analysis into the early 17th century [75]. Other fundamental ecological indicators are the basic presence or absence of a species in an ecosystem, the trends in functional groups, or changes in trophic habits and positions [70]. Future development of ecological indicators will move towards standardization and evaluation of indicators, and the development of reference levels (targets, thresholds, limits) to inform the management of marine resources and ecosystems.

Social indicators are expected to play a very important role in the future for the assessment and management of coastal and marine systems and their changes over time. Such indicators can be qualitative or quantitative and are used to describe the status of social systems, and their dynamics and processes. While status indicators measure, for example, the current perception of target species, their availability or the social networks of resource exploitation, process indicators assess specific actions, changes or functions over a defined time period, such as participation, conflict resolution or institutional change [76]. Social process indicators can be used to evaluate participation and decision-making in the management of coastal and marine resources by assessing metrics such as the number of meetings, the levels of participation, and the character of social networks involved.

Indicators are expected to play a far greater role in the analysis of social–ecological systems in the future. They allow for regular measurements of key ecological, socio-economic, and social–ecological processes to better understand system changes and their underlying causes [76].

5.3 Utilizing modelling approaches

A major research issue in this field is how to integrate historical knowledge and ecosystem modelling, including past and future applications and simulations. The future will bring developments in modelling techniques which increase the capability to hindcast ecosystem dynamics and ensure forecasting possibilities [70]. More comparative analysis of qualitative and quantitative models must be developed since the ability to model past marine ecosystems is crucial to understanding the present, and to predict the future. These analyses will be challenged by the urgent need to consider ecological, economic and social dimensions of change [70].

Qualitative and quantitative modelling techniques [77–81] have already been adapted or modified to model past marine food webs. For example, qualitative modelling was used in the Gulf of Maine to identify distinct and sequential phases in the trophic structure of kelp forests [82]. The authors used archaeological, historical, ecological, and fisheries data to document a serial targeting and depletion of abundant top consumers that led to functional loss of trophic levels, creating trophic cascades that changed the structure and function of the ecosystem [83]. In the Adriatic Sea, qualitative modelling has similarly been used to describe the historical change of marine food webs using ten historical periods that started in the pre-human period before ~100 000 BC to the global expansion of humans in AD 1950–2000 [84].

An interesting approach to modelling past ecosystems has been developing under the "Back to the Future" approach [85]. This approach aims at evaluating historic ecosystems as tools to define possible restoration goals and to design rebuilding strategies for ecosystems. Two remarkable applications were developed for Canadian marine ecosystems of Newfoundland [86] covering the years of 1450, 1900, 1985 and 1995, and in British Columbia [87], covering the periods of 1750, 1900, 1950 and 2000. Both studies highlighted the general depletion of marine resources since the first European contact and the important changes in the structure and functioning of food webs through time.

5.4 Application of molecular tools

Molecular genetics has fundamentally changed our understanding of marine ecology. Investigations have demonstrated extensive adaptive change in marine populations in response to natural and anthropogenic drivers. Research has also shown that the estimated population sizes of several species are much smaller than census sizes, which is highly relevant for management and conservation [88]. A notable example of the use of molecular techniques in historical ecology studies is Roman and Palumbi's study on North Atlantic humpback whales (*Megaptera novaeangliae*), fin whales (*Balaenoptera physalus*) and minke whales (*Balaenoptera acutorostrata*), in which they used molecular markers to estimate the pre-exploitation abundance of these species. Their results indicated that current population numbers are far below their original size [89], (although, in a recent overview, Palumbi was careful to outline the indefinite time parameter of their study and a number of uncertainties [90]). A genetic approach was also used to estimate pre-whaling abundance of Eastern Pacific gray whales. The results substantially differed from population reconstructions based on catch records, which had considered the population to be fully recovered [91]. Both examples clearly demonstrate the relevance of molecular tools in defining baselines for management of marine species, and the recovery of depleted populations.

Molecular tools also have an enormous potential to contribute to the discussion on detangling natural and anthropogenic drivers of change. A recent study using genetic data from the harbor porpoise (*Phocoena phocoena*) in the Black Sea was able to reconstruct the demographic expansion of this species some 5,000 years ago as a result of natural environmental changes, as well as its population collapse due to recent anthropogenic activity [92].

Molecular analyses are increasingly used in research on biological invasions, one of the most serious human-induced threats to marine ecosystems [49]. This has improved our ability to make inferences regarding invasion histories, enabled resolution of some of the long-standing questions regarding the cryptogenic status of marine species and provided means to recover the patterns of community structure of the ocean's biota.

5.5 Advancing methods for written documents and oral history

The provenance of documents traditionally frames modes of inquiry in the history discipline. Port records, log books, tax ledgers etc. are typical sources of quantitative information, while interviews, written memories or newspaper coverage can provide both quantitative and qualitative information.

Witness testimonies, especially, are increasingly used to acquire information on past and contemporary marine environments and fisheries. Through testimonies, individuals or groups share their perceptions and opinions about past events or experiences. Oral history helps to gather these accounts of the past, and to make use of traditional and local environmental knowledge that resource users have accumulated [93]. If oral histories have been recorded or written down, the time span for which this kind of data can be used extends far beyond living memory. The scientific literature on the use of oral history has grown rapidly over the last three decades [94]. But after Saenz-Arroyo's pioneering study on shifting baselines in fishermen's memories [95], the work by Palomares and colleagues on the establishment of abundance data from historical narratives [58], and the application of a fuzzy logic approach to data gathered from interviews [96], the methodological progress seems to have slowed, although the value of oral history has been clearly demonstrated [59].

An important aspect which certainly deserves further attention is the creation of quantitative data. Qualitative information is typically rich and insightful but difficult to control for selectivity and representativity, a problem at the very heart of the historical discipline which emphasises the critical work of uncovering problems of bias. A basic social science approach – the coding of qualitative data – can provide an important contribution in this respect. Coding comprises of the search for emerging topics and key words in documents or interview transcriptions, and their description with a unique code. This allows, for example, quantitative frequency analysis of topics.

In recent years, discourse analysis, meta-analysis and digital representation have provided historians with a wide range of new methodological and technical tools to grapple with vastly increased masses of data. "Big Data" tools which organise masses of data for geographical, temporal and discursive analysis has the potential to bring historical interpretation to a new level of breadth and precision. Marine environmental history has not yet fully embraced the potential of such new methodologies. In particular there is great potential in the application of database and geographical information tools for the study of disparate oral and documentary data.

5.6 Using a gender lens

Marine resource exploitation in general, and fisheries in particular, are often perceived to be a male domain [97]. This is largely the result of the socially constructed roles of men and women within societies: while men are typically regarded as providers, women take care of the home and family. It has also to do with the traditional perspective that fishing refers to the catching of fish with specific gears, such as lines and nets. Gleaning from shorelines and reefs has rarely been acknowledged as fishing [98]. Additionally, it has often been assumed that fisheries largely operate in the public domain, a usually male dominated sphere. The rather female dominated private domain is mostly not in the focus of attention. But especially in subsistence and other small-scale fisheries, much of the administration and logistics including financial issues, as well as the processing happens in the household and through family networks. An analytical focus on public and formal practices misses women's roles as well as a considerable part of what it takes to organize and put into practice a fishing operation [99].

This has two major implications: ignoring the role of woman in fisheries can lead to a substantial underestimation of fishing pressure, especially in coastal areas. It also leads to an underestimation of the social and economic contributions that women provide in fisheries, especially in processing and other value-adding activities [100]. A gender lens can increase understanding of the history of marine resource exploitation. Differentiating roles, responsibilities, access and opportunities of men and woman will provide a more complete picture about access to and control of marine resources [101].

Since the first consolidated publication on women in fisheries by Nadal-Klein and Davis [102], the literature has increased rapidly. While some of these publications have focussed on the role of women in marine resource collection [103,104], pre- and post-harvesting activities [105], or their role in governing and managing marine resources [106,107], others are rather conceptual contributions such as Yodanis's work on the social construction of gender in fishing communities [108]. Recent work has also high lightened women's in depth knowledge on species and environmental changes. For example, oral histories in New Zealand indicate the important role that women, often accompanied by children, once played in gathering seafood from shallow, easily accessible shores sixty to seventy years ago [109]. Women's knowledge confirmed that fish and shellfish were very abundant and could be reliably caught or gathered with little effort.

Another line of work has looked at masculinity and its linkages to fishing. For example, Fabinyi [110] has described how illegal fishing activities in the Philippines provide young men with a higher social status, while Allison has argued that physical settings and the distinct culture of fishing societies shape similar "marine masculinities" in fishing communities in different regions of the world [111]. In a recent publication, Schwerdtner Máñez & Pawelussen [99] call for a gendered perspective on the different roles of both men and women in order to better understand the history of marine resource exploitation over time.

Coverage of Historical Analyses

Under the auspices of HMAP, twelve regional and three species-based projects were developed between 2000 and 2012. These uncovered a wealth of historical information that has been published in more than 200 peer-reviewed scientific publications. However, even such a concerted effort could by no means attain a global coverage in the investigation of humankind's interactions with the sea. To understand the history of human interactions with life in the oceans on global and smaller scales historical research needs to be extended to new geographical areas.

Costello and colleagues have identified a number of areas that received little or no attention during the Census of Marine Life, such as the southern and eastern Mediterranean Sea; estuaries, coastal areas and coral reefs of the Indian Ocean; large regions of South America and the Indian Ocean; and many other habitats including ice-bound waters, ocean trenches, and some deep sea areas [49]. In terms of historical investigations, this list of gaps is even longer.

South America is one of the regions which have so far received very little attention. Although there are an increasing number of studies on the environmental history of the continent, these have had an almost exclusively terrestrial focus. One of the noteworthy exceptions is the work by Perri, who examined the Spanish overexploitation of the oyster beds around the island of Cubagua in what is today Venezulea [112]. Other work covers the ecological history of the Peruvian guano industry from 1800–1973, and its relation to the development of world's largest industrial fishery, that of anchoveta [113]. More recently, Eddy et al. have studied the change in lobster biomass over 400 years of human fishing activity on the Chilean Juan Fernández archipelago [114]. But the vast majority of the continent's coasts and marine areas remain untouched by historical studies.

For the African continent, the situation is similar. An exception is South Africa, where a number historical studies have been undertaken, including work on long-term trends on commercially exploited species [115,116], and work on the history of harbors and fisheries [117]. In an overview of the history of marine and estuarine studies in South Africa, the author argued that South Africa has some of the best data in the world on the biodiversity of rocky shores, coastal waters and estuaries exists, as well as large collections of many taxa [118]. Clearly, further historical work could take advantage of these rich collections. There are also some historical studies on the island of Rodriguez, including research on shifting baselines [119,120]. But for the rest of the continent, we are not aware of any studies in marine environmental history or historical ecology.

The polar regions certainly represent another region with a rich history in marine resource exploitation: whaling, seal hunting, and search for fish have brought people for centuries into these areas. The history of whaling has received some attention, for example in a paper by Hacquebord who analyzed the impacts of eliminating thousands of whales and walruses on other marine mammals [121]. Other studies have assessed ecological changes or population trends resulting from the mass removals of pinnipeds and whales [122]. Whaling records have also been used to establish declines in Antarctic sea-ice extent [123], and the influence of environmental factors on whaling operations has also been studied [124]. However, much research could still be done in these regions to expand our understanding of how they have been changed by resource extraction over time.

An area for which we have very little information is the third dimension of the oceans: the deep sea [125]. Whereas the coastal areas of most developed countries have been exploited for several centuries, the deep sea waters were little known by fishermen and marine scientists alike just 100 years ago. In the meantime deep sea investigations such as the Dutch Snelllius Expedition of the 1920s, and multiple Danish and Norwegian led deep sea investigations traversed the Atlantic, Indian and Pacific Oceans from the turn of the 20th century until the 1960s. *The Carlsberg Foundation's Oceanographic Expedition Round the World* from 1928–30 is an example of how global biogeography was charted well in advance of real impacts from resource exploitation in the deep seas of the World's oceans [126]. Today many of these areas are exploited for multiple marine resources. Such early endeavours deserve careful scrutiny to provide reference points for present day ocean management.

Another focus of attention should be on historically data-rich areas. For example, much documented evidence is available for large parts of maritime Southeast Asia, including the coral triangle. This region is the centre of global marine biodiversity, and the exploitation and trade of its marine resources has at least partly been documented by its former colonial regimes. Especially for the former Dutch East Indies, but also for what today is Papua New Guinea, a wealth of historical material has been collected in Dutch and German libraries and archives, and its processing has just begun [127]. Japan and China are also countries with a long tradition of record keeping, and strong connections to the sea. Working with these data might also improve our understanding of early drivers of resource exploitation, which created networks and trade routes that are still important today [41].

Outlook

Archaeological and historical research has still much to contribute to an improved understanding of the development of marine and coastal ecosystems, and their interactions with humans over time. It helps not only to identify patterns and trends, but also to quantify amounts of living marine resource exploitation. While the aim of this paper has been to establish a new research agenda, we would also like to point out that new institutional arrangements are required in order to successfully implement it. To complement individual projects, which end when their funding runs out, we suggest the establishment of a global research network for marine historical research: the Oceans Past Initiative (OPI). This could become a venue where researchers worldwide interested in historical studies could meet and discuss relevant issues virtually. OPI could provide an umbrella, under which already completed, currently running, and also planned projects and initiatives could be linked, and make their results available for decision-makers and the interested public. Such a network could also commit itself to coordinate resources and provide useful information to the research community, such as by circulating news about grant calls and funding opportunities, information on recent papers in the field, announce successful projects, and distribute information for students on courses related to marine historical research. A first major step into this direction has been undertaken through the establishment of the Oceans Past Platform (OPP), a COST action supported by the European Union which aims to measure and understand the significance and value to societies of living marine resource extraction and production to help shape the future of coasts and oceans.

Acknowledgments

We would like to thank Joseph Christensen, Meryl Williams, James Barrett and an anonymous reviewer, who provided helpful comments on earlier versions of this paper.

Author Contributions

Analyzed the data: KSM PH LB MC AMD H BP MT. Contributed reagents/materials/analysis tools: KSM PH LB MC AMD HO BP MT. Contributed to the writing of the manuscript: KSM PH LB MC AMD HO BP MT.

References

1. Pauly D (1995) Anecdotes and the shifting baseline syndrome of fisheries. Trends Ecol Evol 10: 430.
2. Holm P, Marboe AH, Poulsen B, MacKenzie BR (2010) Marine Animal Populations: A New Look Back in Time. Life in the World's Oceans. Wiley-Blackwell. 1–24.
3. Finney BP, Alheit J, Emeis K-C, Field DB, Gutiérrez D, et al. (2010) Paleoecological studies on variability in marine fish populations: a long-term perspective on the impacts of climatic change on marine ecosystems. J Mar Syst 79: 316–326.
4. Halpern BS, Walbridge S, Selkoe KA, Kappel CV, Micheli F, et al. (2008) A global map of human impact on marine ecosystems. Science 319: 948–952.
5. Carpenter SR, Bennett EM, Peterson GD (2006) Editorial: Special Feature on Scenarios for Ecosystem Services. Ecol Soc 11: 32.
6. Menge BA (2000) Top-down and bottom-up community regulation in marine rocky intertidal habitats. J Exp Mar Bio Ecol 250: 257–289.
7. Seitz RD, Lipcius RN (2001) Variation in top-down and bottom-up control of marine bivalves at differing spatial scales. ICES J. Mar. Sci 58: 689–699.
8. Philippart CM, Beukema J, Cadée G, Dekker R, Goedhart P, et al. (2007) Impacts of Nutrient Reduction on Coastal Communities. Ecosystems 10: 96–119.
9. Zeldis J, Howard-Williams C, Carter CM, Schiel DR (2008) ENSO and riverine control of nutrient loading, phytoplankton biomass and mussel aquaculture yield in Pelorus Sound, New Zealand. Mar Ecol Prog Ser 371: 131–142.
10. Powell Jr AM, Xu J (2011) Abrupt climate regime shifts, their potential forcing and fisheries impacts. Atmospheric and Climatic Sciences 1: 33–47.
11. Nelson GC, Bennett E, Berhe AA, Cassman K, DeFries R, et al. (2006) Anthropogenic Drivers of Ecosystem Change: an Overview. Ecol Soc 11: 29.
12. Campana SE, Gagne JA, MacLaren JW (1995) Elemental fingerprinting of fish otoliths using ID-ICPMS. Mar Ecol Prog Ser 122: 115–120.
13. Roelke LA, Cifuentes LA (1997) Use of stable isotopes to assess groups of king mackerel, Scomberomorus cavalla, in the Gulf of Mexico and southeastern Florida. Fisheries Bulletin 95: 540–551.
14. Campana SE, Thorrold SR (2001) Otoliths, increments and elements: keys to a comprehensive understanding of fish populations? Can J Fish Aquat Sci 58: 30–38.
15. Orton DC (2014) Archaeology as a tool for understanding past marine resource use and its impact. In: K. Schwerdtner Máñez and B Poulsen, editors. Perspectives on Oceans Past: A Handbook of Marine Environmental History. Springer Publishers. (In press).
16. Hufthammer AK, Hoie H, Folkvord A, Geffen AJ, Andersson C, et al. (2010) Seasonality of human site occupation based on stable oxygen isotope ratios of cod otoliths. J Archaeol Sci 37: 78–83.

17. Orton DC, Morris J, Locker A, Barrett JH (2014) Fish for the city: meta-analysis of archaeological cod remains and the growth of London's northern trade. Antiquity 88: 516–530.
18. Behrenfeld MJ, O'Malley RT, Siegel DA, McClain CR, Sarmiento JL, et al. (2006) Climate-driven trends in contemporary ocean productivity. Nature 444: 752–755.
19. Southward AJ, Boalch GT B, Maddock L (1998) Fluctuations in the herring and pilchard fisheries of Devon and Cornwall linked to change in climate since the 16th century. J Mar Biol Assoc U.K. 68: 423–445.
20. Alheit J, Hagen E (1997) Long-term climate forcing of European herring and sardine populations. Fish Oceanogr 6: 130–139.
21. Conversi A, Fonda Umani S, Peluso T, Molinero JC, Santojanni A, et al. (2010) The Mediterranean Sea Regime Shift at the End of the 1980s, and Intriguing Parallelisms with Other European Basins. PLoS ONE 5: e10633.
22. Klais R, Tamminen T, Kremp A, Spilling K, Olli K (2010) Decadal-Scale Changes of Dinoflagellates and Diatoms in the Anomalous Baltic Sea Spring Bloom. PloS ONE 6: e21567.
23. Drinkwater KF (2006) The regime shift of the 1920s and 1930s in the North Atlantic. Prog Oceanogr 68: 134–151.
24. Österblom H, Hansson S, Larsson U, Hjerne O, Wulff F, et al. (2007) Human-induced Trophic Cascades and Ecological Regime Shifts in the Baltic Sea. Ecosystems 10: 877–889.
25. Blamey LK, Howard JAE, Agenbag J, Jarre A (2012) Regime-shifts in the southern Benguela shelf and inshore region. Prog Oceanogr 106: 80–95.
26. Lybolt M, Neil D, Zhao J, Feng Y, Yu K-F, et al. (2010) Instability in a marginal coral reef: the shift from natural variability to a human-dominated seascape. Front Ecol Environ 9: 154–160.
27. Roff G, Clark TR, Reymond CE, Zhao J-X, Feng Y, et al. (2013) Palaeoecological evidence of a historical collapse of corals at Pelorus Island, inshore Great Barrier Reef, following European settlement. Proc Biol Sci 280: 10.1098/rspb.2012.2100.
28. Ojaveer H, Mackenzie BR (2007) Historical development of fisheries in northern Europe–Reconstructing chronology of interactions between nature and man. Fisheries Research 87: 102–105.
29. Erlandson JM, Rick TC (2008) Archeology, marine ecology and human impacts on marine environments. In: T. C Rick and J. M Erlandson, editors. Human impacts on ancient marine ecosystems. Berkely and Los Angeles, California: University of California Press. 1–19.
30. Bailey G, Barrett JH, Craig O, Milner N (2008) Historical ecology of the North Sea basin. In: T. C Rick and J. M Erlandson, editors. Human impacts on ancient marine ecosystems. Berkely and Los Angeles, California: University of California Press. 215–242.
31. Marean CW, Bar-Matthews M, Bernatchez J, Fisher E, Goldberg P, et al. (2007) Early human use of marine resources and pigment in South Africa during the Middle Pleistocene. Nature 449: 905–908.
32. Cortés-Sánchez M, Morales-Muñiz A, Simón-Vallejo MD, Lozano-Francisco MC, Vera-Peláez JL, et al. (2011) Earliest Known Use of Marine Resources by Neanderthals. PLoS ONE 6: e24026.
33. Rick TC, Erlandson JM, Braje TJ, Estes JA, Graham MH, et al. (2008) Historical ecology and human impacts on coastal ecosystems of the Santa Barbara Channel region, California. In: T. C Rick and J. M Erlandson, editors. Human impacts on ancient marine ecosystems: a global perspective. University of California Press. 77–101.
34. Lotze HK (2007) Rise and Fall of fishing and marine resource use in the Wadden Sea, southern North Sea. Fisheries Research 87: 208–218.
35. Halpern BS, Walbridge S, Selkoe KA, Kappel CV, Micheli F, et al. (2008) A global map of human impacts on marine ecosystems. Science 319: 948–952.
36. Barnes DK, Galgani F, Thompson RC, Barlaz M (2009) Accumulation and fragmentation of plastic debris in global environments. Philos Trans R Soc Lond B Biol Sci 364: 1985–1998.
37. Rockström J, Steffen W, Noone K, Persson A, Chapin FS, et al. (2009) A safe operating space for humanity. Nature 461: 472–475.
38. Butcher JG (2004) The Closing of the Frontier: A History of the Marine Fisheries of Southeast Asia c. 1850–2000. Leiden: KITLV Press.
39. Lotze HK, Worm B (2009) Historical baselines for large marine animals. Trends Ecol Evol 24: 254–262.
40. Richards JF (2006) The Unending Frontier. An Environmental History of the Early Modern World. London: University of California Press.
41. Schwerdtner Máñez K, Ferse SCA (2010) The history of Makassan trepang fishing and trade. PLoS ONE 5.
42. Monte-Luna PD, Lluch-Belda D, Serviere-Zaragoza E, Carmona R, Reyes-Bonilla H, et al. (2007) Marine extinctions revisited. Fish Fish 8: 107–122.
43. Dulvy NK, Sadovy Y, Reynolds JD (2003) Extinction vulnerability in marine populations. Fish Fish 4: 25–64.
44. Lotze HK, Lenihan HS, Bourque BJ, Bradbury RH, Cooke RG, et al. (2006) Depletion, degradation, and recovery potential of estuaries and coastal seas. Science 312: 1806–1809.
45. Lotze HK, Coll M, Magera MA, Ward-Paige C, Airoldi L (2011) Recovery of marine animal populations and ecosystems. Trends Ecol Evol 26: 595–605.
46. Jackson JBC, Kirby MX, Berger WH, Bjorndal KA, Botsford LW, et al. (2001) Historical Overfishing and the Recent Collapse of Coastal Ecosystems. Science 293: 629–638.
47. Baum JK, Myers RA, Kehler DG, Worm B, Harley SJ, et al. (2003) Collapse and conservation of shark populations in the Northwest Atlantic. Science 299: 389–392.
48. Myers RA, Worm B (2003) Rapid worldwide depletion on predatory fish communities. Nature 423: 280–283.
49. Costello MJ, Coll M, Danovaro R, Halpin P, Ojaveer H, et al. (2010) A census of marine biodiversity knowledge, resources and future challenges PLoS ONE 5: e12110.
50. Perkol-Finkel S, Airoldi L(2010) Loss and Recovery Potential of Marine Habitats: An Experimental Study of Factors Maintaining Resilience in Subtidal Algal Forests at the Adriatic Sea. PLoS ONE 5: e10791.
51. Thurstan RH, Roberts CM (2010) Ecological Meltdown in the Firth of Clyde, Scotland: Two Centuries of Change in a Coastal Marine Ecosystem. PLoS ONE 5: e11767.
52. Lalas C (2008) Recolonialisation of Otago, southern New Zealand, by fur seals and sea lions: unexpected patterns and consequences. In: B Clarkson, P Kurian, T Nachowitz and H Rennie, editors. Conservation-Vision Conference. Hamilton: University of Waikato.
53. Chilvers BL, Wilkinson IS (2008) Philopatry and site fidelity of New Zealand sea lions (*Phocarctos hookeri*). Wildlife Research 35: 463–470.
54. Pikitch EK, Santora C, Babcock EA, Bakun A, Bonfil R, et al. (2004) Ecosystem-Based Fishery Management. Science 305: 346–347.
55. FAO Fisheries Department(2003) The ecosystem approach to fisheries. FAO Technical Guidelines for Responsible Fisheries. Rome.
56. Kittinger JN, Pandolfi JM, Blodgett JH, Hunt TL, Jiang H, et al. (2011) Historical reconstruction reveals recovery in Hawaiian coral reefs. PLoS ONE 6: e25460.
57. Neubauer P, Jensen OP, Hutchings JA, Baum JK (2013) Resilience and Recovery of Overexploited Marine Populations. Science 340: 347–349.
58. Palomares MLD, Mohammed E, Pauly D (2006) Maria Lourdes D. Palomares, Elizabeth Mohammed, and Daniel Pauly on European expeditions as a source of historic abundance data on marine organisms: A case study of the Falkland Islands. Environ Hist 11: 835–847.
59. Coll M, Carreras M, Ciércoles C, Cornax M-J, Gorelli G, et al. (2014) Assessing Fishing and Marine Biodiversity Changes Using Fishers' Perceptions: The Spanish Mediterranean and Gulf of Cadiz Case Study. PLoS ONE 9: e85670.
60. Olsen SB, Page GG, Ochoa E (2009) The Analysis of Governance Responses to Ecosystem Change: A Handbook for Assembling a Baseline. LOICZ Reports & Studies. Geesthacht: GKSS Research Center.
61. Poulsen B, Holm P, MacKenzie BR (2007) A long-term (1667–1860) perspective on impacts of fishing and environmental variability on fisheries for herring, eel, and whitefish in the Limfjord, Denmark. Fish Res 87: 181–195.
62. Zerner C (1994) Through a green lense: The construction of customary environmental law in Indonesia's Maluku Islands. Law Soc Rev 28: 1079–1121.
63. Cinner JE, Basurto X, Fidelman P, Kuange J, Lahari R, et al. (2012) Institutional designs of customary fisheries management arrangements in Indonesia, Papua New Guinea, and Mexico. Marine Policy 36: 278–285.
64. Gahzoul J (2007) Recognising the complexities of ecosystem management and the ecosystem service concept. Gaia 16: 215–221.
65. Worm B, Barbier EB, Beaumont N, Duffy JE, Folke C, et al. (2006) Impacts of Biodiversity Loss on Ocean Ecosystem Services. Science 314: 787–790.
66. MacDiarmid AB, Law CS, Pinkerton M and Zeldis J (2013) New Zealand marine ecosystem services. In: J. R Dymond, editor editors. Ecosystem services in New Zealand – conditions and trends. Lincoln, New Zealand: Manaaki Whenua Press.
67. Smith I (2005) Retreat and resilience: fur seals and human settlement in New Zealand. In: G Monks, editor editors. The exploitation and cultural importance of sea mammals. Cambridge: Oxbow Books. 6–18.
68. Caroll EL (2006) The demographic and genetic bottleneck of the New Zealand southern right whale. School of Biological Sciences. The University of Auckland, New Zealand.
69. Nicol S, Bowie A, Jarman S, Lannuzel D, Meiners KM, et al. (2010) Southern ocean iron fertilisation by baleen whales and Antarctic krill. Fish Fish 11: 203–209.
70. Coll M, Lotze HK (forthcoming) Ecological indicators and food-web models as tools to study historical changes in marine ecosystems In: K. Schwerdtner Máñez and B. Poulsen, editors. Perspectives on Oceans Past: A Handbook of Marine Environmental History. Amsterdam: Springer Publishers.
71. Broughton JM (1997) Widening diet breadth declining foraging efficiency and prehistoric harvest pressure: ichthyofaunal evidence from the Emeryville Shellmound, California. Antiguity 71: 845–862.
72. Broughton JM (2002) Prey spatial structure and behavior affect archaeological tests of optimal foraging models: examples from the Emeryville Shellmound vertebrate fauna. World Archeol 34: 60–83.
73. Josephson E, Smith TD, Reeves RR (2008) Depletion within a decade: The American 19th-century North Pacific right whale fishery. In: D. J Starkey, P Holm and M Barnard, editors. Oceans Past: Management insights from the history of marine animal populations. London: Earthscan Research Edition. 133–147.
74. Ferretti F, Myers RA, Serena F, Lotze HK (2008) Loss of large predatory sharks from the Mediterranean Sea. Conserv Biol 22: 952–964.

75. Poulsen B (2008) Dutch Herring: An Environmental History, c. 1600–1860. Amsterdam: Aksant.
76. Glaser M, Christie P, Diele K, Dsikowitzky L, Ferse S, et al. (2012) Measuring and understanding sustainability-enhancing processes in tropical coastal and marine social–ecological systems. Curr Opin Environ Sustain 4: 300–308.
77. Christensen V, Walters C (2004) Ecopath with Ecosim: methods, capabilities and limitations. Ecol Modell 72: 109–139.
78. Dambacher JM, Li HW, Rossignol PA (2003) Qualitative predictions in model ecosystems. Ecol Modell 161: 79–93.
79. Fulton EA (2010) Approaches to end-to-end ecosystem models. J Mar Syst 81: 171–183.
80. Plagányi ÉE, Punt AE, Hillary R, Morello EB, Thébaud O, et al. (2012) Multispecies fisheries management and conservation: tactical applications using models of intermediate complexity. Fish Fish.
81. Williams RJ, Martinez ND (2000) Simple Rules Yield Complex Food Webs. Nature 404: 180–183.
82. Steneck RS, Vavrinec J and Leland AV (2004) Accelerating Trophic-Level Dysfunction in Kelp Forest Ecosystems of the Western North Atlantic. Ecosystems 7: 323–332.
83. Pauly D, Christensen V, Daalsgard J, Froese R, Torres FC (1998) Fishing down marine food webs. Science 279: 860–863.
84. Lotze HK, Coll M, Dunne J (2011) Historical changes in marine resources, food-web structure and ecosystem functioning in the Adriatic Sea. Ecosystems 14: 198–222.
85. Pitcher TJ (2005) Back-to-the-future: a fresh policy initiative for fisheries and a restoration ecology for ocean ecosystems. Philos Trans R Soc Lond B Biol Sci 360: 107–121.
86. Pitcher T, Heymans JJ, Vasconcellos M (2002) Ecosystem models of Newfoundland for the time periods 1995, 1985, 1900, 1450. Fisheries Centre Research Report.
87. Ainsworth C, Pitcher T, Heymans J, Vasconcellos M (2008) Reconstructing historical marine ecosystems using food web models: Northern British Columbia from Pre-European contact to present. Ecol Modell 216: 354–368.
88. Hauser L, Carvalho GR (2008) Paradigm shifts in marine fisheries genetics: ugly hypotheses slain by beautiful facts. Fish Fish 9: 333–362.
89. Roman J, Palumbi SR (2003) Whales Before Whaling in the North Atlantic. Science 301: 508–510.
90. Palumbi SR (2011) Whales, logbooks, and DNA. In: J. B. C Jackson, K Alexander and E Sala, editors. Shifting baselines: the past and the future of ocean fisheries. Washington DC: Island Press. 163–174.
91. Alter SE, Rynes E, Palumbi SR (2007) DNA evidence for historic population size and past ecosystem impacts of gray whales. Proc Natl Acad Sci U S A 104: 15162–15167.
92. Fontaine MC, Snirc A, Frantzis A, Koutrakis E, Öztürk B, et al. (2012) History of expansion and anthropogenic collapse in a top marine predator of the Black Sea estimated from genetic data. Proc Natl Acad Sci 109: E2569–E2576.
93. Thurstan RH, Buckley SM (2014) Oral histories: informing natural resource management using perceptions of the past. In: K. Schwerdtner Máñez and B Poulsen, editors. Perspectives on Oceans Past: A Handbook of Marine Environmental History. Amsterdam: Springer Publishers. (In Press).
94. Johannes RE, Neis B (2007) The value of anecdote. In: N Haggan, B Neis and I. G Baird, editors. Fishers Knowledge in Fisheries Science and Management. Paris: UNESCO. 41–58.
95. Saenz-Arroyo A, Roberts CM, Torre J, Carino-Olvera M, Enriquez-Andrade RR (2005) Rapidly shifting environmental baselines among fishers of the Gulf of California. Proc R Soc Lond Ser B-Biol Sci 272: 1957–1962.
96. Ainsworth CH, Pitcher TJ, Rotinsulu C (2008) Evidence of fishery depletions and shifting cognitive baselines in Eastern Indonesia. Biol Conserv 141: 848–859.
97. Choos PS, Nowak BS, Kusakabe K, Williams MJ (2008) Guest Editorial: Gender and Fisheries. Development 51: 176–179.
98. Kleiber D, Harris LM, Vincent ACJ (2014) Improving fisheries estimates by including women's catch in the Central Philippines. Can J Fish Aquat Sci: 1–9.
99. Schwerdtner Máñez K, Pawelussen A (2014) Fish is women's business too – looking at marine resource use through a gender lens. In: K. Schwerdtner Mánez and B Poulsen, editors. Perspectives on Oceans Past: A Handbook of Marine Environmental History. Springer Publishers.
100. Harper S, Zeller D, Hauzer M, Pauly D, Sumaila UR (2013) Women and fisheries: Contribution to food security and local economies. Marine Policy 39: 56–63.
101. Williams MJ (2008) Why look at fisheries through a gender lens? Development 51: 180–185.
102. Nadal-Klein J, Davis DL (1988) To work and to weep. In: I. o. S. a. E Research and M. U. o Newfoundland, editors. Social and Economic Papers. St. John's, Newfoundland: Institute of Social and Economic Research.
103. Hauzer M, Dearden P, Murray G (2013) The fisherwomen of Ngazidja island, Comoros: Fisheries livelihoods, impacts, and implications for management. Fish Res 140: 28–35.
104. Vunisea A (2005) Womens changing roles in the subsistence fishing sector in Fidji. In: I Novaczek, J Mitchell and J Vietayaki, editors. Pacific Voices: Equity and Sustainability in Pacific Island Fisheries. Suva, Fiji: Institute of Pacific Studies, University of South Pacific. 89–105.
105. Bennett E (2005) Gender, fisheries and development. Marine Policy 29: 451–459.
106. McCay B (1993) Fisherwomen, fisheries policy, and maritime anthropology. Reviews in Antropology 22: 77–89.
107. Meltzoff SK (1995) Marisquadoras of the Shellfish Revolution: The Rise of Women in Co-management on Illa de Arousa, Galicia. Journal of Political Ecology 2: 20–38.
108. Yodanis CL (2000) Constructing gender and occupational segregation: a study of women and work in fishing communities. Qual Sociol 23: 267–290.
109. Maxwell K, MacDiarmid AB (2014) Oral histories of marine fish and shellfish state and use in the Hauraki Gulf and along the Otago-Catlins coast 1940–2008. New Zealand Aquatic Environment and Biodiversity Report. (In press).
110. Fabinyi M (2007) Illegal fishing and masculinity in the Philippines. A look at the Calamianes Islands in Palawan. Philipp Stud 55: 509–529.
111. Allison EH (2013) Maritime Masculinities – and why they matter for management. People and the Sea VII. Amsterdam.
112. Perri M (2009) Ruined and Lost': Spanish Destruction of the Pearl Coast in the Early Sixteenth Century. Environ Hist Camb 15: 129–161.
113. Cushmann GT (2003) The lords of guano: Science and the management of Peru's marine environment, 1800–1973. PhD thesis Austin: University of Texas.
114. Eddy TD, Gardner JPA, Pérez-Matus A (2010) Applying Fishers' Ecological Knowledge to Construct Past and Future Lobster Stocks in the Juan Fernández Archipelago, Chile. PLoS ONE 5: e13670.
115. Griffiths MH (2000) Long-term trends in catch and effort of commercial linefish off South Africa's Cape Province: snapshots of the 20th century. Afr. J. Mar. Sci. 22: 81–110.
116. Attwood CG, Farquhar M (1999) Collapse of linefish stocks between Cape Hangklip and Walker Bay, South Africa. Afr. J. Mar. Sci. 31: 415–432.
117. McKenzie MJ (1946) The historical development of harbours and fisheries in South Africa.
118. Day JA (2000) Marine and estuarine studies in South Africa: a historical perspective. Transactions of the Royal Society of South Africa 55: 101–105.
119. Bunce M, Rodwell LD, Gibb R, Mee L (2008) Shifting baselines in fishers' perceptions of island reef fishery degradation. Ocean Coastal Manage 51: 285–302.
120. Oliver PG, Lynch T (2004) A historical perspective of the marine biota of Rodrigues. Journal of Natural History 38: 2927–2935.
121. Hacquebord L (2001) Three centuries of whaling and walrus hunting in Svalbard and its impact on the Arctic ecosystem. Environ Hist Camb: 169–185.
122. Emslie SD, Patterson WP (2007) Abrupt recent shift in δ13C and δ15N values in Adélie penguin eggshell in Antarctica. Proc Natl Acad Sci U S A 104: 11666–11669.
123. William K (1997) Abrupt mid-twentieth-century decline in Antarctic sea-ice extent from whaling records. Nature 389: 57–60.
124. Sanger CW (1991) Environmental factors affecting 17th–19th century whaling in the Greenland Sea. Polar Rec 27: 77–86.
125. Ramirez-Llodra E, Brandt A, Danovaro R, De Mol B, Escobar E, et al. (2010) Deep, diverse and definitely different: unique attributes of the world's largest ecosystem. Biogeosciences 7: 2851–2899.
126. Tåning ÅV (1929) The Danish Circumnavigation Expedition. Science 69: 191–194.
127. Boomgaard P (2005) Resources and people of the sea in and around the Indonesian Archipelago, 900–1900 In: P. Boomgaard and D. Henley, editors. Muddied Waters Historical and contemporary perspectives on management of forest and fisheries in island Southeast Asia Leiden: KITLV Press Koninklijk Instituut voor Taal-, Land- en Volkenkunde. 98–119.

22

Detection of Salt Marsh Vegetation Stress and Recovery after the Deepwater Horizon Oil Spill in Barataria Bay, Gulf of Mexico Using AVIRIS Data

Shruti Khanna[1]*, Maria J. Santos[2], Susan L. Ustin[1], Alexander Koltunov[1], Raymond F. Kokaly[3], Dar A. Roberts[4]

1 Center for Spatial Technology and Remote Sensing (CSTARS), Department of Land, Air and Water Resources, University of California Davis, Davis, California, United States of America, **2** Center for Spatial and Textual Analysis (CESTA), Stanford University, Stanford, California, United States of America, **3** United States Geological Survey, Denver, Colorado, United States of America, **4** Department of Geography, University of California Santa Barbara, Santa Barbara, California, United States of America

Abstract

The British Petroleum Deepwater Horizon Oil Spill in the Gulf of Mexico was the biggest oil spill in US history. To assess the impact of the oil spill on the saltmarsh plant community, we examined Advanced Visible Infrared Imaging Spectrometer (AVIRIS) data flown over Barataria Bay, Louisiana in September 2010 and August 2011. Oil contamination was mapped using oil absorption features in pixel spectra and used to examine impact of oil along the oiled shorelines. Results showed that vegetation stress was restricted to the tidal zone extending 14 m inland from the shoreline in September 2010. Four indexes of plant stress and three indexes of canopy water content all consistently showed that stress was highest in pixels next to the shoreline and decreased with increasing distance from the shoreline. Index values along the oiled shoreline were significantly lower than those along the oil-free shoreline. Regression of index values with respect to distance from oil showed that in 2011, index values were no longer correlated with proximity to oil suggesting that the marsh was on its way to recovery. Change detection between the two dates showed that areas denuded of vegetation after the oil impact experienced varying degrees of re-vegetation in the following year. This recovery was poorest in the first three pixels adjacent to the shoreline. This study illustrates the usefulness of high spatial resolution airborne imaging spectroscopy to map actual locations where oil from the spill reached the shore and then to assess its impacts on the plant community. We demonstrate that post-oiling trends in terms of plant health and mortality could be detected and monitored, including recovery of these saltmarsh meadows one year after the oil spill.

Editor: Aaron Alain-Jon Golden, Albert Einstein College of Medicine, United States of America

Funding: This work has been partially funded by an NSF Macroecosystems program for a Rapid grant (#1058134) and Chevron Corp. (#201118134). The authors project for Chevron is primarily to provide advice on when to use different types of remote sensing data, mostly to identify land cover types using different types of statistical classification methods. The objective of their funding from Chevron has been to identify cost effective data types and methods of analysis that can be applied under a range of environmental conditions. The funders had no role in study design, data collection and analysis, decision to publish, or preparation of the manuscript.

Competing Interests: The authors have declared that no competing interests exist.

* E-mail: shrkhanna@ucdavis.edu

Introduction

Ecosystems located in regions where oil and gas development occurs are highly susceptible to direct and indirect impacts of petroleum extraction and refining [1]. Coastal ecosystems have been disproportionately affected, because a majority of extraction sites are in coastal areas [2], as are the locations for ship transport and refineries. The Mississippi Delta contains the largest area of coastal wetlands in the United States but also supports one of the most extensive petroleum extraction operations in the world, exposing these wetlands to the impacts of oil contamination since the early 1900s [1]. The British Petroleum Deepwater Horizon (BP-DWH) Oil Spill in the Gulf of Mexico is the biggest coastal oil spill in US history and one of the five largest spills in the world by volume [3]. The impact of the BP-DWH spill on Gulf ecosystems is still under active examination [4,5], but a long history of small and large oil spills in these regions has resulted in numerous studies on both impacts and recovery [6,7,8,9,10].

The salt marshes of Louisiana lost about 6500 hectares of wetland area per year between 1985 to 1989 due to subsidence, erosion, scouring, and other reasons [7]. Oil contamination can increase erosion and salt marsh loss due to the oil-induced plant mortality. The longer the residence time of the oil in the wetland, the greater is the impact of oil and slower is the recovery [2,11,12]. All weights of crude oil are highly toxic to plants through direct impacts on plant metabolism and indirectly through disruption of plant-water relationships, and reduced oxygen exchange between atmosphere and soil [1,7,8,13]. Oil coating the leaves can obstruct or prevent gas exchange [1,8,14]. When leaves are oiled, mortality is much higher, especially in *Spartina* species, which are otherwise relatively resistant to oil impacts [1,7]. Oil also affects the microbial community and nutrient cycling in the soil, which can have a negative impact on plant health [8].

Remote sensing has proven to be a valuable tool for detection and mapping of marine oil spills [15,16,17,18,19,20,21,22]. In terrestrial environments, reliable detection typically requires highly detailed spectral information provided by imaging spectrometers [18,19]. While the presence of oil reduces soil reflectance in the visible part of the electromagnetic spectrum [23], other factors can lead to the same effect, for example, high organic matter content in soil [24]. The oil signal is characterized by absorption features in the near-infrared (NIR) and shortwave-infrared (SWIR) regions, due to overtones and combinations of C-H and C-O vibrational absorptions [20,23,25,26,27,28] (Figure 1). In particular, the NIR 1730 nm and SWIR 2300 nm bands are best suited for detecting oil on bare soil as they do not overlap with absorption features of soil background materials [5,23,27,29]. The 2300 nm absorption feature can be confused with carbonate absorption in soil [30,31], but high salinity soils found in saltmarshes are acidic with very low carbonate content [32]. Both absorption features of oil overlap with cellulose-lignin absorption features found in the spectra of dead or senescent plants, often termed non-photosynthetic vegetation (NPV). However, the 2300 nm absorption has a unique spectral shape compared to NPV spectral features [5,33]. As plants senesce and lose their leaves, the soil fraction in the pixel spectrum increases, and oil under the canopy can be detected.

The physiological changes due to plant stress can be measured by specific changes in the reflectance spectrum. As plants become increasingly stressed, they lose pigments, water, and ultimately leaves, all of which can be measured through changes to their spectral properties [34,35,36,37,38]. Plant stress can be detected by several different broadband and narrowband vegetation indexes that track plant pigment concentration [34,37], water content [8,39,40,41] and plant cover [38,42,43]. The spectrum of healthy photosynthetic vegetation has a characteristic "red edge" where the reflectance rises sharply between 680 nm (red) and 780 nm (near infrared) wavelengths. The red edge is the long wavelength edge of the chlorophyll absorption. The properties of the red edge change rapidly as the plant becomes more stressed [44,45,46,47,48,49,50].

Multiple studies since the BP-DWH spill have mapped the oil spill on the ocean surface [51,52,53,54,55] but very few studies have looked at the impact of this spill on wetland vegetation [4]. Mishra et al. [4] assessed the impact of the BP-DWH spill in East Bird's Foot, Louisiana using field spectra and Landsat imagery. However, the broad spatial and spectral resolution of multi-spectral data prevented the assessment of changes in the narrow absorption features necessary for direct oil detection. Hence the study did not examine the spatial proximity of vegetation stress to oil contamination. Large areas of vegetation "change" mapped with 30 m pixels could be due to other factors, e.g., phenological differences between years. The spectral resolution of spectroscopy data acquired at high spatial resolution is necessary to confirm the presence of oil and to attribute whether observed stress is due to oil or other stress factors [5,44]. The Airborne Visible Infrared Imaging Spectrometer (AVIRIS) measures the full spectrum between 380 to 2500 nm in 224 sequential bands, each approximately 10 nm wide. It has been successfully used to map both oil contamination and vegetation stress [5,44,56].

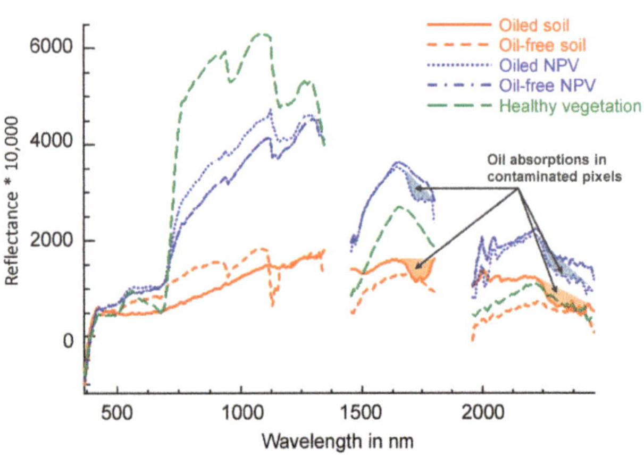

Figure 1. AVIRIS spectra from oiled and oil-free pixels. AVIRIS spectra for oiled and oil-free soil, oiled and oil-free non-photosynthetic vegetation (NPV), and healthy photosynthetic vegetation. Oil absorptions (centered at 1720 nm and 2300 nm) are visible in spectral signatures of oiled pixels. Bad band values are not displayed.

The objective of this study is to determine the impact of the BP-DWH Oil Spill on the tidal salt marshes in Barataria Bay, LA, one of the areas heavily impacted by oil reaching the shoreline. We were interested in assessing: 1) the magnitude of any oil-induced stress on inland vegetation and 2) evidence for recovery or continued decline of contaminated vegetation communities the following year. We analyzed AVIRIS imagery flown over saltmarshes to determine the extent and the level of plant stress in proximity to oil contamination. Band indexes were used to determine the effect of oil on plant stress, as measured by changes in plant water and chlorophyll content. Finally, we used change detection between images collected approximately a month after the oil spill made landfall (September 2010) and one year later (August 2011) to examine whether marsh vegetation continued to decline or recovered from the immediate impact of the oil spill.

Materials and Methods

Field Data Collection

Field work was conducted by the United States Geological Survey (USGS) in Barataria Bay on July 10, 2010 and again on August 12–13, 2010 [5]. At each 2×2 m survey point, vegetation species composition, canopy condition, presence of oil, and penetration of oil into the marsh were noted. Locations of survey points were selected based on access to the marsh through gaps in protective booms. During the August survey, oil detected in a previous image acquired on July 31, 2010 was used to select areas of extensive oiling [5,57]. Forty of the survey points were located in the Bay Jimmy region, which was analyzed for this study.

Data Preprocessing

AVIRIS image data were collected over the oil impacted regions of the Gulf of Mexico coastal wetlands from May to October of 2010 and from May to December of 2011. Flightlines that had all navigation and georectification information available and were at least 70% cloud free were selected for further processing. We focused on Barataria Bay (area 86 km^2; Figure 2) because it was severely impacted by the oil spill. Data collected on two dates at different altitudes were chosen to study the progression of wetland impact and recovery: 1) September 14, 2010 (pixel resolution: 3.5 m), and 2) August 15, 2011 (pixel resolution: 7.7 m). We selected the September 2010 imagery because most of the oil had already come ashore and we hypothesized that the time delay between the arrival of oil and the response of the plant community would have passed. The August 2011 imagery was selected because it was closest in seasonality to the 2010 imagery. We analyzed four flightlines in September 2010 and two flightlines in August 2011 that covered the most heavily impacted region of Barataria Bay (Figure 2). The flightlines were atmospherically corrected using ACORN 6, mode 1.5 (ImSpec LLC, Seattle) to apparent surface reflectance. The September 2010 images were georectified to 1 m National Agricultural

Imagery Program (NAIP) color infrared images collected in 2010. Even though we were planning to compare the September 2010 with the August 2011 imagery, we did not radiometrically calibrate the two images together, because our comparisons were based on index values. Indexes are relatively insensitive to differences in absolute reflectance since they compare relative differences between band values and are frequently normalized.

The AVIRIS image data used in this study are publicly available from the Jet Propulsion Laboratory (JPL) archive. Field data collected by USGS will be made available upon request.

Image co-registration. NASA Jet Propulsion Laboratory (JPL) provides AVIRIS images with georeferencing information derived from inertial navigation data and GPS. Nonetheless, images georeferenced based on this information often suffer from

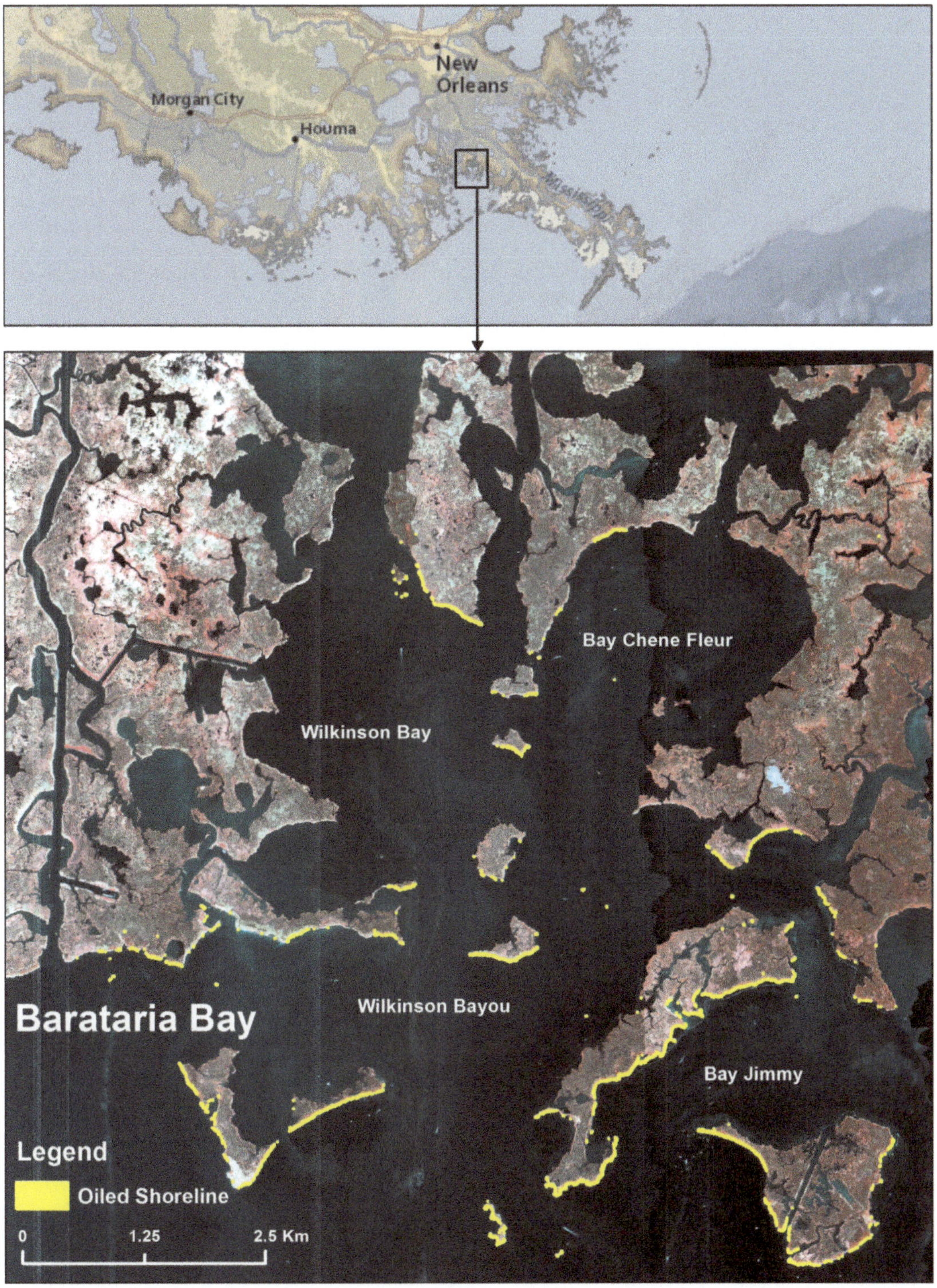

Figure 2. Color-infrared view of study site showing oiled shoreline. Location of Barataria Bay in the Mississippi Delta in the Gulf of Mexico, AVIRIS images of Barataria Bay depicted in color infrared bands with oiled shoreline shown in yellow.

residual misalignment by a few pixels or more. With 3.5 and 7.7 m pixels of AVIRIS imagery used in our study, this misalignment translated into displacements of the order of dozens of meters, whereas surface (including vegetation) conditions near the shore were known to exhibit sharp gradients on the scale of just a few meters. Therefore, we co-registered the AVIRIS images using an automated image registration technique [58] that combines robust band-wise compensation for radiometric differences in images [59] with an iterative gradient-based video-sequence alignment method by Irani [60], under the affine image motion model. The method does *not* change the band values of either image. We resampled images from both dates to 1.1m and subset the two 2011 flightlines to the four 2010 flightlines (half the swath width because pixel size is smaller). Then all four sets of images were coregistered. As a result of the image co-registration, the residual pixel misregistration was markedly reduced, allowing more accurate analysis of the changes in vegetation conditions. The base images were selected to be the September 2010 images and the August 2011 images were co-registered to the base images. Areas of large or systematic change in the scene (e.g., cloud masses, shorelines at different tidal stages, or eroded shorelines) were excluded from the image motion estimation. The tidal stage of the August 2011 image was higher than that of the September 2011 image hence excluding the main shoreline from the coregistration area allowed only within marsh features to be used for coregistration. Thus, the difference in shoreline edge did not impact the accuracy of coregistration.

Landcover classification and oil mapping

Kokaly et al. [5] successfully mapped oil contamination along the shoreline after the BP-DWH using AVIRIS images acquired on July 31, September 14 and October 2 of 2010 over Barataria Bay. They used hydrocarbon absorption features centered at 1720 nm and 2300 nm as inputs to the USGS's Material Identification and Characterization Algorithm (MICA) to map oil presence [5].

To compare inter-annual change for potential recovery or further damage, we needed to do all our analyses on the co-registered images, hence we mapped oil contamination using the continuum removal technique [41] over the two oil absorption features (Figure 1) in four AVIRIS flightlines over Barataria Bay acquired in September 2010, and did not use the oil maps produced by Kokaly et al. [5]. However, we used Kokaly et al. [5] field measured oil penetration data to validate our oil maps. We then compared the oiled coastline between the two maps to quantify the agreement between them.

The images were classified into six classes: water, soil, dry plant material or non-photosynthetic vegetation (NPV), photosynthetic vegetation, oiled soil and oiled NPV (Figure 3). We used a binary decision tree based on vegetation indexes, angle indexes (Table 1), and the depth of the oil absorption continuum removals to produce a classification map for the six classes following the methods in Khanna *et al.* [61]. In this case, the classification map was used to restrict vegetation stress analyses to the land class (i.e., every class except water). Clouds and cloud shadows were excluded from the analyses.

While we had sufficient field data to independently assess the accuracy of our oiled pixels, we did not have enough field data to train and validate our landcover map. To enrich our data set, and since we were mapping classes that are easily discernible by an observer, we collected training and validation data by visually interpreting photosynthetic vegetation, soil, water and NPV from the color infrared images. We randomly divided the September 2010 dataset into training and validation datasets and used the training dataset to train our classifier. Then we applied the same classifier to images from both dates. We assessed the accuracy of the oiled vs. oil-free pixels with data collected in the field using kappa statistics, overall classification accuracy and omission and commission errors [62,63]. Accuracy for all classes was assessed using the visually interpreted validation data by calculating the same metrics.

Comparing oiled vs. oil-free shoreline in September 2010

The boundary of the land class from the September 2010 image classification was used to produce a vector layer of the shoreline. This is because the September 2010 images were acquired at low-tide hence the shoreline generated by using this dataset represents the mean low tide water line. This shoreline was further simplified by deleting small polygons less than 500 m^2 (small depressions within the islands filled with water) resulting in a smoother shoreline. We then calculated a "distance to shore" for every land pixel by computing the Euclidian distance and direction to shore in ArcGIS 10 (ESRI, Redlands, CA). We chose this approach instead of using a pre-existing shoreline vector layer because we wanted to compare the exact distances from the shoreline within our dataset, irrespective of how well the image was geo-registered.

Selection of pixels in oiled and oil-free shoreline zones. We selected pixels that were within 60 m of the oiled shoreline because, based on the USGS field data, we did not expect the oil to impact vegetation further inland. Further, to ensure that the pixels included were located in the section of the shoreline facing the ocean, we selected pixels with a direction perpendicular to the nearest shoreline between 0°–80° or 280°–360° (where 0° is North). This was done because tidal forces act differently on the ocean-facing southern shore compared to the east-west or northern shore. Oiled areas were predominantly located on the southern shoreline since the oil came in with the tide. Finally, only oil-free shorelines within 140 m of an oiled shoreline were used for comparison in order to maximize the similarity of environmental conditions between oiled and oil-free zones, except for the variable of interest, "oil". Any bias would likely be in the direction of including oiled pixels in the oil-free area analysis which would result in minimizing differences between oiled and oil-free areas. Oil-free shoreline inland pixels were included following the same criteria as for oiled shorelines.

Vegetation index trajectories in oiled vs. oil-free shoreline zones. Plant physiological responses to stress often manifest as changes in chlorophyll and water content [64]. We selected seven indexes that track changes in chlorophyll content and/or leaf area of the plant (Normalized Difference Vegetation Index: NDVI, modified NDVI: mNDVI), change in plant condition (green, senescent, dead – Angle at NIR: ANIR, Angle at Red: ARed), and change in plant water content (Normalized Difference Infrared Index: NDII, Water Absorption at 980 nm: ADW1, Water Absorption at 1240 nm: ADW2) (Table 1). No single index is optimal for all conditions, so consistent responses across these diverse indexes would add robustness to our results. To determine the extent of oil-induced stress on interior marsh vegetation, we compared index values along oiled shorelines with those along oil-free shorelines for the September 2010 imagery. We used an analysis of variance to test whether there were significant differences between the indexes in oiled and oil-free shorelines, and then subdivided the data set by distance zones from the shoreline (Figure 3). Zones corresponded to the pixel size of the September 2010 images so that zone 1 was the first 3.5 m pixel next to the shore, zone 2 was the second pixel from the shore, etc. We tested for pair-wise differences in oiled versus oil-free shores in each zone (e.g., oil versus oil-free pixels at zone 1, oil versus oil-free pixels at zone 2, etc.) using a Tukey-HSD test [65]. This test

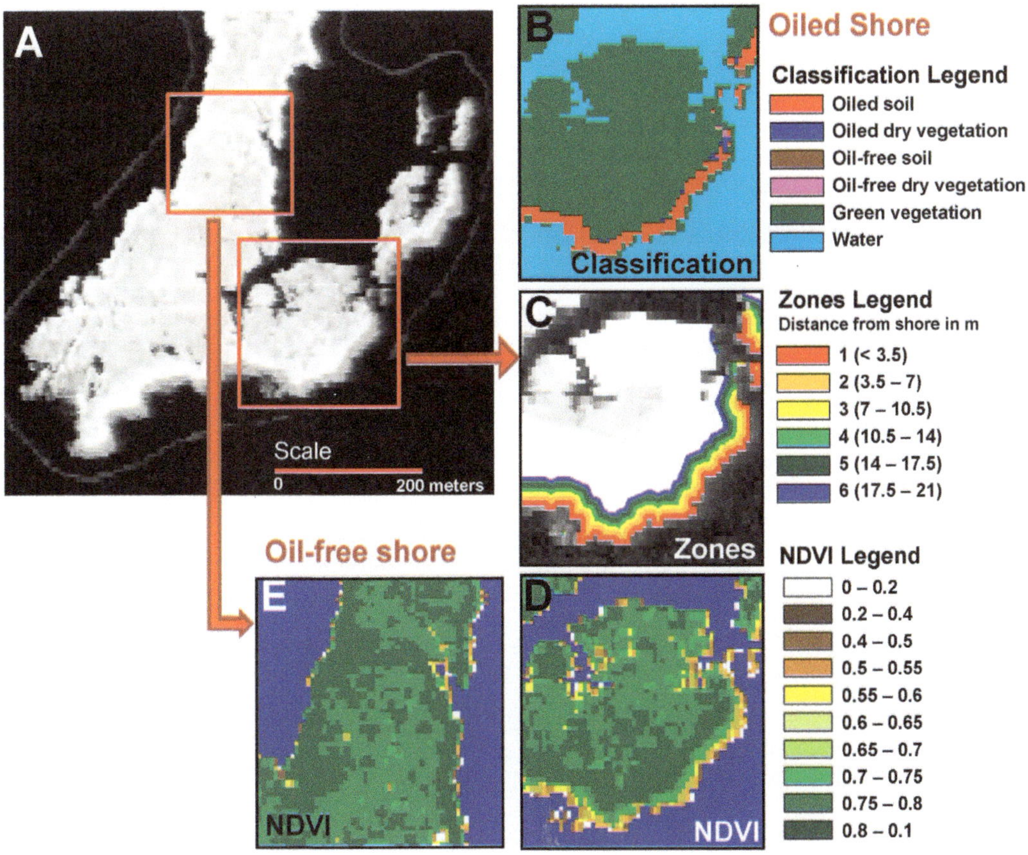

Figure 3. NDVI image, classified image and zones used to analyze oil impact. (a) Gray scale NDVI image of a subset of Barataria Bay, with a close up of an oiled section of the shoreline showing (b) the classified image, (c) the zones as we move away from the shoreline, and (d) the NDVI image showing the low-NDVI band of pixels right next to the oiled shoreline. For comparison, (e) shows the NDVI profile for an oil-free section of the shoreline.

showed the significance of all the possible pair-wise combinations of oiled versus oil-free zone comparisons that went into the overall ANOVA. Pixels were always compared within the same zone. We chose to compare corresponding zones because salt marsh species distributions follow the characteristics of flooding and elevation along shorelines. We observed a similar differentiating pattern in the remote sensing data, from lower index values in the intertidal region to a band of higher index values (robust dense vegetation) just beyond the intertidal and then slightly lower values again in the inner marsh. By grouping data by distance from the shorelines, we attempted to minimize changes in species types and density and their distribution patterns.

We tested the dependent variables, the index images, for autocorrelation between pixels using the Moran's I statistic. We found that the pixels were significantly autocorrelated until a distance of 21 m (Moran's I = 0.48, p-value <<0.001). Despite

Table 1. Vegetation indexes used for stress detection due to oil contamination.

Inputs	Formula	Relevance	References
Normalized Difference Vegetation Index (NDVI)	$\frac{R_{861}-R_{649}}{R_{861}+R_{649}}$	Index of green plant cover and LAI	[42]
Modified NDVI (mNDVI)	$\frac{R_{754}-R_{705}}{R_{754}+R_{705}}$	Sensitive to plant health	[47,48]
Normalized Difference Infrared Index (NDII)	$\frac{R_{861}-R_{1663}}{R_{861}+R_{1663}}$	Sensitive to plant water content	[39,40]
Angle at NIR (ANIR)	λ: 675, 1092, 1662 nm	Angle index sensitive to phase change in plants	[43]
Angle at Red (ARed)	λ: 550, 675, 1092 nm	Angle index sensitive to plant pigments	Adapted from [43]
Water Absorption at 980 nm (ADW1)	$½ * (R_{783} + R_{1054}) - R_{977}$	Quantification of water absorption features	Adapted from [67]
Water Absorption at 1200 nm (ADW2)	$½ * (R_{1054} + R_{1254}) - R_{1187}$	Quantification of water absorption features	Adapted from [67]

this high correlation we decided not to subsample our data to include only independent pixels because of several reasons. We expected the first few pixels along the shoreline to be sensitive to oil and by choosing pixels far enough to be spatially independent, we could have missed detecting the impact of oil altogether. Moreover, the tests we used to determine if there was an impact are known to be robust to violations of the assumption of independence between samples [66]. Finally, a conservative approach can be undertaken to reduce the significance level from 0.05 to 0.01 and if significant differences are still found then despite the violation of the independence assumption, we are reassured that the differences are significant.

Post-classification comparison. Using the classification maps, wherein each pixel was assigned to a single class, we determined the number of photosynthetic vegetation pixels, and soil and NPV pixels within each zone along oiled and oil-free shorelines. If the impact of oil was discernible within a zone, we expected to see a smaller number of the pixels classified as photosynthetic vegetation along the oiled shoreline compared to pixels along the oil-free shoreline.

For both the analyses using index-trajectories and classification maps, we expected the vegetation along oiled and oil-free shorelines to be at the same phenological state barring the impact of oil because we were looking at a single date.

Detecting recovery of stressed vegetation

Index values relative to distance from nearest oiled pixel. We used this analysis to determine potential continued decline or recovery from oil contamination a year later. Using our oil contamination map derived from the classification of the September 2010 imagery, we calculated the "distance to nearest oiled pixel" to determine the effect of proximity to oil contamination on plant stress. To test for significant differences in vegetation indexes we used an ANOVA with all the data (all zones combined) and a post-hoc Tukey-HSD test to determine how far from the oiled pixel we could see significant differences.

We used a linear regression for index values for the first four pixels away from the oiled pixels. At distances closer to an oiled pixel (up to 3 pixels away), we found a linear relationship between index values and distance from an oiled pixel. An asymptote was reached at distances farther away. Then we tested whether the slope of the linear regression was significantly different in August 2011 compared to September 2010 using a t-test and identified the direction of the change in slope. We expected a lower slope and weakened correlation by August 2011, which would indicate vegetation recovery, that is, the index values are no longer influenced by distance from oil. We should clarify that there was negligible oil in the August 2011 imagery. For both dates, "distance to nearest oiled pixel" relates to the original contamination in 2010. The only assumption we made was that there was no significant change in the location of oil contamination (like oil being carried further inland) after September 2010.

Post-classification change detection. We assessed whether there was continued vegetation decline or recovery a year later by looking at the percentage of pixels depleted of photosynthetic vegetation in September 2010 that were re-vegetated by August 2011. We also examined whether there was an effect of the position of the pixel relative to the shoreline in the extent of recovery, in other words, did recovery vary with distance from shore?

We were confident enough in our sub-pixel coregistration to do a pixel-to-pixel comparison of the classifications from both years. We quantified the number of pixels in each zone along oiled shorelines that were classified as soil or NPV in September 2010 but classified as photosynthetic vegetation in August 2011.

Although the 2010 imagery was collected in September and the 2011 imagery in August, thus during the same seasonal period, we *did* expect to see some differences in the phenological state of the wetland because of natural variation from year to year due to precipitation and temperature differences. However, because we were looking at a classified image rather than an index image, the data is categorized, not continuous. This minimizes small differences in plant vigor. As long as the plant is not in a completely different phenological state (e.g. green in 2010 to completely dry in 2011), the class of the pixel should not change. On the other hand, any recovery (new photosynthetic vegetation) in areas of denuded or dying vegetation will result in a class change. Thus recovery using this method does *not* indicate a return to "before oil" state but just that recovery has begun and there is fresh growth. This method is also not sensitive to declines in vegetation cover or health in areas adjacent to heavily oiled shorelines, that is, secondary oiling impacts.

Results

Classification Accuracy

Both 2010 and 2011 images were classified into six classes: soil, NPV, oiled soil, oiled NPV, photosynthetic vegetation and water (see Figure S1 for the full decision tree). The overall classification accuracy for the six classes in September 2010 was 98% and the Kappa coefficient was 0.97 ($n = 427$). For August 2011, the overall accuracy was 87.3% and Kappa was 0.8 ($n = 3108$). Table 2 shows confusion matrices for both dates using validation data distinct from training data used to build the classifier.

Kokaly et al. [5] found that contamination due to the BP-DWH oil spill covered narrow zones along the shoreline. Our oil maps showed that average penetration into the marsh was 7.5 m while maximum penetration was up to 40 m. The accuracy of the classification of oiled pixels was assessed independently using data collected in the field by Kokaly et al. [5,57]. When the classification map was collapsed down to two classes, "oil present" (oiled soil and oiled NPV) and "oil absent" (water, soil, NPV, and photosynthetic vegetation), the overall accuracy was 95% and the Kappa coefficient was 0.88 ($n = 40$, Table 3). Two of the field survey points had oil contamination but were classified as oil-free. Kokaly et al. [5] noted these areas to be lightly oiled with minimal penetration into the marsh. Comparison with the oil contamination map produced by Kokaly et al. [5] showed that 90.6% of the oiled shoreline mapped by Kokaly et al. [5] was classified as oiled in our map while 5.1% of the shoreline mapped as oiled in this study was classified as oil-free by Kokaly et al. [5]. Kokaly et al. [5] used multiple-date images (July, September and October, 2010) to map oiled pixels in all cover classes while our study concentrated on mapping oil in only the September, 2010 images.

Comparing oiled vs. oil-free shoreline for September 2010

Vegetation index trajectories. In September 2010, index values were significantly different between oiled and oil-free shorelines, and index trajectories in the first four oiled shoreline zones (0 to 14 m) were significantly lower than in the corresponding oil-free shoreline zones (Figure 4, see also Tables S1 and S2). The ANIR index shows the opposite pattern to the other indexes, because lower values for this index indicate photosynthetic vegetation (the angle at NIR gets narrower in healthy vegetation).

Post-classification comparison. The impact extended four pixels inland (0–14 m from shore). The percentage of pixels

Table 2. Confusion matrices for September 2010 and August 2011 classifications.

	Ground Reference Data					
September 2010 Confusion Matrix						
Classified as	Oiled NPV	Green Veg.	Water	Oiled Soil	Total	User's Accuracy
Oiled NPV	**107**	**1**	**0**	**1**	**109**	**98.2**
Green Veg.	**3**	**139**	**0**	**2**	**144**	**96.5**
Water	**0**	**0**	**94**	**0**	**94**	**100.0**
Oiled Soil	**1**	**0**	**0**	**79**	**80**	**98.8**
Total	**111**	**140**	**94**	**82**	**427**	
Producer's Accuracy	**96.4**	**99.3**	**100.0**	**96.3**		**98.1**
August 2011 Confusion Matrix						
	Water	Green Veg.	NPV			
Water	**1185**	**1**	**0**		**1186**	**99.9**
Green Veg.	**0**	**1158**	**395**		**1553**	**74.6**
NPV	**0**	**0**	**369**		**369**	**100.0**
Total	**1185**	**1159**	**764**		**3108**	
Producer's Accuracy	**100.0**	**99.9**	**48.3**			**87.3**

Confusion matrices for classification of September 2010 and August 2011 imagery using user-interpreted data. In September 2010, there were no oil-free non-photosynthetic vegetation (NPV) and oil-free soil pixels and in August 2011, there were no oiled pixels in validation data.

classified as photosynthetic vegetation was lower in the first four zones (<60%) in oiled shorelines relative to the corresponding zones in oil-free shorelines (>90%), consistent with observations on the index value trajectories (Figure 5). By the fifth pixel from the shore, the difference in percent of photosynthetic vegetation along oiled vs. oil-free shoreline was much narrower (83.8% vs. 93.2% or 9.4% difference). Beyond the first five pixels, the percentage of photosynthetic vegetation pixels along oiled and oil-free shorelines was similar and above 90%. Pixels in the second and third zone away from the shoreline were the most affected by oil (had the lowest percentage of photosynthetic vegetation pixels) followed by pixels adjacent to the shoreline.

Detecting recovery of stressed vegetation

Distance to nearest oiled pixel. In September 2010, index values were lowest in pixels closest to oiled pixels and increased as distance to the nearest oiled pixel increased (Figure 6, see also Tables S3 and S4). The impact of oil was observed up to 7 m from the nearest oiled pixel (up to two pixels away). All index values significantly increased as distance to the nearest oiled pixel increased in the first three pixels away from an oiled pixel (Table 4), with the exception of ANIR (which showed the opposite trend in values, thereby indicating the same pattern in plant stress). However, the amount of variance explained by this variable was not very high (R^2 between 0.11 to 0.42, Table 4).

Table 3. Confusion matrix for oiled vs. oil-free pixels in September 2010.

	Ground Reference Data			
Classified as	**Oil-free**	**Oiled**	**Total**	**User's Accuracy**
Oil-free	**10**	**2**	**12**	**83.3**
Oiled	**0**	**28**	**28**	**100.0**
Total	**10**	**30**	**40**	
Producer's Accuracy	**100.0**	**93.3**		**95.0**

Confusion matrices for classification of oiled pixels in September 2010 using field data collected by Kokaly et al. [5].

By August 2011, less than 1% of the variance was explained by the linear relationship between proximity to oil and index values ($R^2<0.05$ for all indexes, Table 4). The slopes were also significantly lower in August 2011 than those in September 2010 for all indexes (p-value <0.001, Table 4). Figure 6 illustrates the steep slope in the first three zones in September 2010 relative to August 2011.

Post-classification change detection. When comparing September 2010 and August 2011 classifications, we found that recovery or re-vegetation of pixels classified as dead or dying vegetation in 2010 was poorest closer to the shoreline, with a lower percentage of re-vegetated pixels, improving with increased distance from the shore. By the fourth pixel, almost all pixels that were oiled and denuded of photosynthetic vegetation in September 2010 had recovered and were classified as photosynthetic vegetation in August 2011 (Figure 7).

Discussion

Impact of oil contamination

The indexes we used in this study are indicators of important biophysiological properties of plants such as chlorophyll concentration, leaf area, plant water, and plant phenology [34,35,36,37,38]. Loss of pigment, water, and leaves are all symptoms of plant stress. We found that vegetation was more stressed along oiled shoreline in September 2010 compared to oil-free shoreline. This was true within every zone for at least the first four zones adjacent to the shoreline. Thus, the oil-impacted zone along the shoreline was 14 m. Kokaly et al. [5] noted that mean oil penetration in the region was 11 m. Our results show that the

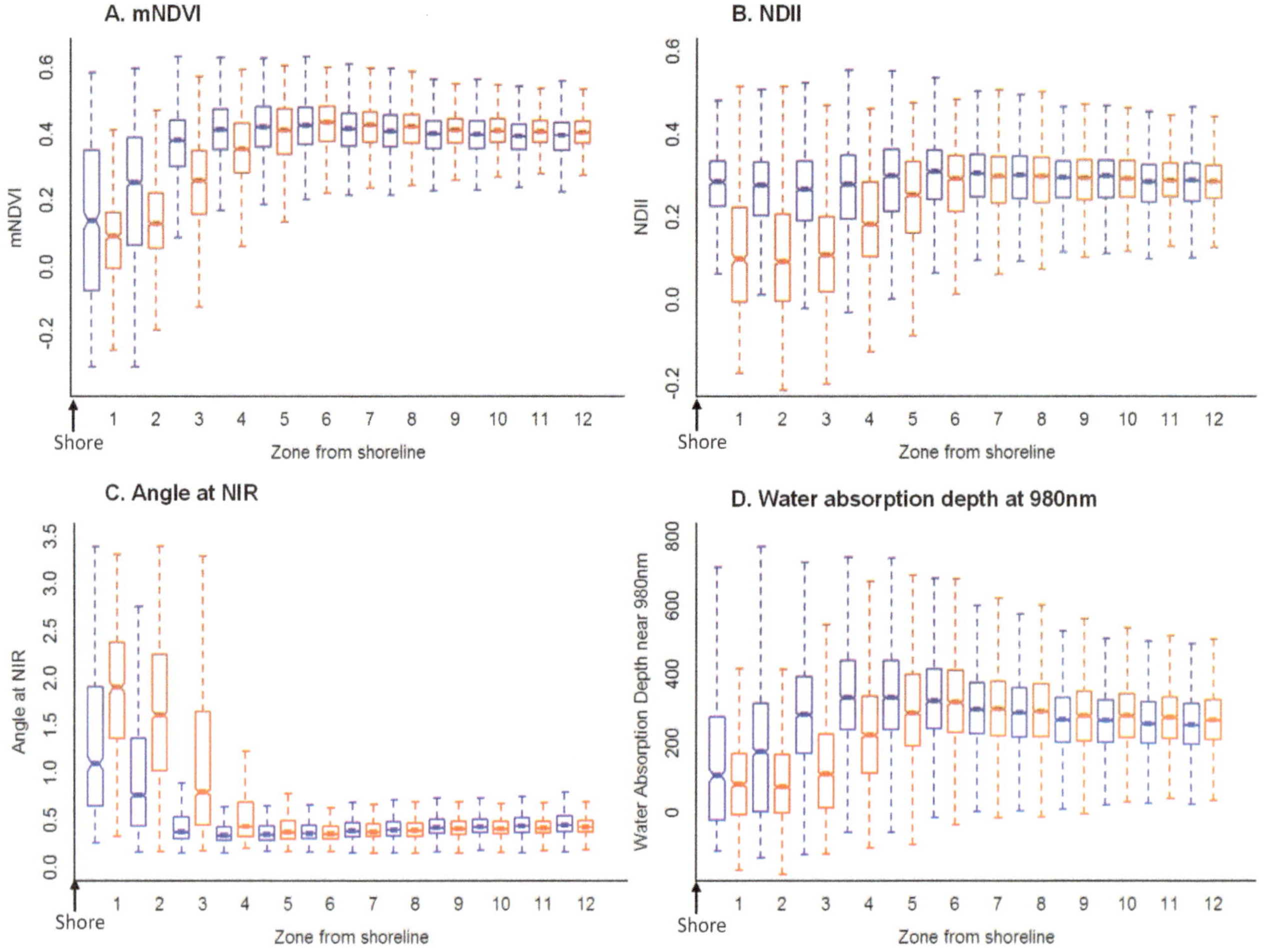

Figure 4. Index trajectories from shore to inland along oiled and oil-free shoreline. Box plots of index values for pixels vs. zone from shore ("1" being the first pixel from the shore and "12" being the twelfth) for oiled (red) and oil-free (blue) shoreline for the September 2010 dataset for (a) mNDVI, (b) NDII, (c) Angle at NIR, and (d) Absorption depth of water centered near 980nm. Distributions with non-overlapping notches are significantly different from each other. The notch represents the 95% confidence interval, the box extends from the first to the third quartile and the dashed lines extend to the data point which is 1.5 times the interquartile range.

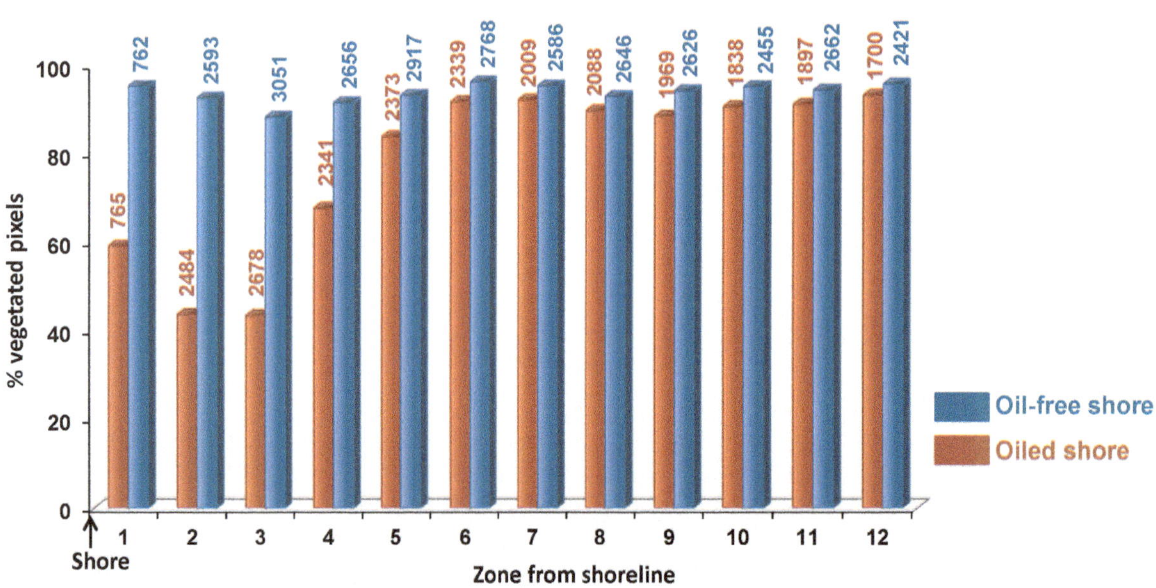

Figure 5. Photosynthetic vegetation cover from shore to inland along oiled and oil-free shoreline. Percentage of photosynthetic vegetation pixels in oiled vs. oil-free shoreline zones in September 2010 ("1" being the first pixel from the shore and "12" being the twelfth). Total number of pixels for each zone for oiled and oil-free shoreline is shown above the bars.

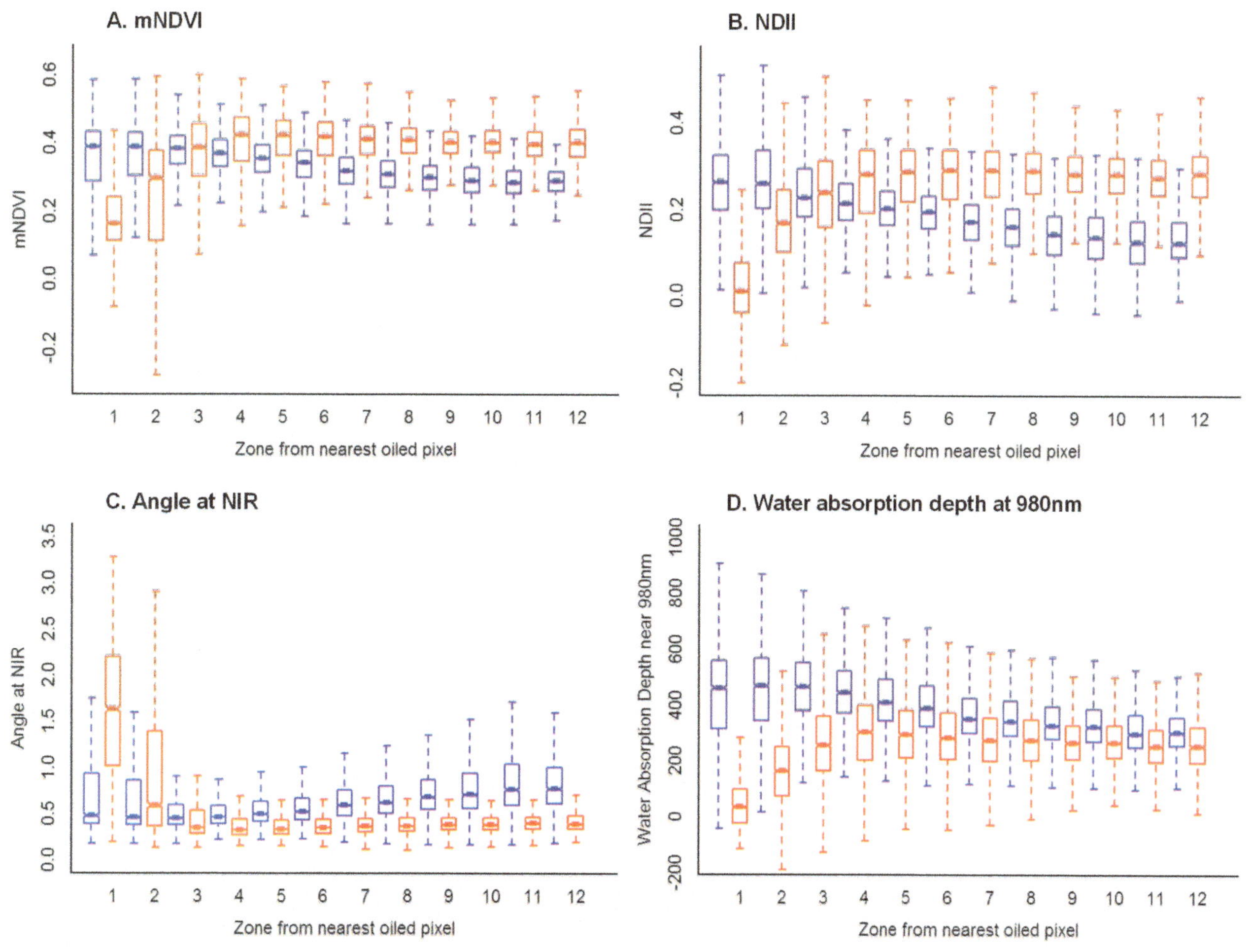

Figure 6. Index trajectories w.r.t. distance to the nearest oiled pixel in September 2010 and August 2011. Box plots of index values for pixels vs. distance from nearest oiled pixel ("1" being an oiled pixel and "12" being 12 pixels away from the nearest oiled pixel) in September 2010 (red) and August 2011 (blue) for (a) mNDVI, (b) NDII, (c) Angle at NIR, and (d) Absorption depth of water centered near 980 nm. Distributions with non-overlapping notches are significantly different from each other. The notch represents the 95% confidence interval, the box extends from the first to the third quartile and the dashed lines extend to the data point which is 1.5 times the interquartile range.

actual impact of oil on vegetation extends beyond the region where oil penetration is visible in the image.

We also observed in the field that when oil is carried inland through tidal action, it can contaminate the soil under the canopy while the plant canopy surface has no oil. Plant mortality is much higher when leaves are coated by oil than when the oil is present below the canopy only in the soil layer [1,7,12]. Therefore, we expected the impact of oil to decrease as we moved inland from the shoreline. This was confirmed by the general trajectory of index values with respect to the shoreline. Index values were lowest (indicating high plant stress) in the zone next to the shoreline but increased as we moved away from the shoreline, stabilizing at about 14 m from the shore. While significant differences were found only in the first four zones (five for water absorption bands), this does not mean that there was no impact beyond this zone. Some places where oil had penetrated much further inland likely showed impact far beyond 14 m but since the analysis pooled data from the entire study area, localized areas that may have been affected further inland did not result in a significant difference.

When we examined index behavior with respect to distance to the nearest oiled pixel, the response can be divided into two segments: the first 2–3 pixels from the nearest oiled pixel have an approximately linear response (index values increase with increasing distance) while pixels farther than that seem independent of distance from shore (see Figure 6). This is observed in September 2010 along the oiled shorelines by the significantly positive slopes of mNDVI, NDII, and ADW1 (negative slope for ANIR) as distance from oil increases. Plant stress extended, on average, 7 m inland from the closest oiled pixels. This gradient in impact was also likely due to the same reasons described above. The sub-canopy oil did not seem to have a measurable impact on the photosynthetic ability of the plants beyond this distance.

Comparison of classified pixels along oiled and oil-free shorelines showed that the lowest percentage of pixels containing photosynthetic vegetation occurred in the first three pixels away from the shoreline. This pattern is likely due to the fact that plants near the shoreline had their leaves and stems completely flooded by oil, hence plant mortality was very high [57]. Zonal bands, 1–2 pixels away from the shoreline (3.5 to 10.5 m) had a lower percentage of photosynthetic vegetation pixels compared to the pixels adjacent to the shore. We suspect this was because many of the oil absorbing floating booms that were used to prevent oil from reaching the shore may have washed onto the marshes and settled

Table 4. Regression analysis of September 2010 and August 2011 indexes.

Index	September 2010			August 2011			Sep 2010 vs. Aug 2011: T value**
	Slope coefficient	Coefficient of determination (R^2)	F value*	Slope coefficient	Coefficient of determination (R^2)	F value*	
NDVI	0.06	0.11	1780	0.02	0.01	122	17.9
mNDVI	0.07	0.20	3692	0.01	0.00	53	43.1
NDII	0.08	0.42	10360	−0.01	0.02	324	87.5
ANIR	−0.37	0.29	5733	−0.08	0.01	145	−37.1
ARed	0.49	0.34	7521	−0.08	0.02	286	73.0
ADW1	80.5	0.27	5386	7.77	0.00	2949	40.3
ADW2	119.8	0.28	5597	26.58	0.01	151	34.6

*All regressions were significant (p-value <0.001)
**All 2010 vs. 2011 slopes were significantly different (p-value < 0.001)
Regression analysis of September 2010 and August 2011 indexes data with respect to "distance from the nearest oiled pixel" (n = 14,372).

at the high tide line, within the 1–2 pixels from the shoreline, corresponding to the zone where plant mortality was highest (personal observation in October, 2010). The booms that were transported into the marsh were generally saturated with oil and thereby increased the intensity and residence time of oil contamination near the high tide line which can lead to higher impact on the plant community [2,11,12]. Moreover, frequency of tidal action decreases as we move away from the shore. This also leads to longer residence time for the oil. Another possible reason for greater impact in the pixels 3.5 to 10.5 m from shore is that water levels were higher when the oil came ashore and as the water receded, it deposited more oil inland than in the zone adjacent to the shore.

Detecting recovery of stressed vegetation

Our results indicate that tidal marsh communities were beginning to recover in August 2011 because their spectral index values were no longer influenced by proximity to oil contamination. *S. alterniflora*, the dominant salt marsh species along the shoreline in Barataria Bay, is relatively robust to oil contamination in terms of above-ground biomass, stem density and nutrient uptake [1,7,12]. *S. patens* is relatively more sensitive to oil but is found further inland in the brackish marsh. Recovery in above-ground biomass in both species is quick. DeLaune and Wright [12] note that *S. alterniflora* recovers within a year even when the leaves have been oiled and the canopy has died.

A comparison of the classifications of oiled and oil-free shorelines in September 2010 imagery indicated that the first 10.5 m along oiled shorelines had a high percentage of oiled bare soil and oiled NPV pixels (as high as 50%) potentially indicating areas that were denuded of vegetation as a result of oil-induced plant stress. By 24 m inland from shore, the number of oiled bare soil and NPV pixels was negligible (<1%). In August 2011, recovery was the poorest in pixels adjacent to the shore (65% of oiled soil and NPV pixels contained photosynthetic vegetation in the following year), followed by the next two zones, less than 7 m from shore (more than 80% of the oiled soil and NPV pixels in the second zone from the shore contained photosynthetic vegetation in the following year) and by the fourth zone (14 m from shore), re-vegetation was almost 100% supporting the conclusion that much of the inland marsh had regenerated. Hester and Mendelssohn [7] observed in a five year study of a Louisiana brackish marsh recovering from oil contamination that failure to re-vegetate was more often due to lack of elevation (resulting in waterlogging stress) than directly due to the oil. Elevation is lowest in pixels adjacent to the shoreline, and additionally, it is subject to erosion as a result of denuded vegetation. This probably led to poor recovery in pixels next to the shoreline.

The difference in the range of index values beyond the impact zone between September 2010 and August 2011 was likely due to 2011 being a much drier year than 2010, and not oil impacts. Precipitation in 2011 (754 mm) was only one-third of 2010 levels (2233 mm; National Climate Data Center) resulting in lower overall index values and a drier interior marsh relative to the near-shore marsh. This interpretation is further supported by the absence of large storms that would have been powerful enough to bring tidal water farther into the marsh.

This study clearly showed that the impact of oil characterized using vegetation indexes extended no further than a couple of pixels from a contaminated pixel. While Mishra et al. [4] found oil impacts to be far more widespread (>400 km^2), the broad spatial and spectral resolution of the Landsat imagery they used prevented the direct detection of oil. Instead, they compared the 2009 growing season to the 2010 growing season to determine

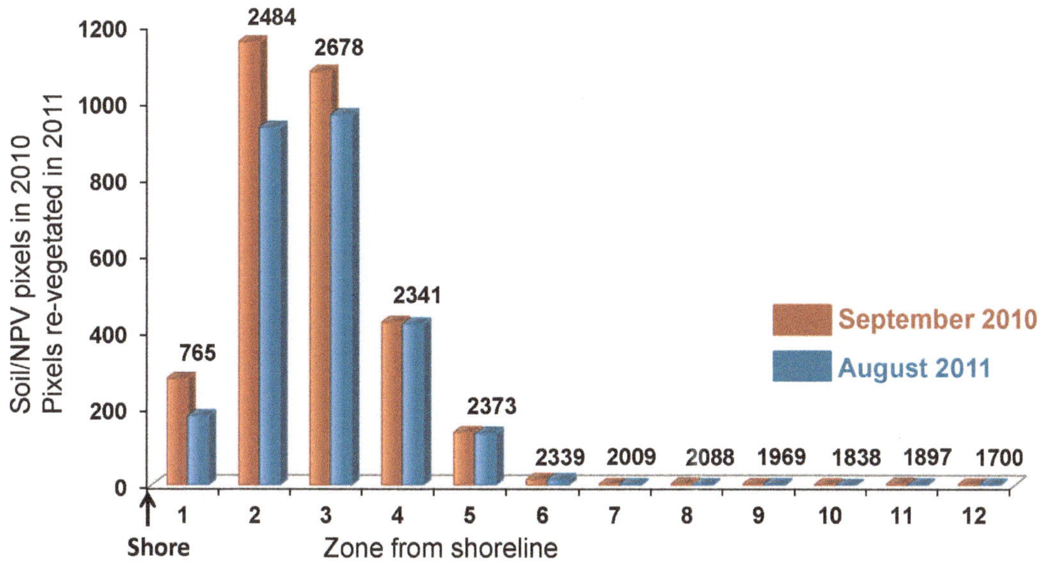

Figure 7. Recovery of photosynthetic vegetation from September 2010 to August 2011. The number of oiled soil & NPV pixels in September 2010 that recovered (regrowth of vegetation) a year later in August 2011 in pixels next to the shore ("1" being the first pixel from the shore and "12" being the twelfth). Total number of pixels for each zone is shown above the bars.

extent of oil impact. The large areas of vegetation change mapped with 30 m pixels could be due to other reasons such as precipitation differences, natural variation in vegetation phenology from year to year, etc. The impacted areas included salt marshes (e.g. Barataria Bay), brackish and freshwater marshes (e.g. Bird's Foot) and mangroves (e.g. Chandeleur Islands). Some of these areas are far more fragmented by oil and gas infrastructure than others. Species composition also changes considerably among different vegetation types. Thus they may show considerable variation in oil impact.

There was a loss of 15% to 30% vegetated pixels in 2011 that changed to water from photosynthetic vegetation, residue and soil classes. According to local tide tables for these two dates and the time that the images were flown, the tidal stage in August 2011 was about 0.12 m higher than in September 2010 (tides.rodn-reel.com). In September 2010, time of data collection coincided with the low tide (0.18 m), while in August 2011, data were collected only slightly below high tide levels (0.30 m). Hence, the increase in water pixels in August 2011 was in part due to tidal differences between the two dates.

Conclusions

The full impact of the BP-DWH oil spill on the marshland, mangroves and wildlife of the region is still under study. We used AVIRIS imagery collected over Barataria Bay in September 2010 and August 2011 to evaluate whether vegetation health in these marshes continued to decline or experienced recovery in the following year. We successfully assessed marsh impacts of the oil spill finding that the marsh showed negative impacts due to oil exposure but that some recovery occurred within one year. The impact was contingent on distance to oil contamination. We also found that sub-canopy oil did not seem to have a discernable effect on marsh communities beyond a couple of pixels. While this is encouraging news for marsh health, we don't know whether different plant communities respond differently. Further we cannot address how this immediate impact cascades through the ecosystem.

This analysis is based on two AVIRIS snapshots in time, September 2010 and August 2011. Recovery in the marsh is likely still ongoing. Moreover, the strong storm surge brought in by hurricane Isaac in 2012 might have remobilized oil and brought it further inland. More research is needed to determine long-term impacts of the oil spill. However, this study illustrates that high spatial resolution imaging spectroscopy serves as a powerful tool to establish the link between oil contamination and plant stress and to document plant die-off and recovery in the marsh after the disaster.

Supporting Information

Figure S1 Decision tree for classification of AVIRIS imagery. Decision tree showing multiple variables used at each node to separate one target class from the rest of the classes. Oiled pixels were classified using continuum removal (CR) over two oil absorptions at 1720 nm and 2300 nm in only soil and non-photosynthetic vegetation (NPV) pixels. R_x indicates reflectance in the "x" region of the electromagnetic spectrum.

Table S1 Mean index values w.r.t distance to shore in September 2010. Mean values for seven indexes vs. distance to shore for oiled and oil-free shorelines for the September 2010 dataset (index acronyms are listed in table 1).

Table S2 Index analysis for oiled vs. oil-free zones w.r.t distance to shore in September 2010. Analysis of variance between oil versus oil-free shorelines in September 2010 and per zone comparisons (index acronyms are listed in table 1). Degrees of freedom = 185,887.

Table S3 Mean index values w.r.t distance to nearest oiled pixel. Mean values for seven indexes vs. distance to nearest oiled pixel for oiled shoreline in September 2010 and a year later in August 2011.

Table S4 Index analysis for Sept. 2010 vs. Aug. 2011 oiled zones w.r.t distance from nearest oiled pixel. Analysis of variance between pixels at different distances from an oiled pixel in September 2010, and Tukey HSD tests for pair-wise comparisons (index acronyms are listed in Table 1). Degrees of freedom = 114,371.

Acknowledgments

We thank Dr. Diane Wickland, program manager for Terrestrial Ecology at NASA, for providing access to the AVIRIS database and to the AVIRIS team for preprocessing support. We thank Dr. Michael Whiting and Dr. Jose Zarate for collecting calibration data in the field, Mr. George Scheer for systems support, and Mr. Sinzee Tran for help in additional pre-processing of the data at CSTARS. We thank Ms. Keely Roth from UCSB and the staff at USGS, Baton Rouge, LA for collecting and sharing field data, and Mr. Seth Peterson from UCSB for pre-processing some of the AVIRIS imagery. We particularly thank Brady Couvillion for reviewing the manuscript and offering insightful comments.

Author Contributions

Analyzed the data: SK MJS. Contributed reagents/materials/analysis tools: SLU RFK AK. Wrote the paper: SK. Pre-prossessed image data for analysis: SK DAR. Did image coregistration: SK. Did image analysis: AK SK RK. Collaborated on statistical data analysis: SK MJS SLU. Contribution to critical review of the manuscript in final approval: SK MJS SLU AK RFK DAR.

References

1. Ko J-Y, Day JW (2004) A review of ecological impacts of oil and gas development on coastal ecosystems in the Mississippi Delta. Ocean & Coastal Management 47: 597–623.
2. Gundlach ER, Hayes MO (1978) Vulnerability of coastal environments to oil spill impacts. Marine Technology Society Journal 12: 18–27.
3. Moss L (2010) The 13 largest oil spills in history. Mother Nature Network.
4. Mishra DR, Cho HJ, Ghosh S, Fox A, Downs C, et al. (2012) Post-spill state of the marsh: Remote estimation of the ecological impact of the Gulf of Mexico oil spill on Louisiana Salt Marshes. Remote Sensing of Environment 118: 176–185.
5. Kokaly RF, Couvillion BR, Holloway JM, Roberts DA, Ustin SL, et al. (2013) Spectroscopic remote sensing of the distribution and persistence of oil from the Deepwater Horizon spill in Barataria Bay marshes. Remote Sensing of Environment 129: 210–230.
6. DeLaune RD, Smith CJ, Patrick Jr WH, Fleeger JW, Tolley MD (1984) Effect of oil on salt marsh biota: Methods for restoration. Environmental Pollution Series A, Ecological and Biological 36: 207–227.
7. Hester MW, Mendelssohn IA (2000) Long-term recovery of a Louisiana brackish marsh plant community from oil-spill impact: vegetation response and mitigating effects of marsh surface elevation. Marine Environmental Research 49: 233–254.
8. Pezeshki SR, Hester MW, Lin Q, Nyman JA (2000) The effects of oil spill and clean-up on dominant US Gulf coast marsh macrophytes: a review. Environmental Pollution 108: 129–139.
9. Lin Q, Mendelssohn IA (1996) A comparative investigation of the effects of south Louisiana crude oil on the vegetation of fresh, brackish and salt marshes. Marine Pollution Bulletin 32: 202–209.
10. Mendelssohn IA, Hester MW, Sasser C, Fischel M (1990) The effect of a louisiana crude oil discharge from a pipeline break on the vegetation of a Southeast Louisiana brackish marsh. Oil and Chemical Pollution 7: 1–15.
11. Jackson JBC, Cubit JD, Keller BD, Batista V, Burns K, et al. (1989) Ecological effects of a major oil spill on Panamanian coastal marine communities. Science 243: 37–44.
12. DeLaune RD, Wright AL (2011) Projected impact of deepwater horizon oil spill on U.S. gulf coast wetlands. Soil Science Society of America Journal 75: 1602–1612.
13. Kenworthy WJ, Durako MJ, Fatemy SMR, Valavi H, Thayer GW (1993) Ecology of seagrasses in northeastern Saudi Arabia one year after the Gulf War oil spill. Marine Pollution Bulletin 27: 213–222.
14. Smith CJ, DeLaune RD, Patrick WH (1981) A method for determining stress in wetland plant communities following an oil spill. Environmental Pollution Series A, Ecological and Biological 26: 297–304.
15. Brekke C, Solberg AHS (2005) Oil spill detection by satellite remote sensing. Remote Sensing of Environment 95: 1–13.
16. Cross AM (1992) Monitoring marine oil pollution using AVHRR data: observation of the coast of Kuwait and Saudi Arabia during January 1991. International Journal of Remote Sensing 13: 781–788.
17. De Domenico L, Crisafi E, Magazzù G, Puglisi A, La Rosa A (1994) Monitoring of petroleum hydrocarbon pollution in surface waters by a direct comparison of fluorescence spectroscopy and remote sensing techniques. Marine Pollution Bulletin 28: 587–591.
18. Fingas MF, Brown CE (1997) Review of oil spill remote sensing. Spill Science & Technology Bulletin 4: 199–208.
19. Goodman R (1994) Overview and future trends in oil spill remote sensing. Spill Science & Technology Bulletin 1: 11–21.
20. Howari FM (2004) Investigation of hydrocarbon pollution in the vicinity of United Arab Emirates coasts using visible and near infrared remote sensing data. Journal of Coastal Research 20: 1089–1095.
21. Lin B, An J, Carl B, Zhang H (2004) Neural networks in detection and identification of littoral oil pollution by remote sensing. In: Yin F-L, Wang J, Guo C, editors. Advances in Neural Networks – ISNN 2004: Springer Berlin / Heidelberg. pp. 977–982.
22. Volckaert FAM, Kayens G, Schallier R, Jacques TG (2000) Aerial surveillance of operational oil pollution in Belgium's maritime zone of interest. Marine Pollution Bulletin 40: 1051–1056.
23. Cloutis EA (1989) Spectral reflectance properties of hydrocarbons: Remote-sensing implications. Science 245: 165–168.
24. Ben-Dor E, Inbar Y, Chen Y (1997) The reflectance spectra of organic matter in the visible near-infrared and short wave infrared region (400–2500 nm) during a controlled decomposition process. Remote Sensing of Environment 61: 1–15.
25. Bianchi R, Cavalli RM, Marino CM, Pignatti S, Poscolieri M. Use of airborne hyperspectral images to assess the spatial distribution of oil spilled during the Trecate blow-out (Northern Italy); 1995. pp. 352–362.
26. Gaffey SJ, McFadden LA, Nash D, Pieters CM (1993) Ultraviolet, visible, and near-infrared reflectance spectroscopy: Laboratory spectra of geologic materials. In: Pieters CM, Englert PAJ, editors. Geochemical Analysis: Elemental and Mineralogical Composition. Cambridge: Cambridge University Press. pp. 43–78.
27. Hörig B, Kühn F, Oschütz F, Lehmann F (2001) HyMap hyperspectral remote sensing to detect hydrocarbons. International Journal of Remote Sensing 22: 1413–1422.
28. Leifer I, Lehr WJ, Simecek-Beatty D, Bradley E, Clark R, et al. (2012) State of the art satellite and airborne marine oil spill remote sensing: Application to the BP Deepwater Horizon oil spill. Remote Sensing of Environment 124: 185–209.
29. Kühn F, Oppermann K, Hörig B (2004) Hydrocarbon Index — An algorithm for hyperspectral detection of hydrocarbons. International Journal of Remote Sensing 22: 1413–1422.
30. Ben-Dor E, Banin A (1990) Near-Infrared Reflectance Analysis of Carbonate Concentration in Soils. Applied Spectroscopy 44: 1064–1069.
31. Gaffey SJ (1986) Spectral reflectance of carbonate minerals in the visible and near infrared (O.35-2.55 microns); calcite, aragonite, and dolomite. American Mineralogist 71: 151–162.
32. Gerrard AJ (1981) Soils and landforms: An integration of geomorphology and pedology. London, UK: George Allen & Unwin (Publishers) Ltd.
33. Kokaly RF, Asner GP, Ollinger SV, Martin ME, Wessman CA (2009) Characterizing canopy biochemistry from imaging spectroscopy and its application to ecosystem studies. Remote Sensing of Environment 113: S78–S91.
34. Ustin SL, Gitelson AA, Jacquemoud S, Schaepman M, Asner GP, et al. (2009) Retrieval of foliar information about plant pigment systems from high resolution spectroscopy. Remote Sensing of Environment 113: S67–S77.
35. Houborg R, Soegaard H, Boegh E (2007) Combining vegetation index and model inversion methods for the extraction of key vegetation biophysical parameters using Terra and Aqua MODIS reflectance data. Remote Sensing of Environment 106: 39–58.
36. Zarco-Tejada PJ, Rueda CA, Ustin SL (2003) Water content estimation in vegetation with MODIS reflectance data and model inversion methods. Remote Sensing of Environment 85: 109–124.
37. Peñuelas J, Gamon JA, Griffin KL, Field CB (1993) Assessing community type, plant biomass, pigment composition, and photosynthetic efficiency of aquatic vegetation from spectral reflectance. Remote Sensing of Environment 46: 110–118.
38. Jurgens C (1997) The modified normalized difference vegetation index (mNDVI) a new index to determine frost damages in agriculture based on Landsat TM data. International Journal of Remote Sensing 18: 3583–3594.
39. Hardisky MA, Klemas V, Smart RM (1983) The influence of soil-salinity, growth form, and leaf moisture on the spectral radiance of *Spartina-alterniflora* canopies. Photogrammetric Engineering and Remote Sensing 49: 77–83.
40. Hunt ER, Rock BN (1989) Detection of changes in leaf water content using near-infrared and middle-infrared reflectances. Remote Sensing of Environment 30: 43–54.
41. Clark RN, Roush TL (1984) Reflectance spectroscopy - Quantitative analysis techniques for remote sensing applications. Journal of Geophysical Research 89: 6329–6340.
42. Tucker CJ (1979) Red and photographic infrared linear combinations for monitoring vegetation. Remote Sensing of Environment 8: 127–150.

43. Khanna S, Palacios-Orueta A, Whiting ML, Ustin SL, Riano D, et al. (2007) Development of angle indexes for soil moisture estimation, dry matter detection and land-cover discrimination. Remote Sensing of Environment 109: 154–165.
44. Li L, Ustin SL, Lay M (2005) Application of AVIRIS data in detection of oil-induced vegetation stress and cover change at Jornada, New Mexico. Remote Sensing of Environment 94: 1–16.
45. Van Der Meer F, Van Dijk P, Van Der Werff H, Yang H (2002) Remote sensing and petroleum seepage: a review and case study. Terra Nova 14: 1–17.
46. Yang H, Zhang J, Van der Meer F, Kroonenberg SB (1998) Geochemistry and field spectrometry for detecting hydrocarbon microseepage. Terra Nova 10: 231–235.
47. Gitelson A, Merzlyak MN (1994) Spectral reflectance changes associated with autumn senescence of *Aesculus-hippocastanum* L. and *Acer-platanoides* L. leaves - Spectral features and relation to chlorophyll estimation. Journal of Plant Physiology 143: 286–292.
48. Gitelson AA, Merzlyak MN (1997) Remote estimation of chlorophyll content in higher plant leaves. International Journal of Remote Sensing 18: 2691–2697.
49. Horler DNH, Dockray M, Barber J (1983) The red edge of plant leaf reflectance. International Journal of Remote Sensing 4: 273–288.
50. Milton NM, Eiswerth BA, Ager CM (1991) Effect of phosphorus deficiency on spectral reflectance and morphology of soybean plants. Remote Sensing of Environment 36: 121–127.
51. Hu C, Weisberg RH, Liu Y, Zheng L, Daly KL, et al. (2011) Did the northeastern Gulf of Mexico become greener after the Deepwater Horizon oil spill? Geophysical Research Letters 38: L09601.
52. Liu P, Li X, Qu JJ, Wang W, Zhao C, et al. (2011) Oil spill detection with fully polarimetric UAVSAR data. Marine Pollution Bulletin 62: 2611–2618.
53. Liu Y, Weisberg RH, Hu C, Zheng L. Combining numerical ocean circulation models with satellite observations in a trajectory forecast system: a rapid response to the Deepwater Horizon oil spill; 2011 May 4, 2011. pp. 80300K.
54. Ramsey III E, Rangoonwala A, Suzuoki Y, Jones CE (2011) Oil detection in a coastal marsh with polarimetric Synthetic Aperture Radar (SAR). Remote Sensing 3: 2630–2662.
55. Svejkovsky J, Lehr WJ, Muskat J, Graettinger G, Mullin J (2012) Operational utilization of aerial multispectral remote sensing during oil spill response: Lessons learned during the Deepwater Horizon (MC-252) spill. Photogrammetric Engineering and Remote Sensing in press.
56. Merton R. Monitoring community hysteresis using spectral shift analysis and the red-edge vegetation stress index; 1998 12–16 January; Pasadena, California, USA. NASA, Jet Propulsion Laboratory.
57. Kokaly RF, Heckman D, Holloway J, Piazza S, Couvillion B, et al. (2011) Shoreline surveys of oil-impacted marsh in southern Louisiana, July to August 2010. U.S. Geological Survey. 124 p.
58. Koltunov A, Ustin SL, Quayle B, Schwind B (2012) GOES Early Fire Detection (GOES-EFD) system prototype. ASPRS 2012 Anuual Conference. Sacramento, CA.
59. Koltunov A, Ben-Dor E, Ustin SL (2008) Image construction using multi-temporal observations and Dynamic Detection Models. International Journal of Remote Sensing 30: 57–83.
60. Irani M (2002) Multi-frame correspondence estimation using subspace constraints. International Journal of Computer Vision 48: 173–194.
61. Khanna S, Santos MJ, Ustin SL, Haverkamp PJ (2011) An integrated approach to a biophysiologically based classification of floating aquatic macrophytes. International Journal of Remote Sensing 32: 1067–1094.
62. Rosenfield GH, Fitzpatrick-Lins K (1986) A coefficient of agreement as a measure of thematic classification accuracy. Photogrammetric Engineering and Remote Sensing 52: 223–227.
63. Story M, Congalton RG (1986) Accuracy assessment - A user's perspective (map interpretation). Photogrammetric Engineering and Remote Sensing 52: 397–399.
64. Carter GA, Knapp AK (2001) Leaf optical properties in higher plants: linking spectral characteristics to stress and chlorophyll concentration. American Journal of Botany 88: 677–684.
65. Zar JH (2010) Biostatistical Analysis. Englewood CliffsNJ, USA: Prentice-Hall. 944 p.
66. Dale MRT, Fortin MJ (2002) Spatial autocorrelation and statistical tests in ecology. Eco-science 9: 162–167.
67. Nagler PL, Daughtry CST, Goward SN (2000) Plant litter and soil reflectance. Remote Sensing of Environment 71: 207–215.

In Situ Investigation the Photolysis of the PAHs Adsorbed on Mangrove Leaf Surfaces by Synchronous Solid Surface Fluorimetry

Ping Wang[1], Tun-Hua Wu[2], Yong Zhang[3]*

1 School of Environmental Science and Public Health, Wenzhou Medical University, Wenzhou, China, **2** School of Information and Engineering, Wenzhou Medical University, Wenzhou, China, **3** State Key Laboratory of Marine Environmental Science (Xiamen University), Environmental Science Research Center of Xiamen University, Xiamen, China

Abstract

An established synchronous solid surface fluorimetry (S-SSF) was utilized for *in situ* study the photolysis processes of anthracene (An) and pyrene (Py) adsorbed on the leaf surfaces of *Kandelia obovata* seedlings (*Ko*) and *Aegiceras corniculata* (L.) Blanco seedlings (*Ac*). Experimental results demonstrated that the photolysis of An and Py adsorbed on the leaf surfaces of two mangrove species under the laboratory conditions, followed first-order kinetics with their photolysis rates in the order of *Ac*>*Ko*. In addition, with the same amount of substances, the photolysis rate of An adsorbed on the same mangrove leaf surfaces was much faster than the adsorbed Py. In order to investigate further, the photolysis processes of An and Py in water were also studied for comparison. And the photolysis of An and Py in water also followed first-order kinetics. Moreover, for the same initial amount, the photolysis rate of the PAH in water was faster than that adsorbed on the leaf surfaces of two mangrove species. Therefore, photochemical behaviors of PAHs were dependent not only on their molecular structures but also the physical-chemical properties of the substrates on which they are adsorbed.

Editor: Jie Zheng, University of Akron, United States of America

Funding: National Natural Science Foundation of China (21177102, 21075102 and 21207103), Public Benefit Project of Zhejiang Province (2012C31025), Natural Science Foundation of Zhejiang Province (LY13H180012), Scientific Research Fund of Zhejiang Provincial Education Department (Y201222932), Project of Wenzhou Municipal Science and Technology (G20110004) and Science and Technology Innovation Action of College Student of Zhejiang Province (2013R413010). The funders had no role in study design, data collection and analysis, decision to publish, or preparation of the manuscript.

Competing Interests: The authors have declared that no competing interests exist.

* E-mail: fruitful@xmu.edu.cn

Introduction

Polycyclic aromatic hydrocarbons (PAHs) are well-known environmental persistent pollutants listed as priority pollutants by European Union and the U.S. Environmental Protection Agency due to their teratogenic, carcinogenic and mutagenic [1,2]. The persistence of PAHs in the environment poses a potential threat to human health through bioaccumulation and biomagnifications via food chains [3]. Arising from both natural and anthropogenic sources, PAHs are ubiquitous in different natural phases such as plant, soil, sediment, water and air [4–6]. Thus it is of great importance to investigate the transport and transformation of PAHs in the environment. Many PAHs, especially with four or more rings, are generally recalcitrant to be biodegraded in the environment because of their low water-solubility [7]. These compounds are more likely to be affected by sedimentation and photooxidation [8]. Therefore, studies on the photolysis processes of PAHs are extremely important in order to evaluate the fate and transformation of these hazardous compounds in the environment, and have being attracted great attention [9,10]. At present, most studies are focus on the photolysis of PAHs in liquid medium (e.g., water, ethyl alcohol and acetone) or adsorbed on solid particles (e.g., fly ash, carbon black silica gel, porous glass, clay sand and active aluminum) [11–13]. It is well known that vegetation plays a key role in the environmental fate of PAHs [1]. Only a few studies involve the photolytic behaviors of PAHs adsorbed on vegetation [7,14,15].

Mangroves are various kinds of trees growing along the coastlines of tropical and sub-tropical regions. Because of their special ecological functions, mangroves are considered to be reservoir of lipophilic contaminations, including PAHs from various sources [16–18]. Mangrove leaves are common covered with large surface areas and thick waxy layers, which can accumulate the lipophilic PAHs in atmosphere [18]. Ke et al have investigated the fate of PAHs contamination in a mangrove swamp after an oil spill accident. They have believed that photolysis is one of the important way to remove the PAHs adsorbed on the mangrove leaves [19]. Thus it is of great significance to study the photochemical behaviors of PAHs adsorbed on mangrove leaves. However, recent studies on the photolysis of PAHs adsorbed on plant leaves are mostly entirely destructive sample extraction, which can not realize *in situ* monitoring the photolysis of PAHs on the leaves. In addition, the amount of PAHs volatilization from the leaf surfaces into the atmosphere and entering into the inner leaf tissues could not be ignored during the long duration experiment [14,15]. To overcome the limitations of the traditional extraction procedures, Wild and his co-workers have used the two-photon microscopy to track directly some PAHs adsorbed on the plant leaves [20,21]. In our previous studies, the two-photon microscopy has also been utilized for visualization three typical PAHs into the inner tissues of two mangrove species [22,23]. However it is very expensive to purchase and maintain that instrument, which could be difficult to be widely used. Optical fiber with high light focalization, low weight and small size could be used for *in situ* investigation of pollutants adsorbed on solid substrates [24–27].

In our previous studies, a fiber-optic fluorimetry has been used for direct investigation the photolysis of some PAHs adsorbed on the leaf surfaces of three mangrove species [7,28]. Nevertheless, the photolysis behaviors of PAHs adsorbed on the mangrove leaf surfaces have not been fully explored. Questions without answers are still forthcoming. What are the main transformation pathways of the PAHs adsorbed on the mangrove leaf surfaces? Is photolysis significant for the fate of PAHs in the environment? Are the photolysis mechanisms of different kinds of PAHs adsorbed on mangrove leaves similar? What are the photolytic products of the PAHs? How to get the photolytic products from the mangrove leaf surfaces? These questions are of great importance for us to learn more about the environmental behaviors of the adsorbed PAHs on mangrove leaves. Moreover, a synchronous solid surface fluorimetry (S-SSF) combined with an optical fiber and a fluorescence spectrophotometer has also been established for *in situ* determination of the PAHs adsorbed on mangrove leaves in our previous work [27]. The S-SSF method has many obvious advantages: it was easy, rapid and inexpensive to purchase, maintain and operate. Additionally, the original existing forms and the distribution of the PAHs adsorbed on the mangrove leaf surfaces could be easily determined in situ, which would be impossible for traditional methods, such as GC-MS, GC, and HPLC, which need an extraction of sample before analysis. Furthermore, no complex pretreatment involving a large amount of organic solvent is needed, and it takes less than 1 minute for the determination of one sample. To further demonstrate the tremendous scope of this method, the S-SSF was also applied for *in situ* investigation the photolysis of typical PAHs adsorbed on the mangrove leaf surfaces under the laboratory experimental conditions. In this work, anthracene (An) and pyrene (Py) were selected as the model PAHs, the viviparous seedlings of *Ko* in salt-resistance and the non-viviparous seedlings of *Ac* in salt-tolerance were studied as the typical mangrove species. In addition, with the same initial amount the photolysis of An and Py in water was also studied for comparison.

Materials and Methods

Apparatus and compounds generation

The spectra of PAHs were obtained utilizing a Cary Eclipse (Varian, Habor City, California) equipped with a 150-W xenon flash lamp. A fiber optic accessory with solid sample tip was installed on the fluorospectrophotometer and aligned as described in the user instructions supplied with the accessory. An optical fiber was used for fluorescence measurements. The instrumental parameters were as follows: excitation and emission slits were set at 20 and 10 nm, scan speed was performed as 120 $nm.min^{-1}$, photomultiplier tube voltage was 600 V, and an ultrasonic cleaning device (Model KQ3200, power 150W, frequency 40 kHz) was also used to extract the leaf-wax on mangrove leaf surfaces and the wax content of the two specific mangrove leaves was quantified by the method in our previous work [18,27].

A high pressure mercury lamp equipped with the optical fiber (CHF-XM500 W, Beijing Trusttech, Co., Ltd., China) was utilized as the light source for the photolysis study of PAHs adsorbed on the leaf surfaces of two mangrove species. In order to keep the emit light intensity steady during the experiment, the light intensity on the leaf surfaces was controlled by adjusting the height between the mangrove leaf surfaces and the optical fiber probe of the mercury lamp.

The stock solution of An and Py (Aldrich, purity>99%, USA) were prepared by dissolving the solutes in acetone in the brown volumetric flask, respectively. These solutions were then ultrasonicated for 30 min to aid solubilization and stored at 4°C in darkness to avoid possible photodegradation. Working solutions of An and Py were prepared by dilution of the stock solutions in acetone before use, respectively.

Plant preparation

Mature hypocotyls of *Ko* and one-year-old *Ac*, were collected from a mangrove forest growing in Cao Putou village, Longhai city, China (longitude: 24°29′3″; latitude: 118°5′59″; altitude: 0 m above sea level). This mangrove forest is wild and does not belong to any individuals or organizations, thus no specific permissions were required for these activities. In the experiment, only a few mangrove seedlings were collected, which neither involved endangered or protected species nor vertebrate studies. The seedlings of the two mangrove species (Figure 1) with similar length and fresh weight were then planted on sand in pots partially submerged in nutrient Hoagland solution [27]. The sample preparations were all based on the methods of our previous studies [22,23,27]. The leaves of the two mangrove species (Figure 1) with similar length and fresh weight were collected from these mangrove seedlings cultivated for about ninety days. After the collection of the leaves, experiments were carried out as soon as possible. The picked fresh mangrove leaves were carefully rinsed with distilled water to remove surface silt. After air drying, six circles of 0.5 cm radius were drawn on the upper surfaces of each leaf with a pencil [27]. The distribution of the sample location was shown in Figure 1. The size was the same as the light circle formed by the fiber optical probe. With the use of a micropipette, a certain amount of An (44.5 $ng.spot^{-1}$ and 445 $ng.spot^{-1}$) and Py (50.5 $ng.spot^{-1}$ and 505 $ng.spot^{-1}$) acetone solutions were applied as homogeneous layers to the circles of upper leaf surfaces respectively. After evaporation of the acetone from the leaf surfaces at room temperature, a series of similar-sized spots were formed, each spot indicating a sample location (Figure 1).

Photolysis of PAHs on the leaf surfaces of two mangrove species

The leaves of *Ko* and *Ac* with PAHs adsorbed on them were put under the mercury lamp whose light is guided by an optical fiber to avoid heating effects on the photolysis processes caused by the mercury lamp. By adjusting the height of the fiber optical probe of mercury lamp, the diameter size formed by the mercury lamp was fixed as the same as the length of the leaf blade during the course of the experiment. And the center of the light spot was directly faced the center of the leaf vein. Thus, the light intensity accepted by each sample location of the leaves is basically the same. In addition, the proper illumination intensity on the leaf surfaces was controlled at $3.53(\pm 0.07)\times 10^4$ lx, which was determined by a ZDS-10 illuminometer (Shanghai, China) in the photolysis experiments. After a certain period of illumination, the leaves of *Ko* and *Ac* were then placed under the fiber optic probe of the fluorospectrophotometer, respectively. And the relative S-SSF intensities of the PAH adsorbed on the leaves were directly determined. To avoid the interference from the scattered light, an angle between the fiber optic probe of the fluorospectrophotometer and the tested mangrove leaves was kept at 45° during the experiment. Four mangrove leaves of each species were involved, yielding 24 locations and 72 measurements for each species. The

Figure 1. The leaves and seedlings of *Ac* and *Ko* and the distribution of sample location on leaf surface.

value of relative S-SSF intensity presented represents the average behavior of the adsorbed PAH on different locations of each mangrove species. The mean values of the data were utilized to express the final results. And the data processing was based on the method of our previous studies. Statistical analysis for the variation of the fluorescence intensity obtained was performed using the statistics package for social science (SPSS) 13.0 for Windows. The significant differences in the results were determined using a one-sample t-test at the 95% confidence level ($p<0.05$ means that a remarkable difference existed) [7,28].

Photolysis of PAHs dissolved in water

As mentioned above, most studies are focus on the photolysis of PAHs in liquid medium (e.g., water, ethyl alcohol and acetone). The water as the homogeneous media is very simple. Thus, the mechanisms of the photolysis of PAHs in water are relatively simple and much work has been done in relevant researches. The photolytic behaviors of the PAHs adsorbed on the leaf surfaces have been found very common in the nature. And the leaf as the heterogeneous substrate is very complicated. Thus, few studies involve the photolysis processes of PAHs adsorbed on leaf surfaces. In addition, the photolysis mechanisms of the PAHs adsorbed on mangrove leaves have been hardly studied. Therefore, in order to further study the photolysis mechanisms of the PAH adsorbed on mangrove leaf surfaces, the photolysis of the same initial amount of PAH dissolved in water was also investigated. A self-made photodegradation reactor in our previous studies was used to study the photolysis processes of the PAHs dissolved in water [27]. With the use of a micropipette, a certain amount of An (44.5 $ng.spot^{-1}$) and Py (50.5 $ng.spot^{-1}$) aqueous solutions were added into the reactor which was directly put under the mercury lamp, respectively. After a certain period of illumination, the reactor was then put into the fluorescence spectrophotometer. And the relative fluorescence intensities of An and Py in aqueous solution were obtained, respectively. Because of the small volume of the reactor, working solutions of An and Py need not to be stirred and could realize *in situ* determination of the PAHs during the photolysis processes. These experiments were repeated three times.

Data processing

It is reported that two important parameters (C_t and C_0) should be determined in study of the photolysis processes of the PAH. The C_t means the concentration of PAH at time t during the UV irradiation, and C_0 is the initial concentration of PAH. In addition, when the concentration of PAH is proportional to its fluorescence intensity, C_t and C_0 could be replaced respectively by F_t (the fluorescence intensity of PAH at time t) and F_0 (the initial fluorescence intensity of PAH) [30]. Many studies have shown that the photolysis of PAHs adsorbed on the plant leaves always follow first-order kinetics [14,15]. Thus, the first-order reaction rates (k) and half lives ($t_{1/2}$) for the photolysis of the PAHs could be obtained by the following equations [29].

$$\ln\left(\frac{F_0}{F_t}\right)=kt \quad (1)$$

$$t_{1/2}=\frac{\ln 2}{k} \quad (2)$$

Results and Discussion

S-SSF spectra of An and Py adsorbed on the mangrove leaf surfaces

In this work, the S-SSF established in our previous studies was utilized for *in situ* determination of An and Py adsorbed on the leaf surfaces of *Ko* and *Ac*, respectively [27]. Because the mangrove leaf is the complicated and heterogeneous media, the peak area spectrophotometry was not used for determination of the PAHs adsorbed on mangrove leaf surfaces. The wavelength of 387 nm and 343 nm was selected as the fluorescence spectra for the quantification of An and Py, respectively. It is reported that photolytic products of the PAH might be the main interference for *in situ* study of the fluorescence signals during photolysis [31]. Therefore, the S-SSF spectra of An and Py adsorbed on the leaf surfaces of two mangrove species should be investigated over time. Figure 2 showed the S-SSF spectra of An and Py adsorbed on the *Ko* leaf surfaces taken at different time during the irradiations. And the decrease of the relative S-SSF intensities was observed with time. As can be seen from Figure 2, the shape of the spectrum and width of the half-wide spectral band were no considerable differences over time. Thus, the photolytic products could not interfere with the detection of the An and Py adsorbed on the mangrove leaf surfaces. Moreover, the autofluorescence of the uncontaminated mangrove leaves was very weak at the selected spectrum, which could not affect the detection of the fluorescence signals either (Figure 2). There was also a similar trend of the photolysis of An and Py adsorbed on *Ac* leaf surfaces. Consequently, the S-SSF was acceptable to be utilized as an *in situ* method for the photolysis study of An and Py adsorbed on two specific mangrove leaf surfaces. Thus under laboratory conditions, photolysis might play a major role in the fate of PAHs adsorbed on mangrove leaf surfaces under UV irradiation.

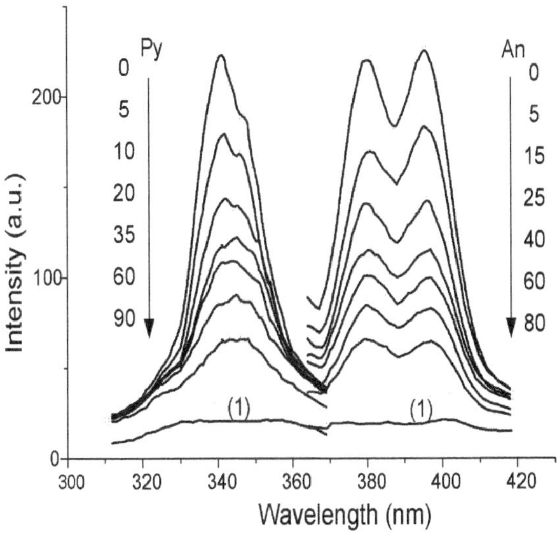

Figure 2. S-SSF spectra of the adsorbed An and Py on the *Ko* leaf surface. The order of the S-SSF intensity changes vs. time (min) is indicated by arrow. 1, uncontaminated *Ko* leaf.

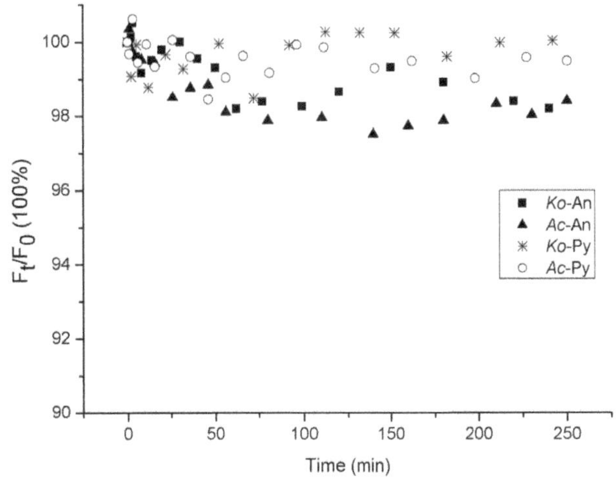

Figure 3. Variations of F_t/F_0 vs time of the adsorbed An (44.5 ng.spot^{-1}) and Py (50.5 ng.spot^{-1}) on the leaf surfaces of two mangrove species without illumination.

Study the photolysis processes of the An and Py adsorbed on the leaf surfaces of two mangrove species

It has been shown that volatilization might be one of the significant pathways for PAHs loss from certain environment surfaces [32]. According to our previous studies, volatilization could be a major loss for the PAH adsorbed on leaf surfaces within six hours [33]. In this experiment, the illumination time was less than four hours. Thus, the decrease of fluorescence signals of the PAHs adsorbed on mangrove leaf surfaces might be primarily due to the volatilization from the leaf surface into the atmosphere, and the quantities of the PAHs entering into the inner leaf tissues could be neglected in a short-term experiment. In this work, the heating effects of mercury lamp whose light is guided by an optical fiber could be avoided. And results had shown that the temperature of the location placing the leaves was similar with room temperature. Thus, it is reasonable that the effects of evaporation for PAHs loss without illumination could be used as blank control. Therefore, the variations of F_t/F_0 without illumination were investigated over time during the photolysis processes of the PAHs adsorbed on the leaf surfaces of two mangrove species (Figure 3). The purpose of the non-irradiation control group is to investigate the effect of evaporation on the PAH disappearances. As can be deduced from Figure 3, the relative S-SSF intensities of the An and Py adsorbed on the leaf surfaces of two mangrove species decreased very little with the extension of time. Thus, the decrease of relative S-SSF intensities of the An and Py adsorbed on mangrove leaf surfaces might be all caused by photochemical reaction, and the quantities of An and Py volatilization from the leaf surfaces into the atmosphere could be negligible during a short-term experiment.

It could also be deduced from Figure 3 that the decreasing extent of the relative S-SSF intensities of the adsorbed An was a little more than that of the adsorbed Py on the leaf surface of the same mangrove species with the extension of time. It is well known that the octanol-water partition coefficient (K_{ow}) of the An is of lower value than the Py. More An might volatilize into the atmosphere under the identical experimental conditions [34]. In addition, for the same initial amount of PAH, the decrease of the relative S-SSF intensities of the PAH adsorbed on the *Ko* leaf surfaces was a little less than on *Ac* leaf surfaces (Figure 4). In this experiment, the leaf-wax content of *Ko* (7.63 mg.g^{-1}) was much higher than the *Ac* (4.98 mg.g^{-1}). Thus, the interactions between the adsorbed PAH and *Ko* leaf surfaces might be much stronger, which make the adsorbed PAH on *Ko* leaf surface difficult to volatilize [21,27].

As mentioned above, the photolysis processes of An and Py adsorbed on the leaf surfaces of two mangrove species were firstly investigated in the course of time, and the experimental results were shown in Figure 4. As can be seen from Figure 4, the curves demonstrated the photolysis processes of each PAH were multi-decays. And within a certain period of time, the photolysis processes for both An and Py adsorbed on the leaf surfaces of two mangrove species followed first-order reaction kinetics with their photolysis rates in the order of *Ac*>*Ko* (Figure 5). Table 1 showed the kinetic results of the photolysis of An and Py adsorbed on the leaf surfaces of two mangrove species. It was shown from Table 1 that the photolysis half-lives of the lower initial amount of An (44.5 ng.spot^{-1}) and Py (50.5 ng.spot^{-1}) ranged from 14.8 min to 19.2 min and 32.5 min to 43.9 min, respectively. In addition, the photolysis half-lives of the higher initial amount of An (445 ng.spot^{-1}) and Py (505 ng.spot^{-1}) ranged from 25.3min to 38.9 min and 76.2 min to 101.9 min, respectively. Consequently, for both An and Py in the same experimental conditions, the higher the quantity of the PAH adsorbed on the mangrove leaf surfaces, the slower the photolysis rate of the PAH. Because each sample area on the leaf surface was fixed in the present study, the PAH adsorbed on the mangrove leaf surfaces might be of multi-layers, and the higher the initial amount of the adsorbed PAH, the thicker the PAH covered on the leaf surfaces [35]. Therefore, some excited states or polymer molecules of PAH might be formed on the leaf surfaces and absorb the photons of UV light. And the higher the quantity of the PAH adsorbed on the leaf surfaces, the more the excited states or polymer molecules of PAH formed. In conclusion, the number of photons striking the leaf surfaces decreased, and thereby decrease the photolysis rate of the PAH adsorbed on mangrove leaf surfaces [36–39]. However, the equations obtained in this work were not generalized. There are some limiting factors. Because of the multi-layers of the PAHs adsorbed on mangrove leaf surfaces, the UV photons accepted by the adsorbed PAHs on a different layer should be different, and the photolysis processes of the adsorbed PAHs should also be

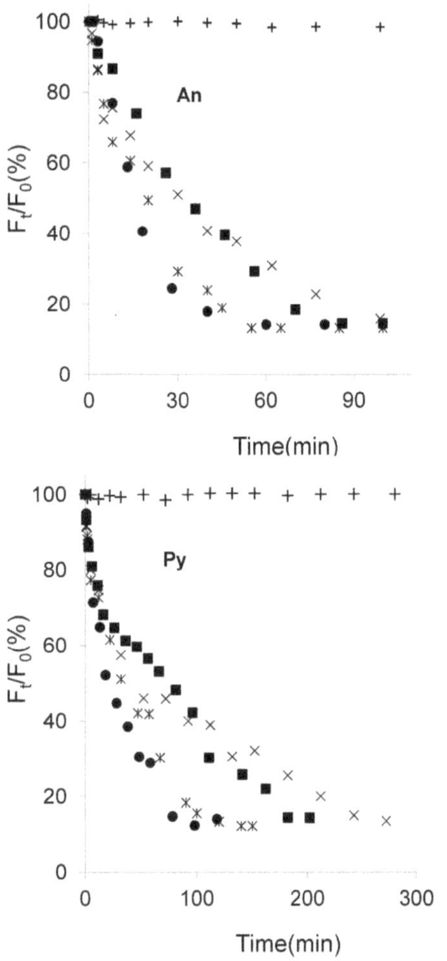

Figure 4. Photolysis processes of different initial amount of An (44.5 ng.spot^{-1}, 445 ng.spot^{-1}) and Py (50.5 ng.spot^{-1}, 505 ng.spot^{-1}) adsorbed on the leaf surfaces of two mangrove species.

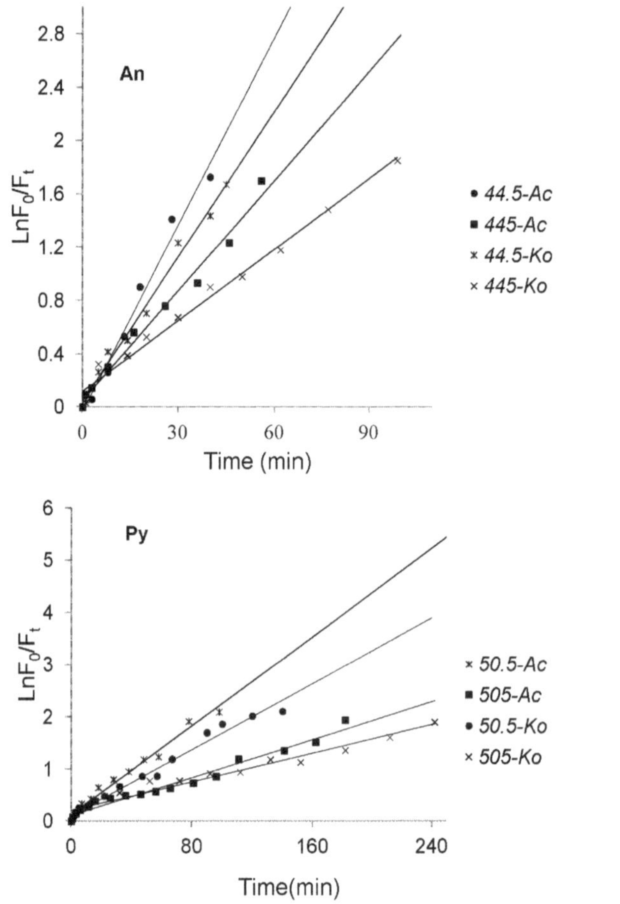

Figure 5. First-order kinetic plot of results of different initial amount of An (44.5 ng.spot^{-1}, 445 ng.spot^{-1}) and Py (50.5 ng.spot^{-1}, 505 ng.spot^{-1}) adsorbed on the leaf surfaces of two mangrove species (n = 9).

different. However, only a small amount of PAHs were selected in this work, the number of the layers could be small. Thus, the differences could be acceptable for *in situ* study the photolysis processes of PAHs adsorbed mangrove leaf surfaces in a short time. It was also from Table 1 that for the same amount of substances, the photolysis rate of Py adsorbed on the leaf surfaces of the same mangrove species was much lower than that of An. Thus, it was reasonable that the molecular structure might be one of the important factors for the photochemical behaviors of PAH [40,41]

It is supposed that the photolysis of PAHs on plants predominantly take place in the cuticle and especially the cuticular wax coating on the leaf surfaces [19]. Previous studies found that the wax or lipid on the plant leaves could absorb the UV photons, which might create the light filtering effect. Thus the number of the UV photons striking the PAH adsorbed on mangrove leaf surfaces might decrease, and thereby decrease its photolysis rate [14,15]. Because the leaf-wax content of *Ko* (7.63 mg.g^{-1}) was much higher than the *Ac* (4.98 mg.g^{-1}) in this experiment, the light filtering effects of *Ko* leaves might be stronger than *Ac* leaves. Consequently, the photolysis rate of the same initial amount of PAH adsorbed on *Ko* leaf surfaces was much slower than that adsorbed on *Ac* leaf surfaces (Table 1). In addition, the relative S-SSF intensities of the each PAH had little change when the amount of each PAH degraded to about 12% of the initial amount (Figure 4). In other words, the photolysis processes of the residual PAH adsorbed on the mangrove leaf surfaces could not occur with the illumination time prolonging. Xu et al and Schuler et al also found the same phenomena in their studies. However, they did not give the possible reasons [42,43]. Combined with others and our present studies, a little amount of PAH might enter into the inner leaf tissues and could not absorb the UV photons because of the light filtering effects of mangrove leaf surfaces. On the other hand, the generated photolytic products of the PAH in the upper layers might cover on the residual PAH in the low layers during illumination. Thus, the photolysis rates of the PAHs in the low layers might decrease. As time goes on, the photolysis of the residual PAH in the low layers might not occur [14,15]. Certainly, further studies are also needed.

Comparisons of the photolysis of An and Py adsorbed on mangrove leaf surfaces and in water

According to previous reports, the more polar the solvent is, the faster the photolysis processes of PAH [43,44]. Thus, the photolysis rate of PAH appears to be highly dependent on the substrate on which they are adsorbed. In order to further study the

Table 1. Kinetic results of the photolysis of An and Py adsorbed on the leaf surfaces of two mangrove species (n = 9) and in water (n = 9).

Mangrove	PAHs (ng.spot^{-1})		Calibration carve	Correlation coefficient	K	$t^{1/2}$ (min)
Ko	An	44.5	$y^b = 3.62\times10^{-2}x^a + 0.043$	0.9952	3.62×10^{-2}	19.2
		445	$y = 1.78\times10^{-2}x + 0.115$	0.9935	1.78×10^{-2}	38.9
	Py	50.5	$y = 1.58\times10^{-2}x + 0.114$	0.9899	1.58×10^{-2}	43.9
		505	$y = 6.80\times10^{-3}x + 0.208$	0.9802	6.80×10^{-3}	101.9
Ac	An	44.5	$y = 4.68\times10^{-2}x - 0.036$	0.9908	4.68×10^{-2}	14.8
		445	$y = 2.74\times10^{-2}x + 0.050$	0.9926	2.74×10^{-2}	25.3
	Py	50.5	$y = 2.13\times10^{-2}x + 0.125$	0.9865	2.13×10^{-2}	32.5
		505	$y = 9.10\times10^{-3}x + 0.109$	0.9898	9.10×10^{-3}	76.2
water	An	44.5	$y = 5.22\times10^{-2}x + 0.042$	0.9960	5.22×10^{-2}	13.3
	Py	50.5	$y = 2.88\times10^{-2}x + 0.088$	0.9937	2.88×10^{-2}	24.1

[a]stands for the illumination time of the PAH adsorbed on mangrove leaf surface.
[b]stands for the value of $\ln(F_0/F_t)$.

possible photolysis mechanisms of the PAHs adsorbed on mangrove leaf surfaces, the photolysis processes of An (44.5 ng, 0.25 nmol) and Py (50.5 ng, 0.25 nmol) in water were also investigated for comparison (Figure 6). The fluorescence method for determination of the An and Py in water has been established in our previous studies [29]. As can be seen from Figure 6, the PAH in water could be degraded continually without stop under

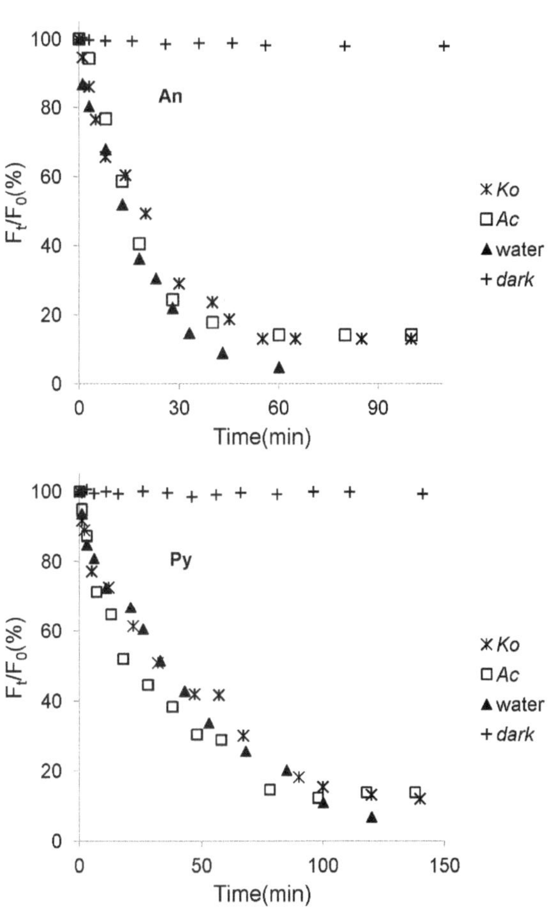

Figure 6. Photolysis Processes of the same initial amount of the An (44.5 ng.spot^{-1}) and Py (50.5 ng.spot^{-1}) adsorbed on two typical mangrove leaf surfaces and in water.

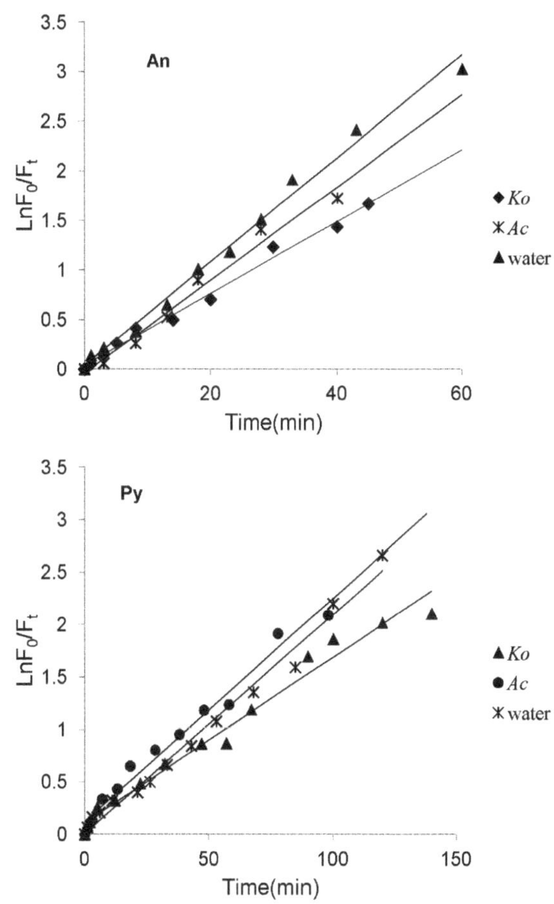

Figure 7. Variation of F_t/F_0 vs time during photolysis of the same initial amount of the An (44.5 ng.spot^{-1}) and Py (50.5 ng.spot^{-1}) adsorbed on two typical mangrove leaf surfaces and in water.

the same light intensity. It was different with the photolysis processes of the PAH adsorbed on mangrove leaf surfaces. Figure 7 showed the variations of F_t/F_0 vs time during the photolysis processes of An and Py adsorbed on the leaf surfaces of two mangrove species and in water. From the Figure 7, it was evident that the photolysis of An and Py dissolved in water also followed first-order reaction kinetics, respectively. With same amount of substances, the kinetic results of the photolysis of An and Py adsorbed on the leaf surfaces of two mangrove species and in water were shown in Table 1. According to the data of Table 1, the photolysis half-lives for An adsorbed on *Ko*, *Ac* leaf surfaces and in water were 19.2, 14.8 and 13.3 min, respectively, while the photolysis half-lives for Py adsorbed on *Ko*, *Ac* leaf surfaces and in water were 43.9, 32.5 and 24.1 min, respectively. These results indicated that the photolysis rate of the same PAH was much faster in water than on mangrove leaf surfaces. Mangrove leaf surfaces are covered by thick cuticle consisting of many long chain saturated and unsaturated fatty acids [18]. Consequently the polarity of mangrove leaf surfaces might be weaker than that of water. Thus, it was reasonable that the photolysis rate of the same PAH adsorbed on mangrove leaf surfaces was slower than in water. Moreover, some of the lipoidal compounds on plant leaves could adsorb and reflect the UV light [26,27,36–38]. And the number of UV photons striking to the PAH adsorbed on the mangrove leaf surfaces decreased, and thereby slowed down the photolysis rate of the adsorbed PAH. For this reason, the photolysis mechanism of the same PAH adsorbed on mangrove leaf surfaces and in water might be different, further studies are therefore needed.

Conclusions

Using the established S-SSF method, the photolysis processes of An and Py adsorbed on the leaf surfaces of two mangrove species were *in situ* studied for the first time. Meanwhile, the photolysis for both An and Py adsorbed on the leaf surfaces of two mangrove species followed first-order reaction kinetics with their photolysis rates in the order of *Ac*>*Ko*. In addition, the higher the quantity of the PAH adsorbed on the mangrove leaf surfaces, the slower the photolysis rate of the PAH. Moreover, for the same amount of substances, the photolysis rate of Py adsorbed on the same mangrove leaf surfaces was much lower than that of the An, and the photolysis rate of the same PAH in water was much faster than on mangrove leaf surfaces. Thus, photochemical behaviors of PAHs were dependent not only on their molecular structures but also the physical-chemical properties of the substrates on which they are adsorbed.

The S-SSF method is simple, accurate and easy operating for *in situ* investigation the photolysis of the PAHs adsorbed on mangrove leaf surfaces. However, much more studied need to be carried out in the near future. For example, PAHs usually exist in a multi-component in real environment, and this S-SSF might be a good way to be used for determination of multi-component PAHs simultaneously. It could be prospected that with the development of the S-SSF methods, it would be helpful for us to understand more the environmental behavior of PAHs in real environment, and it would also provide us a new way to study on mechanisms of phytoremediation of PAHs in mangrove wetland or other contaminated mediums.

Author Contributions

Conceived and designed the experiments: YZ. Performed the experiments: PW. Analyzed the data: PW YZ. Contributed reagents/materials/analysis tools: THW YZ. Wrote the paper: PW.

References

1. Gorshkov AG (2008) Determination of polycyclic aromatic hydrocarbons in the needles of a Scotch pine (*Pinus sylvestris L.*), a biomonitor of atmospheric pollution. J Anal Chem 8: 805–811.
2. Laender FDJ, Hammer J, Hendriks AJ, Jassent CR (2011) Combining monitoring data and modeling identifies PAHs as emerging contaminants in the Arctic. Environ Sci Technol 45: 9024–9029.
3. Augusto S, Máguas C, Matos J, Pereira MJ, Branquinho C (2010) Lichens as an integrating tool for monitoring PAH atmospheric deposition: a comparison with soil, air and pine needles. Environ Pollut 158: 483–489.
4. Wu NY, Huang HL, Zhang SZ, Zhu YG, Christie P, et al. (2009) Phenanthrene uptake by *Medicago sativa L.* under the influence of an arbuscular mycorrhizal fungus. Environ Pollut 157: 1613–1618.
5. Gao YZ, Ling WT, Wong MH (2006) Plant-accelerated dissipation of phenanthrene and pyrene from water in the presence of a nonionic-surfactant. Chemosphere 63: 1560–1567.
6. Wang XT, Miao Y, Zhang Y, Li YC, Wu MH, et al. (2013) Polycyclic aromatic hydrocarbons (PAHs) in urban soils of the megacity Shanghai: occurrence, source apportionment and potential human health risk. Sci Total Environ 447: 80–89.
7. Chen L, Zhang Y, Liu BB (2010) *In situ* simultaneous determination the photolysis of multi-component PAHs adsorbed on the leaf surfaces of living *Kandelia candel* seedlings. Talanta 83: 324–331.
8. Pagni RM, Dabestani R (2005) Recent developments in the environmental photochemistry of PAHs and PCBs in water and on solids. Env Chem 2: 193–219.
9. Tham YW, Sakugawa H (2007) Preliminary study of the photolysis of fluorene in rainwater. Bull of Environ Contam Toxicol 79: 670–673.
10. Huang XD, Dixon DG, Greenberg BM (1993) Impact of UV radiation and photo-modification on the toxicity of PAHs to the higher plant *Lemna gibba* (Duckweed). Environ Toxicol Chem 12: 1067–1077.
11. Korfmacher WA, Wehry EL, Mamantov G, Natusch DFS (1980) Resistance to photochemical decomposition of polycyclic aromatic hydrocarbons vapor adsorbed on coal fly ash. Environ Sci Technol 14: 1094–1097.
12. Robert AY, Arlene AG, Wehry EL, Mamantov G (1986) Photochemical transformation of pyrene and benzo[a]pryene vapor deposited on eight coal stack ashes. Environ Sci Technol 20: 86–90.
13. Reza D, Keith JE, Michael ES (1995) Photodecomposition of anthracene on dry surface: products and mechanism. J Photoch Photobio A 86: 231–239.
14. Wang DG, Chen JW, Zhen X, Qiao XL, Huang LP (2005) Disappearance of polycyclic aromatic hydrocarbons sorbed on surfaces of pine [*Pinua thunbergii*] needles under irradiation of sunlight: volatilization and photolysis. Atmospheric Environment 39: 4583–4591.
15. Niu JF, Chen JW, Martens D, Quan X, Yang FL, et al. (2003) Photolysis of polycyclic aromatic hydrocarbons adsorbed on spruce [*Picea abies (L.) Karst.*] needles under sunlight irradiation. Environmental Pollution 123: 39–45.
16. Oyo-Ita OE, Offem JO, Ekpo BO, Adie PA (2013) Anthropogenic PAHs in mangrove sediments of the Calabar River, SE Niger Delta, Nigeria. Appl Geochem 28: 212–219.
17. Ramdine G, Fichet D, Louis M, Lemoine S (2012) Polycyclic aromatic hydrocarbons (PAHs) in surface sediment and oysters (Crassostrea rhizophorae) from mangrove of Guadeloupe: Levels, bioavailability, and effects. Ecotoxicol Environ Saf 79: 80–89.
18. Wang P, Du KZ, Zhu YX, Zhang Y (2008) A novel analytical approach for investigation of anthracene adsorption onto mangrove leaves. Talanta 76:1177–1182.
19. Ke L, Wong TWY, Wong YS, Tam NF (2002) Fate of polycyclic aromatic hydrocarbon (PAH) contamination in a mangrove swamp in Hong Kong following an oil spill. Mar Pollut Bull 45: 339–347.
20. Wild E, Dent J, Barber JL, Thomas GO, Jones KC (2004) A novel analytical approach for visualizing and tracking organic chemicals in plants. Environ Sci Technol 38: 4195–4199.
21. Wild E, Dent J, Barber JL, Thomas GO, Jones KC (2005) Direct observation of organic contaminant uptake, storage, and metabolism within plant roots. Environ Sci Technol 39: 3695–3702.
22. Wang P, Zhang Y, Wu TH (2010) Novel method for in situ visualization of polycyclic aromatic hydrocarbons in mangrove plants. Toxicol Environ Chem 92: 1825–1829.
23. Wang P, Wu TH, Zhang Y (2012) Monitoring and visualizing of PAHs into mangrove plant by two-photon laser confocal scanning microscopy. Mar Poll Bull 64: 1654–1658.

24. Niessner R, Panne U, Schroeder H (1991) Fiber-optic sensor for the determination of polynuclear aromatic hydrocarbons with time-resolved laser-induced fluorescence. Anal Chim Acta 255: 231–243.
25. Rogers KR, Poziomek EJ (1996) Fiber optic sensors for environmental monitoring. Chemosphere 33:1151–1174.
26. Panne U, Niessner R (1993) A fiber-optical sensor for polynuclear aromatic hydrocarbons based on multidimensional fluorescence. Sensors Actuators B Chem 13: 288–292.
27. Wang P, Wu TH, Wang XD, Zhang Y (2012) Novel method for in situ investigation of PAH adsorption onto mangrove leaves. J Coast Res 28: 499–504.
28. Chen L, Wang P, Liu JB, Liu BB, Zhang Y, et al. (2011) In situ monitoring the photolysis of fluoranthene adsorbed on mangrove leaves using fiber-optic fluorimetry. J Fluoresc 21: 765–773.
29. Xiao X (2009) Photodegradation of Dissolved Anthracene and Pyrene in a Mixture and the Effect of Surfactant on the Photodegradation of Dissolved Anthracene or Pyrene [D]. [Ph.D. thesis] Xiamen: Xiamen University. 18 p.
30. Lehto KM, Vuorimaa E, Lemmetyinen H (2000) Photolysis of polycyclic aromatic hydrocarbons (PAHs) in dilute aqueous solutions detected by fluorescence. J Photoch Photobio A 136: 53–60.
31. Li YQ (1991) A study on synchronous and derivative synchronous fluorescence spectrometry and a preliminary investigation on fluorescence spectra of chlorophyll in vivo of plant Leaves. [Ph.D. thesis]. Xiamen: Xiamen University. 34 p.
32. Komp P, McLachlan MS (2000) The kinetics and reversibility of the partitioning of polychlorinated biphenyls between air and ryegrass. Sci Total Environ 250: 63–71.
33. Wang P, Zhu YX, Lu HL, Zhang Y (2010) Visualizing localizations and movement of anthracene in *Kandelia candel* (L.) Druce leaves by fluorescence microscopy. J Coast Res 3: 549–554.
34. Mackay D, Shiu WY, Ma KC (1997) Illustrated Handbook of Physical-Chemical Properties and Environmental Fate for Organic Chemicals: Pesticide Chemicals. Lewis Publishers. 259 p
35. Chen GZ, Huang XZ, Xu JG (1990) Fluorescence Analysis. Beijing: Beijing Science and Technology Press. 198 p.
36. Fioressi S, Arce R (2003) Excited states and intermediate species of Benzo[e]pyrene photolyzed in solution and adsorbed on surfaces. J Phys Chem A 107: 5968–5975.
37. Reyes CA, Medina M, Crespo-Hernandez C, Cedeno MZ, Arce R, et al. (2000) Photochemistry of pyrene on unactivated and activated silica surfaces. Environ Sci Technol 34: 415–421.
38. Worrall DR, Williams SL, Wilkinson F (1997) Electron transfer reactions of anthracene adsorbed on silica gel. J Phys Chem B 101: 4709–4716.
39. Fioressi S, Arce R (1999) The effect of physical and chemical properties of the surface on the photochemistry and spectroscopy of adsorbed benzo[e]pyrene Polycyclic Aromatic Compounds, 14: 285–294.
40. Feilberg A, Nielsen T (2001) Photodegradation of nitro-PAHs in viscous organic media used as models of organic aerosols. Environ Sci Technol 35: 108–113.
41. Mallakin A, Dixon DG, Bruce MG (2000) Pathway of anthracene modification under simulated solar radiation. Chemosphere 40: 1435–1441.
42. Xu Z (2004) Study on the Chemical and Photolychemical Transformation of the Some Organic Pollutants in Aatmosphere. [Ph.D. thesis]. Dalian: Dalian University of Technology. 23 p.
43. Schuler F, Schmid P, Schlatter C (1998) Photodegradation of polychlorinated dibenzo-p-dioxins and dibenzofurans in cuticular waxes of laurel cherry (*prunus laurocerasus*). Chemosphere 36: 21–34.
44. Kahan TF, Donaldson DJ (2007) Photolysis of polycyclic aromatic hydrocarbons on water and ice surfaces. J Phys Chem A 111: 1277–1285.

24

Bmp Suppression in Mangrove Killifish Embryos Causes a Split in the Body Axis

Sulayman Mourabit, Michael W. Moles, Emma Smith, Ronny van Aerle, Tetsuhiro Kudoh*

Biosciences, College of Life & Environmental Sciences, University of Exeter, Exeter, United Kingdom

Abstract

Bone morphogenetic proteins (Bmp) are major players in the formation of the vertebrate body plan due to their crucial role in patterning of the dorsal-ventral (DV) axis. Despite the highly conserved nature of Bmp signalling in vertebrates, the consequences of changing this pathway can be species-specific. Here, we report that Bmp plays an important role in epiboly, yolk syncytial layer (YSL) movements, and anterior-posterior (AP) axis formation in embryos of the self-fertilizing mangrove killifish, *Kryptolebias marmoratus*. Stage and dose specific exposures of embryos to the Bmp inhibitor dorsomorphin (DM) produced three distinctive morphologies, with the most extreme condition creating the splitbody phenotype, characterised by an extremely short AP axis where the neural tube, somites, and notochord were bilaterally split. In addition, parts of caudal neural tissues were separated from the main body and formed cell islands in the posterior region of the embryo. This splitbody phenotype, which has not been reported in other animals, shows that modification of Bmp may lead to significantly different consequences during development in other vertebrate species.

Editor: Michael Schubert, Laboratoire de Biologie du Développement de Villefranche-sur-Mer, France

Funding: SM was funded by a PhD studentship from the Natural and Environmental Research Council (NERC) in the UK. The funders had no role in study design, data collection and analysis, decision to publish, or preparation of the manuscript.

Competing Interests: The authors have declared that no competing interests exist.

* E-mail: t.kudoh@exeter.ac.uk

Introduction

Patterning of the DV axis is regulated by a differential spatial activation of Bmp signalling during early vertebrate embryogenesis [1,2]. In zebrafish embryos, from the blastula to the gastrula stage, Bmp promotes the development of ventrally derived tissues such as epidermis, posterior spinal cord, posterior somites, and blood, in the ventral side of the embryo; whereas Bmp inhibited areas in the dorsal side give rise to anterior neural cell fates and the notochord. A graded decrease of Bmp from the ventral side to the dorsal side produces laterally derived structures such as trunk somites and the neural tube [1,3,4,5].

In zebrafish, mutations in genes of this signalling pathway lead to a dorsalized phenotype, i.e. an expansion of dorsal-lateral regions of the gastrula at the expense of ventrally derived structures. Milder phenotypes of mutations in genes involved in Bmp signalling display a reduction in the ventral tail fin as seen in the recessive phenotypes *mini fin/tolloid* and *lost-a-fin/alk8* [6,7,8]. With increasing severity, ventrally derived tissues such as blood and tail are not apparent, and dorsal tissues such as the notochord are expanded. For instance, the homozygous *snailhouse/bmp7* phenotype is characterised by a shortened anterior-posterior axis, which twists around itself posteriorly like a coiled snail shell [6,9,10]. In the most severe homozygous phenotypes, such as *swirl/bmp2b* and *somitabun/smad5*, anterior somites expand dramatically constricting the yolk and causing it to burst [6,10,11].

Although the Bmp signalling pathway is highly conserved in vertebrates, modified Bmp activity can have different consequences depending on the morphological and genetic characteristics of the species. For example, whereas zebrafish chordino mutants, which have a null mutation in the gene coding for the Bmp-antagonist Chordin, have a significantly smaller brain [12], Chordin knock-out mice do not show such clear reduction in brain size, possibly due to a differential redundancy of another Bmp-antagonist, Noggin [13].

To examine the evolutionary conserved and divergent role of the Bmp signalling pathway and to further investigate the stage specific role of Bmp in vertebrate embryos, we examined Bmp inhibition in the self-fertilizing mangrove killifish *K. marmoratus*. This species is a unique hermaphroditic vertebrate that reproduces primarily by internal self-fertilization, essentially creating clonal lines with homozygous progeny. Mangrove killifish are also androdioecious, meaning that populations are composed of males and hermaphrodites [14]. This reproductive system involves three distinct phenotypes: primary males, hermaphrodites, and secondary males. Primary males (i.e. developed with unisexual male gonads) are rare in most of the wild populations [15]. There are currently 21 distinct clonal lines available for research, of which 11 have been demonstrated to be truly isogenic [16]. The self-fertilizing ability of mangrove killifish makes this species very interesting for mutant screening as zygotic mutant phenotypes appear in F2 embryos, one generation earlier than other animals such as zebrafish and medaka [17]. To establish the species as a novel model for developmental biology, we have recently analysed developmental stages and also developed basic embryological techniques for gene expression, cell labelling, imaging, and chemical treatment [18,19]. However, the role of key signalling pathways during early development in *K. marmoratus* has not yet been investigated.

Here, we show that a reduction of Bmp activity in *K. marmoratus* achieved by addition of the specific inhibitor DM resulted in significantly different morphological defects compared with the phenotypes previously reported in zebrafish [6,20]. It is noted that mangrove killifish and zebrafish shared a common ancestor over 250 million years ago, making this is a broad evolutionary comparison.

Results

Stage Specific Inhibition of Bmp Induces a Split Body Axis in *K. marmoratus* Embryos

To investigate the role of the Bmp signalling pathway during early embryonic development in *K. marmoratus*, we treated the embryos with the Bmp inhibitor DM [20]. It has been reported that the *snailhouse* phenotype can be produced in zebrafish with early DM exposures (10 μM) [20]. However the authors have not reported severer phenotypes such as the *swirl/bmp2b* mutant phenotype, which presents a more elongated embryo at the bud to early somitogenesis stage and also embryonic lethality during somitogenesis, suggesting that the dose used was not strong enough to suppress all Bmp signalling. Therefore, in the present study, we have used a higher dose of DM (100 μM) to examine severer loss of function of Bmp. Using this dose, zebrafish embryos show the *swirl* mutant phenotype (Cruz et al., in preparation). In addition, to investigate the stage specific role of Bmp, *K. marmoratus* embryos were treated with DM at various stages ranging from cleavage to gastrula.

A phenotype resembling *snailhouse* was observed in *K. marmoratus* by exposing embryos to DM from the late blastula stage (Figure 1C) (see [18] for *K. marmoratus* developmental stages), displaying a shortened and curled tail (Figure 1 C3 arrowhead) by 3 days post-fertilization (dpf). Embryos treated with the same concentration but starting from late epiboly (Figure 1D) produced a milder phenotype characterised by its bent tail (Figure 1 D3 arrowhead). However, DM exposures from the 32-cell stage (Figure 1B) produced a distinctive phenotype, hereby referred to as splitbody, characterised by its short body (Figure 1 B1), morphologically undifferentiated head region (Figure 1 B2 arrowhead), split body axis (Figure 1 B2 arrow) and cell clumps (hereby referred to as cell islands) in the posterior region of the embryo (Figure 1 B3 arrowhead). This splitbody phenotype has not been reported in other model species including zebrafish, *Xenopus*, chick and mouse.

Bmp is Essential for Normal Epiboly Progression in *K. marmoratus*

To determine the cause of a divided AP axis in the splitbody phenotype, we first examined differences in early development between wild type and embryos exposed to 100 μM DM at the 32-cell stage (Figure 2). During gastrulation, Bmp inhibition was shown to clearly delay epiboly progression. At 1 dpf, when control embryos reached *c.* 70% epiboly (Figure 2 A1), DM treated embryos were delayed with epiboly covering *c.* 30% of the yolk (Figure 2 A2). Similarly at 2 dpf, control embryos entered the otic vesicle formation stage (Figure 2 B1), whilst exposed embryos lagged behind with epiboly covering *c.* 90% of the yolk (Figure 2 B2). Such a significant delay in epiboly, resulting from Bmp signalling inhibition, has not been reported in zebrafish embryos [6].

Prior to gastrulation, the embryo is composed of 4 layers, the enveloping layer, deep cells, the yolk syncytial layer (YSL) and the yolk [21]. It is known that in zebrafish, during late gastrulation, delays in movements of the deep cells do not equate to delays in the YSL [22]. Furthermore, research in *Fundulus* has shown that the movements of the YSL are independent from the blastoderm, as the YSL continues its epibolic migration if the blastoderm is removed [23]. Thus, in order to determine if inhibition of Bmp signalling also delayed YSL movements, yolk syncytial nuclei (YSN) were stained by sytox green injection at the late blastula stage. Embryos were then exposed to 200 μM DM, a concentration capable of mimicking the splitbody phenotype (see the next section and Figure 3). Embryos were observed the next day, and whilst controls reached the eye formation stage with YSN spread throughout the yolk (Figure 2 C1), both the YSN and the blastoderm of DM exposed embryos were delayed at mid-epiboly (Figure 2 C2). This data suggests that the YSL is also affected by the inhibition of Bmp, as YSN were moving relative to the blastoderm margin and displayed the same level of delay.

Laterally Derived Structures and the Notochord are Divided in Splitbody

The severe delay of epiboly movements observed in splitbody embryos suggests that laterally derived structures are unable to merge at the end of epiboly, leading to the formation of two body axes. To confirm this hypothesis, we examined the spatial arrangement of the neural tube and somites in splitbody (Figure 3&4). These two tissues may be derived from lateral gastrula domains, as reported in zebrafish [3]. We used Hoechst staining for the body contour, *sox3* and *ntl in situ* hybridization for the neural tube and the notochord respectively, and MF-20 immunofluorescence staining for the somites.

Hoechst and *sox3* staining confirmed a split in the body axis and the neural tube for all embryos exposed to both 100 μM DM at the 32-cell stage and 200 μM DM at the late blastula stage. Two different phenotypes were observed, with some individuals showing an opened end of the two neural tube strands (Figure 3 B1, 2 & D1, 2 arrowheads), whilst others had a closed end with both strands joining in their most posterior section (Figure 3 C1, 2 & E1, 2 arrowheads). Significantly more individuals with the former phenotype were observed in the 100 μM DM treated embryos (19/20 individuals) compared to the 200 μM DM treated embryos (12/20; $P = 0.02$). Furthermore, the cell islands seen in splitbody were observed by Hoechst staining (Figure 3 B-E3 arrowheads), and displayed *sox3* positive staining suggesting that the island is partly composed of neural plate cells (Figure 3 B-E4 arrowheads).

At the late gastrula stage, the axial mesoderm of normal embryos stained by *ntl* was observed in the dorsal axis (Figure 4 A1 arrowhead), whereas in DM treated embryos these cells appeared to stay separated in the lateral domains (Figure 4 A2 arrowhead). At day 4 in control embryos, the tip of the notochord was stained by *ntl* (Figure 4 B2 arrowhead). On the other hand, splitbody embryos possessed two *ntl* stained tips (Figure 4 C2 arrowheads), suggesting that the axial mesoderm cells separated in the lateral domains of the gastrula embryo were split into the two body axes. This separation of lateral structures was further demonstrated by the staining of somite muscles in embryos 4 dpf using the myosin antibody MF-20. If epiboly occurs correctly, somites form pairs either side of the neural tube of the developing embryos (Figure 4 D1–3 arrowheads). In the splitbody phenotype, somites were unpaired and appeared divided in the two strands of the embryonic body axis (Figure 4 E1 Hoechst staining showing the clear split of the body axis; E2, 3 arrowheads, somites are present in both strands of the divided body). In the cell islands, *ntl* and MF-20 did not show any staining suggesting these cells do not contain notochord, muscle, and undifferentiated mesoderm.

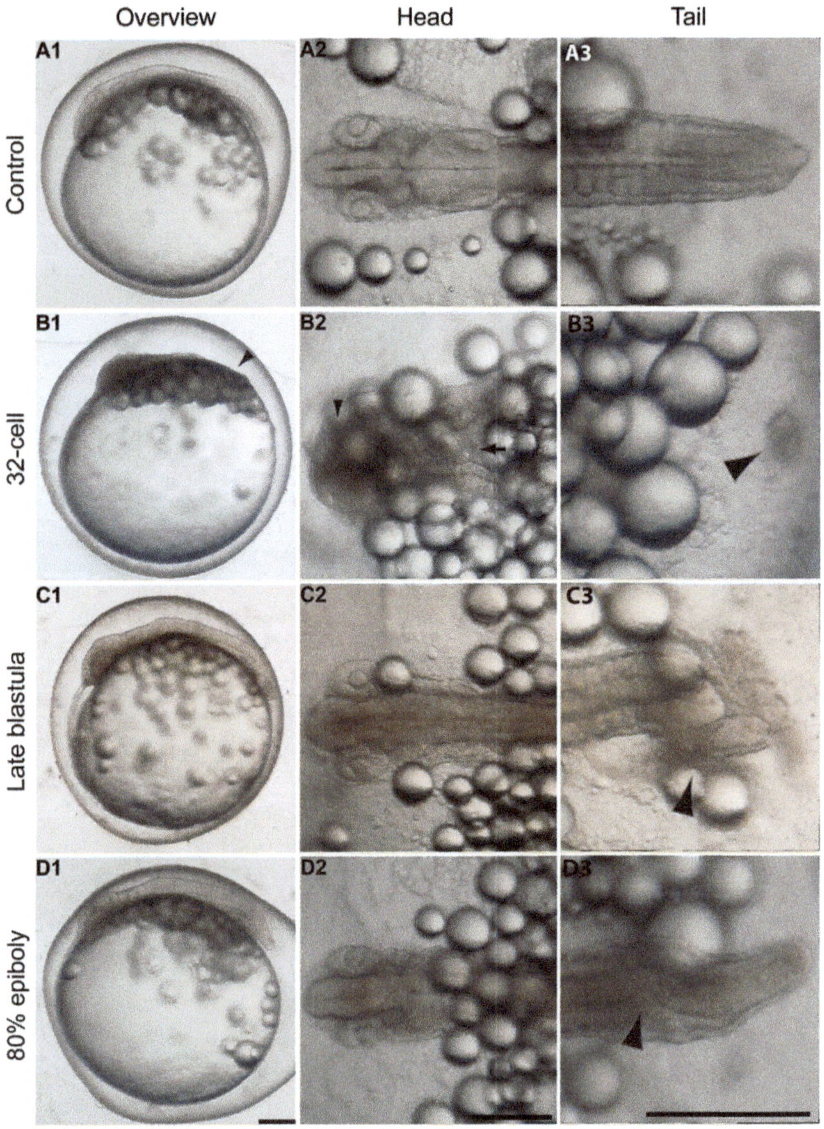

Figure 1. Stage specific inhibition of Bmp in *K. marmoratus*. Embryos were exposed to 100 μM dorsomorphin at the 32-cell (**B**), late blastula (**C**) and 80% epiboly (**D**) stages of development. Photographs of the embryos were taken 3 days post-fertilization. **A1–3**: Control (n = 20/20). **B1–3**: splitbody (phenotype variation details in Figure 3), this phenotype is characterised by absence of a distinct tail region (**B1** arrowhead), morphologically undifferentiated head region (**B2** arrowhead) and split body axis (**B2** arrow), and cell islands in the posterior region (**B3** arrowhead). **C1–3**: Curled tail (n = 8/10), this phenotype resembles *snailhouse* seen in zebrafish and is characterised by its curled tail (**C3** arrowhead). **D1–3**: Bent tail (n = 12/12, this phenotype primarily displayed a bent tail (**D3** arrowhead). Overview images are lateral views and head/tail images are dorsal views of the embryos. Scale bars: 250 μm.

DM Dose Dependence of the Splitbody Phenotype

Given that embryos treated to 100 μM DM from 32-cell and 200 μM DM from late blastula displayed a similar splitbody phenotype, we hypothesised that the 100 μM dose took longer to fully suppress Bmp signalling, but as embryos were treated earlier in development Bmp was fully suppressed by the mid-blastula transition and produced splitbody. To explore this hypothesis, we examined the level of Bmp signalling activity by measuring phosphorylation of Smad1/5 with Western blotting. Embryos were exposed to 100 μM from 32-cell as well as 100 and 200 μM DM from late blastula. These were then frozen at the late gastrula stage and used for Western Blotting. Quantification of densitometry results was obtained from 3 independent experiments, normalised to total Smad and indicated as fold increase over the resting control condition (Figure 5A, Mean ± SE). The representative Western Blot of the 3 independent experiments (Figure 5B) shows the levels of total Smad1/5/8 and phospho-Smad1/5 at late gastrula from the dose and stage specific treatments. These data demonstrate that all three treatments equally suppress phospho-Smad1/5 by late gastrula. These results confirmed that DM effectively suppressed Bmp signalling during gastrulation, but also suggested that zygotic Bmp is key for normal epiboly movements, as the 100 μM treatment only produced splitbody if applied earlier in development.

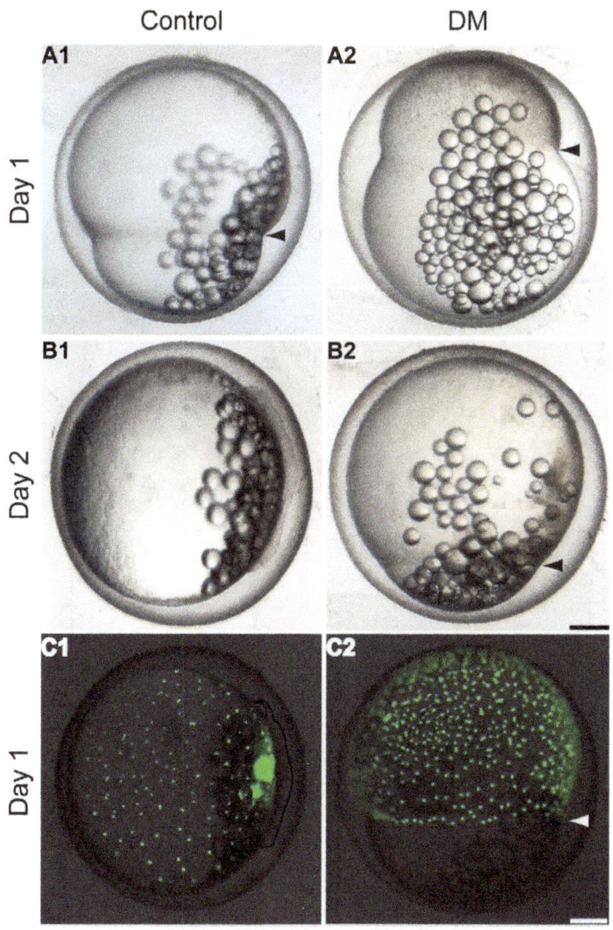

Figure 2. Bmp inhibition delays epiboly progression in *K. marmoratus*. Epiboly coverage was recorded at day 1 (**A**) and day 2 (**B**) post-fertilization (dpf) in embryos exposed to 100 μM dorsomorphin (DM) at the 32-cell stage. Progression of the yolk syncytial layer (YSL) during gastrulation was assessed via staining of yolk syncytial nuclei (YSN) using Sytox Green. The green fluorescent YSN were observed 1 dpf (**C**). **A1, 2**: As control embryos reach *c.* 70% epiboly (**A1** arrowhead, n = 10/10), DM treated embryos are delayed with epiboly covering *c.* 30% of the yolk (**A2** arrowhead, n = 10/10). **B1, 2**: Controls reach the otic vesicle formation stage (**B1**, n = 10/10) whilst exposed embryos are lagging behind around 90% epiboly (**B2** arrowhead, n = 10/10). **C1, 2**: Shortly after epiboly closure, control embryos enter the eye formation stage (**C1**, n = 10/10) (embryo and the eye are outlined) and YSN are spread all over the yolk. On the other hand DM exposed embryos are still mid-epiboly and fluorescent YSN are observed near the blastoderm margin (**C2** arrowhead, n = 10/10), demonstrating that YSN are also delayed by inhibition of Bmp signalling. All images are lateral views of the embryos. Scale bars: 250 μm.

Discussion

Here, we report a defect in the merging of laterally/ventrally derived structures at the end of epiboly and a division of the notochord, as a result of Bmp signalling inhibition. Both neural tube and somites were unable to completely merge at the end of gastrulation due to delayed epiboly and YSL movements, thus producing the splitbody phenotype.

The YSL plays an important role in epiboly movements of the blastoderm. Studies in *Fundulus* have shown that although epiboly can take place after removal of the blastoderm, the latter is unable to perform this morphogenetic movement without the YSL [23,24]. Here, we demonstrate that the delayed blastoderm was accompanied by a delay in movements of the YSN. It is unclear whether the delay in movements of the YSL is triggered by the setback of the blastoderm, but the relative independence of the YSL from the blastoderm discussed previously suggests that Bmp plays a crucial role in movements of the YSL during gastrulation.

It is known from zebrafish research that the Bmp gradient of the zebrafish gastrula regulates convergent extension (CE), as these morphogenetic movements are absent ventrally whilst lateral tissues display increased CE until the dorsal side where convergence weakens and extension stays strong [25]. This pattern corroborates with and is driven by the ventrally-high, laterally-gradient, and dorsally-low distribution of Bmp signalling during gastrulation [25]. It is thought that this Bmp gradient in the zebrafish gastrula regulates CE movements by applying a reverse gradient of cell adhesion in which Bmp signalling negatively affects the Cadherin-dependent adhesion of lamellipodia from mesodermal cells to adjacent cells. This creates a spatial gradient of cell-cell adhesion that directs the lamellipodia-driven cell migration of lateral regions towards the dorsal side, as cell movement via lamellipodia is increasingly disabled as Bmp signalling increases [26]. Here, splitbody *K. marmoratus* embryos experienced low to absent levels of Bmp throughout the gastrula thus disturbing CE movements. In the absence of strong convergence and a gradient of cell-cell adhesion, lumps of cells could be isolated from the main body axis. In addition, the weakened extension movements explain the extremely shortened body axis of the splitbody phenotype.

The variation observed in splitbody, where some individuals exhibited an opened double body axis and others a closed one, resulted from the stage and dose specific exposure of *K. marmoratus* embryos to DM, with the more mild treatments producing the *snailhouse* equivalent. Splitbody was only produced when embryos were treated early in development (100 μM at 32-cell stage or 200 μM at late blastula) to ensure a delay of epiboly and thus a split in the body axis. These results suggested that zygotic Bmp is key for normal epiboly movements, as the 100 μM treatment only produced splitbody if applied earlier in development, thus giving DM the time to fully suppress Bmp by the onset of epiboly. On the other hand, it may also be possible that maternal Bmp acts as an activator to the zygotic signalling pathway; as such the lower dose earlier in development could affect the late blastula to gastrula stages. Recent studies have shown the temporal importance of Bmp signalling in patterning DV tissues along the AP axis [2]. Our data demonstrate that Bmp signalling is also crucial for the correct timing of epiboly closure and thus the formation of a single anterior-posterior body axis in the mangrove killifish. Such results have not been reported in zebrafish, suggesting that the suppression of Bmp may have different consequences during development in other fish species.

The mangrove killifish produces eggs roughly two times bigger than zebrafish. Consequently the ratio between the diameter of the blastoderm margin and the yolk is 1:1.4 for killifish Stage 10 embryos and 1:1 in zebrafish embryos at an equivalent stage (shield stage) [18,21]. As the blastoderm is under extreme tension to move over the yolk during epiboly, the larger yolk of *K. marmoratus* embryos may increase the stretch required for the sheet of deep cells to reach the mid-gastrula point, which may enhance the epiboly defect less obvious in zebrafish. Furthermore, the differential genetic background of this species may result in varying levels of severity of the patterning defect caused by Bmp suppression. Both a genome project and a mutant screen are currently ongoing for the mangrove killifish [17,27], and will help further uncover the molecular mechanisms and phenotypic

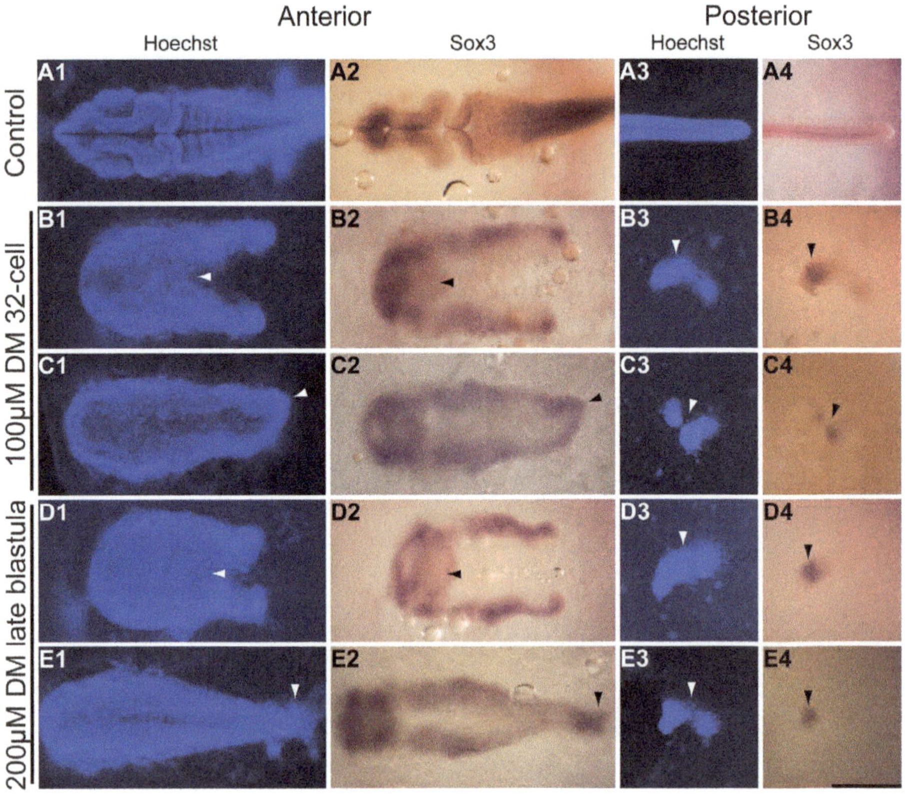

Figure 3. The neural tube is separated in embryos of the splitbody phenotype. *K. marmoratus* embryos were exposed to 100 µM dorsomorphin (DM) at the 32-cell stage (**B, C**), and 200 µM DM at the late blastula stage (**D, E**) of development. These embryos were then fixed 3 days post-fertilization and used for *in situ* hybridization using a *sox3* probe (stains all neural tissue) (**A–E2, A–E4**) and Hoechst staining (a blue fluorescent DNA stain) (**A–E1, A–E3**) in order to examine body contour and split neural tube (**A–E1, 2**), and the nature of the posterior isolated cell lumps or cell islands (**A–E3, 4**). **A1–4**: Control embryo (n = 20/20). **B, D**: Splitbody individual with an opened end of the body axis and neural tube split (**B1, 2** arrowheads n = 19/20, and **D1, 2** arrowheads n = 12/20). Splitbody individuals with a closed end, as both strands of the body axis and neural tube join in their most posterior region (**C1, 2** arrowheads n = 1/20, and **E1, 2** arrowheads n = 8/20). All DM embryos presented here generated cell islands (**B-E3** arrowheads) with distinct *sox3* positive staining (**B-E4** arrowheads). All images are dorsal views of the embryos. Scale bar: 250 µm.

variation of the loss of function of Bmp signalling between different species.

It has been demonstrated that DM may inhibit other receptors of the Alk family [28], therefore it might be possible that part of the splitbody phenotype is due to inhibition of other pathways. However, if DM inhibits Alk receptors transducing Nodal/Activin, the mesoderm marker *ntl* would have been suppressed, which was not the case. The Bmp specificity of the chemically induced phenotype is also supported by the phenotype of DM treated zebrafish embryos, as these are identical to *bmp2b* and *bmp7* mutants ([10,29]; Cruz et al., in preparation). Therefore in the context of early development it is likely that the main target of inhibition causing the splitbody phenotype is the Bmp pathway.

In summary, by using a novel model animal we were able to find a very different morphological phenotype when blocking Bmp signalling. In zebrafish, most embryos affected by a severe suppression of Bmp die by late somitogenesis, whereas mangrove killifish embryos at an equivalent stage all survived and displayed a very unique splitbody phenotype, enabling analysis for both early and late phenotypes resulting from a severe reduction of Bmp signalling. These data provide new insights for the conserved and species-specific roles of Bmp at the blastula, gastrula, and somitogenesis stages, demonstrating that the mangrove killifish is a very useful model for studying the roles of Bmp and other key signalling pathways in early development.

Experimental Procedures

Experimental Animals

K. marmoratus of the Hon9 clonal lineage were obtained from an existing stock at the University of Exeter (UK). Hermaphroditic individuals were kept individually in 1.5 L plastic containers (25°C, 14 psu (practical salinity unit), 12:12 h light:dark photoperiod) and were fed daily *ad libitum* on *Artemia* nauplii. Brackish water was made using demineralised water and marine salts (Tropic Marin, Germany). Eggs were collected from aquaria filter pads placed in the containers (Pondmaster filter foams), and provided a substrate for oviposition. Both control and DM treated embryos were reared under the same conditions as adult individuals. Embryonic stages were determined using the staging series in [18]. All Protocol used were permitted by the UK Home Office guidance to Animals Scientific Procedures Act (Project License No. PPL 30/2360).

Experimental Protocols

Imaging. Micrographs were taken using a Nikon Digital Sight DS-U2 camera mounted on a Nikon SMZ1500 microscope

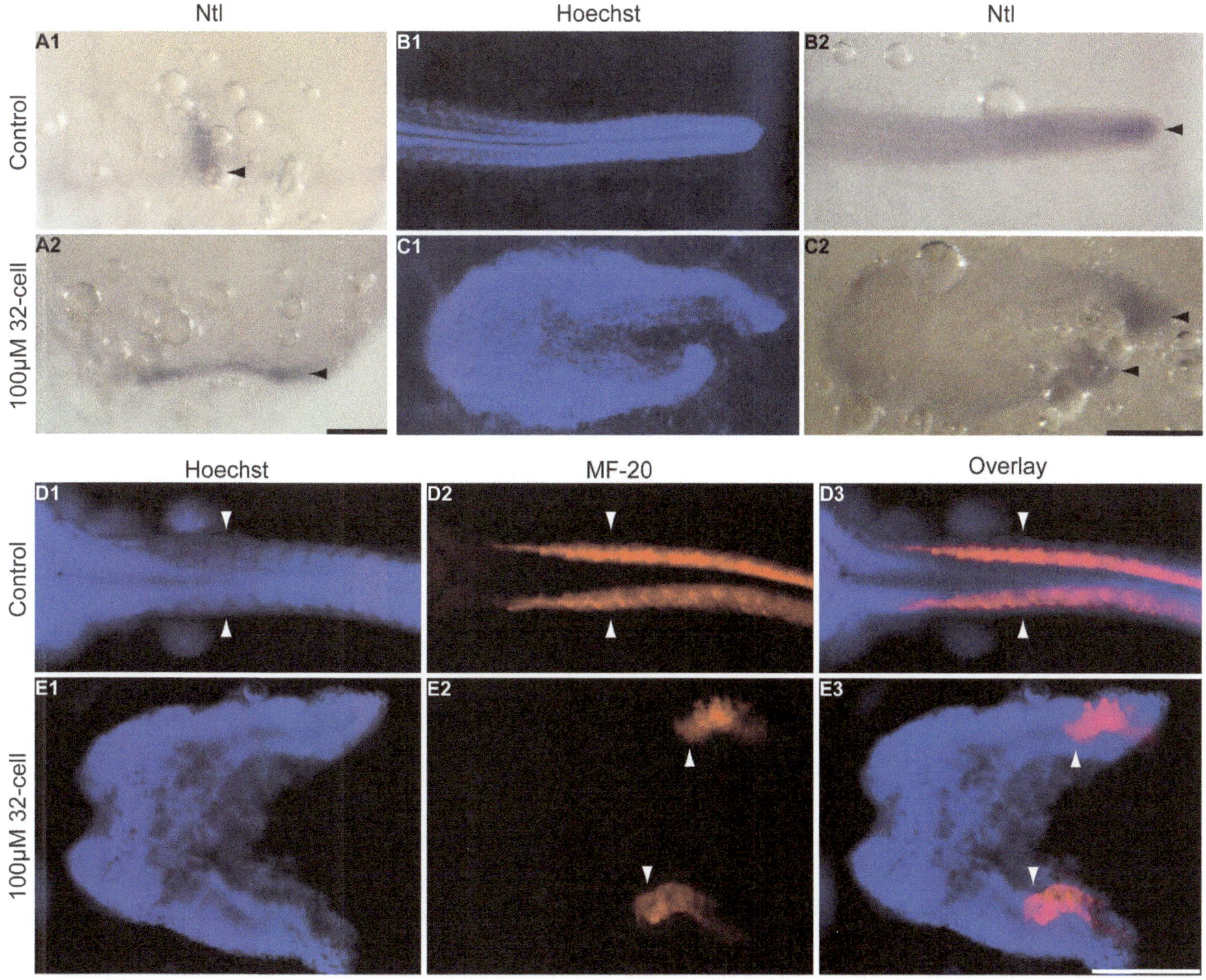

Figure 4. Somites and the notochord are divided in the splitbody phenotype. *K. marmoratus* embryos were exposed to 100 μM dorsomorphin at the 32-cell stage and fixed 1 and 4 days post-fertilization in order to stain the notochord by *in situ* hybridization using a medaka *ntl* probe (**A, B, C**), and somites using the myosin antibody MF-20 (**D, E**). In control embryos at late gastrula, *ntl* stained axial mesoderm in the dorsal axis (**A1** arrowhead, n = 10/10), whereas in DM treated embryos these cells appeared to have stayed in the lateral domains (**A2** arrowhead, n = 10). At day 4, *ntl* stained the notochord in the tip of the tail for control embryos (**B2** arrowhead, n = 10/10), whereas splitbody embryos had the tips of both body axes stained with *ntl* (**C2** arrowheads, n = 10/10). For control embryos, somites are formed as pairs arranged either side of the neural axis (**D1–3** arrowheads, n = 10/10). In the splitbody phenotype, somites were unpaired and separated in the two body axes (**E1** Hoechst staining showing the body split; **E2, 3** arrowheads, somites are present in both axes, n = 10/10). Photographs were taken at late gastrula for **A**, and 4 days post-fertilization for **B–E**. Images in **A** are lateral views and for **B–E** dorsal views of the embryos. Scale bars: 250 μm.

and an Olympus XC10 camera mounted on an Olympus SZX16 microscope. Imaging of live and fixed *K. marmoratus* embryos was performed using the Agarose bed and methyl cellulose techniques respectively, described in detail by [18,19].

Dorsomorphin exposures. Stock solutions for dorsomorphin (DM) (6-(4-(2-(1-Piperidinyl)ethoxy]phenyl]-3-(4-pyridinyl)-pyrazolo(1,5-*a*]pyrimidine dihydrochloride, Tocris 3093) were prepared as 10 mM dissolved in demethyl sulfoxide (DMSO) and diluted in 14 psu brackish water to final concentrations. Embryos at the 32-cell, late blastula and 80% epiboly stages were exposed to different concentrations of DM (100 μM and 200 μM).

Whole-mount *in situ* hybridization. *K. marmoratus sox3* was cloned by PCR using the following primers: forward GAGTGTGTGAGTGATCACTGA, and reverse TCTGA-GAGTGGGACGTGATGG. Primer design was based on *sox3* sequence information that was obtained from a *de novo K. marmoratus* transcriptome assembly (Illumina RNA-seq sequencing) conducted in our laboratory. *K. marmoratus Sox3* cDNA was sequenced and deposited to Genbank (KF887913).

The PCR product was inserted into the pGEM-T Easy vector (Promega). *Escherichia coli* colonies containing this plasmid were cultured and the plasmid DNA was then purified using a QIAprep Spin Midiprep (QIAGEN). The plasmids were digested with *PstI*, and Digoxigenin-labelled RNA probe was synthesised by T7 RNA polymerase (Roche). The medaka *ntl* plasmids [30] were digested with *SalI*, and the Digoxigenin-labelled RNA probe was synthesised by T3 RNA polymerase (Roche).

Live *K. marmoratus* embryos with chorions were placed in 1.5 ml Eppendorf tubes (5 embryos/tube). After removing brackish water with a pipette, 1 ml of 4% paraformaldehyde (PFA) (14 psu brackish water, 20 mM HEPES buffer, pH adjusted to 7) was added for fixation and kept at room temperature for 4 days.

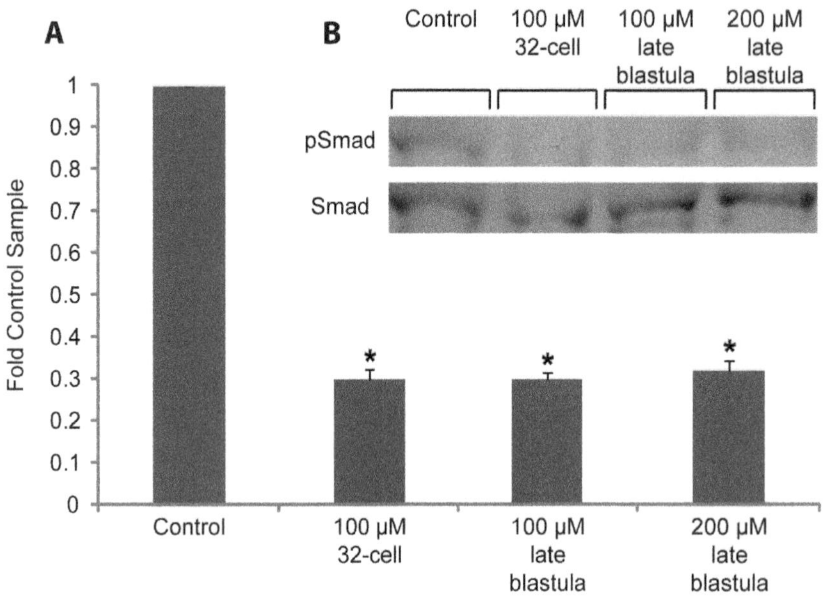

Figure 5. Dorsomorphin inhibits phosphorylation of Smad1/5. Bmp signalling activity was quantified by measuring phosphorylation of Smad 1/5 at the late gastrula stage. Embryos were exposed to 100 µM from 32-cell as well as 100 and 200 µM DM from late blastula. These were then frozen at the late gastrula stage and used for Western Blotting. **A:** Quantification of densitometry results obtained from 3 independent experiments (Mean ± SE), normalised to total Smad and indicated as fold increase over the resting control condition. All 3 treatments were significantly different from the control as indicated by asterisks ($P<0.001$), but no significant differences were observed between the treatments. **B:** Representative Western Blot of 3 independent experiments showing the levels of total Smad1/5/8 and phospho-Smad1/5 at late gastrula from the dose and stage specific treatments. These data demonstrate that all 3 treatments equally suppress phospho-Smad1/5 by late gastrula.

Following fixation, these embryos were washed with 1 ml phosphate buffer saline (10 minutes) then manually dechorionated and dehydrated in 1 ml 100% methanol at −20°C for one hour (they can be stored at this step for several weeks). These embryos were then used for whole-mount *in situ* hybridization, performed according to the method described by [31], with modifications. A full protocol is available in the supplemental information section (Supporting Information S1).

Differences between the phenotypes observed after DM treatment (Figure 3) were determined using Fischer Exact Tests (Systat Software, San Jose, CA). Differences between groups were considered to be significant when $P<0.05$.

Hoechst and immunofluorescent staining. For MF-20 antibody staining (Hybridoma Bank), fixed embryos stored in methanol (see above for conditions), were rehydrated in PBSTx (PBS+0.5% Tritonx, Sigma) and further permeabilised using Proteinase K (PK). Control and DM treated embryos at 3 dpf were treated to 10 µg/ml PK for 5 minutes, and day 4 embryos for 10 minutes. These embryos were then washed in PBSTx to stop the digestion and re-fixed with 4% PFA for an hour at room temperature. Embryos were put in blocking solution for 3 hours at room temperature (1% skimmed milk and 1% DMSO in PBSTx), and then incubated in primary antibody overnight (1:20 monoclonal mouse antibody MF-20 in blocking solution). The next day, the primary antibody was thoroughly washed off (four 30 minute washes in PBSTx), and incubated in Alexa Fluor 546 goat anti-mouse IgG secondary antibody overnight (Invitrogen A11003). Finally, the secondary antibody was thoroughly washed off.

For Hoechst staining, *sox3* or MF-20 stained embryos were incubated in a Hoechst solution for 30 minutes (0.5 µg/ml in PBSTx). The solution was then thoroughly washed off and embryos were ready for imaging.

Microinjection. Microinjection of sytox green (Invitrogen S7020) into the yolk syncytial layer (YSL) was performed following the procedure described by [18,19]. Sytox green (0.5 mM) was injected in the YSL at the late blastula stage and fluorescent yolk syncytial nuclei were photographed at 1 dpf in control and DM treated embryos.

Western blotting. Embryos at the late blastula stage were lysed in cold 2× lysis buffer (4% SDS, 20% glycerol, 125 mM Tris-HCl pH 6.8, 50 µg/ml BPB, 10% β-Mercaptoethanol) at 5 embryos/400 µl lysis buffer. Lysates were clarified by centrifugation (14.5 Krpm for 5 minutes) and the supernatants were heated at 70°C for 10 minutes then analysed by SDS-PAGE. Western Blots for Smad1/5/8 (1:200; Santa Cruz Biotechnology sc-6031-R) and Phospho-Smad1/5 (1:1000: New England Biolabs 9516S) were performed according to the manufacturer's instructions (blocking solution, for pre-blocking and dilution of all the antibodies, was composed of 2% bovine serum albumin in Tris buffer saline (20 mM Tris-Hcl pH 7.5 and 150 mM NaCl). Differences in phospho-Smad1/5 densitometry results between experimental groups were determined using One Way ANOVA, followed by pair-wise comparisons between DM-treated embryos and the controls using the Holm-Sidak method (Systat Software, San Jose, CA). Differences between groups were considered to be significant when $P<0.05$.

Acknowledgments

We thank Matthias Carl for the medaka *ntl* probe, and Carlos Cruz and Máté Varga for their helpful discussions and comments on the manuscript. We are grateful to technicians of the University of Exeter fish facilities for animal husbandry.

Author Contributions

Conceived and designed the experiments: SM TK. Performed the experiments: SM MWM ES. Analyzed the data: SM RVA TK. Contributed reagents/materials/analysis tools: SM MWM ES RVA TK. Wrote the paper: SM TK.

References

1. Neave B, Holder N, Patient R (1997) A graded response to BMP-4 spatially coordinates patterning of the mesoderm and ectoderm in the zebrafish. Mech Develop 62: 183–195.
2. Tucker JA, Mintzer KA, Mullins MC (2008) The BMP signalling gradient patterns dorsoventral tissues in a temporally progressive manner along the anteroposterior axis. Dev Cell 14: 108–119.
3. Kimmel CB, Warga RM, Schilling TF (1990) Origin and organization of the zebrafish fate map. Development 108: 581–594.
4. Kudoh T, Concha ML, Houart C, Dawid IB, Wilson SW (2004) Combinatorial Fgf and Bmp signalling patterns the gastrula ectoderm into prospective neural and epidermal domains. Development 131: 3581–3592.
5. Schier AF, Talbot WS (2005) Molecular genetics of axis formation in zebrafish. Annu Rev Genet 39: 561–613.
6. Mullins MC, Hammerschmidt M, Kane DA, Odenthal J, Brand M, et al. (1996) Genes establishing dorsoventral pattern formation in the zebrafish embryo: The ventral specifying genes. Development 123: 81–93.
7. Connors SA, Trout J, Ekker M, Mullins MC (1999) The role of tolloid/mini fin in dorsoventral pattern formation of the zebrafish embryo. Development 126: 3119–3130.
8. Mintzer KA, Lee MA, Runke G, Trout J, Whitman M, et al. (2001) Lost-a-fin encodes a type IBMP receptor, Alk8, acting maternally and zygotically in dorsoventral pattern formation. Development 128: 859–869.
9. Dick A, Hild M, Bauer H, Imai Y, Maifeld H, et al. (2000) Essential role of Bmp7 (snailhouse) and its prodomain in dorsoventral patterning of the zebrafish embryo. Development 127: 343–354.
10. Schmid B, Furthauer M, Connors SA, Trout J, Thisse B, et al. (2000). Equivalent genetic roles for bmp7/snailhouse and bmp2b/swirl in dorsoventral pattern formation. Development 127: 957–967.
11. Hild M, Dick A, Rauch GJ, Meier A, Bouwmeester T, et al. (1999) The *smad5* mutation somitabun blocks Bmp2b signaling during early dorsoventral patterning of the zebrafish embryo. Development 126: 2149–2159.
12. Schulte-Merker S, Lee KJ, McMahon AP, Hammerschmidt M (1997) The zebrafish organizer requires chordino. Nature 387: 862–863.
13. Bachiller D, Klingensmith J, Kemp C, Belo JA, Anderson RM, et al. (2000) The organizer factors Chordin and Noggin are required for mouse forebrain development. Nature 403: 658–661.
14. Tatarenkov A, Lima SMQ, Taylor DS, Avise JC (2009) Long-term retention of self-fertilization in a fish clade. Proc Natl Acad Sci 106: 14456–14459.
15. Turner BJ, Fisher MT, Taylor DS, Davis WP, Jarrett BL (2006) Evolution of 'maleness' and outcrossing in a population of the self-fertilizing killifish, *Kryptolebias marmoratus*. Evol Ecol Res 8: 1475–1486.
16. Tatarenkov A, Ring BC, Elder JF, Bechler DL, Avise JC (2010) Genetic composition of laboratory stocks of the self-fertilizing fish *Kryptolebias marmoratus*: A valuable resource for experimental research. PLoS ONE 5: e12863.
17. Moore GL, Sucar S, Newsome JM, Ard ME, Bernhardt L, et al. (2012) Establishing developmental genetics in a self-fertilizing fish (*Krytolebias marmoratus*). Integr Comp Biol 52: 781–791.
18. Mourabit S, Edenbrow M, Croft DP, Kudoh T (2011) Embryonic development of the self-fertilizing mangrove killifish *Kryptolebias marmoratus*. Dev Dyn 240: 1694–1704.
19. Mourabit S, Kudoh T (2012) Manipulation and imaging of *Kryptolebias marmoratus* embryos. Integr Comp Biol 52: 761–768.
20. Yu PB, Hong CC, Sachidanandan C, Babitt JL, Deng DY, et al. (2008) Dorsomorphin inhibits BMP signals required for embryogenesis and iron metabolism. Nat Chem Biol 4: 33–41.
21. Kimmel CB, Ballard WW, Kimmel SR, Ullmann B, Schilling TF (1995) Stages of embryonic-development of the zebrafish. Dev Dyn 203: 253–310.
22. Kane DA, Hammerschmidt M, Mullins MC, Maischein HM, Brand M, et al. (1996)The zebrafish epiboly mutants. Development 123: 47–55.
23. Trinkaus JP (1951) A study of the mechanism of epiboly in the egg of *Fundulus heteroclitus*. J Exp Zool 118: 269–319.
24. Betchaku T, Trinkaus JP (1978) Contact relations, surface-activity, and cortical microfilaments of marginal cells of enveloping layer and of yolk syncytial and yolk cytoplasmic layers of *Fundulus* before and during epiboly. J Exp Zool 206: 381–426.
25. Myers DC, Sepich DS, Solnica-Krezel L (2002) Bmp activity gradient regulates convergent extension during zebrafish gastrulation. Dev Biol 243: 81–98.
26. von der Hardt S, Bakkers J, Inbal A, Carvalho L, Solnica-Krezel L, et al. (2007) The Bmp gradient of the zebrafish gastrula guides migrating lateral cells by regulating cell-cell adhesion. Curr Biol 17: 475–487.
27. Kelley JL, Yee MC, Lee C, Levandowsky E, Shah M, et al. (2012) The possibility of *de novo* assembly of the genome and population genomics of the mangrove Rivulus, *Kryptolebias marmoratus*. Integr Comp Biol 52: 737–742.
28. Vogt J, Traynor R, Sapkota GP (2011) The specificities of small molecule inhibitors of the TGFβ and BMP pathways. Cell Signal 23: 1831–1842.
29. Kishimoto Y, Lee KH, Zon L, Hammerschmidt M, Schulte-Merker S (1997) The molecular nature of zebrafish swirl: BMP2 function is essential during early dorsoventral patterning. Development 124: 4457–4466.
30. Araki K, Okamoto H, Graveson AC, Nakayama I, Nagoya H (2001) Analysis of haploid development based on expression patterns of developmental genes in the medaka *Oryzias latipes*. Dev Growth Differ 43: 591–599.
31. Kudoh T, Tsang M, Hukriede NA, Chen XF, Dedekian M, et al. (2001) A gene expression screen in zebrafish embryogenesis. Genome Res 11: 1979–1987.

Permissions

All chapters in this book were first published in PLOS ONE, by The Public Library of Science; hereby published with permission under the Creative Commons Attribution License or equivalent. Every chapter published in this book has been scrutinized by our experts. Their significance has been extensively debated. The topics covered herein carry significant findings which will fuel the growth of the discipline. They may even be implemented as practical applications or may be referred to as a beginning point for another development.

The contributors of this book come from diverse backgrounds, making this book a truly international effort. This book will bring forth new frontiers with its revolutionizing research information and detailed analysis of the nascent developments around the world.

We would like to thank all the contributing authors for lending their expertise to make the book truly unique. They have played a crucial role in the development of this book. Without their invaluable contributions this book wouldn't have been possible. They have made vital efforts to compile up to date information on the varied aspects of this subject to make this book a valuable addition to the collection of many professionals and students.

This book was conceptualized with the vision of imparting up-to-date information and advanced data in this field. To ensure the same, a matchless editorial board was set up. Every individual on the board went through rigorous rounds of assessment to prove their worth. After which they invested a large part of their time researching and compiling the most relevant data for our readers.

The editorial board has been involved in producing this book since its inception. They have spent rigorous hours researching and exploring the diverse topics which have resulted in the successful publishing of this book. They have passed on their knowledge of decades through this book. To expedite this challenging task, the publisher supported the team at every step. A small team of assistant editors was also appointed to further simplify the editing procedure and attain best results for the readers.

Apart from the editorial board, the designing team has also invested a significant amount of their time in understanding the subject and creating the most relevant covers. They scrutinized every image to scout for the most suitable representation of the subject and create an appropriate cover for the book.

The publishing team has been an ardent support to the editorial, designing and production team. Their endless efforts to recruit the best for this project, has resulted in the accomplishment of this book. They are a veteran in the field of academics and their pool of knowledge is as vast as their experience in printing. Their expertise and guidance has proved useful at every step. Their uncompromising quality standards have made this book an exceptional effort. Their encouragement from time to time has been an inspiration for everyone.

The publisher and the editorial board hope that this book will prove to be a valuable piece of knowledge for researchers, students, practitioners and scholars across the globe.

List of Contributors

Joseph Tzanopoulos and Douglas C. MacMillan
Durrell Institute for Conservation and Ecology, School of Anthropology and Conservation, University of Kent, Canterbury, Kent, United Kingdom

Nicola K. Abram
Durrell Institute for Conservation and Ecology, School of Anthropology and Conservation, University of Kent, Canterbury, Kent, United Kingdom
HUTAN/Kinabatangan Orang-utan Conservation Programme, Kota Kinabalu, Sabah, Malaysia
Living Landscape Alliance, Reading, Berkshire, United Kingdom
Borneo Futures initiative, People and Nature Consulting International, Jakarta, Indonesia

Isabelle Lackman
HUTAN/Kinabatangan Orang-utan Conservation Programme, Kota Kinabalu, Sabah, Malaysia

Panteleimon Xofis
Technological Education Institute of Kavala, Department of Forestry and Management of Natural Environment, Drama, Greece

Marc Ancrenaz
HUTAN/Kinabatangan Orang-utan Conservation Programme, Kota Kinabalu, Sabah, Malaysia
Borneo Futures initiative, People and Nature Consulting International, Jakarta, Indonesia
North England Zoological Society, Chester, Cheshire, United Kingdom

Robin Chung
C H Williams Talhar and Wong (Sabah) Sdn Bhd, Sandakan, Sabah, Malaysia

Lucy Peter
Danau Girang Field Centre, c/o Sabah Wildlife Department, Kota Kinabalu, Sabah, Malaysia

Robert Ong
Forest Research Centre, Sabah Forestry Department, Sandakan, Sabah, Malaysia

Benoit Goossens
Organisms and Environment Division, School of Biosciences, Cardiff University, Cardiff, Wales, United Kingdom
Sabah Wildlife Department, Kota Kinabalu, Sabah, Malaysia

Laurentius Ambu
Sabah Wildlife Department, Kota Kinabalu, Sabah, Malaysia

Andrew T. Knight
Department of Life Sciences, Imperial College London, Ascot, Berkshire, United Kingdom
Department of Botany, Nelson Mandela Metrropolitan University, Port Elizabeth, Eastern Cape, South Africa

Michael J. Osland and Richard H. Day
U.S. Geological Survey, National Wetlands Research Center, Lafayette, Louisiana, United States of America

Jack C. Larriviere and Andrew S. From
Five Rivers Services, LLC, U.S. Geological Survey, National Wetlands Research Center, Lafayette, Louisiana, United States of America

Anna McConville and Michael J. Mahony
School of Environmental and Life Sciences, University of Newcastle, Callaghan, New South Wales, Australia

Bradley S. Law
Forest Science Centre, Department of Primary Industries, Beecroft, New South Wales, Australia

Alan Jordan and Melinda A. Coleman
New South Wales Fisheries, Department of Primary Industries, PO Box 4321, Coffs Harbour 2450, New South Wales, Australia

Brendan P. Kelaher
New South Wales Fisheries, Department of Primary Industries, PO Box 4321, Coffs Harbour 2450, New South Wales, Australia

National Marine Science Centre & Centre for Coastal Biogeochemistry Research, School of Environment, Science and Engineering, Southern Cross University, Coffs Harbour, New South Wales, Australia

Allison Broad, Andrew R. Davis and Matthew J. Rees
Institute for Conservation Biology & Environmental Management, School of Biological Sciences, University of Wollongong, New South Wales, Australia

Takaomi Arai
Institute of Oceanography and Environment, Universiti Malaysia Terengganu, Kuala Terengganu, Terengganu, Malaysia

Carole Di-Poi, Jennyfer Lacasse and Nadia Aubin-Horth
Département de Biologie & Institut de Biologie Intégrative et des Systémes (IBIS), UniversitéLaval, Québec, Queébec, Canada

Sean M. Rogers
Department of Biological Sciences, University of Calgary, Calgary, Alberta, Canada

Charitha Pattiaratchi
School of Environmental Systems Engineering, University of Western Australia, Perth, Australia
Oceans Institute, University of Western Australia, Perth, Australia

Julia Reisser
School of Environmental Systems Engineering, University of Western Australia, Perth, Australia
Oceans Institute, University of Western Australia, Perth, Australia
Wealth from Oceans Flagship, Commonwealth Scientific and Industrial Research Organisation, Perth, Australia

Jeremy Shaw
Centre for Microscopy, Characterisation and Analysis, University of Western Australia, Perth, Australia

Gustaaf Hallegraeff
Institute for Marine and Antarctic Studies, University of Tasmania, Hobart, Australia

Maira Proietti
Instituto de Oceanografia, Universidade Federal do Rio Grande, Rio Grande, Brazil

David K. A. Barnes
British Antarctic Survey, Natural Environment Research Council, Cambridge, United Kingdom

Michele Thums
Oceans Institute, University of Western Australia, Perth, Australia
Australian Institute of Marine Science, The UWA Oceans Institute, Perth, Australia

Chris Wilcox and Britta Denise Hardesty
Wealth from Oceans Flagship, Commonwealth Scientific and Industrial Research Organisation, Perth, Australia
Marine and Atmospheric Research, Commonwealth Scientific and Industrial Research Organisation, Hobart, Australia

Lucy G. Gillis, Tjeerd J. Bouma and Peter M. J. Herman
Spatial Ecology Department, Royal Netherlands Institute for Sea Research (NIOZ), Yerseke, Zealand, The Netherlands

Alan D. Ziegler
Geography Department, National University of Singapore (NUS), Singapore, Singapore

Dick van Oevelen
Ecosystems Studies Department, Royal Netherlands Institute for Sea Research (NIOZ), Yerseke, Zealand, The Netherlands

Cecile Cathalot
Laboratoire Environnement Profond (LEP), French Research Institute for Exploitation of the Sea, Polouzane, Brittany, France

Jan W. Wolters
Department of Biology, University of Antwerp, Antwerp, Flanders, Belgium

James J. La Clair
Xenobe Research Institute, San Diego, California, United States of America

Steven T. Loveridge, Karen Tenney and Phillip Crews
Department of Chemistry and Biochemistry, University of California Santa Cruz, Santa Cruz, California, United States of America

Mark O'Neil-Johnson
Lead Discovery and Rapid Structure Elucidation Group, Sequoia Sciences, Inc., St. Louis, Missouri, United States of America

Eli Chapman
Department of Pharmacology and Toxicology, College of Pharmacy, University of Arizona, Tucson, Arizona, United States of America

Paulo M. Brito
Department of Zoology, Universidade Estadual do Rio de Janeiro, Rio de Janeiro, Brazil

Cesar R. L. Amaral
Department of Zoology, Universidade Estadual do Rio de Janeiro, Rio de Janeiro, Brazil
Department of Ecology, Universidade Estadual do Rio de Janeiro, Rio de Janeiro, Brazil

Dayse A. Silva and Elizeu F. Carvalho
Department of Ecology, Universidade Estadual do Rio de Janeiro, Rio de
Janeiro, Brazil

Ari J. S. Ferreira, Rania Siam, Ahmed Moustafa, Ahmed Sayed, Mohamed A. Ghazy, Hazem Sharaf, Amged Ouf Alyaa M. Abdel-Haleem, Eman Ramadan and Hamza El-Dorry
Department of Biology and the Science and Technology Research Center, School of Sciences and Engineering, The American University in Cairo, Cairo, New Cairo, Egypt

João C. Setubal
Departamento de Bioquímica, Instituto de Química, Universidade de São Paulo, São Paulo, Brazil

Felipe S. Chambergo
Escola de Artes, Ciências e Humanidades, Universidade de São Paulo, São Paulo, Brazil

Adam S. Dawe, Intikhab Alam, Heikki Lehvaslaiho, John A. C. Archer, Boris R. Jankovic and Vladimir B. Bajic
Computational Bioscience Research Center, King Abdullah University of Science and Technology, Thuwal, Kingdom of Saudi Arabia

AndréAntunes
Institute for Biotechnology and Bioengineering, Centre of Biological Engineering, University of Minho, Portugal

Ulrich Stingl
Red Sea Research Center, King Abdullah University of Science and Technology, Thuwal, Kingdom of Saudi Arabia

Mitchell Sogin
Josephine Bay Paul Center, Marine Biological Laboratory, Woods Hole, Massachusetts, United States of America

Anca G. Delgado, Dae-Wook Kang, Katherine G. Nelson, Devyn Fajardo-Williams, Joseph F. Miceli III, Hansa Y. Done, Sudeep C. Popat and Rosa
Swette Center for Environmental Biotechnology, Biodesign Institute, Arizona State University, Tempe, United States of America

Krajmalnik-Brown
Swette Center for Environmental Biotechnology, Biodesign Institute, Arizona State University, Tempe, United States of America
School of Sustainable Engineering and the Built Environment, Arizona State University, Tempe, United States of America

Laetitia Helene Marie Schmitt
Centre for Health Economics/Department of Economics and Related Studies, University of York, Heslington, York, United Kingdom

Cecile Brugere
Stockholm Environment Institute, University of York, Heslington, York, United Kingdom

Sébastien Personnic, Charles F. Boudouresque, Delphine Thibault-Botha, Mireille Harmelin-Vivien, Jean-Christophe Poggiale, Thierry Thibaut and Sandrine Ruitton
Aix-Marseille University, Mediterranean Institute of Oceanography (MIO), Universitéde Toulon, CNRS/INSU, IRD, UM 110, Marseille, France

Patrick Astruch and Patrick Bonhomme
GIS Posidonie, Pytheas Institute, Aix-Marseille University, Marseille, France

Enric Ballesteros
Centre d'Estudis Avanc¸ats de Blanes - CSIC, Blanes, Spain

Sylvain Blouet
Aire marine protégée de la côte Agathoise, site natura 2000, Agde, Franc

Denise Bellan-Santini5, Florent Renaud
Aix-Marseille University, Institut Méditerranéen de Biodiversitéet d'Ecologie (IMBE), UMR 7263, Station Marine d'Endoume, Marseille, France

Eric Feunteun
Museum National d'Histoire Naturelle, UMR 7208, Station Marine de Dinard, France

Gérard Pergent and Christine Pergent-Martini
Equipe Ecosystémes Littoraux, FRES 3041, University of Corsica, Corte, France

Jérémy Pastor
Universitéde Perpignan, Via Domitia, Centre de Formation et de Recherche sur les Environnements Méditerranéens, UMR 5110, Perpignan, France

Michał Kowalewski
Florida Museum of Natural History, University of Florida, Gainesville, Florida, United States of America

Rosa Doménech and Jordi Martinell
Institut de Recerca de la Biodiversitat and Departamentd'Estratigrafia, Paleontologia i Geociéncies Marines, Facultat de Geologia, Universitat de Barcelona, Barcelona, Spain

Jane E. Bryan
School of Geography and Environmental Studies, University of Tasmania, Hobart, Tasmania, Australia
UPNG Remote Sensing Centre, University of Papua New Guinea, Waigani, National Capital District, Papua New Guinea

Barbara Lokes and Geraldine Aoro
UPNG Remote Sensing Centre, University of Papua New Guinea, Waigani, National Capital District, Papua New Guinea

Philip L. Shearman
UPNG Remote Sensing Centre, University of Papua New Guinea, Waigani, National Capital District, Papua New Guinea
Research School of Biology, Australian National University, Canberra, Australian Capital Territory, Australia

Gregory P. Asner and David E. Knapp
Department of Global Ecology, Carnegie Institution for Science, 260 Panama Street, Stanford, California, United States of America

Chairmandurai Aravindraja, Dharmaprakash Viszwapriya and Shunmugiah Karutha Pandian
Department of Biotechnology, Alagappa University, Karaikudi, Tamil Nadu, India

Martin Aberhan
Museum für Naturkunde, Leibniz Institute for Evolution and Biodiversity Science, Berlin, Germany

Wolfgang Kiessling
Museum für Naturkunde, Leibniz Institute for Evolution and Biodiversity Science, Berlin, Germany
GeoZentrum Nordbayern, Paläoumwelt, Universität Erlangen Nürnberg, Erlangen, Germany

Fiorenza Micheli and Francesco Ferretti
Hopkins Marine Station, Stanford University, Pacific Grove, California, United States of America

Benjamin S. Halpern
National Center for Ecological Analysis and Synthesis, Santa Barbara,
California, United States of America, and the Center for Marine Assessment and Planning, University of California Santa Barbara, Santa Barbara, California, United States of America

Shaun Walbridge
ESRI, Redlands, California, United States of America

Saul Ciriaco
Miramare Marine Protected Area, Trieste, Italy

Simonetta Fraschetti
Laboratory of Marine Biology, Universitádel Salento, Lecce, Italy

Rebecca Lewison
Coastal and Marine Institute Laboratory, San Diego State University, San Diego, California, United States of America

Leo Nykjaer
European Commission, Joint Research Centre, Institute for Environment and Sustainability, Ispra, Italy

Andrew A. Rosenberg
Union of Concerned Scientists, Cambridge, Massachusetts, United States of America

Lingxia Wang and Wei Chen
Key Laboratory of Ministry of Education for Genetics, Breeding and Multiple Utilization of Crops, Fujian Agriculture and Forestry University, Fuzhou, Fujian, China

School of Life Sciences, Fujian Agriculture and Forestry University, Fuzhou, Fujian, China

Xiao Liu, Meng Liang, Yiyong Chen, Yongxiang Lin and Jianhong Xing
School of Life Sciences, Fujian Agriculture and Forestry University, Fuzhou, Fujian, China

Fanglin Tan and Li Huang
Fujian Academy of Forestry, Fuzhou, Fujian, China

Wenyu Liang
School of Life Sciences, Ningxia University, Yinchuan, Ningxia, China

Kathleen Schwerdtner Máñez
Department of Social Sciences, Leibniz Center for Tropical Marine Ecology, Bremen, Germany
Asia Research Center, Murdoch University, Murdoch, Western Australia, Australia

Poul Holm
Trinity Long Room Hub Arts and Humanities Research Institute, Trinity College Dublin, Dublin, Ireland

Louise Blight
World Wildlife Fund (WWF) -Canada, Vancouver, British Columbia, Canada

Marta Coll
Renewable Marine Resources Department, Institute of Marine Science, Barcelona, Spain
Laboratoire Écosystémes Marins Exploités, Séte Cedex, France

Alison MacDiarmid
Marine Ecology, National Institute of Water and Atmospheric Research, Kilbirnie, Wellington, New Zealand

Henn Ojaveer
Estonian Marine Institute, University Tartu, Tartu, Estonia

Bo Poulsen
Faculty of Social Sciences, Aalborg University, Aalborgø, Denmark

Malcolm Tull
Murdoch Business School, Murdoch University, Murdoch, Western Australia, Australia

Shruti Khanna, Susan L. Ustin and Alexander Koltunov
Center for Spatial Technology and Remote Sensing (CSTARS), Department of Land, Air and Water Resources, University of California Davis, Davis, California, United States of America

Maria J. Santos
Center for Spatial and Textual Analysis (CESTA), Stanford University, Stanford, California, United States of America

Raymond F. Kokaly
United States Geological Survey, Denver, Colorado, United States of America

Dar A. Roberts
Department of Geography, University of California Santa Barbara, Santa Barbara, California, United States of America

Ping Wang
School of Environmental Science and Public Health, Wenzhou Medical University, Wenzhou, China

Tun-Hua Wu
School of Information and Engineering, Wenzhou Medical University, Wenzhou, China

Yong Zhang
State Key Laboratory of Marine Environmental Science (Xiamen University), Environmental Science Research Center of Xiamen University, Xiamen, China

Sulayman Mourabit, Michael W. Moles, Emma Smith, Ronny van Aerle, Tetsuhiro Kudoh
Biosciences, College of Life & Environmental Sciences, University of Exeter, Exeter, United Kingdom

Index